HUMAN

Stephen Gaukroger presents an original account of the development of empirical science and the understanding of human behavior from the mid-eighteenth century. Since the seventeenth century, science in the west has undergone a unique form of cumulative development in which it has been consolidated through integration into and shaping of a culture. But in the eighteenth century, science was cut loose from the legitimating culture in which it had found a public rationale as a fruitful and worthwhile form of enquiry. What kept it afloat between the middle of the eighteenth and the middle of the nineteenth centuries, when its legitimacy began to hinge on an intimate link with technology? The answer lies in large part in an abrupt but fundamental shift in how the tasks of scientific enquiry were conceived, from the natural realm to the human realm.

At the core of this development lies the naturalization of the human, that is, attempts to understand human behavior and motivations no longer in theological and metaphysical terms, but in empirical terms. One of the most striking features of this development is the variety of forms it took, and the book explores anthropological medicine, philosophical anthropology, the 'natural history of man', and social arithmetic. Each of these disciplines re-formulated basic questions so that empirical investigation could be drawn upon in answering them, but the empirical dimension was not conceived very differently in each case, with the result that the naturalization of the human took the form of competing, and in some respects mutually exclusive, projects.

Stephen Gaukroger, who was educated at the University of London and the University of Cambridge, is Emeritus Professor of History of Philosophy and History of Science at the University of Sydney. He is author of nine books in the history of science and history of philosophy, including an intellectual biography of Descartes, as well as translations of the works of Descartes and Arnauld. *The Emergence of a Scientific Culture: Science and the Shaping of Modernity, 1210–1685*, was published in 2006, and *The Collapse of Mechanism and the Rise of Sensibility: Science and the Shaping of Modernity, 1680–1760*, was published in 2010. His work has been translated into Arabic, Chinese, French, German, Italian, Portuguese, Russian, and Serbian.

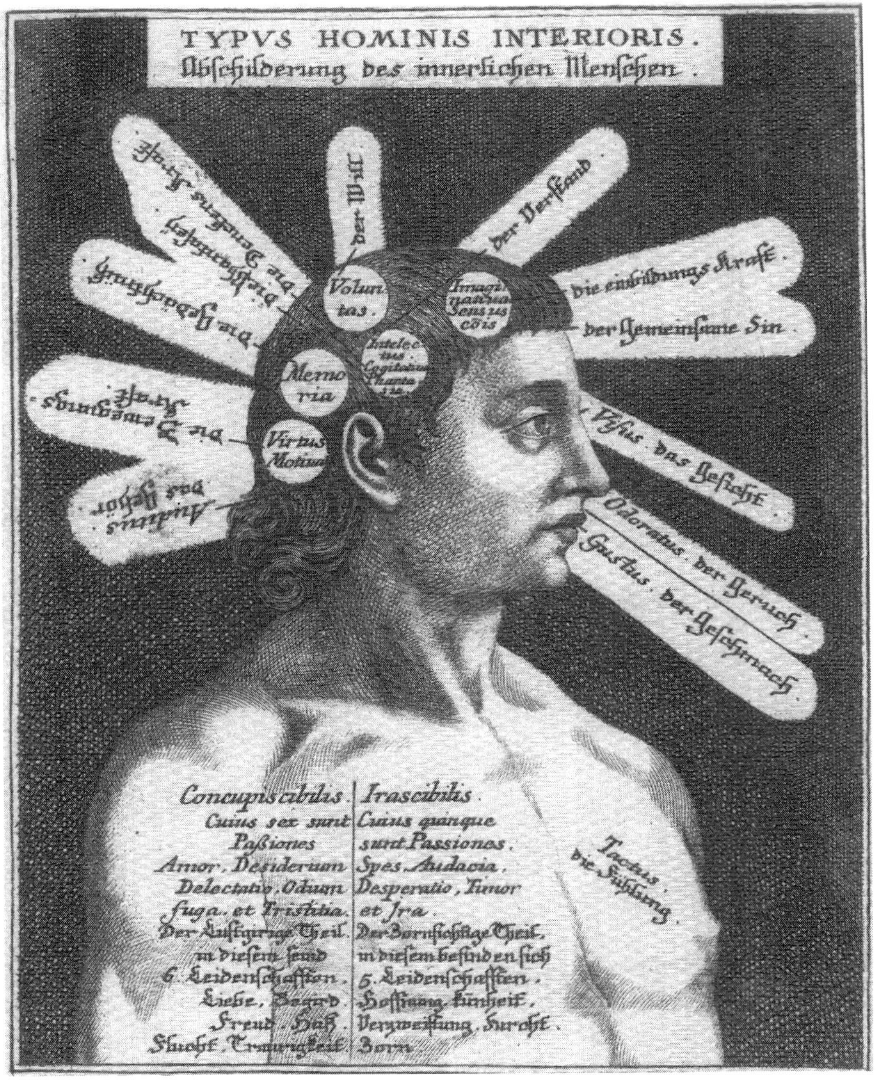

'The Inner Man', *Ichnographia Emblematica Triplicis Vitæ* (Augsburg, 1779).

The Natural and the Human

Science and the Shaping of Modernity, 1739–1841

STEPHEN GAUKROGER

OXFORD
UNIVERSITY PRESS

OXFORD
UNIVERSITY PRESS

Great Clarendon Street, Oxford, OX2 6DP,
United Kingdom

Oxford University Press is a department of the University of Oxford.
It furthers the University's objective of excellence in research, scholarship,
and education by publishing worldwide. Oxford is a registered trade mark of
Oxford University Press in the UK and in certain other countries

First published 2016
First published in paperback 2018

Published in the United States of America by Oxford University Press
198 Madison Avenue, New York, NY 10016, United States of America

British Library Cataloguing in Publication Data
Data available

Library of Congress Cataloging in Publication Data
Data available

ISBN 978–0–19–875763–4 (Hbk.)
ISBN 978–0–19–880160–3 (Pbk.)

Preface

Over the last two decades, I have been preoccupied with the question of how the values of a scientific culture came to predominate in the West, something that was part of a transformation of Western culture that began in the thirteenth century. In two earlier volumes—*The Emergence of a Scientific Culture: Science and the Shaping of Modernity, 1210–1685* and *The Collapse of Mechanism and the Rise of Sensibility: Science and the Shaping of Modernity 1680–1760*—I set out to uncover the complex and unprecedented processes by which scientific values were established, vindicated, and reinforced from then up to the middle of the eighteenth century. In continuing this project into the late eighteenth and early nineteenth centuries, my attention has turned to the human sciences, and I have been taken back to the political and intellectual interests of my student days in the 1970s, interests that I was fortunate enough to share, in conversation and in regular reading groups, with Beverley Brown, Terry Counihan, Herman Feddema, Lynn Ferguson, John Forrester, Mike Gane, Barry Hindess, Paul Hirst, Colin McCabe, Jill McGreal, Jeff Minson, Mike O'Pray, Cesia Prytys, Grahame Thompson, Keith Tribe, Ken Umpleby, and Karel Williams. Although I haven't exactly returned to these interests here—the nature of politics has changed so radically since then as to make that inconceivable—they have helped over the years to shape my sense of what the point and value of intellectual enquiry is.

The *termini* of the book are Hume's *Treatise of Human Nature* (1739), and Feuerbach's *Das Wesen des Christentums* (1841). I have allowed myself some flexibility with the timespan, but these two works mark crucial developments, the former firing the opening shots in the naturalization of the human, the latter bringing the project to a climax with the extension of the naturalization of the human to the naturalization of Christianity.

A long-term endeavour such as that of which this volume is part incurs many intellectual debts. In working on the present volume, I would particularly like to thank: Frédérique Aït-Touati, Francesco Borghesi, Stefanie Buchenau, Alan Chalmers, Conal Condren, Peter Cryle, Ofer Gal, Peter Harrison, Daniela Helbig, Ian Hunter, Mogens Laerke, Jennifer Milam, Dominic Murphy, Dalia Nassar, Joanna Picciotto, Annette Pierdziwol, Paul Redding, Jessica Riskin, Eric Schliesser, John Schuster, Katharina Schwarz, Anik Waldow, Catherine Wilson, Charles Wolfe, and John Zammito. As ever, my deepest thanks go to my wife, Helen Irving, and our children, Cressida and Hugh.

Material from the book has been presented at talks at the University of Aberdeen, Institute of Advanced Study University of Durham, École Normale Supérieure Lyon, École Normale Supérieure Paris, the Free University Berlin, University of Ghent, University of Graz, University of Helsinki, Humboldt

University Berlin, University of Lisbon, New York University, University of Oxford, University of Palermo, Princeton University, University of Queensland, the Rotman Institute at the University of Western Ontario, Shanghai Jiao Tong University, University of Sydney, and the University of Western Sydney.

Work on the project has been pursued primarily at the University of Sydney where, from the mid-1990s, I have received very generous support from the Australian Research Council, including two five-year Professorial Research Fellowships, without which it would probably not have been feasible to undertake something on this scale.

Contents

Introduction

This is a book about the emergence of the human sciences between the mid-eighteenth and the mid-nineteenth centuries, but within a broad agenda. The development of science in the West has, until recently, been quite unlike that in those other societies in which there were successful scientific traditions.[1] Science in these latter cases effectively came to an end once a set of original aims had been achieved. In the West, by contrast, it underwent a unique form of cumulative development in the early-modern era, one in which it was consolidated through the integration into and shaping of a culture. But in the eighteenth century, science was cut loose from the legitimating culture in which it had found a public rationale as a fruitful and worthwhile form of enquiry. What kept it afloat between the middle of the eighteenth and the middle of the nineteenth centuries, when it began to form an intimate link with technology? Why didn't it follow the boom–bust pattern characteristic of those other cultures in which there were successful scientific traditions that simply came to an end when basic problems had been resolved? I have set out to show that the answer lies largely in an abrupt but fundamental shift in how the tasks of scientific enquiry were conceived, from the natural realm to the human realm.

Hume provides us with an appropriate starting point. The subtitle of his *Treatise of Human Nature* (1739), which describes the work as 'An Attempt to Introduce the Experimental Method of Reasoning into Moral Subjects', heralds the emergence of a novel programme of accounting for human understanding and behaviour, one based on what were seen as scientific principles.[2] A distinctive

[1] In the two earlier volumes in this series—*The Emergence of a Scientific Culture: Science and the Shaping of Modernity, 1210–1685* (Oxford, 2006), and *The Collapse of Mechanism and the Rise of Sensibility: Science and the Shaping of Modernity 1680–1760* (Oxford, 2010)—I used the term 'natural philosophy' rather than 'science'. But 'science' was gradually replacing 'natural philosophy' in the second half of the eighteenth century, and I shall talk about 'science' and 'scientific' here when discussing eighteenth-century developments, although without the connotations of an exclusively professionalized form of activity, associated with the term 'scientist', introduced in the 1830s (but not generally accepted until the end of the century). In every European language other than English, 'science' covers the natural sciences, social sciences, and the humanities. Pre-nineteenth-century uses of the word in English are closer to the broader sense common to other modern European languages, and indeed remained so in the definitions given in dictionaries, such as Johnson's dictionary, as late as the 1870s.

[2] The idea of modelling moral reasoning on natural philosophy was not without precedent: notably Francis Hutcheson, *An Essay on the Nature and Conduct of the Passions and Affections, with*

feature of this programme was its normative edge, for, from the middle decades of the eighteenth century onwards, large-scale projects were initiated to establish a connection between science and civilization: science became an essential part of the civilizing process. As Condorcet put it in a reception speech to the Académie Française in 1782: 'Experience also proves that in all countries where the physical sciences have been cultivated, barbarism in the moral sciences has been more or less dissipated and at least error and prejudice have disappeared.'[3] This was certainly not a new thought, and by no means one confined to *philosophes*. In a speech given in September 1759 for example, Linnaeus, who was no *philosophe*, remarks that 'only the Sciences distinguish Wild people, Barbarians and Hottentots, from us', and 'the sciences are thus the light that will lead people who wander in the darkness'.[4] These attempts to bind science and civilization shaped the broad understanding of both science and civilization, and the connection between the two, from that time onwards. In this book, we shall be concerned with how and why this came about by focusing on an unprecedented form of understanding the human condition, the naturalization of the human, and its correlate, the humanization of nature.

In broad terms, what we encounter from the middle of the eighteenth century is a new type of consolidation of European scientific culture, by which I understand a culture in which cognitive values generally become subordinated to scientific ones, and where, correspondingly, science is expected to provide an archetypal path to cognitive success not just for scientific disciplines but for any discipline with cognitive aspirations, such as theology, metaphysics, political theory, political economy, law, and history. Nothing like this had occurred in any other culture that we can identify as having successful long-term scientific programmes: classical Greece and the Hellenistic Greek diaspora; Arab-Islamic North Africa/Near East/Iberian peninsula in the ninth, tenth, and eleventh centuries; thirteenth- and fourteenth-century Paris and Oxford; or China from the twelfth to the fourteenth centuries. In none of these cases did science become integrated into a culture so that the values of that culture became identified with those of science. And in none of the cases did science escape a pattern of peaking and then decline, and undergo the kind of uninterrupted and cumulative growth that we witness in the West from the early-modern period onwards. Before the

Illustrations on the Moral Sense (London, 1728), and George Turnbull, *The Principles of Moral Philosophy: An Enquiry into the Wise and Good Government of the Moral World* (London, 1740). The difference is that Hutcheson and Turnbull use natural-philosophical reasoning to bolster pre-established accounts of human nature, and particularly morality, as evidenced in their providentialism, whereas Hume used it much more as a genuine exploratory device.

[3] Bibliothèque de l'Institut de France, MS 855, f. 7: quoted in Keith Michael Baker, *Condorcet: From Natural Philosophy to Social Mathematics* (Chicago, 1975), 75. The idea of a moral regeneration that naturally accompanies cultivation of the sciences was a theme in a number of writers: it is particularly stressed, for example, in Charles Louis Dumas, *Principes de physiologie, ou introduction à la science expérimentale, philosophique et médicale de l'homme vivant* (3 vols, Paris, 1800–3).

[4] Quoted in Lisbet Koerner, *Linnaeus: Nature and Nation* (Harvard, MA, 1999), 94.

middle decades of the nineteenth century, when scientific and technological developments finally began to converge in an intensive way, it was this cultural consolidation of the scientific enterprise, rather than any particular scientific discoveries, that established the character of science in the West. It did more than simply allow a consolidation of science, however: it shaped its goals and aspirations, it shaped views of just what science could and could not achieve, and it largely determined what place science found for itself among other disciplines.

The emergence of a scientific culture is a modern phenomenon, but a number of pre-modern developments provided the conditions that made it possible, and enabled it to flourish. I explored these developments in the first volume in this series, *The Emergence of a Scientific Culture.* They date back as far as the thirteenth century, when theologians attempted to discover a more satisfactory philosophical basis for Christian theology. The solution to the theological problems was to replace what was effectively the Platonist theology of the Church Fathers with Aristotelianism. In stark contrast to Platonism, Aristotle had insisted that all knowledge derives from sense perception, so that, within Aristotelianism, natural philosophy—which comprised the study of the sensible world—was effectively the starting point for enquiry generally. Aristotelianism raised natural philosophy from the status of a marginal exercise wholly subordinate to theology and metaphysics, to that of the entry point for a philosophical understanding of the world. In theory, natural philosophy and Christian theology had an equal standing, and they dealt with different subject matters, but there were issues on which they nevertheless came into conflict—such as whether the universe had always existed, and the possibility of the personal immortality of the soul—and by the sixteenth century these had become problems that bore on the legitimacy and autonomy of both Christian theology and natural philosophy. The difficulties were compounded when Aristotelian natural philosophy came to be undermined in areas such as the theory of the heavens (astronomy), the theory of vision (optics), and—with the rise of atomism of various kinds—matter theory. By this time, the systematic structure of natural philosophy had been reinforced, partly because late sixteenth-century scholastic commentators had matched it up very closely with a systematic theology, and partly because, although the new natural-philosophical theories that were successfully coming to replace Aristotelianism were often piecemeal, there was an inevitable assumption that the best way to deal with Aristotelianism was to offer something that mirrored the breadth and generality of its claims.

As a result, a number of the new natural-philosophical programmes that emerged in the early and middle decades of the seventeenth century can be described as doctrinal. That is to say, they aspired to be both comprehensive and exhaustive; they envisaged a single source of truth, and tended to build on a set of general principles; they offered an exclusive 'world view'; and what they aimed at above all was the establishment of a single and permanent body of truth, securing this in part by a regimenting of, and reconciling of, particular truths, where appropriate by grounding them in the basic principles. Certainly not all work in natural philosophy was of this kind, but by the middle decades of the seventeenth century, Descartes' version of mechanism, particularly as set out in his *Principia philosophiae*, provided the general framework for natural-philosophical thinking. In broad terms, mechanism—not to be confused with mechanics—can be characterized in terms of its two main claims. First, macroscopic physical events are reducible to the interactions between micro-corpuscles making up bodies: there is nothing in the natural realm that cannot be accounted for in micro-corpuscularian terms, and this is the ultimate form of all natural explanations.[5] Second, all interactions between these micro-corpuscles are deemed to be a result of exchange of motion resulting from physical contact, and are as a consequence characterizable wholly in terms of mechanics. At the same time that mechanism was providing a framework for natural-philosophical thinking generally, natural philosophy itself was entering into a symbiotic relationship with natural theology, particularly in Protestant England, where there was a concerted attempt to make natural theology commensurate with natural philosophy, so that the already doctrinal nature of Christianity was reinforced, as it became increasingly assimilated to a system of beliefs—beliefs which could, where necessary, be adjudicated by the same means as those in natural philosophy. Natural theology thereby took on some of the aspects of natural philosophy. It became more open-ended, and began to be construed very much as a form of enquiry, for example in the work of Boyle. At the same time, natural philosophy, in the form of mechanism, took on an unprecedented cultural role, for its association with natural theology meant that it was now a partner in a project that went far beyond simply understanding the natural world. The kind of understanding of our place in the world to which theology had traditionally laid claim was now no longer its exclusive preserve: natural theology sharing in the benefits of natural philosophy meant natural philosophy sharing in the benefits of natural theology.

But the mechanist model of explanation that provided the basis for understanding just what natural philosophy or science was had effectively collapsed by the early decades of the eighteenth century, a question to which the second volume in this series, *The Collapse of Mechanism and the Rise of Sensibility*, was devoted. One set of problems, in particular, went to the core of mechanism's claims to explanatory power. In the 1660s, questions arose about the price to be paid for the grand unified project that mechanism represented in natural philosophy. The possibility began to

[5] Micro-corpuscularianism is a broader notion than atomism, for atoms were conceived as discrete pieces of matter separated by empty space, whereas micro-corpuscularism embraces both atoms in empty space and discrete pieces of matter in a material fluid medium, which may be distinguished from this surrounding fluid medium purely in terms of motions relative to one another (as in Descartes' account). These differences mattered in the seventeenth-century developments discussed in *The Emergence of a Scientific Culture*, but by the middle of the eighteenth century they are less important.

be raised that, as well as explanation in terms of micro-structure, there was also a legitimate form of explanation of phenomena in terms of how they were related to other phenomena. Indeed, the latter—non-reductionist and non-universalist—forms of explanation took on an increased significance as, from the 1680s onwards, it became increasingly difficult to see how the mechanist project might be realized, with mechanical explanation degenerating into promissory notes when it came to recalcitrant phenomena such as gravitation, electricity, magnetism, and chemical reactions.

Since it was mechanism that provided the model of general understanding both for those who wanted to effect a synthesis of natural philosophy and Christian theology, as well as for those who thought that mechanism could progressively usurp at least many of the claims of Christianity, its demise was momentous. Mechanism was the only comprehensive modern system of natural philosophy, and as a result its collapse had significant consequences for how one established a connection between an account of the natural world and an account of our place in it. While there had been some advocates of mechanism who tried to draw what they took to be the consequences of a mechanist model for religious, moral, and political questions, notably Hobbes and Spinoza, such an approach was marginal. The main thrust of mechanist thinking was to establish a mutually reinforcing relationship between natural philosophy and Christian theology, and Descartes and Boyle both proceeded in this way for example, despite profound differences in how they carried out the project. They took an understanding of our relation to the natural world from Christianity and mapped it on to the account of the world provided by mechanism, with a view to establishing an integrated picture of the world and our place in it (as it was conceived in Christian thinking) or, to put it in more modern terms, our relationship to it.[6]

This is precisely what was no longer possible. If, as Locke and his followers insisted, the unificatory project to which mechanism subscribed was based on a fundamental misunderstanding of the nature of physical enquiry, then our account of the world would come not in the form of a general theory, as Aristotelianism and Cartesianism had done, but rather in the form of relatively self-contained areas of inquiry—such as chemistry, electricity, measurement of the shape of the earth, and developmental biology—which had no intrinsic or obvious connections with one another. These provided successful and novel accounts of various physical phenomena, but they did not offer anything like the kind of general account that could be supplemented to provide an understanding of the world and our relation to it. In short, with the demise of

[6] I use the word 'picture' advisedly, for picturability was crucial to understanding in seventeenth-century natural philosophy. See Frédérique Aït-Touati and Stephen Gaukroger, *Le monde en images: Voir, représenter, savoir, de Descartes à Leibniz* (Paris, 2015); and Dennis Sepper, *Understanding Imagination: The Reason of Images* (Dordrecht, 2013). Cf. Sachiko Kusukawa, *Picturing the Book of Nature: Image, Text, and Argument in Sixteenth-Century Human Anatomy and Medical Botany* (Chicago, 2012).

mechanism, scientific developments could no longer necessarily be seen as part of a coordinated enterprise. They were relatively piecemeal, for there was no general picture into which they fit, of the kind that had been provided by mechanism. Correspondingly, they lost the kind of purchase on legitimacy and relevance that the seventeenth-century natural philosophers were able establish for them. There were, it is true, those who did try to generate an understanding of the human condition, as it were, from an account of physical science, more specifically from an account of the nature of matter. The French materialists, notably La Mettrie, offered something along these lines, but quite apart from the question of the intrinsic implausibility of their proposals, the project was really an outdated throwback to the Hobbesian and Spinozean ones of the seventeenth century. For those who took seriously the collapse of the seventeenth-century conception of natural philosophy, epitomized in mechanism, no revision of this conception was going to suffice.

The demise of mechanism forces us to confront a question that dominated *The Emergence of a Scientific Culture* and *The Collapse of Mechanism and the Rise of Sensibility*, and will continue to preoccupy us here: how was the scientific enterprise validated, such that it was able to occupy a central place in Western culture and gradually shape all cognitive values around its own? The decisive factor in the West was not the level of scientific innovation, because the other examples of long-term scientific activity that I have mentioned were successful in this respect. Rather, it was the consolidation of the scientific enterprise at a general level, one that placed a value not so much on particular scientific theories or results as on science as such. In the early-modern era, the picture of how this came about, at least up to the early decades of the eighteenth century, is reasonably clear. The connection, initially made by theologians, of narrowly conceived natural-philosophical values with broadly conceived theological ones, meant that scientific values generally took on a unique prominence, which early-modern natural philosophers built on and transformed. More specifically, the general vindication of scientific enquiry resulted in large part from a number of complex symbiotic exchanges between natural philosophy and Christian doctrine, from which both emerged transformed. Natural philosophy was transformed into something that contributed to a task which had previously been the sole responsibility of Christian doctrine, namely understanding our place in nature. For this, it was crucial that natural philosophy have some overall coherence; that it be a unified enterprise, somewhat as Christian theology was.

Here we come face to face with the core problem. By the middle of the eighteenth century, if scientific practice was to have any connection with broader issues, the path to securing this could no longer be to proceed in the way that mechanism had up to the early decades of the eighteenth century. Something very different would be needed. And if it were not provided, there was a strong chance that Western science would increasingly turn out to look like the earlier cultures with strong scientific programmes which underwent a few decades of success before interest in science gradually began to decline, as the original aims of specific programmes came to be realized. Remember also that we are not dealing here with something that had any significant practical impact before the middle decades of the nineteenth century, so legitimacy and relevance could not be established in general, practical terms. Indeed, we can locate the problem period more precisely as being that between the collapse of mechanism in the early decades of the eighteenth century and the emergence of a new intimate relationship between science and technology in the mid-nineteenth century. Before this period, what secured the legitimacy of science was its integration with a Christian world view, as a result taking on roles that had previously been the preserve of Christianity. This source of legitimacy was not wholly abandoned between the mid-eighteenth and the mid-nineteenth centuries, and indeed in areas such as natural history attempts were made to foster it from time to time,[7] but there was no well-defined 'science' in general which could continue to contribute to, and benefit from, this relationship. By contrast, after this period, the legitimacy of science, now becoming more integrated as a coherent whole, turned in large part on its usefulness, at least until the horrors of technologically driven warfare prompted attempts to dissociate it from technology in the wake of the First World War, and to establish a new identity for itself.

I have set out in this book to explore what happened in the period between these two developments. Was there a hiatus, during which the West began to mirror other cultures? There is no evidence of such a hiatus. We certainly do not witness a decline in interest in science, for example. Quite the contrary, there are a number of new attempts to define what science is and what it means to be scientific, above all in the burgeoning new fields of the 'moral' or 'human' sciences. What I have set out to show is that the legitimacy of science was established anew in a new arena, and that this brought about a reversal of legitimatory priorities. In the first place, it was not the natural sciences—which had been regarded as the core of scientific activity not just for mechanists but for all natural philosophers up to the middle decades of the eighteenth century, especially the areas of mechanics and astronomy—but the newly emerging human or moral sciences that now came to the fore, providing the ground on which new scientific conceptions were forged, in the process displacing traditional religious, humanist, and other approaches. It is in the moral sciences—defined by Condorcet as 'all those sciences that have as the subject of their researches either

[7] The *Bridgewater Treatises* (8 vols, London, 1834–7) are often taken as a significant late work in this genre, but it was no longer possible by this stage to provide the kind of overarching physico-theological structure for knowledge at which earlier physico-theological works had aimed. The volumes acted instead merely as vehicles for instruction in science in a way that did not threaten religious orthodoxy. See Jonathan Topham, 'Science and Popular Education in the 1830s: The Role of the *Bridgewater Treatises*', *British Journal for the History of Science* 25 (1992), 397–430.

the human mind in itself, or the relations of men one to another'[8]—that questions of legitimacy are pressed. Second, the move to the moral sciences changes the way in which the understanding of the natural realm and the understanding of our relation to this realm are connected. Seventeenth-century natural philosophy took the first—understanding of the natural realm—as its own domain, and the second—understanding of our relation to this realm—as something provided by Christianity (or less commonly civic humanism of various kinds), or in extreme cases (Hobbes, for example, at least as he was commonly conceived) took understanding our relation to the natural realm to be nothing over and above understanding the natural realm. But what happens with the emergence of the human sciences is that considerations of our relation to the natural realm now shape conceptions of the natural realm itself.

In short, a new scientific culture did emerge, but it was one quite different from anything that had come before. One of its principal sources was Lockean sensationalism, which was itself an attempt to make sense of the experimental natural philosophy tradition, and consequently was intimately tied in with the development that did more than anything else to undermine the explanatory ambitions of mechanism. Locke had stressed the importance of sensation as a source of knowledge, and his followers, particularly those in France, extended this into a more general notion of sensibility—something that drew together questions in epistemology, physiology, medicine, and aesthetics. What resulted was a rethinking of the sources of our knowledge of the world, which developed into a new understanding of our relation to the natural world. In particular, French Lockeans such as Diderot argued that no cognitive relation to the world was possible without sensibility. Without sensibility, our cognitive faculties would just furnish us with experiences: it is sensibility that provided access to the world as something independent of us. Yet this dimension of experience was something alien to the mechanist model of natural philosophy, with its very limited biomechanical conception of perceptual cognition. Consequently, any account of our relation to the natural world now needed something more than just natural philosophy, at least as conceived up to this point. What I have set out to show is that, by the second half of the eighteenth century, considerations of our relation to the natural world started to shape conceptions of what kind of account of the natural world we should be seeking. This is the exact reverse of the mechanist programme, which starts with an account of the world and then supplements it with an account of our place in and our relation to this world. Instead, what now happens is that considerations of our relationship to the world shape what we want out of science, thereby providing it with a new cultural standing.

The new scientific culture of the second half of the eighteenth century was effected primarily by means of a process of naturalization of the human, that is,

[8] 'Éloge de M. Bucquet': *Œuvres de Condorcet*, ed. A. Condorcet-O'Connor and F. Arago (12 vols, Paris, 1847–9), ii. 410.

the formulation in empirical terms of questions about the human realm that had up to that point taken a non-empirical form.[9] This changed conceptions of what science was and what it should be doing. Above all, to the extent to which it immersed the human in the natural world, it also played a role in the humanization of the world, within which one needs to include both the natural and the supernatural realms.[10] It not only made their investigation subject to human interests, but helped to transform nature from a cold inert realm into an active living one in which the human being found a natural place. By the last decades of the eighteenth century, the naturalization of the human and the humanization of nature existed in a complex interplay which had significant consequences for scientific, religious, and philosophical questions: as Kant put it in his lectures on anthropology, 'nature exists for the sake of the human, the human being is the purpose of nature'.[11] In short, the naturalization of the human, with its complementary humanization of nature, was a broad and consequential development, and it forms the core of our investigation into why a scientific culture in the West not only didn't collapse in the second half of the eighteenth century, but was renewed in a novel way.

In order to appreciate just what was involved in the naturalization of the human, significant orientation is necessary, and this is what I set out to provide in Part I. In Chapter 1, I begin by exploring a number of (often overlapping) issues that shaped the debates about the nature of the human between the middle of the eighteenth century and the beginning of the nineteenth. At the core of these is a complex and often acrimonious debate over whether the key to understanding human behaviour consists in the understanding of rationality. Philosophical, religious, and scientific questions were explicitly caught up in these debates. None of the issues was straightforward, and among the proponents of reason, for example, there was a sharp divide between those who considered that the rationality of individual behaviour was the proper object of study, and those who focused on the rationality of collective behaviour. By contrast, among those who rejected the idea that reason was sufficient, it was typically assumed that sensibility underlay our cognitive states as well as our affective ones, and indeed that the two

[9] This characterization of naturalism is different from most of those offered in contemporary philosophy, which tie naturalism to continuity with the physical sciences (or, in areas such as ethics, to evolutionary biology). See for example the helpful account in Jack Ritchie, *Understanding Naturalism* (Stocksfield, 2008). I hope it will be clear in what follows why this is inadequate for my purposes, but also why my definition captures something crucial to naturalism as commonly conceived.

[10] The separation between natural and supernatural is not a perennial one: two separate realms were distinguished only in the thirteenth century. Prior to that, while Christian theology remained Platonist, they were conceived in terms of continuous gradations. The strict separation, which comes with Aquinas, radically changed conceptions of 'our place in nature'.

[11] Immanuel Kant, *Gesammelte Schriften* (29 vols, Berlin, 1900 onwards), xxv. 470; *Lectures on Anthropology* (Cambridge, 2012), 48. (Note that I cite English translations where available, but do not always follow them to the letter: I have occasionally offered my own readings of particular sentences or phrases, without comment, where necessary for clarity or precision.)

were inseparable. This development put the autonomy of reason itself into question. But the problem had a number of dimensions, for reason and sensibility were aligned with two other issues: that of systematic versus non-systematic understanding, where the questions turned on the legitimacy of non-systematic forms of understanding; and propositional understanding (knowing that something is the case) and non-propositional forms of understanding (coming to terms with the world in terms of aspirations, anxieties, raw beliefs, desires, etc.), where the questions turned on whether the two could be reconciled. Finally, I examine the replacement of the search for certainty, which had played such an important part in the establishment of mechanics as a model for scientific enquiry, with the elaboration of procedures for the minimization and avoidance of error: rigour and precision now became separated from certainty.

Chapter 2 looks at conceptions of the nature of matter, traditionally one of the key ingredients in natural understanding. From the middle decades of the eighteenth century onwards, the understanding of science and its tasks was significantly altered. Above all, it was no longer thought of as something self-contained, but rather as an ingredient in a project of accounting for our relation to the natural realm. The naturalization of the human and the humanization of nature of necessity worked with basic assumptions about nature, but traditional assumptions had been abandoned in the seventeenth century with the advent of mechanism, only for mechanistic ones to be themselves abandoned by the middle of the eighteenth century. Consequently, it is important that we understand the new conceptions of nature that underlay developments in the second half of the eighteenth century. What I want to establish above all is the existence of a radical change in the understanding of basic natural processes. In the middle of the eighteenth century, mechanism had relinquished any role in those disciplines at the cutting edge of developments in the physical and life sciences.[12] In the case of chemistry, a complex experimental-phenomenological approach yielded significant progress in the understanding of chemical processes, but it remained remote from other areas, such as electricity, until late in the century. At the end of the eighteenth century this changed, with the assimilation of electricity to vital forces, the discovery of the ability of electrical currents to effect chemical processes and vice versa, and the construal of basic properties of matter exclusively in terms of forces. These suggested a new factor unifying the physical and the life sciences: the existence of some basic force underlying all natural phenomena. In short, we explore the move from a reductionist account, along mechanist lines, to the attempts to establish autonomous disciples, notably in chemistry and the life sciences, to attempts to provide a new unifying synthesis.

Part II is devoted to our core concern, the naturalization of the human. I identify a number of different naturalizing projects, considered under four

[12] I argued this in detail in Part IV of *The Collapse of Mechanism and the Rise of Sensibility*.

general headings: anthropological medicine, philosophical anthropology, the 'natural history of man', and social arithmetic. A separate chapter is devoted to each of these.

Anthropological medicine, the subject of Chapter 3, was a highly naturalistic endeavour, but it was diametrically opposed to what was the dominant form of naturalization of the organic between the mid-seventeenth century and the early decades of the eighteenth, namely biomechanics. Biomechanics comprised a reduction of vital phenomena to mechanical interactions between inert micro-corpuscles. Anthropological medicine, by contrast, was not reductive. It originated in the 1730s with the physicians of the University of Montpellier, for whom the subject matter of medicine included every aspect of human experience. These *médecins philosophes* aimed to offer a general account of the interconnections between the realms of the physical, the mental/intellectual, and the emotions. Anything external to the body that affected the three realms was observed and recorded, for it had an effect on the well-being of the person. The resolutely naturalistic approach of anthropological medicine to these domains became a dominant way of thinking about the human in France, and it included everything from attempts to temper the effects of excessive sensibility caused by the unnatural environment of modern urban life, to the provision of a psychological basis for political programmes in the thinking of the *idéologues*, where a sensationalist psychology became the route to moral, economic, and social theory. Because advocates of anthropological medicine saw health in terms of harmonious balance, apparently extraneous political and social factors were considered part of the overall balance and needed to be adjusted to secure human happiness. The route to naturalization in this project was, in the first instance, via the naturalization of nervous sensitivity, but there were different forms of naturalization—such as those offered by laboratory medicine—that cut across and were at odds with this, and which consequently had to be excluded. It was not a matter of naturalization *tout court*, but rather one of choosing between competing strategies of naturalization. Anthropological medicine was committed to very strong claims about the physiological basis of human behaviour and it found itself in an increasingly hostile climate in the post-Revolutionary period, contributing to the difficulties that the *médecins philosophes* faced in holding together a programme which had extolled the virtues of naturalization against metaphysical and theological speculation, while at the same time resisting forms of naturalization in physiological research and experimental medicine due to the increasing use of microscopy and chemistry.

Philosophical anthropology, the subject of Chapter 4, was largely an outcome of a reaction to metaphysics. Outside Germany, metaphysics had become marginalized from the beginning of the eighteenth century.[13] Within Germany, by

[13] The term 'Germany', in this period, covers the German-speaking parts of the Holy Roman Empire, although for all intents and purposes we shall be concerned exclusively with Prussia.

contrast, Wolff's version of Leibnizian metaphysics had some hold on university curricula, although by the middle of the century it was largely discredited, and Hume (in his essays and historical writings) and Rousseau began to provide models for the kind of enterprise that might replace metaphysics in the understanding of human behaviour. From the 1760s, Herder was advocating a novel philosophical anthropology which took its bearings from rearguard attempts by metaphysicians to come to terms with sense perception, and sensibility more generally. Sensibility had been thought by a number of philosophers since mid-century, when hostility to metaphysics was at its peak, to underlie our cognitive states, with the consequence that what, in metaphysical treatises for example, had been treated as epistemological questions were now given distinctive psychological and anthropological dimensions. Of particular importance in this context was the way in which a form of empirical psychology came to the fore as a means of explaining our cognitive and affective states. Herder uses this empirical psychology to replace abstract notions of reason. Treating reason in terms of thought, and thought in terms of empirical psychology, he accounts for the 'empirical' aspect of this psychology by exploring the way in which language reflects thought, so that the study of language now becomes an investigation of thought processes as they are mirrored in particular languages. Kant, by contrast, rejected all such attempts at naturalization, on the basis of an understanding of scientific enquiry that took rational mechanics as its model, arguing that philosophical anthropology could not aspire to the standing of 'knowledge'.

Anthropological medicine and philosophical anthropology both offer programmes for cultivation of the faculties, and shaping of the personality. By contrast, the second two forms of naturalization, which I look at in Chapters 5 and 6, are more descriptive than prescriptive, a feature that played a significant role in their subsequent domination of nineteenth-century thinking about human behaviour. Above all, they put in place profoundly challenging reassessments of the nature and sources of human behaviour.

The 'natural history of man', the subject of Chapter 5, was in many respects the forerunner of what came to be known as physical anthropology. Although philosophical anthropology and the natural history of man were not clearly distinguished in the eighteenth and early nineteenth centuries, they proceeded in entirely different ways from one another. I take the natural history of man to comprise comparative anatomy (what will in the course of the nineteenth century form the basis for evolutionary theories); what in the eighteenth century was termed 'physical geography' (in more modern terms a mix of human geography and physical geography), which attempts to identify the environmental factors that shape human behaviour; human variation, such as racial variation, and its significance; and the external marks of character, notably physiognomy, craniology, and phrenology. Crucial to this comparative exercise are the questions of whether there is any continuity between apes and humans, what the standing of seemingly half animal/half human feral children is, whether there is a continuity between human races or whether they have separate origins, whether physical differences between men and women extend beyond their reproductive organs, and more generally whether significant differences are due to environmental or intrinsic factors. 'Environmental' factors are increasingly explored in the course of the eighteenth century in terms of comparative geography, which considers regional variations and their causes, and comparative history, which works primarily in terms of stages of historical development, from a primitive to modern civilization. Both of these played an important role in anthropological understanding, especially the second, which provided richer resources. But whereas physical geography offered a set of possible determinants of behaviour which one could either accept or reject, the programmes behind the identification of historical stages varied radically, some identifying socio-economic factors as driving the historical sequences, others cultural and aesthetic factors, yet others religious ones, and so on. In consequence, claims about the influence of such factors depended on larger, intractable issues.

Chapter 6 deals with a fourth form of naturalization, social arithmetic, which treats questions of human behaviour not in individual terms or in terms of universal attributes—as Christian teaching and the metaphysical and humanist traditions had done for example—but in terms of collective or aggregate properties. These are distinctive features of collective activity that are not mirrored at the individual level, and cannot be arrived at by abstraction from individual behaviour. I trace the investigation of collective properties back to the beginning of the eighteenth century, to Mandeville's notorious tract *The Fable of the Bees*, which provoked many of the questions concerning the relationship between individual behaviour and collective outcomes that were subsequently to dominate social and political thought. Above all, Mandeville stressed the existence of a discrepancy between the reliance of economic and political well-being on the collective consequences of self-interest, and the moral condemnation of self-interest at an individual level. The use of social arithmetic as a means of naturalization of the human required a redefinition of morality, provoking one of the most profound transformations of human self-understanding of the modern era. In particular, Bentham offered a system of morality in which individual behaviour is modelled on quantitative procedures for assessing the morality of collective outcomes. At the same time, attempts to account for a range of aspects of everyday life—from health and population to law and politics—are subjected to measurement as a means of accounting for various kinds of human behaviour. The development that most sums up the magnitude of the shift is the way in which the rational man comes to be replaced, as an object of study in the nineteenth century, by Quetelet's idea of the average man.

Finally, in Chapter 7, we turn to the naturalization of non-propositional forms of understanding: coming to terms with the world by means of such things as aspirations, desires, fears, raw beliefs, anxieties, responsibilities, and goals. Some coherence can be introduced into these otherwise disparate ways of engaging with the world by treating them under the rubric of myth. This category includes both religious and aesthetic forms of understanding our relation to the world, and one large-scale theme that will occupy us is the move from a religious to an aesthetic-humanist understanding. There are four developments that we shall be concerned with. The first was an influential argument, in Kant and the Kantians, about the limits of a philosophical account of theological questions, which leaves open the need for empirical—primarily historical—exploration. The second development was the independent emergence of a form of historical enquiry, one that made no concessions to 'sacred history', into the development of Christianity, especially into the circumstances under which its doctrines were formulated, substituting secular history, its critics complained, for faith in the sacred. At the same time there emerged a critical assessment of the role of Jesus, transforming him into a moral teacher. Third, there was a comparative account of the evolution of religions, ranking them from primitive to the most developed. The basic assumption behind such accounts was that at the apex of this development came Christianity. This led to the fourth development: the questioning of whether Christianity did in fact represent the final stage in spirituality. Feuerbach, for example, argued that the evolution did not in fact cease with Christianity but went one step further, culminating in a form of humanism. These developments highlight the way in which the naturalization of the human and the humanization of nature come to be part of the same programme by the nineteenth century, as Feuerbach's naturalization of the spiritual takes the form of a humanization of the spiritual. This became in many respects a defining feature of modern thought in the West. But these developments did not mean that the understanding of the spiritual effectively became reduced to knowledge of the world. In the final section of the chapter, we see how this understanding could remain in the mythical realm, but needed to be given some new coherence, perhaps in the form of a reworking of Christianity, or alternatively in terms of an aesthetic-humanist conception of our aims and aspirations.

PART I

I
The Dichotomies of Understanding

On 26 July 1772, Mlle Julie de Lespinasse, *salonnière* of one of the greatest Parisian *salons* from the mid-1760s to her death in 1776, wrote a letter of advice to Condorcet, whose career she and d'Alembert were nurturing, on how he might reinforce his credentials as a candidate for the position of next permanent secretary of the Académie des Sciences. Condorcet was well-respected in scientific and mathematical circles, and this was now, at d'Alembert's urging, a necessary credential. He lacked one crucial qualification, however. Previous secretaries had been known for their literary skills. Fontenelle, the most powerful and prestigious of them, whose forty-four-year tenure of the secretariat had raised the position to one of exceptional prestige, made his name initially in literature rather than natural philosophy. His contributions ranged from poetry, journalism, histories of mythology, and an opera libretto, to what can be considered the first piece of 'popular science', the highly successful *Entretiens sur la pluralité des mondes* of 1686. Working in a *salon* culture as much as in the Académies, it was Fontenelle who introduced the work of the Académie des Sciences, and natural philosophy more generally, to a broad reading public, by incorporating it into the Republic of Letters. Such skills were considered crucial to the secretary, and in particular, it was his duty to compose the *éloges* of academicians, which treated them in terms previously reserved for monarchs and generals. In her letter, Mlle de Lespinasse asks Condorcet whether he knows that his 'fellow academicians' are asking whether he 'has enough talent' to replace Jean-Paul Grandjean de Fouchy, whose ineffectual tenure of the position had seen the Académie fall behind in a number of respects. The talent she particularly has in mind is a literary one. She advises him to 'write a few pieces to be published before St. Martin's Day. I hear that you have something on printing. You would do well to look over it and publish it.'[1]

The topic of Condorcet's essay on printing was the effect of the birth of printing on the development of the sciences, but it was quickly refocused as a history of the Académie des Sciences. The history was to begin with an account of the sciences before the advent of printing, then proceed to their development before the foundation of the Académie, concluding with an account of their

[1] Julie de Lespinasse, *Lettres inédites de Mlle de Lespinasse*, ed. Charles Henry (Paris, 1887), 79–80.

progress from the founding of the Académie up to the present, with biographies of members of the Académie. It was this project, in itself hardly novel or adventurous, that was to grow, over a period of twenty-one years, into the most important statement of the idea that the history of civilization was the history of the progress of science. The association between science and civilization set out in the *Esquisse d'un tableau historique des progrès de l'esprit humain,* which was published posthumously in 1795, was to prove a pervasive and resilient idea.[2] Although the work certainly had precedents, most notably d'Alembert's genealogy of reason in the preliminary Discours of the *Encyclopédie*,[3] and Turgot's December 1750 lecture to the Sorbonne, 'Sur les progrès successifs de l'esprit humain',[4] it was Condorcet's defence of the inevitability of progress that would quickly become the default notion of civilization in the early decades of the nineteenth century, and it has remained so for many thinkers since that time, at least up to the second half of the twentieth century, displacing religion, the humanities, politics, and the arts in any claims they might have made to this role. The key argument is that progress in the physical sciences brings with it progress in the moral sciences and in practical applications: mastery of one's physical environment through the physical sciences leads to mastery of one's social and political environment by means of the moral sciences. The association of science and civilization consolidates a social role and standing for science that establishes it at the core of modern culture.

It has traditionally been assumed that the centrepiece of this association, which is distinctive of the Enlightenment, is the primacy of reason. In fact, the situation is more complex. What comes to the fore in the course of the eighteenth century is what I have termed naturalization: a concerted attempt to argue that questions that had been treated as conceptual or a priori were in fact more fruitfully pursued in empirical terms. As we shall see, this empirical investigation took a range of forms—medical, historical, physiological, anatomical, environmental, economic, and so on—and there was some disagreement on what the appropriate form was in many cases. The point is that what was at issue here was not reason versus religion, a common trope in studies of the Enlightenment, but reason versus empirical enquiry, in the form of a concern over what resources should be deployed in treating particular questions, and what we are going to be particularly concerned with are questions about the sources and nature of human behaviour. Metaphysics, theology, and to some extent civic humanism, had construed such

2 As Baker notes, Condorcet 'was a central figure in that widespread attempt to apply scientific thinking to all aspects of social affairs that was so marked a feature of late eighteenth and early nineteenth centuries, particularly in France. Saint Simon and Comte claimed him as the principal precursor of their positive sociology; Quetelet fulfilled his program for a social mathematics; Laplace, Poisson, and Cournot took seriously his dream of a rational science of decision-making.' (*Condorcet*, viii.)

3 See my discussion in *The Collapse of Mechanism and the Rise of Sensibility*, 427–37.

4 Anne Robert Jacques Turgot, *Oeuvres de Turgot*, ed. Eugène Daire (2 vols, Paris, 1844), ii. 597–611.

questions as falling exclusively under rational, as opposed to empirical, enquiry. This was now a point of contention. But the primacy of reason was also contested in a number of other ways. Some of these, such as the debates over reason versus sensibility, were connected directly with the question of empirical resources, for sensibility was easily naturalized in a way that reason could not be.[5] In other cases, such as disputes over the relative standing of reason and religious faith, there is no direct connection, but there is a probing of the nature and limits of reason that bears on the general issue of the role and limits of reason in enquiry. The primacy of reason was fiercely contested on a number of grounds in the period between the middle of the eighteenth century and the early decades of the nineteenth, and it was in the context of an unprecedented questioning of the nature and limits of reason that a new and intimate connection between 'science and civilization' was established. The human sciences formed a point of contact between science and civilization of a wholly new kind, and reshaped the understanding of both.

THE PLACE OF REASON

In Book I of his *Treatise Concerning Human Nature*, Hume offered an empirical exploration of the basic principles of human understanding, that is, one he envisaged as being based on observation, by contrast with general speculative principles. The core of observation, he tells us, lies in sensory or emotional experiences, either in the direct lively form, or in the fainter form of memories. The ideas that result from these are combined in three different ways. First, there are those cases where we experience two things as being routinely contiguous, and by association we acquire a disposition to think of the one when we think of the other. Second, there are resemblances between objects and, by association, when we think of an object we come to think of others that resemble it. Third, the way in which our minds are organized reflects our experience of a causal order that we perceive in the world, so that our thoughts run along tracks shaped by our (associative) experience.

Taken as claims describing various aspects of cognitive powers in animals, these would have been widely accepted. Although they might have been contested at the level of fine detail, there is nothing especially novel or radical in them. Similarly with the extension of such considerations to rudimentary human cognition, of the kind characteristic of the 'lower' faculties, and manifest perhaps in learning processes in infants, before reason locks in. We can find such an approach in Descartes, for example. In his account of cognition in the *Traité de*

[5] It is even possible to see sensibility in materialist terms, though I shall not be arguing that here. For an account of sensibilist materialism, see Charles Wolfe, 'Sensibility as Vital Force or Property of Matter in Mid-Eighteenth Century Debates', in Henry Martin Lloyd, ed., *The Discourse of Sensibility: The Knowing Body in the Enlightenment* (Dordrecht, 2013), 147–70.

l'Homme, he had explored animal cognition, partly as something of interest in its own right, but also partly as a way of clarifying what features of cognitive operations were purely corporeal and hence mechanistically characterizable, so that one could then ask what the introduction of a rational mind added to our abilities, and in what way it transformed them.[6] The aim of such an enterprise is to understand how animal cognition is effected and how far we share in this, with a view to establishing the boundary between what might be characterized as animal cognition, and properly human cognition.

Hume's project is far more radical than this, however. This is not because he sets out to establish that there is in fact no boundary between the animal/lower faculties and the distinctively human ones. Both the insistence on boundaries and the rejection of boundaries can be considered as equally speculative from a Humean point of view. Rather, what is at issue for Hume is a methodological question, which he couches in terms of a contrast between a speculative approach in which one comes to one's subject matter with already formed principles, dictated by metaphysical or other considerations, and a Newtonian 'natural history' that is free of such attempts at guidance on the basis of (assumed) prior understanding. Newton's account of gravitation is explicitly the model in the latter case. Rather than attempting to probe underlying causes, in the form of a theory of the behaviour of the micro-corpuscles that are deemed to constitute bodies, one instead tries to provide a comprehensive description of the phenomenon, showing how it underlies a range of events which had hitherto been considered entirely separate and unconnected (such as bodies falling to the ground and planets occupying stable orbits around the sun), and capturing their lawlike behaviour. Newton's achievement, construed in terms of its prevalent eighteenth-century Lockean interpretation, was presented as the triumph of factual investigation over speculative metaphysics: the latter being associated, in this case, with Cartesianism.

Hume's attempt to change the direction of philosophical enquiry had a particular resonance in Germany where, by contrast with France, there had still been a residual commitment to metaphysics in the middle of the century.[7] In his lectures on logic dating from the early 1770s, Kant distinguishes 'dogmatic' from 'critical' philosophy. The dogmatic philosophers identified are Descartes, Malebranche, Leibniz, and Wolff, 'the last of whom, through his industry,

[6] See Stephen Gaukroger, 'Descartes' Theory of Perceptual Cognition and the Question of Moral Sensibility', in John Cottingham and Peter Hacker, eds, *Mind, Method and Morality: Essays in Honour of Anthony Kenny* (Oxford, 2010), 230–51.

[7] On the background to this question, see Peter Petersen, *Geschichte der aristotelianischen Philosophie im protestantischen Deutschland* (Leipzig, 1921); Hans Maier, 'Die Lehre der Politik an den deutschen Univeritäten vornehmlich vom 16. bis 18. Jahrhundert', in Dieter Oberndorfer, ed., *Wissenschaftliche Politik: Eine Einführung in Grundfragen ihrer Tradition und Theorie* (Freiburg, 1967), 59–116; and Ian Hunter, *Rival Enlightenments: Civil and Metaphysical Philosophy in Early Modern Germany* (Cambridge, 2011).

produced a *systema* of philosophy'.[8] Among the critical philosophers, Hume figures prominently in Kant's account, although only Locke is identified here. Summing up the influence of the two approaches, Kant tells us that 'the critical philosophy thrives most, and in this the English have the greatest merit. For the most part the dogmatic method has fallen into disuse in all sciences; even morals is not expounded dogmatically any more, but more often critically.'[9] In aligning himself with 'critical' philosophy in these lectures, Kant is not using the term as he will after his 'critical' turn in the 1780s. The distinction drawn here between critical philosophy and dogmatic philosophy is different, and it is both fundamental and complex. It reflects a number of deep divisions in European intellectual culture in the second half of the eighteenth century. In its starkest terms, the choice can be seen as that between modelling oneself on Wolff—the systematic metaphysician cloistered in the university and writing in a dry style exclusively for a university audience—and Rousseau—the 'free agent' who sees little if any value in the education provided by universities, and who writes in a personal and engaging style for men and women of letters.

In the German case, *Bildung*—the notion of cultivating a particular kind of educated, cultured, and enlightened *persona* for oneself—is crucial.[10] The division is not merely between different styles of pursuing philosophy and science, but above all between different *personae* for the practitioner of these forms of enquiry. The contrast between the systematic metaphysics or *Schulphilosophie* of the Wolffians—typified for its critics in the *Stubengehlerte*, the antisocial and argumentative learned man who spent his all his time alone in his study—and the advocates of *Popularphilosophie* is not a simple one between content and presentation, but something that explicitly went to the heart of what it meant to be a philosopher.[11] A number of images were used by the *Popularphilosophen*. The problem with the professional philosopher, writes Feder, is that he is 'locked in his

[8] The 'Blomberg Logic': Kant, *Gesammelte Schriften*, xxiv. 37; translated in Immanuel Kant, *Lectures on Logic* (Cambridge, 1992), 24. Cf. Wolff: 'I finally brought everything together into an order such that one thing could be made understandable and known through another, and such that a *systema* could be made out of the foundational science, because all theories stand together like parts in a human body.' *Ausführliche Nachricht von seinen eigenen Schriften, die er in deutscher Sprache von den verschiedenen Teilen der Weltweisheit ans Licht gestellet* (Frankfurt am Maine, 1733), 229 (¶78).

[9] *Gesammelte Schriften*, xxiv. 37; *Lectures on Logic*, 24–5.

[10] See, for example, Hans Weil, *Die Entstehung des deutsches Bildungsprinzips* (Bonn, 1930). The literature is vast: for a guide to it, see Rudolph Vierhaus, 'Bildung', in O. Brunner, W. Conze, and R. Koselleck, eds, *Geschichtliche Grundbegriffe* (5 vols, Stuttgart, 1972–89), i. 508–51. For a recent account of the history of the notion, see Julian Hamann, '*Bildung* in German Human Sciences: The Discursive Transformation of a Concept', *History of the Human Sciences* 24 (2011), 48–72. Cf. Eckhardt Fuchs, 'Nature and *Bildung*: Pedagogical Naturalism in Nineteenth-Century Germany', in Lorraine Daston and Fernando Vidal, eds, *The Moral Authority of Nature* (Chicago, 2004), 155–81.

[11] See Doris Bachmann-Medick, *Die ästhetische Ordnung des Handelns: Moralphilosophie und Ästhetik in der Popularphilosophie des 18. Jahrhunderts* (Stuttgart, 1989); Gert Ueding, 'Populärphilosophie', in R. Grimminger, ed., *Deutsche Aufklärung bis zur Französischen Revolution 1680–1789* (Munich, 1984), 605–34. More generally on what it meant to be a philosopher in this period in Germany, see Werner Schneiders, 'Der Philosophiebegriff des philosophischen Zeitalters:

closet, buried under his books, withdrawn from the world' and preoccupied solely with 'people in his head'.[12] Garve compares the academic philosopher to the traveller who, concerned only with reaching his destination, takes the shortest route there, without any interest in the fields and farmlands that he traverses, whereas the *Popularphilosoph* is like someone who enjoys and enlightens himself on his stroll, taking in everything around him with no particular destination in mind.[13]

Popularphilosophie was in some respects a continuation of a Wolffian programme, one aimed at a cultivation of knowledge and virtue, even though the audience at which it was directed was different, as were the means by which it was achieved. Its aims were the cultural and ethical cultivation of members of the educated classes, as opposed to a university audience, and the means of cultivation lay not in metaphysics, as it did for Wolff and *Schulphilosophie*, but in areas such as history, aesthetics, and pedagogy, with a strong commitment to a literary culture, broad learning, and the art of conversation.[14] Although a few of the *Popularphilosophen* were professors of philosophy, most were not, and many earned a living outside the university system: Tetens was briefly a professor of physics and philosophy, then mathematics and philosophy, before becoming a financial official in the Danish civil service; Wieland left the university system after a short spell and earned his living as a novelist; Mendelssohn was a bookkeeper in, and then co-owner of, a silk factory; Nicolai was a bookseller and publisher;[15] and so on. Given this lack of adherence to any philosophical school, of the kind one finds in the university system at the time, it is not surprising that there was no consensus on substantial philosophical issues among the *Popularphilosophen*. As van der Zande points out, 'all the great debates of their times found them on both sides of the issues, whether it was about Lavater's theories of physiognomy, about the philosophy of Spinoza, about the question of deism in view of Lessing's publication of the Wolfenbüttel fragments, or about the supposed threat from obscurantist forces after the Edict of Wöllner

Wandlungen im Selbstverständnis der Philosophie von Leibniz bis Kant', in Rudolph Vierhaus, ed., *Wissenschaften im Zeitalter der Aufklärung* (Göttingen, 1985), 58–92.

12 Johann Georg Heinrich Feder, *Untersuchungen über den menschlichen Willen* (4 vols, Göttingen, 1779–93), i. 350.

13 See for example his disquisition on 'the art of thinking': Christian Garve, 'Einige Beobachtungen über die Kunst zu denken', in *Versuche über verschiedene aus der Moral, der Litteratur und dem gesellschaftlichen Leben* (5 vols, Breslau, 1792–1802), ii. 245–430.

14 This is evident for example in the journal *Der Philosoph für die Welt*, the first volume of which was published in 1775 in Leipzig by Johann Jakob Engel, to be followed by a second in 1777. It was the epitome of *Popularphilosophie*, containing, for example, literary pieces by Engel himself, a cautionary discussion of Helvétius' materialism, debates on Lessing's *Emilia Galotti* and on Goethe's *Werther*, more philosophical pieces by Mendelssohn and Benjamin Franklin, and an essay by Kant on racial differences.

15 On the cultural importance of publishing at this time, and Nicolai's role in it, see Pamela Eve Selwyn, *Everyday Life in the German Book Trade: Friedrich Nicolai as Bookseller and Publisher in the Age of Enlightenment, 1750–1810* (University Park, Penn., 2000).

became effective'.[16] But there was one theme that did unite the *Popularphilosophen*: a commitment to an anthropology, a science or philosophy of man. As the *Popularphilosoph* Karl von Irwing put it, 'presently one can incorporate almost our whole philosophy, and not without profit, into the science of man'.[17]

What is at issue here is not a question of Enlightenment versus Counter-Enlightenment, for Wolff and his followers were very much the standard bearers of the German *Aufklärung*, especially in its early phase. Neither is it just a question of the standing of metaphysics. Science too is subjected to severe criticism in *Popularphilosophie*, basically along lines dictated by Rousseau. In his pre-critical days, for example, Kant had few doubts about the dangers of pursuing metaphysics and science, and his characterization of them is in terms of something unhealthy. In an essay of 1764 he includes metaphysical speculation in an account of mental illness and its cure.[18] And elsewhere he tells us that science 'cannot make up for its own discontents', and its practitioners 'become increasingly harmful to themselves and useless to the public'; science itself 'tears us away from pleasures'.[19] Newton, in other contexts Kant's great hero (along with Rousseau), is described as becoming 'an ape or an angel instead of a human being', never satisfying himself, becoming childish, and ending his days as someone who was laughed at.[20]

The concerns that emerge here are not specific to Germany, and Hume and Diderot were raising related worries. To take these concerns seriously, as British, French, and German thinkers in the second half of the eighteenth century certainly did, was not only to question the ability of science to provide a general model of enquiry, but also, potentially, to call into doubt the very legitimacy of the scientific enterprise, at least as it had been construed up to this point. To understand what can only be described as a crisis of confidence in the ability of science to articulate, embody, or share broader cultural values, we need to separate out some core issues and to try to understand how they came to be connected in the way that they were.

There are three levels on which this dilemma works: that of propositional versus non-propositional forms of understanding; that of systematic versus non-systematic conceptions of understanding; and that of reason versus sensibility. The respective terms of the three dichotomies tended to go together. Advocacy of the primacy of reason was associated with the advocacy of comprehensive systematic understanding, and in turn with the idea that understanding had to take

[16] Johan van der Zande, 'In the Image of Cicero: German Philosophy between Wolff and Kant', *Journal of the History of Ideas* 56 (1995), 419–42: 424.

[17] Karl Franz von Irwing, *Erfahrungen und Untersuchungen über den Menschen* (2 vols, Berlin, 1777–9), preface.

[18] 'Versuch über die Krankheiten des Kopfes': *Gesammelte Schriften*, ii. 271.

[19] 'Metaphysik Herder' (i.e. Herder's notes on Kant's metaphysics courses from the 1760s): *Gesammelte Schriften*, xxviii. 892.

[20] Ibid., 894.

the form of propositional understanding. Correlatively, advocacy of the primacy of sensibility tended to be associated with the advocacy of non-systematic understanding. But such forms of non-propositional understanding as coming to terms with the world and our place in it through our desires, anxieties, or aspirations, did not seem in any way to be a replacement for propositional understanding. In short, there appeared to be a fundamental dichotomy in our understanding of the world.

We can distinguish three broadly different kinds of response to the dichotomy: those that privilege propositional understanding; those that privilege non-propositional understanding; and those that see the solution in balancing the two. It will be helpful to get a sense of what motivates these three kinds of response, before we examine the larger issues of reason versus sensibility, and systematic versus non-systematic understanding.

On the first view, the claim is that non-propositional forms of understanding have to be 'propositionalized' in some way if they are to count as genuine forms of understanding. Failure to achieve this means that they lose their status as forms of understanding. This is because, on such an approach, propositional understanding is the only possible genuine or legitimate form of understanding. There are different versions of this kind of approach, and we shall be considering the Wolffian defence of reason, which is a defence of exclusively propositional understanding, later. A different kind of approach which ends up with largely the same conclusions, albeit couched in a more radical form, is that associated with some of the ideas inspiring the French Revolution, and the key figure here is Condorcet, permanent secretary of the Académie des Sciences, and the only one of the *philosophes* to play a significant role in the events following 1789.

Condorcet's ambition, as Baker has put it, was 'to guide the French Revolution by transforming societal choice into the rational decision-making of the idealized republic of science'.[21] He set out to provide 'stronger combinations, more precise means of calculation' because 'all of the truths recognized by the enlightened have been confused in a mass of uncertain changing opinions', so that it is necessary to 'fetter men to reason by the precision of ideas, by the rigour of proofs, so as to put the truths presented to them beyond the sophistry of interest'.[22] Whereas in the Wolffian tradition—and that of the German Enlightenment, the *Aufklärung*—reason is embodied in metaphysics, in the French *philosophe* tradition reason or rationality is embodied in science. The connection between the social and political domain and that of science had been established and set out in detail mid-century in terms of the association of science and civilization in d'Alembert's

[21] Baker, *Condorcet*, ix. Cf. two earlier works: Gilles-Gaston Granger, *La mathématique sociale du Marquis de Condorcet* (Paris, 1956); and Roshdi Rashed, *Condorcet: Mathématique et société* (Paris, 1974).

[22] Marie Jean Antoine Nicolas de Caritat, Marquis de Condorcet. *Esquisse d'un tableau historique des progrès de l'esprit humain* (Paris, 1795), 542.

preliminary 'Discours' to the *Encyclopédie*.[23] D'Alembert reworked the theme of the transition from superstition to rational belief that had been pursued in writers such as Fontenelle from the late seventeenth century onwards, providing a progressive genealogy in which various stages in the triumph of reason can be discerned. We are offered a vindication of the project of the *Encyclopédie* in the distinctively Baconian genre of a legitimating genealogy. What is at issue is primarily a question of establishing a historical sequence in which one can follow a progression that starts with the origins of knowledge and traces a process of growth—while uncovering and analysing various false starts—which can be shown to culminate in the present, so that the present, as represented by the *Encyclopédie*, provides both a secure vantage point for scrutinizing the past and a secure starting point for further enquiry. In his 1753 *Essai sur la société des gens de lettres et des grands*, d'Alembert adds a political dimension to this account, telling us that Louis XIV gradually conquered the traditional predilection of the nobility for ignorance, and natural philosophers and others engaged in intellectual pursuits were taken out of the solitude to which they had been accustomed and raised to the standing of 'greats'.[24]

What d'Alembert sets out to describe in his account is the progress of civilization from superstition to a scientific understanding of the world, and the *Essai* shows how this transition can be reinforced by a raising of the standing of those to whom we owe the greatest debt in this respect. What is at issue here is not a reworking of Plato's advocacy of 'philosopher-kings', for the contribution of philosophers and scientists is less direct than that. But we do have a notion of a rational ordering of society, and with d'Alembert's protégé Condorcet, the model for the reordering is that of a rationally organized science.

This model is set out in his *Esquisse d'un tableau historique des progrès de l'esprit humain*, published in 1795, a year after Condorcet's death in prison. Like d'Alembert's 'Discours', the *Esquisse* was originally intended as an introduction to a larger (unfinished) work. The Tenth Epoch, on the 'Future Progress of Mankind', begins:

> If man can predict, almost with certainty, those appearances of which he understands the laws; if, even when the laws are unknown to him, experience of the past enables him to foresee, with considerable probability, future appearances; why should we suppose it a chimerical undertaking to delineate, with some degree of truth, the picture of the future destiny of mankind from the results of its history? The only foundations of faith in the natural sciences is the principle, that the general laws, known or unknown, which regulate the phenomena of the universe, are regular and constant; and why should this principle,

[23] Denis Diderot and Jean le Rond d'Alembert, *Encyclopédie ou Dictionnaire raisonné des sciences, des arts et des métiers par une société des gens de Lettres, mis en ordre et publié par Diderot et quant à la Partie mathématique par d'Alembert* (2nd edn, 40 vols, Geneva, 1777–9), i. xxxii–lix.

[24] Jean Baptiste d'Alembert, *Œuvres* (5 vols, Paris, 1821–2), iv. 338–43.

applicable to the other operations of nature, be less true when applied to the development of the intellectual and moral faculties of man?[25]

The crucial thing to note about the claims regarding predictability here is that they apply not to individuals as such but to collective properties of larger-scale human organizations and societies. That is to say, questions of human behaviour are treated neither in individual terms nor in terms of universal attributes—as Christian teaching and the humanist tradition had done for example—but in terms of collective properties, that is, distinctive features of collective activity which are not mirrored at the individual level, and cannot be arrived at by abstraction from individual behaviour. The eighteenth-century shift to the study of collective properties will occupy us in Chapter 6. For the moment, it is sufficient to note that Condorcet's focus is on collective properties.

The 'Introduction' to the *Esquisse* takes its bearings from a Condillac-inspired sensationalist psychology—virtually a *sine qua non* of French intellectual life in the second half of the eighteenth century, as we shall see—and he makes the standard move of associating sensations with feelings of pleasure and pain, and transforming these into 'durable sentiments'. These latter enable us to form ties of interest and duty with our fellow creatures. But to leave things at this individual level, Condorcet tells us, is to pursue metaphysics, whereas he has something different in mind:

But if we consider this development in its results, relative to the mass of individuals co-existing at the same time on a given space, and follow it from generation to generation, it then exhibits a picture of the progress of human intellect. This progress is subject to the same general laws, observable in individual development of our faculties; being the result of that very development considered at once in a great number of individuals united in society.[26]

Here we can begin to appreciate the immense distance between what Condorcet is doing and a metaphysical project such as that of Wolff. Despite the fact that they are both committed to the idea that 'reason' is the sole key to understanding human behaviour, their objects of study are completely different. Wolffian metaphysics is concerned to establish the grounding of human behaviour—whether cognitive, affective, or moral—in reason, whereas Condorcet's concern is with what account we can give of the rationality of collective behaviour.[27] This leads him to deal with various voting paradoxes, such as that where the ratification

[25] Condorcet, *Esquisse*, 327. I have used the English translation which appeared in the same year as the French edition: *Outlines of an Historical View of the Progress of the Mind* (London, 1795), 316.

[26] Condorcet, *Esquisse*, 2–3; *Outlines*, 3.

[27] The gradual separation of 'reason' (considered as the highest human faculty, and something with general application) and 'rationality' (considered as formal reasoning independent of personality and context, and usually taking the form of algorithms) came to a peak in the middle of the twentieth century: see Paul Erikson et al., *How Reason Almost Lost its Mind: The Strange Career of Cold War Rationality* (Chicago, 2013).

of a constitution can produce different results if each clause must be ratified by a majority of assemblies and the whole constitution must be ratified by a majority of assemblies;[28] with the conditions under which it is legitimate to subject citizens to a law which has not received unanimous approval; and with calculating the maximum acceptable error in collective judicial proceedings.[29] Such questions are quite different from any of the considerations raised in the Wolffian metaphysical tradition, but they are critical to the kind of rationality at stake in Condorcet.

By contrast, the second kind of response denies that rationality is what is at stake either in the individual case or the collective one. Rousseau, who privileges non-propositional forms of understanding, is concerned with both kinds of case, above all in his treatment of how one moves from individual wills to the general will. But for Rousseau, primitive 'natural' forms of understanding are taken to have been perverted by civilization and the educational precepts by which it is perpetuated. Reason is associated with a rigid dogmatism, something evident in his account of education, where the idea of an authority figure who imparts knowledge according to a pre-established scheme is replaced with a process whereby the child's natural capacities are activated in a process of autonomous discovery.[30] This focus on natural capacities shifts the centre of gravity of the discussion from reason to sensibility. Sensibility in this context can be thought of as a very specific form of non-propositional understanding, so to this extent Rousseau is rejecting propositional understanding in favour of a form of non-propositional understanding. By contrast, a different kind of strategy to elevate sensibility is that pursued by Diderot, who argues that sensibility underlies our cognitive abilities. Since these cognitive abilities were traditionally included under 'reason', what is being claimed in this case is that a form of non-propositional understanding does not replace, but underlies, propositional understanding.[31] Nevertheless, the autonomy of propositional understanding is clearly threatened by such a strategy.

[28] Condorcet, 'Sur la nécessité de faire ratifier la constitution par les citoyens' (1789): *Œuvres*, ix. 411–13. What he shows is that majority preferences are intransitive.

[29] Condorcet, *Essai sur l'application de l'analyse à la probabilité des décisions rendues à la pluralité des voix* (Paris, 1785).

[30] See, in particular, Jean-Jacques Rousseau, *Émile, ou de l'éducation* (The Hague, 1762).

[31] I will be restricting my attention here to the French Lockean tradition, but it is worth noting that a conclusion mirroring this was reached on wholly different grounds in the German natural law tradition. Johann Jacob Schmauß, for example, in his *Vorstellung des wahren Begriffs von einem Recht der Nature* (Göttingen, 1748), argued that no one corresponding to the idea of man as presented in natural law theory can be found in history. History shows reason to be a rare occurrence, and the discrepancy between philosophy and history in this respect indicates that there is something wrong with philosophy. The emotions and the will are the natural features of human beings, Schmauß argues. Reason is a product of development: it comes after, and rests on, the emotions. See the discussion in Peter Hanns Reill, *The German Enlightenment and the Rise of Historicism* (Berkeley, 1975), 57–8.

A third option is that offered by Hume. Rather than attempting to come to a satisfactory conception of understanding by choosing between the alternatives, he opens up the possibility of conceiving of understanding in terms of a balance between propositional and non-propositional forms of understanding. At the beginning of *An Enquiry Concerning Human Understanding*, he proposes a dilemma with respect to the nature of understanding. He writes:

But the mind requires some relaxation, and cannot always support its bent to care and industry. It seems, then, that nature has pointed out a mixed kind of life as most suitable to the human race, and secretly admonished them to allow none of these biases to *draw* too much, so as to incapacitate them for other occupations and entertainments. Indulge your passion for science, says she, but let your science be human, and such as may have a direct reference to action and society. Abstruse thought and profound researches I prohibit, and will severely punish, by the pensive melancholy which they introduce, by the needless uncertainty in which they involve you, and by the cold reception which your pretended discoveries shall meet with, when communicated. Be a philosopher, but amidst all your philosophy, be still a man.[32]

Simplifying somewhat, Hume came to the view that the most developed form of systematic general understanding, metaphysics, led to paradoxical conclusions which were so contrary to 'common life' that we could not accept them as true. Yet to the extent to which we desire to develop a critical understanding of the world and our moral and other practices, we cannot renounce such forms of philosophical enquiry. We are faced with a mode of life and thought that leads to dangerous or ridiculous phantasies, but which is part of the thoughtful existence distinctive of civilization, by contrast with an unreflective mode which he associated with barbarism. Neither 'reason' alone nor 'sensibility' alone can reliably guide our choice here: it is a matter of judging between the competing demands of the two.[33] Correlatively, Hume would also have rejected the notion advocated by d'Alembert and Condorcet that history was the history of reason. In his detailed account of the development of constitutionalism in Britain in his *History of England*, for example, he is at pains to stress that the British constitution was not the work of conscious design: rather its provisions were largely unintended results of painful and unwilling adjustments forced by over a century of political chaos.[34]

The range of positions is, then, considerable. To probe more deeply we need to consider two basic dichotomies that make up the fault lines of thought about science and philosophy in the second half of the eighteenth century: systematic versus non-systematic knowledge, and reason versus sensibility.

[32] David Hume, *Enquiries concerning the Human Understanding and Concerning the Principles of Morals*, ed. L. A. Selby-Bigge, 2nd edn (Oxford, 1962), 9.

[33] See Gaukroger, *The Collapse of Mechanism and the Rise of Sensibility*, 438–52.

[34] See, for example, the discussion in Donald W. Livingston, *Philosophical Melancholy and Delirium: Hume's Pathology of Philosophy* (Chicago, 1998), ch. 2.

SYSTEMATIC KNOWLEDGE

'Nothing makes me sicker than that arch-error of the Germans, namely the building of systems', writes Herder in 1767.[35] The dichotomy between systematic and non-systematic forms of understanding was absolutely basic after the early decades of the eighteenth century, but the issues go back to classical antiquity. The origins of philosophy (including natural philosophy) as a discipline lie in Plato, whose characterization of philosophy was set out in explicit contrast to the activity of the sophists. Plato drew on the systematic nature of philosophical enquiry, characterizing the failing of the sophists to lie in their seeking simply to win arguments or show off their ingenuity, whereas the truth that the philosopher sought requires systematic connections. The implication of this view is that it is the systematic nature of philosophy that prevents the decontextualized form of argument for its own sake that characterized the sophists' practice.[36] This systematic nature of the project was later reinforced with attempts, beginning in the thirteenth century, to reconcile Christian theology with Aristotelian natural philosophy.[37]

To understand the long-term significance of this, it is important to grasp a very distinctive feature of Christianity. Before the modern era, Christian theologians treated all religious differences in terms of differences in doctrine, and they construed Islam and Judaism, for example, as forms of heresy, rather than as different religions, for there was no concept of 'religions' in the plural. In the course of the seventeenth and eighteenth centuries, this changed. There now developed the view that there were different religions, although these were distinguished on a doctrinal basis. It is in the seventeenth century that we witness the first attempts to distinguish other religions from Christianity,[38] an exercise

[35] Herder to Scheffner, 31 October 1767: Johann Gottfried Herder, *Briefe: Gesamtausgabe 1763–1803*, ed. Wilhelm Dobbek and Günter Arnold (10 vols, Weimar, 1977), i. 92. Cf. his remark in his *Journal meiner Reise im Jahr 1769*: 'Above all I want to resist the temptation of the Germans to try to develop everything out of nominal explanations of what follows or what cannot follow.' Herder, *Werke in zehn Bänden*, ed. Günther Arnold et al. (9 vols in 10 parts, Frankfurt am Maine, 1985–2000), ix–2. 110. What Herder is opposed to is system-building, not systematic thinking: see Marion Heinz, *Sensualisticher Idealismus: Untersuchungen zur Erkenntnistheorie des jungen Herder* (Hamburg, 1994).

[36] See the discussion in Gaukroger, *The Emergence of a Scientific Culture*, 229–39. We are not concerned here with how accurate Plato's characterization of the sophists' practices were.

[37] By the seventeenth century, the notion of 'right reason' was being used extensively by Anglican divines to indicate an alignment of reason with Christian teaching: there was (mere) reason and there was sound or proper reason. See Lotte Mulligan, '"Reason", "Right Reason", and "Revelation" in Mid-Seventeenth-Century England', in Brian Vickers, ed., *Occult and Scientific Mentalities in the Renaissance* (Cambridge, 1984), 375–402; and idem, 'Robert Boyle, Right Reason, and the Meaning of Metaphor', *Journal of the History of Ideas* 55 (1994), 235–57.

[38] See Peter Harrison, *'Religion' and Religions in the English Enlightenment* (Cambridge, 1990), and Ernst Feil, *Religio: Die Geschichte eines neuzeitlichen Grundbegriffs vom Frühchristentum bis zum Reformation* (Göttingen, 1986). Note that as well as different religions there was a concern with

renewed in the nineteenth century, when there was a concerted move to offer comprehensive classifications of world religions.[39] The point I want to stress is that different religions were distinguished exclusively on a doctrinal basis, on the model of identifying heresies: irrespective of what these other religions might have considered their own religious identity to consist in, they were assimilated to sets of core doctrinal beliefs. There are distinctive core beliefs in other religions of course, but Christianity is quite unique in its construal of the identity of a religion as lying in its beliefs on theological questions. For Christianity, it is necessary and sufficient to be a Christian that one hold particular theological beliefs. No moral qualities are required, for example: one can be excommunicated for propounding heretical doctrines, but not for behaving immorally. Moreover, it was assumed—despite worries from the seventeenth century onwards about whether Confucianism was a religion in any recognizable sense or whether China was, *per impossibile*, a nation of atheists—that this criterion was shared with the other world religions. But it was not shared. In Islam, Buddhism, Confucianism, Hinduism, and Judaism, all kinds of things other than beliefs play crucial roles—dietary restrictions and fasting, for example, or daily rituals governing hygiene, or such practices as meditation and chanting. Note that whereas post-Reformation Christianity takes doctrines about the nature of God as its core (his existence, his tripartite nature, the Incarnation, etc.),[40] such a concern may be completely absent in what are identified as other religions. Confucianism, for example, does not posit the existence of a God, and what is distinctive of it for its practitioners is rather the transition from chaos to order. Even when a God is posited, belief in the existence of that God may not be a necessary condition for belonging to that religion, as in Judaism.[41] In short, the doctrinal focus of Christianity, which had become established by the fourth century,[42] is something peculiar to Christianity. This is not to deny that some of these other 'religions' have significant doctrinal

different sources of heresy. John Robertson has compared the 1692 atheism trial of Basilio Giannelli and Giacinto Cristofato in Naples with the 1696 Scottish trial of Thomas Aikenhead for atheism, noting that whereas in the Naples case 'the prosecutors were much concerned by non-Christian accounts of the origin and development of the world, and particularly by ideas associated with Epicureanism', in Scotland 'heterodox ideas about nature and the eternity of the world were secondary in importance to disrespect for Scripture and the denial of the Trinity': *The Case for Enlightenment: Scotland and Naples 1680–1760* (Cambridge, 2005), 95.

[39] See Tomoko Masuzawa, *The Invention of World Religions* (Chicago, 2005). Cf. Talal Asad, *Genealogies of Religion: Discipline and Reasons of Power in Christianity and Islam* (Baltimore, 1993).

[40] I say 'post-Reformation' because medieval Christianity was not quite so straightforward, with rituals playing a more significant role in shaping its identity: see Eamon Duffy, *The Stripping of Altars: Traditional Religion in England, 1400–1580* (New Haven, 1992).

[41] On the complex issues surrounding the meaning of 'religion', see Wilfred Cantwell Smith, *The Meaning and End of Religion* (London, 1978). There is a case to be made that the term is misleading, to the extent that it encourages the view that the different world 'religions' have something in common. See Stephen Prothero, *God is Not One* (New York, 2010).

[42] See Diarmaid MacCulloch, *A History of Christianity* (London, 2009), ch. 6. Cf. Guy Stroumsa, *Barbarian Philosophy: The Religious Revolution of Early Christianity* (Tübingen, 1999).

aspects.[43] It is just that everything does not hinge on doctrine, as it comes to do in Christianity. Their identities go beyond whatever doctrines they may subscribe to. From the point of view of our present concerns, what is distinctive about Christianity is not so much the content of the doctrines it espouses, but the fact that it defines itself in terms of doctrines.

The upshot of this was that building up and protecting a body of doctrine was a fundamental feature of Christianity, at least from the fourth century onwards. It was an indispensable ingredient in defining its identity. Doctrinal consistency was at a premium, and this was achieved by means of systematic theology, for many purposes the core of Christianity from the Patristic thinkers onwards, and something that shaped its sense of authority, manifested in a striking way in its concern with heresy, for example, and its naturally proliferating sectarianism. What was at issue in the thirteenth-century scholastic reconciliation of Christianity and Aristotelian natural philosophy, as well as in the subsequent seventeenth-century attempts (e.g. among mechanists) to align Christian theology and natural philosophy, was not a question of the relation between 'science and religion', but rather a question of bringing a Christian doctrine together with natural-philosophical tenets formulated in terms of a natural-philosophical doctrine. There were many forms of natural philosophy that were not doctrinal—such as natural history, as well the various forms of mechanics and positional astronomy placed under the rubric of the 'subordinate sciences' in the Aristotelian account—but these were not, and (at least as they stood before the modern era) could not have been, part of the package.

In the light of this, it is hardly surprising that the scholastic reconciliation of Christianity and Aristotelian natural philosophy proceeded along a trajectory that was both doctrinal—in that it set out to establish and build up a permanent body of truth—and systematic. Not only was metaphysics, which was considered the most fundamental philosophical doctrine, intrinsically systematic from the thirteenth century onwards, but the scholastic development of Aristotelian natural philosophy was predominantly a matter of its reformulation in more systematic terms, something that came to a head in the scholastic textbooks of the late sixteenth and early seventeenth centuries.[44] Nor was this exclusively a feature of

[43] In some cases, they may even have taken doctrinal questions seriously as a reaction to Christianity. There is a case to be made, for example, that in the wake of the attempts of Justin Martyr to distinguish Christianity from Judaism in doctrinal terms—one of the formative early attempts to establish what Christian orthodoxy consisted in—the rabbinical movement responded by moving in a doctrinal direction. See Daniel Boyarin, *Border Lines: The Partition of Judeo-Christianity* (Philadelphia, 2004), chs 2 and 3.

[44] See Charles H. Lohr, 'Metaphysics', in Charles B. Schmitt, Quentin Skinner, and Eckhard Kessler, eds, *The Cambridge History of Renaissance Philosophy* (Cambridge, 1988), 537–638. The legal model of enquiry also played an important role here: see Gaukroger, *The Emergence of a Scientific Culture*, 141–6. For details, see Harold J. Berman, *Law and Revolution: The Formation of the Western Legal Tradition* (Cambridge, Mass., 1983), and idem, *Law and Revolution* ii.: *The Impact of the Protestant Reformations on the Western Legal Tradition* (Cambridge, Mass., 2003).

scholasticism. On the contrary, the dominant natural philosophy of the mid-seventeenth century, namely mechanism as set out in Descartes' *Principia philosophiæ*, aspired to much the same degree of doctrinal integrity and systematicity as scholastic philosophy, and it was the claim of mechanism to encompass the whole of the natural realm in terms of a systematic fundamental reduction that was taken to indicate its explanatory power and its suitability as *the* natural philosophy.

By the early decades of the eighteenth century, the situation had changed radically, and there was a strong reaction to systematic explanation, coming to a head in the third quarter of the century. There did, however, persist two programmes, one in mechanics and one in metaphysics, that promoted the ideals of systematicity. 'Rational mechanics'—a project that Huygens, Newton, and Leibniz helped to initiate, and which, using the advanced mathematical techniques devised above all by the Bernoullis, was brought to fruition mid-century in the work of d'Alembert and Euler—adhered to strict standards of mathematical rigour and axiomatic organization. Its practitioners conceived it as a model of physical enquiry, and their hope was that once the mechanical core had been developed and firmly established, one could work outwards to areas such as gravity, chemistry, electricity, and optics, reducing them to an axiomatically structured mechanics and reworking them into a systematic form. But the mathematical sophistication of rational mechanics was not mirrored in its physical relevance, and it became effectively insulated from the main body of physical sciences, which were the successors to the seventeenth-century 'experimental natural philosophy' tradition, and, in areas like chemistry and electricity, were concerned largely with making sense of a mass of complicated experimental results. Rational mechanics could provide no guidance at all here, any more than could micro-corpuscularianism. In consequence, it became distanced from what were the cutting-edge developments in the sciences. Indeed, what can be regarded as the epitome of rational mechanics, Lagrange's *Mécanique analytique* (1788), is really just a work of systematization through a rewriting of mechanics as developed by Euler and others in more elegant algebraic form.[45]

As regards metaphysics, in the first half of the eighteenth century, after Leibniz, Wolff was the only significant advocate of a comprehensive systematic

[45] Joseph Fourier, in his 1829 *éloge* of Lagrange, notes that 'all his mathematical compositions are remarkable for a singular elegance, by the symmetry of forms and generality of methods, and if one may speak thus, by the perfection of analytical style', and Jean Delambre in his introduction to the 1867 edition of his works notes that, in his work in astronomy, Langrange 'made of these problems, simple, common and already resolved, the same use that is made by other analysts of questions of pure curiosity, that they furnish examples of calculus and occasions to develop new analytical artifices'. Both quoted in Craig G. Fraser, 'Lagrange's Analytical Mathematics, its Cartesian Origins and Reception in Comte's Positive Philosophy', *Studies in History and Philosophy of Science* 21 (1990), 243–56: 243. In this respect, Lagrange's work was surpassed only in Hilbert's axiomatization of classical mechanics: see Leo Corry, *David Hilbert and the Axiomatization of Physics (1898–1918): From* Grunglagen der Geometrie *to* Grundlagen der Physik (Dordrecht, 2010).

metaphysics.[46] What he offered was a basically scholastic metaphysics reformulated along Leibnizian lines. It was the contemporary archetype of *Schulphilosophie*, and was directly opposed to Thomasian and pietist thought, which were considered to have abandoned the claims of reason.[47] Wolff's influence was largely confined to a few German states however,[48] and even there it was often under attack, not just from within the university establishment, which was its natural home, but also from the powerful Frederick I of Prussia, who expelled Wolff from his kingdom in 1723. It is true that his successor, Frederick II, subsequently reinstalled him at Halle and appointed him vice-chancellor of the university in 1740, but his influence had waned considerably by that time, and his classes, once the highlight of the academic term, were now poorly attended. Many German scholars were contemptuous of him by this stage, the historian Johann Köhler in 1736 characterizing him as someone who had nothing better to do than sit around playing with monads and speculating about pre-established harmony.[49]

Wolff's system was a simplified and revised version of a scholastic/Leibnizian metaphysics. Scholastic metaphysics had always been highly systematic, and Leibnizian metaphysics had systematic aspirations, although it had never received a comprehensive formulation. Wolff began by rewriting logic (syllogistic) along Leibnizian lines, and he conceived the aim of logic to be the provision of a means whereby confused concepts were analysed and made clear and distinct. This procedure was then applied to the whole of philosophy, including natural theology. Philosophy was conceived as a discipline that investigated why things are as they are, including the distinctively Leibnizian project of accounting for why things are possible if they are possible, and why they are actual if they are actual.[50] The model was explicitly that of mathematics (although rational mechanics could equally have acted as a model along the same lines), where necessary truths were

[46] With the exception of his *Théodicée*, Leibniz's works were known only in fragmentary form before the posthumous publication of the *Nouveaux essais sur l'entendement humain* in 1765, and the appearance of some of his more systematic works in the edition of Louis Dutens in 1768. So Wolff (who died in 1754) and his followers effectively had the field to themselves before the mid-1760s. On the relation between Wolff's philosophy and that of Leibniz, see Walther Arnsperger, *Christian Wolffs Verhältnis zu Leibniz* (Weimar, 1897); Charles Corr, 'Christian Wolff and Leibniz', *Journal of the History of Ideas* 36 (1975), 241–62.

[47] See Max Wundt, *Die deutsche Schulphilosophie im Zeitalter der Aufklärung* (Tübingen, 1945).

[48] Wolff did have a French disciple, Jean Deschamps—author of the *Cours abrégé de la philosophie wolfienne* (2 vols, Leipzig and Amsterdam, 1743–7)—although his attempt to promote Wolffian philosophy in France was largely unsuccessful. More generally, Wolff's work was also known through the Marquise de Châtelet's *Institutions de physique* (Paris, 1740). The short anonymous entry on 'Psychologie' in the *Encyclopédie* offered a positive account of Wolff: on the impact of Wolffian psychology in the encyclopedia tradition see Fernando Vidal, 'Psychology in the 18th Century: A View from Encyclopedias', *History of the Human Sciences* 6 (1993), 89–119: 101–5.

[49] Johann David Köhler, *Erneurter Entwurf eines Collegii über den gegenwärtigen Zustand von Europa und die jetzigen Welt-Händel* (Göttingen, 1736), 2–3.

[50] Wolff set out his system at length on a number of occasions. The first complete exposition was *Vernünftige Gedanken von den Kräften des menschlichen Verstandes* (Halle, 1713).

revealed by establishing the systematic connections that propositions have with one another.[51]

It was a distinctive feature of the systematizing projects of rational mechanics and Wolffian metaphysics that the success of the deductive apparatus they deployed rested on the a priori nature of the premises. Euler, for example, went to great lengths to establish the a priori nature of the spatial extension of bodies, their impenetrability, and the principle of inertia.[52] Wolff sought to construct the sciences in a 'rational order', with what he considered the basic laws of thought—the principle of contradiction, the principle of sufficient reason, and the principle of the syllogism—being inviolable and innate to the human mind. On his account, these principles are not merely logical, however, but ontological, in that they govern not just thought but also reality.[53] In this way, a priori truths about the world are established, mirroring Euler's attempt to begin a systematic physical account of the world from a priori truths.

A version of an attempt to offer a systematic account of understanding, drawing on both Wolff and rational mechanics, would be revived in Kant in the 1780s, following an abrupt change of direction in his thought, but in the intervening period (and for some time afterwards among Kant's critics) there was a rejection of the idea that scientific understanding must take the form of systematic understanding. In the physical sciences, this rejection was motivated largely by developments in matter theory, as the mechanical model for explanation of physical phenomena was abandoned as inadequate in the burgeoning fields of electricity and chemistry. The notion of active matter that was developed in these areas was even more pronounced in the life sciences, where developmental physiology, in particular, required a notion of organic matter that bore no relation to the understanding of matter at play in mechanics. But there was also a rejection of the idea of the unique value of systematic understanding in an explicitly philosophical tradition. What is perhaps most remarkable about this rejection of systematic aspirations is the fact that, by mid-century, it was to be found across the religious and philosophical spectrum, from the radical defences of eclecticism and 'democratic' science in the *Encyclopédie* to the 1755 textbook *Philosophiae Naturalis* by the Jesuit Josef Redlhamer, in which he provides a genealogy for Jesuit 'physics' in the history of the experimental sciences, complaining about the

[51] This model had a stylistic impact on Wolff's writings, and the application of 'mathematical method' to philosophy was widely criticized. See Giorgio Tonelli, 'Der Streit über die mathematische Methode in der Philosophie in der ersten Hälfte des 18. Jahrhunderts und die Entstehung von Kants Schrift Über die "Deutlichkeit"', *Archiv für Philosophie* 9 (1959), 37–66.

[52] See, for example, Leonard Euler, *Mechanica sive motus scientia analytice exposita* (1736), in *Leonhardi Euleri opera omnia,* series 2, vols i, ii (Leipzig & Berlin, 1912). See the discussion in Gaukroger, *The Collapse of Mechanism and the Rise of Sensibility*, 304–17.

[53] This is a view that Kant challenged early in his career, a challenge that was to be formative in his later thought. The criticism begins with his 1762 'Die falsche Spitzfindigkeit der vier syllogistischen Figuren erwiesen', in Kant, *Gesammelte Schriften*, ii. 45–62: 59–60. It comes to a head in 1770 in the 'Inaugural Dissertation': *Gesammelte Schriften*, ii. 415–16.

substitution of contentious abstract metaphysical questions for physical ones in medieval thought.[54] How did such a radical change come about?

Note that in exploring this issue, it cannot be a question of attempting to uncover what science or natural philosophy is *really* like in its unadulterated form, as if the task were one of identifying the accretions that shape it into systematic or doctrinal forms. In particular, despite the claims of some of the participants in the eighteenth-century debates, we cannot assume that what happens is that systematic accretions are stripped off to reveal the true, unsystematic nature of natural philosophy. Rather, what happened is that the doctrinal and systematic form which natural philosophy had in the mid-seventeenth century was found to be an obstacle in a growing number of areas: notably pneumatics and optics, and later chemistry and electricity. But in other areas, notably rational mechanics, the systematic nature of the enterprise was crucial to the generation of results. Natural philosophy had to take one form or another, and some worked better than others in particular areas, but not because they reflected the *true* nature of the enterprise. Bearing this in mind, we can distinguish three stages in the development of the question of systematic versus non-systematic forms of knowledge in the seventeenth and eighteenth centuries: from the 1630s to the 1660s, from the 1660s up to about 1740, and the period from around 1740 until the revival of systematic forms of understanding in the 1780s, with the appearance of Kant's 'critical' writings, which offer a renewed doctrinal notion of understanding.

As regards the mid-seventeenth century debates,[55] in order to get a sense of what was at issue, we need only consider the conditions under which Galileo and Descartes, for example, levelled arguments against opponents along anti-systematic lines. It is striking that in both cases what is at issue is the charge of intellectual dishonesty. Plato had used the constraints imposed by systematic thought to counter what he construed as the intellectually dishonest 'argument for argument's sake' approach of the sophists. Galileo and Descartes, by contrast, both explicitly associate intellectual dishonesty with commitment to a system. However, what is involved, in both these latter cases, is a rhetorical ploy, a means of disabling criticism of their own accounts from the point of view of an already constituted systematic understanding, usually Aristotelianism, by accusing those who argue from a systematic position of dogmatism and lack of objectivity. Galileo's attack on Grassi in *Il Saggiatore*, is a case in point. Grassi, in virtue of being a supporter of Aristotelianism, is presented as someone who is unable to argue a case on its merits and so has to rely on a philosophical system, which is

[54] Diderot and d'Alembert, *Encyclopédie*, especially the entries on eclecticism and encyclopedias; Josef Redlhamer, *Philosophiae Naturalis. Pars prima, seu Physica generalis ad praefixam in scholis nostris normam concinnata* (Vienna, 1755). See Gaukroger, *The Collapse of Mechanism and the Rise of Sensibility*, chs 6 and 7, for a full discussion of these questions.

[55] In the scene-setting, and hence very summary, account that I give in this and the following paragraphs I draw on *The Emergence of a Scientific Culture* and *The Collapse of Mechanism and the Rise of Sensibility*, where full discussions can be found.

treated as a form of intellectual dishonesty and a lack of impartiality. Galileo's attack on systems was also motivated by the fear that a Tychonic system was rapidly becoming the default astronomical system, with increasing awareness of the faults of the Ptolemaic one, thereby pre-empting serious consideration of Copernicanism, and he responded by stressing the hypothetical nature of all astronomical systems and the ill-advisedness of attaching oneself to any single system. Nevertheless, there can be no doubt that he himself believed that there was in fact a true system: Copernicanism. Descartes too, in *La recherche de la verité*, argued that someone who had been brought up on a doctrinal and systematic understanding was at least seriously disadvantaged in pursuing natural philosophy, and he lauded the intellectual honesty and abilities of the *honnête homme*, such as the gentleman officer who lacked such an education but was able to rely on his natural untainted faculties. But again there is an underlying agenda, which shapes the polemics. Descartes objected to the Aristotelian system of natural philosophy because he believed it to be false, but its systematic nature played a role in obscuring this, and in the *Meditationes* he advocated a procedure of clearing one's mind of any prior system. This was not proposed as a rejection of systems as such, however, but rather as a prerequisite for starting from first principles in order to grasp the truth of the Cartesian system.

In the 1660s in England, a very different kind of challenge to the idea of a system emerged, one in which genuine questions were raised about whether systematic understanding was always the most suitable and fruitful form of understanding in natural philosophy. The possibility was raised whether there could be non-systematic but nevertheless genuine forms of understanding, and indeed whether the search for systematic forms of understanding might not, on occasion, actually be an obstacle to understanding. This new questioning began in earnest in the 1660s with two controversies: that over Boyle's pneumatics, and that over Newton's prism experiments. Boyle and Newton discovered that in order to account for certain phenomena in a satisfactory way they had to suspend their commitment to corpuscularian micro-reductionism as the sole form of explanation. For the advocate of a systematic mechanism, the ultimate explanations took the form of accounts in terms of underlying microscopic states, so that causation, and with it explanation, were always construed as vertical. By contrast, Boyle and Newton postulated horizontal causal processes—those where cause and effect were on the same level—and defended this as a genuine and independent form of explanation. In effect, they rejected the idea that causes must be restricted to what underlies the phenomena, and in consequence that they must be located at a different level from the phenomena. Rather, their treatment implied that there is a way of understanding at least some phenomena that consists in exploring the causal connections between (as opposed to underlying) them. There is a direct link here between the rejection of reductive explanation and the rejection of explanation in terms of a system. Mechanism—the reduction of macroscopic physical behaviour to the mechanically describable behaviour of the micro-corpuscles making up bodies—offered an exceptionally powerful form of unification, in which all physical phenomena are comprehended within a single system. What was at issue was not systems per se, but whether a particular system—almost universally agreed to have replaced all earlier ones, and to have no viable competitors—had the scope and capability claimed for it, and if not whether it was actually an obstacle to physical enquiry. Boyle and Newton effectively demonstrated the latter in specific fields, but could offer no general defence of non-systematic explanations.

It was Locke who managed to devise a philosophically defensible statement of phenomenal explanation, which was part of a programme of providing a philosophical vindication of experimental natural philosophy, a vindication which allowed that phenomenal explanations were not subordinate to micro-corpuscularian accounts. The defence worked on two levels: it denied that observational procedures are reliable only if and until judged satisfactory by the lights of a foundational micro-corpuscularian account of fundamental natural processes; and it showed that, from an explanatory point of view, it is an open question what kinds of considerations are going to be of fundamental explanatory significance, a question that may be decided in favour of phenomenal explanations in many cases. In the course of the eighteenth century we can find numerous examples of the rejection of a systematic, micro-corpuscularian model of physical explanation in favour of explicitly phenomenal accounts, and in the 1710s and 1720s there are fundamental developments in chemistry (in the construction of the table of affinities) and electricity (the discovery of electrical conductivity) which were achieved by abandoning a micro-corpuscular model.

The next development comes with Hume, initially with his *Treatise of Human Nature* (1739), where the Lockean account is extended into the moral sciences, in the process becoming an explicitly general account of understanding in a way that had been hinted at but never really pursued in Locke. Like Locke, Hume does not deny that understanding can take a systematic form, only that the kind of understanding that can take a systematic form—and he is concerned with metaphysics in this context—cannot be constitutive of our understanding of the world: it can be part of it (and indeed, he believes, by contrast with his later 'common sense' school critics such as Reid, *must* be part of it), but not all of it. Hume offers an account of 'human understanding' whereby it is a judicious balancing of propositional and non-propositional considerations, of considerations of reason and sensibility. To the extent to which natural philosophy offers a model for understanding here, it takes the form of what Hume terms 'natural history', something deriving from the experimental philosophy tradition and its empiricist successor, and which he uses to naturalize enquiry on the one hand, while denying either systematicity or even a unified programme to natural philosophy on the other.

What exactly is this 'natural history'? Natural history in the literal sense is a classificatory discipline, but it acquires more complex senses in Bacon at the beginning of the seventeenth century, and then in Boyle and the defence of 'experimental natural philosophy' in the second half of the seventeenth century. The central contrast for the 'experimental natural philosophy' tradition is between, on the one hand, an experimental form of natural philosophy which identifies and accounts for problematic phenomena at a purely phenomenal level without recourse to any idea of an underlying causal level; and, on the other hand, systematic natural philosophy, which refers explanations to an underlying microscopic level, on the understanding that this is where the real action happens. What is at issue is whether all explanations must take the form of deducing the macroscopic behaviour of bodies from the behaviour of their microscopic constituents. Advocates of experimental natural philosophy—beginning with Boyle in his pneumatics and Newton in his account of spectral colours, and effectively dominating chemistry and the study of electricity by mid-century—urged the legitimacy, and fruitfulness, of drawing connections between phenomena. If we take natural history in this sense, then we can perhaps see how we might conceive of a discipline that can in some way help us to make sense of the relation between reason (a propositional form of understanding) and sensibility (a non-propositional form of understanding), without it having any pretensions to understanding these in terms of a system resting on a small number of basic truths, and without the aspiration to offer some absolute critical perspective, itself immune to criticism.

Hume's use of sceptical arguments to undermine the claims of systematic metaphysics—he takes Malebranchean metaphysics as its most advanced form—came in time to be immensely influential, but equally important was his own practice as an essayist and historian, which had a more immediate impact.[56] It was in this latter capacity that his life and work manifested the other side of the balance between systematic and non-systematic enquiry. The essays, in particular, balanced the sceptically driven metaphysical writings, and they each represented two different *personae*, both equally necessary. Here we have what is most distinctive about Hume's position, namely his refusal to choose between systematic and non-systematic considerations. His combination of a commitment to a sceptically driven form of enquiry as absolutely central to our critical engagement with the world in which we live, on the one hand, and his demonstration of the absurdities to which this leads, on the other, results in Hume adopting a distinctive view of the role and nature of judgement. His view is that understanding consists in the exercise of judgement between considerations that are not exclusively propositional.

[56] It was this aspect of his work that appealed to the *Popularphilosophie* movement. Garve, for example, was particularly impressed by the way in which Hume integrated philosophy and history, and saw this as a model for his own work: see his 'Einige Beobachtungen über die Kunst zu denken', in *Versuche über verschiedene aus der Moral*, ii. 427.

What is distinctive about this judgement is that it is not a 'philosophical' judgement as such. That is to say, it is not a judgement that the philosopher, *qua* philosopher, is best qualified to make. If one wants to judge the 'good life', this is not something that any particular group—theologians or metaphysicians—have any real claim on. But as traditionally conceived, for instance if one thinks of the philosophical tradition from Plato and Aristotle to the Stoics and Epicureans, an answer to the question how to live is precisely the ultimate goal of philosophy, and it is a goal that establishes its autonomy from theology for example. Even worse, metaphysics and other disciplines are compared in terms of what Hume terms 'natural history', which is both historicizing and naturalizing. It is true that this avenue would be taken up by some of Hume's successors, notably in Herder's naturalistic anthropology, but otherwise it met massive resistance. Even more disconcerting for its critics, was the fact that Hume's enquiry issued in a dilemma, which he took to be indicative of the human condition. Such an approach, in which the outcome of a philosophical investigation is a dilemma rather than a solution, was for many philosophers a profoundly unsatisfactory one, indeed a profoundly unphilosophical one.[57] It was as if Hume had failed in a philosophical project and, instead of persisting, had the audacity to hold up this failure as telling us something about the nature of philosophy. What subsequently happened was that many of his successors sought to identify the source of the failure so that it could finally be remedied, and indeed this became the problem that Hume bequeathed to philosophers up to the 1780s, when Kant's 'critical' philosophy presented itself as a properly 'philosophical'—systematic and doctrinal—solution, offering a general account of the relationship between different forms of understanding, one that fixed the limits of reason while at the same time exploring in a systematic way what could be said about matters that fell outside those limits.

In France and Germany, the questions became highly polemicized in the 1750s and 1760s through the influence of Rousseau. In Britain, Rousseau's influence was significantly weaker, but what Rousseau represented for many philosophers—an essayist of wit and urbanity whose philosophical interests went well beyond those subjects that fell under traditional metaphysics, and who represented a philosophical *persona* quite different from the metaphysician—was something that could also be found in Hume, and his 'On Essay Writing' (1742),[58] which stresses the importance of reaching out to a wide public, was an influential work. Although the writings of Smith, Burke, and Ferguson were appearing in German translation in the 1770s, it was Rousseau and Hume who provided the two models of thinkers who had achieved in the essay form what metaphysicians

[57] The otherwise sympathetic reception of Hume in France, for example, was constantly qualified by a rejection of what was taken to be his commitment to paradox: see Michel Malherbe, 'Hume's Reception in France', in Peter Jones, ed., *The Reception of David Hume in Europe* (London, 2005), 43–96.

[58] David Hume, *Essays, Moral and Political* (2 vols, Edinburgh, 1742), ii. 1–8.

could never have achieved in their treatises.[59] Moreover, both dealt with moral, political, educational, and pedagogic questions—those questions that would fall under the 'moral sciences'—in terms of anthropology rather than either metaphysics, or Christian teaching, or civic humanism (as traditionally conceived). Although they drew different conclusions on many of the questions they explored, what was crucial was their way of pursuing the subject, and their questioning of the power of reason to regulate a range of issues from morality and aesthetics to epistemology and religious belief.

The two tend to be drawn together in Germany, where we find the influence of Rousseau balanced to some extent by Hume, especially inasmuch as Hume is associated with the project of a naturalized psychology. In mid-century, British and French ideas were introduced into Berlin and other centres in a context of growing hostility to Wolffian *Schulphilosophie* and a revival of what was seen as Thomasius' eclecticism.[60] Above all, the material from France and Britain was disseminated through journals announcing their subject matter as *schöne Künste und Wissenschaften*.[61] The idea of 'beautiful science' or 'beautiful arts' is a translation of *belles-lettres*, and what is happening in this literature is a shift of learning generally, including the sciences, from an academic culture into an essay culture—one whose audience is the Republic of Letters.[62] Although the shift might be viewed as bringing Germany into alignment with Britain and France (more specifically, bringing Berlin into alignment with Edinburgh and Paris), the relative suddenness and abruptness of the transition, together with the fact that the leading figures—Nicolai, Lessing, and Mendelssohn—were incorporating this shift into an explicit programme of cultural and political reform, mean that the rejection of metaphysics as a vehicle for thought about and engagement with fundamental issues about the nature of the world and our place in it raised the whole question of the value of particular modes of enquiry. What is distinctive about Germany is the extent to which these questions turned on universities and

[59] On Rousseau's reception in Germany see Herbert Jaumann, 'Rousseau in Deutschland, Forschungsgeschichte und Perspektiven', in Herbert Jaumann, ed., *Rousseau in Deutschland: Neue Beiträge zur Erforschung seiner Rezeption* (Berlin, 1994), 1–22. On Hume's reception see Manfred Kuehn, 'The Reception of Hume in Germany', in Peter Jones, ed., *The Reception of David Hume in Europe* (London, 2005), 98–138. Diderot was also an important figure: see Roland Mortier, *Diderot en Allemagne, 1760–1850* (Paris, 1954).

[60] See Werner Schneiders, 'Vernünftiger Zweifel und wahre Eklektik: Zur Entstehung des modern Kritikbegriffs', *Studia Leibnitiana* 17 (1985), 143–61; and Helmut Holzhey, 'Philosophie als Eklektik', *Studia Leibniziana* 15 (1983), 19–29.

[61] The ground had been prepared by the 'moral weeklies', the German contribution to the journals of the Republic of Letters, designed for a bourgeois reading public, which emerged in the 1710s and continued up to mid-century. See Wolfgang Martens, *Botschaft der Tugend: Die Aufklärung im Spiegel der deutschen Moralischen Wochenschriften* (Stuttgart, 1968).

[62] Werner Schneiders, 'Zwischen Welt und Weisheit: Zur Verweltlichung der Philosophie in der frühen Moderne', *Studia Leibnitiana* 15 (1983), 2–18; John H. Zammito, *Kant, Herder, and the Birth of Anthropology* (Chicago, 2002), 8–13; Lorraine Daston, 'The Ideal and Reality of the Republic of Letters in the Enlightenment', *Science in Context* 4 (1991), 367–86.

university education. There had been a massive confessional investment in German universities since the Reformation, making the universities an issue in a way they could never have been in France (where it was the Académies that were at the centre of institutional intellectual life) or England.[63] Wolff had been a vigorous university reformer, setting in train a questioning of the university curriculum that even the anti-Wolffian reformers of the 1750s and 1760s built upon, and he played a role in the establishment of philosophy as an object of study for the educated (*Gebilden*), and not just academic scholars (*Gelehrten*).[64] Yet it was far from clear that philosophy could actually play this role. Distinguishing between 'historical' knowledge, which concerns itself with empirical and causal matters, and 'philosophical' knowledge, which works from basic principles, Wolff rejected the traditional ascendancy of theology, jurisprudence, and medicine, arguing that they were merely 'historical' forms of enquiry, whereas philosophy could raise knowledge to the level of what the scholastics had termed *scientia*: comprehensive understanding based on first principles.[65] But his opponents had no doubt that he had failed in this regard,[66] and that the failure was in large part due to how he conceived the content of philosophy. By the 1750s in Germany,

[63] Scotland, where university education was fostered along Northern European lines rather than English ones, was different from England in this respect. In particular, note that Aberdeen had two universities with the founding (along confessional lines) of a new university, Marischal College, in 1593, to counter King's College, founded in 1495. But this divide was nothing in comparison with the numerous confessionally-instituted universities in Germany. On the German universities in the eighteenth century see Notker Hammerstein, 'Zur Geschichte der deutschen Universitäten im Zeitalter der Aufklärung', in Hellmuth Rössler and Günther Franz, eds, *Universität und Gelehrtenstand 1400–1800* (Limburg, 1970), 145–82; idem, 'Die Universitätsgründungen im Zeichen der Aufklärung', in Peter Baumgarten and Notker Hammerstein, eds, *Universität und Gelehrenstand 1400–1800* (Nendeln, 1978), 263–98; idem, 'Christian Wolff und die Universitäten', in Werner Schneiders, ed., *Christian Wolff 1679–1754* (Hamburg, 1983), 266–77.

[64] See Schneiders, 'Der Philosophiebegriff des philosophischen Zeitalters'; Norbert Hinske, 'Wolffs Stellung in der deutschen Aufklärung', in Werner Schneiders, ed., *Christian Wolff 1679–1754* (Hamburg, 1983), 306–20; Hans Erich Bödeker, 'Von der "Magd der Theologie" zur "Leitwissenschaft": Vorüberlegungen zu einer Geschichte der Philosophie des 18. Jahrhunderts', *Das achtzehnte Jahrhundert* 14 (1990), 19–57.

[65] One of the most comprehensive treatments in this respect was that of Johann Christoph Gottsched, *Erste Gründe der Gesamten Weltweisheit, darinn alle Philosophische Wissenschaften in ihrer natürlichen Verknüpfung abgehandelt werden* (2 vols, Leipzig, 1733–4), where Newtonian physics and a version of Wolffian metaphysics are set out in detail. The former is characterized as dealing with the 'exteriors' of things, whereas metaphysics is needed to illuminate their interiors (¶398, ¶¶664–6). Compare Mendelssohn's defence of the superiority of metaphysics in his *Abhandlung über die Evidenz in metaphysischen Wissenschaften* (Berlin, 1764).

[66] As Terry Pinkard points out, by the 1780s, 'the dominant opinion in Germany (and elsewhere in Europe) was that universities were outmoded, medieval institutions, staffed by tenured professors who taught students useless knowledge, and whose traditions of drunken student revelry were detrimental to the students' moral health; and the conventional wisdom was that it might just be better to abolish the universities and replace them with more forward-looking academies and institutes that would train students in more useful skills.' *German Philosophy 1760–1860: The Legacy of Idealism* (Cambridge, 2002), 86. For detailed accounts, see Notker Hammerstein, 'Die deutschen Universitäten im Zeitalter der Aufklärung', *Zeitschrift für historische Forschung* 19 (1983), 73–89; idem, 'Christian Wolff und die Universitäten', 266–77.

philosophy acted as an ethical and socio-political vehicle for change and progress, of freedom of expression and informed public opinion.[67] Wolff's opponents wanted to replace traditional university education with more socially engaged forms of instruction and debate. Consequently, they considered philosophy as something far too important to be guided by academic philosophers, whose preoccupation with metaphysics and logic rendered them unsuitable for furthering the aims of *Aufklärung*.[68]

These academic philosophers considered that philosophy must not only be systematic, but also 'self-grounded', which was how they conceived of mathematics. Such a view was becoming not only widely rejected but widely ridiculed from mid-century onwards, beginning with Lessing in his 1748 play *Der junge Gelehrte*. The idea that it was rigour that should guide the discussion of fundamental issues was rejected in favour of *schönen* disciplines. This transformation was supported in some respects by the Berlin Akademie.[69] Berlin had no university, and, from the 1740s onwards, the Akademie was very much at the forefront of intellectual life in Berlin (and increasingly elsewhere in Germany), something helped along by the lucrative and prestigious prizes that its regular international essay competitions attracted.[70] The language of communication of the Akademie was French,[71] and French Enlightenment ideas in particular started to flood into German culture with the installation of Maupertuis as its president in 1746.[72]

Concern with questions of audience, as evidenced by the move away from a scholarly university readership to a more general educated one, was crucial. But

[67] The influence of Scottish philosophy is significant here: see Günter Gawlick, 'Über einige Charakteristika der britischen Philosophie des 18. Jahrhundert', *Studia Leibnitiana* 15 (1983), 30–41.

[68] See, in particular, Zammito, *Kant, Herder, and the Birth of Anthropology*, ch. 1, to which I am particularly indebted in what follows.

[69] See Adolf Harnack, *Geschichte der Königlich Preussischen Akademie der Wissenschaften zu Berlin* (4 vols, Berlin, 1900), i., Book 2, ch. 2; and Conrad Grau, *Die Preußische Akademie der Wissenschaften zu Berlin* (Heidelberg, 1993). The Akademie underwent a number of name changes: up to 1744 it was the Societas Regia Scientiarum, then (under Maupertuis) becoming L'Académie Royale des sciences et belles lettres, attesting to the francophilia of Frederick II of Prussia.

[70] On the significance of the prize essay competitions see: Andreas Kraus, 'Die Bedeutung der deutschen Akademien des 18. Jahrhunderts für historische und naturwissenschaftliche Forschungen', in F. Hartmann and R. Vierhaus, eds, *Der Akademiegedanke im. 17 und 18. Jahrhundert* (Bremen, 1977), 139–70; and Cornelia Buschman, 'Die philosophische Preisfragen und Preisschriften der Berliner Akademie der Wissenschaften in 18. Jahrhundert', in W. Förster, ed., *Aufklärung in Berlin* (Berlin, 1989), 165–228.

[71] The standing of French was such that the 1784 Prize Essay question of the Berlin Akademie, announced in 1782, concerned the universality of French, and asked whether it deserved this status, as well as whether this was something permanent. For details see Jürgen Storost, *Langue française—Langue universelle? Die Diskussion über die Universalität des Französichen an der Berliner Akademie der Wissenschaften* (Bonn, 1994).

[72] Nevertheless, despite its importance, Berlin was not the only source of enlightened ideas from outside. The occupation of Königsberg from January 1758 onwards introduced a cosmopolitan culture into the city, for example: see Zammito, *Kant, Herder and the Birth of Anthropology*, 99–104, and, more generally, Kurt Stavenhagen, *Kant und Königsberg* (Göttingen, 1949).

the actual practice of reading was also at issue. There was a very significant shift from the repeated and intensely focused reading of a few chosen texts to a continuous assimilation of what was often the ephemeral material contained in the 'moral weeklies'.[73] The expansion of reading material outside a narrowly prescribed corpus was associated with a move to foster an intellectual maturity in readers, through the values of thinking for oneself and independence of judgement. In his 1784 essay 'Was ist Aufklärung?' Kant begins with a definition that makes this constitutive of enlightenment: '*Enlightenment is the human being's emergence from his self-incurred minority. Minority* is inability to make use of one's own understanding without direction from another. This minority is self-incurred when its cause lies not in lack of understanding but in lack of resolution and courage to use it without direction from another.'[74] Not only the cultivation of observational skills, but drawing on one's own experiences, was crucial here, because what was at stake was not the aspiration to attain some metaphysical system that transcended the individual, but rather something with a personal and psychological focus, as well as a cognitive or epistemological one, namely a process of character formation (*Bildung*). To put such character formation at the centre of understanding has radical consequences for what is considered worth pursuing and what is not.[75] In some early remarks on this question, Kant drew these consequences in no uncertain terms:

> When something is not in keeping with our lifespan, or with our epochs, or with the larger part of humanity, but is ultimately very much a matter of chance and is possible only with the greatest difficulty, it makes no contribution to the happiness and perfection of the human race. How many centuries passed before the real sciences appeared, and how many nations are there in the world that will never possess them?[76]

He continues:

> The poor fit between science and humanity is primarily to be seen in the fact that the vast majority of those who so wish to ornament themselves achieve absolutely no improvement of understanding, but rather a perversion of it, not to mention the fact that for most people science serves only as a vehicle for vanity.[77]

[73] See Rolf Engelsing, *Der Bürger als Leser: Lesergeschichte in Deutschland 1500–1800* (Stuttgart, 1974). Note that publishing in Germany had been paralyzed by the Thirty Years War, and the Leipzig catalogue of books took until 1764 to reach its pre-war level of new publications: see the discussion in Robert Darnton, 'History of Reading', in P. Burke, ed., *New Perspectives on Historical Writing* (University Park, Penn., 1992), 140–67. On the sudden rise of ephemeral literature in Germany, see Jochen Schulte-Sasse, *Der Kritik an der Trivialliteratur seit der Aufklärung: Studien zur Geschichte des modernen Kitschbegriffs* (Münich, 1971).

[74] Immanuel Kant, 'Beantwortung der Frage: Was ist Aufklërung?', *Gesammelte Schriften*, viii. 35; idem, *Practical Philosophy* (Cambridge, 1996), 17.

[75] See Frederick C. Beiser, *The Romantic Imperative: The Concept of Early German Romanticism* (Cambridge, Mass., 2003), ch. 6.

[76] Immanuel Kant, *Bermerkungen in den 'Beobachten über das Gefühl des Schönen und Erhabenen'*, ed. Marie Rischmüller (Hamburg, 1991), 34–5.

[77] Ibid., 35.

The opposition here between 'science and humanity' is in effect that between reason and sensibility, which was in many ways a mirror of that between systematic and non-systematic understanding.

REASON AND SENSIBILITY

What was at stake in the eighteenth-century debate over the respective claims of reason and sensibility was whether reason provided the only resource by which to understand the world and our place in it. Put in these broad terms, it would seem that some moderation of the claims of reason is clearly necessary. There are many phenomena that are captured only very awkwardly by reason. Consider the emotions, for example. There is a case to be made that the range of emotions and their affective power are best displayed not in the form of the kind of literal treatment that is distinctive of reason, but in terms of something more subtle and indirect such as the metaphors, similes, and analogies that are characteristic of literary forms such as the novel. The distinction between cognitive states and affective ones appears to be a natural one, and it would therefore seem appropriate to confine reason to the former. But the distinction between reason and sensibility is not so straightforward, and does not neatly mirror that between cognitive and affective states.

The complexity of the issues becomes apparent once we begin to explore just what 'sensibility' amounted to. The way in which sensibility was thought of in France is instructive here, for although British and German conceptions often have a different focus, there is no doubt that mid-eighteenth century French thought was very influential, particularly in Germany. From the entries on 'sensibilité' in the *Encyclopédie*,[78] we can distinguish four phenomena that fall under the term, as used at the time: aesthetic sensibility, a kind of awareness or appreciation that accompanies perceptual cognition in varying degrees; moral sensibility, an ability to exercise untutored moral discrimination; sensation, something that figures in our cognitive relation to the world; and sensitivity, something that accompanies certain kinds of physiological activity. These are not unconnected phenomena. As Anne Vila has pointed out, 'sensibility was the essential link between the human body and the psychological, intellectual, and ethical faculties': it helped unify the human faculties, as it was 'seen as the root of all human perceptions and reflections, as the innate and active principle of sociability that gave rise to human society, as a kind of sixth sense whose special affective energy was essential to both virtue and to art, and finally, as the paradigmatic vital force whose actions could

[78] Diderot and d'Alembert, *Encyclopédie*, xiii. 781–811. The author of the main entry was Henri Fouquet, who was associated with the Montpellier Faculty of Medicine: see Luis Dulieu, 'Les articles d'Henri Fouquet dans l'*Encyclopédie*', *Revue d'histoire des sciences et de leurs applications* 5 (1952), 18–25.

be detected in every bodily function, be it healthful or morbid'.[79] In particular, the inclusion of sensitivity—a physiological phenomenon—in sensibility[80] adds a distinctive layer of complexity to considerations of the relation between reason and sensibility, for it means that we are considering phenomena that, to a significant extent, mirror the contrast between the spiritual and the material. In France, where the debates were particularly intense, there was a general acceptance that questions of sensibility fell under the purview of medicine, with the result that the physiological nature of sensibility was taken as given. Here, what had earlier been formulated in terms of the mind/body distinction now comes to be reformulated in terms of the relationship between reason and sensibility, with the decision that one came to on these questions playing a determining role in deciding what resources—metaphysical or physiological—it was most appropriate to draw on in investigating questions of our relation to the natural realm.

Rousseau's lively and engaging writings, with their appeal to a wide audience, proved to be the catalyst in these debates, and his account of sensibility sharpened the issues and helped to transform them into the most pressing questions of the day. In short, Rousseau regarded the civilizing process as detrimental to morality, because it denatured and overstimulated sensibility, and in pitting morality against civilization he put sensibility at the centre of the 'human condition'. If Rousseau were right, then the aims and aspirations, and forms of self-understanding, that had traditionally been advocated by philosophers, humanists, theologians, and reformers needed to be rethought at a fundamental level.

In his first significant work, his prize-winning 1750 essay *Discours sur les sciences et les arts*, Rousseau argued that the development of the arts and sciences has corrupted morality.[81] Human beings are by nature good, and indeed good while in their natural state. What marks out the natural state is the dominance of sensibility, above all in the form of a disinclination to witness suffering. But with the development of political and civil society, pride or self-esteem (*amour propre*) takes over. While there was general agreement that the solution must lie in the restoration of sensibility, the situation was complex because there was a sharp division among those who wanted to promote sensibility and those who believed that it could not be promoted in a controlled manner.[82] On the one hand, there were those who considered the cultivation of sensibility as a crucial part of the

[79] Anne C. Vila, *Enlightenment and Pathology: Sensibility in the Literature and Medicine of Eighteenth-Century France* (Baltimore, 1998), 2.

[80] There had been discussions that linked sensation and the action of nerves prior to this but not in the context of a general notion of sensibility. On these earlier developments, see Ronan de Calan, *Généalogie de la sensation: Physique, physiologie et psychologie en Europe, de Fernel à Locke* (Paris, 2012).

[81] Jean-Jacques Rousseau, *Discours qui a remporté le prix a l'Académie de Dijon en l'année 1750. Sur cette Question proposée par la même Académie: Si le rétablissement des sciences et des arts a contribué à épurer les mœurs* (Geneva, 1750). The argument is followed up five years later in *Discours sur l'origine et les fondements de l'inégalité parmi les hommes* (Amsterdam, 1755).

[82] See especially Vila, *Enlightenment and Pathology*.

civilizing process, as the instrument by which morality could be improved, something which the progress of civilization brought in its train. This was a view particularly promoted by physicians associated with the Montpellier Faculty of Medicine and their Parisian successors. By contrast, there were those, such as Rousseau, who regarded the civilizing process as detrimental to morality, not because it marginalized sensibility, but rather because it destroyed the delicate balance that sensibility required. These are questions to which we shall return in Chapter 3. For the moment, I want to focus on the consequences of this concern with sensibility for the understanding of the role of reason.

One paramount question at stake in debates over the standing of sensibility and reason, was whether morality was a matter of sensibility, or one of reason. Wolff and the metaphysicians who took up his system tended to subordinate everything to reason, and although concessions to sensibility begin to emerge mid-century in the work of Alexander Baumgarten, the model for analysing the sensory realm, as we shall see in Chapter 4, is drawn exclusively from that devised for analysing the intellectual realm. In contrast, by the middle decades of the century it was an increasingly shared assumption that the origins of morality lay in sensibility. In Scotland for example, Hutcheson, who was a crucial influence on Diderot among others, had argued in detail that the spontaneity of morality meant that it could not be a matter of reason—i.e. a matter of connecting ideas in the requisite way—but must derive from sensibility.[83] Adam Smith, Hutcheson's one-time student and successor in the Chair of Moral Philosophy at Glasgow, simply dismisses the view that reason might ground morality. In *The Theory of Moral Sentiments* (1759), the 'moral sentiments' to which the title refers are felt—as opposed to reasoned—judgements on the propriety of others' actions. Morality is taken out of an abstract realm of intellectual reflection, and becomes a form of sensibility. Smith writes that 'it is altogether absurd and unintelligible to suppose that the first perceptions of right and wrong can be derived from reason'; rather, they must be traced to 'sense and feeling'.[84] In France, the argument was taken one step further: it was not just the origins of morality that were a question of sensibility, but, reflecting a Lockean strain, epistemology as well ultimately came down to sensibility. Here sensibility was not only autonomous with respect to reason, but held a philosophically central position: sensation/sensibility was argued to be constitutive of our sense of an external world, and the central claim was that we are not in a position to understand the role of reason until we understand how sensibility underpins our cognitive life.

[83] Francis Hutcheson, *An Inquiry into the Original of our Ideas of Beauty and Virtue* (London, 1726); idem, *An Essay on the Nature and Conduct of the Passions and Affections*.

[84] *The Theory of Moral Sentiments*, VII.iii.2.7; *The Glasgow Edition of the Works and Correspondence of Adam Smith* (6 vols, Oxford, 1976–83), i. 320. Cf. Hume, *A Treatise of Human Nature: Being an Attempt to Introduce the Experimental Method of Reasoning in Moral Subjects* (3 vols, London, 1739): 'Morality . . . is more properly felt than judg'd of' (468).

Questions of the connection between sensibility and sensation were introduced into French thought in the 1740s in the work of Condillac and Buffon. Both independently used the image of a human statue (implicitly following the device that Descartes had used in *L'homme*) which is progressively given sensory faculties, in order to explore just what is needed to produce a conscious self-aware human being.[85] Moreover, both used the thought experiment in the same way, and reached much the same conclusions. The image of the statue should be considered in the context of French Lockeanism, for it provided a means of exploring how one might replace an innate ideas tradition in which the qualitative difference between mere sensation and perceptual experience comes with innate ideas. Buffon, Condillac, and Diderot all rejected innate ideas, yet each wanted to register the qualitative difference between merely sensing and having an experience of an external world.[86] Sensibility takes on a key role here, since for each of these thinkers the qualitative difference between merely sensing and having an experience of an external world comes with sensibility, not with reason. We do not infer the existence of an external world from our sensations, they argued: we are aware of its existence by means of an affective relation to the world, not a rational one. The reasoning behind this is that our knowledge of the world comes through sensation rather than purely through intellectual reflection, and that experience of an external world, as opposed to mere sensation, is prior to having and exercising rational faculties, because these rational faculties are, at least primarily (and for French Lockeans exclusively), exercised on the basis of our experience of the external world. It follows from this that sensibility not only forms the link between sensation and ratiocination, but it plays a crucial role in what we are and how we think, a role that reason, for example in the form of a faculty that relies on innate ideas, is simply unable to play.

On the move from sensation to experience of the world, consider Condillac's version of the argument. Condillac asks us to imagine a human statue which we bring alive by attributing various sensory faculties to it, one by one. We can separate each stage and ask what its experience of the world would be like at that stage. Imagine we give the statue the power to smell and place a rose in front of its nose, so that the statue experiences the odour of a rose. There is nothing in this experience that would lead the statue to imagine that that experience had an external source, so the olfactory sensation is not experienced as being that of an external object, but simply as an experiential state. Similarly with the power to hear, to see, to taste. Touch is treated differently by Buffon and Condillac.

[85] Etienne Bonnot de Condillac, *Traité des sensations* (Paris, 1754), Part I; Buffon, in the section 'Des sens en général' in *De l'homme*: George Louis Leclerc, Comte de Buffon, *Œuvres complètes* (34 vols, Paris, 1827), xii. 165–86.

[86] Condillac's work was considered the natural development of Locke's not only in France but also by many in England. His *Essai sur l'origine des connoissances humaines* appeared in English in 1756 with the title: *An Essay on the Origin of Human Knowledge. Being a Supplement to Mr. Locke's Essay on the Human Understanding*.

Buffon, for example, argues that the statue at first believes all its sensations to refer to internal experiences, and it only when a tactile faculty is provided that it realizes, 'in horror', that there is something outside it. Only touch can provide a sense of an external world. Condillac recognized the distinctiveness of tactile sensation, but makes it clear that to gain a general idea of 'sensation', the statue must reflect on the qualities it senses without reference to the five ways in which the bodies are affecting its organs: that is, it must run together all the individual sensations it receives to form a single class. Nevertheless, even a statue with all five of the senses which was able to compare, reflect, remember, and accomplish the other intellectual operations, would not be led, on these grounds alone, to imagine that its states were anything but internal and self-contained. Despite having a grasp of spatial relations, for example, the statue would be solipsistic: its sensory experiences would not project it into, or connect it with, the world. On the contrary, it would remain isolated and self-contained. This question now is how it is possible for us to develop a conception of the world as independent of us. Buffon's view is that only touch can provide a sense of an external world, and with this sense effectively comes a precondition for morality, in that with it comes a recognition of the existence of others, and a love for them replaces the earlier narcissism. Condillac's answer to the question is that a sense of reflection emerges from the sensations, allowing us to distinguish our own body from the sensations themselves, by combining the sensations and making something new out of them, in which the various objects of sensation can be compared under different descriptions.

The crucial point in the present context is that, for both Buffon and Condillac, sensibility and a recognition of an external world come simultaneously. It is axiomatic to the sensationalist project that one begins life with a *tabula rasa*, and the question is how one develops a cognitive, affective, and moral life on this basis. Diderot develops this line of thought most fully, explicitly linking sensation with moral and aesthetic sensibility. He does this through an exploration of the mentality of someone who is deprived of sensory capacities in some way. In his account of blindness, for example, he compares the blind with the sighted in order to explore what this tells us about sensibility in general.[87] A deficient sensibility is primarily a question of an emotional, aesthetic, and moral challenge for Diderot, and because of their impoverished sensibilities, the blind turn their minds inwards and are drawn to thinking in terms of abstractions. The blind offer a crucial case study, because he believes that their abstract manner of experiencing pain in others weakens their sense of sympathy for the suffering of others. His target is not confined to the blind, however. The blind just manifest in a particularly concrete way a general insensibility to the world on the part of those who think and experience it in terms of abstractions. He is particularly concerned with Cartesians

[87] [Denis Diderot], *Lettre sur les aveugles, à l'usage de ceux qui voient* ('London' [Paris], 1749).

in this French context, but Wolff would be a prime contender for this status also. For Diderot, our relation to the world depends very much upon how we arrive at that relation, and one assimilates cognitive information in a process which is always and necessarily social, cultural, and has moral implications, so that what is shaped is not merely a cognitive sensibility but a sensibility in which cognitive, affective, and moral questions are inextricably tied together. What is ultimately at stake is the sensory basis of civic life, where the contrast is between sensibility and solipsistic 'rationalism'.[88] The general question underlying this is that of where the ideas that regulate our lives—our moral, emotional, social, political, and intellectual lives—come from. The approach of those metaphysicians and others who conceive of the world in abstract terms, and as a consequence examine our relation to it in these same terms, now becomes not merely misguided but socially and morally irresponsible.

Diderot's account of the process of our coming to understand the world was focused on the respective roles that sensibility and reason played. It challenged the primacy of reason in this process, but it did not challenge the autonomy of reason as such. But other thinkers—Hume, Rousseau, and Helvétius, for example—did raise the question of the autonomy of reason explicitly. In 1739, in his *Treatise of Human Nature*, for example, Hume describes reason as 'nothing but a wonderful and intelligible instinct in our souls',[89] and goes on to remark that 'reason is, and ought only to be the slave of the passions, and can never pretend to any other office than to serve and obey them'.[90] In Germany, there were a number of different challenges to the autonomy of reason, and we shall be looking in some detail at that of Herder in Chapter 4. A particularly influential criticism, which illustrates in stark terms what was at issue, was that contained in Jacobi's 1785 *Brief über Spinoza*.[91] Jacobi focuses on the relation between the faculties of reason and will.[92] Developing a Humean theme, he argues that reason is the servant of the will. But he goes beyond the view that the will determines the ends of action, providing tasks for reason: it also determines the goals of intellectual enquiry. It is human interests that shape and guide human enquiry, not some abstract notion of 'reason' cut off from these interests and supposedly standing in judgement of them. In his *David Hume über den Glauben* of 1787, he sets out this assessment of reason in more detail, and there are two basic arguments. The first is that reason knows only its own creations, so there is nothing 'in nature', so to speak, to which

[88] See the account in Jessica Riskin, *Science in the Age of Sensibility: The Sentimental Empiricists of the French Enlightenment* (Chicago, 2002), ch. 2.

[89] Hume, *A Treatise of Human Nature*, 179. [90] Ibid., 295.

[91] Friedrich Heinrich Jacobi, *Briefe über die Lehre Spinozas* (Leipzig, 1785). See also the account in the work of Jacobi's close friend Thomas Wizenmann, *Die Resultate der Jacobischen und Mendelsohnschen Philosophie: kritische untersucht* (Leipzig, 1786).

[92] See the discussion in Frederick C. Beiser, *The Fate of Reason: German Philosophy from Kant to Fichte* (Cambridge, Mass., 1987), 85–91. For a detailed account, see Herman Timm, *Gott und die Freiheit: Studien zur Religionsphilosophie*, i. *Die Spinozarenaissance* (Frankfurt, 1974).

our thoughts must correspond if they are to be true. Nature must conform to reason, not the other way around. In itself, this was not such a novel position, and it was something held by Kant and others who sought (in one form or another) to defend reason, although in these cases there was generally a concerted attempt to mitigate any relativist or subjectivist consequences: for example, in Kant's case, by reference to universal and necessary rules. Jacobi rejects any such qualifications. Second, he argues that reason is not an end in itself but is guided by a basic desire for survival which is beyond the control or even comprehension of reason. Reason is not autonomous: it is subject to our needs and desires, and its purpose is to help us realize these. Here, the issues go beyond the balance of reason and sensibility, and to the heart of the question of the standing of reason.

THE CHALLENGE TO REASON

In the course of the 1780s there emerged a fundamental divide between a Kantian conception of the *Aufklärung*, and a Romantic sensibilist alternative to this 'rationalist' idea of Enlightenment. This reaction was epitomized in the *Sturm und Drang* movement, which stressed religious and ethnic tradition, and rejected the cosmopolitanism so distinctive of the *Aufklärung* in favour of a national literary culture. Hamann was a key figure in the early stages of the *Sturm und Drang*. His writings were deliberately unsystematic, bombastic, humorous, and typically written in response to the publications of others, and there was usually a strong element of parody.[93] They were an extreme example of the essay form designed to replace the treatise in the treatment of serious philosophical questions. Indeed it was Hamann who had introduced Kant to the writings of Rousseau and Hume, and who was therefore a key figure in shaping his pre-critical trajectory. However, with the appearance of Kant's *Kritik der reinen Vernunft* (1781), and Hamann's response, *Metakritik über den Purismus der Vernunft* (composed in 1784),[94] their paths diverged radically, and in many respects Kant and Hamann represent the two opposite poles in thinking about rationality in the 1780s, with Schiller attempting to reconcile the two currents in his immensely influential *Über die ästhetische Erziehung des Menschen in einer Reihe von Briefen* in 1795.[95]

The issues that came to a head in the 1780s had been simmering for thirty years.[96] Given the extent to which philosophical thought in Germany was

[93] On Hamann, see Beiser, *The Fate of Reason*, ch. 1; and Michael N. Forster, *After Herder: Philosophy of Language in the German Tradition* (Oxford, 2010), part ii.

[94] It was not published until 1800, but its contents were known to his contemporaries from the time of composition.

[95] The 'letters' were first published in the journal *Die Horen*, which Schiller edited.

[96] See John H. Zammito, '"The Most Hidden Conditions of Men of First Rank"—The Pantheist Current in Eighteenth-Century Germany "Uncovered" by the Spinoza Controversy', *Eighteenth-Century Thought* 1 (2003), 335–68.

embedded in metaphysics, it was perhaps inevitable that questions on the standing of reason turned less on Rousseau than on what was perceived as Hume's dilemma. By contrast with the sensibilist reaction to reason in France, for example, in Germany the threat to reason could not come from outside metaphysics, from sensibility, because it was metaphysics that was deemed to provide the criteria of rationality by which to judge everything else, so nothing outside metaphysics could threaten it. Rather, the threat came from inside, from Hume's internal questioning of metaphysics. The dilemma that German philosophers drew from Hume was a sharp one: either one accepted reason, which meant scepticism, for that was what turned out to be the logical end point of reason, or one was forced to accept an irrational leap of faith. As I have indicated, Hume's own position did not issue in a dilemma of this kind. Rather than choosing between scepticism and blind faith, he saw a balance between propositional understanding, which he associated with metaphysics, and which he considered did indeed lead to scepticism if carried through to the end, and non-propositional forms of understanding (coming to terms with the world in terms of aspirations, anxieties, fears, untutored common-sense beliefs, and traditions, for example), in which the phenomena of sensibility were paramount. We cannot do without either form of understanding, so the idea that there could be a choice would simply be a mistake on Hume's view. Moreover, it was a crucial part of the Humean argument that there was no philosophical formula or rule that captured what form this balance should take. Understanding the 'human condition', which is in effect what is at issue here, is taken out of the realm of philosophical understanding, ultimately becoming a matter of informed personal judgement.[97]

Such a notion was not wholly alien to German culture—the concern with *Bildung*, with cultivation of the self, was a very important part of this culture—but, up to this point, we have considered Hume's solution only in a decontextualized way. We can think of the various, rather more contextualized, forms of naturalization that we shall be exploring in subsequent chapters—anthropological medicine, philosophical anthropology, the natural history of man, and social arithmetic—in Humean terms as ways of taking fundamental questions of the human condition out of the hands of theologians and metaphysicians, and exploring how they might be dealt with in a more satisfactory way. These can certainly be seen as responses to the kind of problem that Hume addressed, even though his name was not invoked in most of the cases. Here, however, we must introduce a crucial qualification. Although I have considered Hume's solution in an abstract form, it is never actually available in this abstract form: it is always overdetermined by a range of other questions. If we contextualize the Humean dilemma in the case of mid-century debates in Germany, we find that the issues are inseparable from various theological and political questions for example, so

[97] See the discussion of Hume in Gaukroger, *The Collapse of Mechanism and the Rise of Sensibility*, 438–52.

that a choice between rational scepticism and an irrational leap of faith maps neatly on to a deeply embedded and complex conflict between various understandings of religion, morality, philosophy, literature, and emancipation. This shapes the form that the questions now take: Hume's dilemma becomes a resource for focusing and sharpening issues which may have only a tangential relation to it.

Hume had a particular theory about the development of religion from a primitive and non-philosophical form to an explicitly philosophical form, that of Christian monotheism. Primitive forms of religion were polytheist, on Hume's reading. They were an integral ingredient in making sense of the world, and he saw them as having significant value. Assessing them by the dictates of reason was something inappropriate on this view. Christian monotheism is a wholly different matter however. Here, in Hume's view, was something that had been shaped philosophically, more specifically in terms of metaphysics, and which made claims of a kind that were subject to the dictates of reason. Hume used the Catholic philosophical theology of Malebranche as his model for a developed Christian theology, and this is indeed explicitly an integral part of a metaphysical system. The end result of such philosophical theology was, in Hume's view, a thoroughgoing scepticism.[98] For those who wanted to defend both the primacy of reason and at the same time to adhere to basic Christian precepts, the problem was how, without treating it in terms of metaphysics—whose ability to secure theological claims had been, and remained, a deeply problematic enterprise—an account of Christianity could be devised which established that Christianity was not a matter of blind faith,[99] and consequently had a form of legitimacy, one different from metaphysical truths and not inferior to them.

When the questions came to a head in the mid-1780s, it emerged that there was an elaborate proposal for a solution to this problem that had been overlooked, in a work that had been published in 1781, Kant's *Kritik der reinen Vernunft*. The *Kritik* was a difficult work, an academic treatise of a kind that was out of fashion, in a genre which Kant himself had rejected in the 1760s and 1770s. It had

[98] Hume was certainly not alone in believing that Christianity had been corrupted by its encounter with philosophy, and one does not have to look too far to find Protestant thinkers arguing along similar lines. The Protestant theologian Johann Lorenz von Mosheim, for example, writes that 'the immoderate love of philosophy did violence to the holy scriptures', and 'inflamed by philosophic zeal, men laboured with extraordinary subtlety to extract what dogmas they thought reasonable from the writings of inspired authors; and if any revelations in the sacred text seemed contrary to philosophical doctrine, they twisted and misinterpreted them desperately'. *De Rebus Christianorum ante Constantinum Magnum Commentarii* (Helmstedt, 1753), 298. Quoted in John G. A. Pocock, *Barbarism and Religion* (6 vols, Cambridge, 1999–2015), v. 205–6. Cf. Bernard Mandeville, *Free Thoughts on Religion, the Church and National Happiness* (London, 1720), who argued that the clergy had taken the mysteries of the Gospel to absurdities by mixing theology with philosophy.

[99] Note that it is Christianity that is at issue, not religion generally, because eighteenth-century Christian commentators had no hesitation in treating non-Christian religious claims in a completely 'rational' way. See Pocock, *Barbarism and Religion*, v. ch. 12.

received little attention since its publication, and was written at such a level of abstraction that it seemed unconnected to the concerns of the day. This changed when a one-time pupil of Kant, and indeed one of his early critics, Karl Leonhard Reinhold, a Viennese ex-Jesuit, underwent what, given it religious overtones, can only be described as a conversion to Kantianism in 1785, and began publishing a series of letters on the *Kritik* from the beginning of 1786.[100] Reinhold translated the 'critical philosophy' from an academic treatise into a series of popular essays, renouncing any claims to systematic exposition or demonstrative argument.

Kant's project in the *Kritik* was to fix the limits of reason and to explore in a systematic way what could be said about matters that fell outside those limits, and this project was crucially tied up in that of providing a solution to the question of the legitimacy of religious belief. But in order to achieve this, what Kant needed to offer was a wholly general and comprehensive account of the limits of reason, of which this was just a part, if a crucial one. The limits of reason for Kant are the limits of metaphysics, but they also turn out to be the limits of scientific understanding, because Kant is adamant throughout his career that science requires grounding in metaphysics.[101] In a short early piece on 'physical monadology' published in 1756 for example, he writes:

> Clear-headed philosophers, who seriously engage in the investigation of nature, are unanimous in their agreement that punctilious care must be taken lest anything concocted with rashness or with arbitrary conjectures should insinuate itself into natural science, or lest anything be vainly undertaken in it without the support of experience and without the mediation of geometry. Certainly, nothing can be thought more useful to philosophy, or more beneficial to it, than this counsel. However, hardly any mortal can advance with a firm step along the straight line of truth without here and there turning aside in one direction or another.[102]

One way to deal with this question is what might be described as the phenomenalist route, associated with Newtonians, especially those influenced by Locke, but Kant rejects this path:

> For this reason there have been some who have observed this law to such a degree that, in searching out the truth, they have not ventured to commit themselves to the deep sea but have considered it better to hug the coast, only admitting what is immediately revealed by the testimony of the senses. And certainly if we follow this sound path, we can exhibit the laws of nature though not the origin and cause of these laws. For those who only hunt out

[100] The letters appeared in instalments in the *Teutsche Merkure*. They were first published in a collected form as *Briefe über die kantische philosophie* (2 vols, Leipzig 1790–2). There is an English translation: *Letters on Kantian Philosophy*, ed. K. Ameriks (Cambridge, 2005). On Reinhold see Beiser, *The Fate of Reason*, ch. 8.

[101] There are changes of detail that will not concern us here: on these, see Martin Schönfeld, *The Philosophy of the Young Kant: The Precritical Project* (Oxford, 2000).

[102] *Metaphysicae cum geometrica iunctae usus in philosophia naturali, cuius Specimen I. continet Monadologium physicam*: *Gesammelte Schriften*, i. 475; *Theoretical Philosophy, 1755–1770* (Cambridge, 1992), 51.

the phenomena of nature are always that far removed from the deeper understanding of first causes. Nor will they ever attain knowledge of the nature of bodies.[103]

For Kant, there is only one conclusion to be drawn from this: 'metaphysics, which many say may be properly absent from physics is in fact its only support; it alone provides illumination'.[104] Consequently, Kant attempts to rewrite Newtonian physics[105] in terms of a new foundational metaphysics, one that draws inspiration from Leibniz and Wolff.

Something like this view carries over into Kant's critical period. His estimation of what metaphysics can and cannot do changes radically, however, and he argues that the basic principles regulating enquiry need to go beyond metaphysics, and indeed to go beyond what he terms 'theoretical' or 'pure' reason.

The 'pure reason' of Kant's 'critique of pure reason' ostensibly refers to metaphysics, because metaphysics deals with things that we cannot examine empirically. Kant's model for metaphysics is a Leibnizian/Wolffian one, in which deductive demonstration is the means by which the discipline progresses. But his model for science is rational mechanics, whose greatest exponent was Euler, and rational mechanics too proceeds by means of deductive demonstration.[106] Rational mechanics was distinctive in that it was promoted, especially by Euler, as an a priori discipline.[107] Euler worked within a tradition of mechanics that had begun at the turn of the century in France, in which Newtonian mechanics was translated into an algebraic form. Although French natural philosophers were initially hostile to the physical content of Newton's *Principia*, especially his account of gravitation, this changed after 1730, and the physical theory pursued in rational mechanics was henceforth Newtonian (as opposed to Cartesian or Leibnizian), even though its algebraic formulation—following the work of the Bernoullis, d'Alembert, Clairaut, Euler, and others—transformed the way in which it was pursued.

In his *Theoria motus corporum solidorum* (1765), for example, Euler set out to present an axiomatic version of this physical theory starting from indubitable definitions and axioms, in what can only be described as a mix of

[103] Ibid. [104] Ibid.

[105] 'Physics' here is basically mechanics. The unified discipline of physics as we now understand it only emerged in the 1830s: see Iwan Rhys Morus, *When Physics Became King* (Chicago, 2005).

[106] On Kant and Euler see Michael Friedman, *Kant and the Exact Sciences* (Cambridge, Mass., 1992), 1–52.

[107] See Gaukroger, *The Collapse of Mechanism and the Rise of Sensibility*, 304–17. The need for a metaphysical grounding of mechanics, along similarly a priori grounds, was being urged at the same time in Emilie du Châtelet, *Institutions de physique* (1st edn, Paris, 1740), a work that was instrumental in reintroducing Leibniz to French natural philosophers. See Harmut Hecht, 'In the Spirit of Leibniz—Two Approaches from 1742', and Dieter Suisky, 'Leonhard Euler and Emilie du Châtelet: On the Post-Newtonian Development of Physics', in Ruth Hagengruber, ed., *Emilie du Châtelet Between Leibniz and Newton* (Dordrecht, 2012), 61–76 and 113–56 respectively. The 1756 Berlin Akademie prize question was 'whether the truth of the principles of statics and mechanics is necessary or contingent'. No winners were declared.

metaphysics—conceived in Leibnizian fashion as the analysis of propositions into clear constituents—and mathematics.[108] He begins with definitions of position, distance, shape, space, time, rest, motion, speed, and direction of motion. Next he provides the basic mathematical and conceptual techniques for characterizing all motions vectorially, and for dealing with motions in terms of reference frames. This provides him with what he needs for a purely kinematic theory, enabling him to move next to dynamics. His first task as far as dynamics is concerned is to distinguish internal from external sources of motion. The internal determinant of a body's state of rest or motion cannot be forces, he argues, and there is only one internal factor, the body's inertial state, that is, its perseverance in a state of (relative) rest or (relative) uniform rectilinear motion: in the absence of external forces the way in which it behaves is due entirely to the inertial state that it is in. While there are important clarifications in Euler's account up to this stage, there is nothing novel. This changes when he turns to the question of the external sources of motion (forces). Up to this point, the discussion has proceeded in conceptual or analytical terms, before being fleshed out mathematically. But once we turn to the question of external forces acting upon a body, contingent facts about the world would seem to enter the picture. Euler sets out to establish that we are still in the realm of conceptual truths, however, something to be decided without recourse to empirical questions. He identifies three basic features of the nature of bodies: they are spatially extended, their behaviour is inertial, and they are impenetrable. These three, he argues, are conceptual truths about the nature of bodies, in that something that lacked any of them would not be a body, and anything possessing all three would of necessity be a body. It is at least plausible that it is essential to what we mean by 'body' that bodies are extended and impenetrable, and that these are conceptual truths about bodies in that we cannot imagine something being a body yet being unextended or penetrable. But the notions of extension and impenetrability are not sufficient in themselves to allow for an adequate characterization of the notions of force and change of motion: inertia is also needed. The problem is that inertia does not look at all like a conceptual truth, for we can manifestly conceive of a body not obeying Newton's law of inertia. Yet the formulation of the foundations of mechanics in terms of conceptual truths was as crucial to Euler's project as the formulation of the foundations of metaphysics in terms of conceptual truths was to Leibniz and the Wolffians. Indeed, in his *Lettres à une princesse*,[109] Euler explicitly engages metaphysical views of the nature of bodies, with a whole letter devoted to Wolff's account of monads,[110] and it is clear that he sees his own account to be in the same genre.

[108] *Theoria motus corporum solidorum seu rigidorum* (1765) in *Leonhardi Euleri opera omnia*, series 2, vols iii, iv (Bern, 1948).

[109] Leonhard Euler, *Lettres à une princesse d'Allemagne sur divers sujets de physique et de philosophie* (3 vols, St Petersburg, 1768–72). References are to the English translation: *Letters of Euler on Different Subjects in Physics, Addressed to a German Princess* (2 vols, London, 1802).

[110] *Letters of Euler*, i. letter LXXIV. Cf. Euler, 'Reflexions sur l'espace et le tems', *Histoire de l'Académie Royale des sciences et belles lettres de Prusse* (Berlin, 1748) vol. 4: 324–33. Note also Euler's

The accounts of Wolff, Euler, and Kant share three assumptions: that mechanics embodies the ideals of rational explanation; that, having established its core on a rigorous basis, one should then, in time, be able to work outwards from this core to other areas of physics, gradually assimilating them to mechanics; and that there is an intimate and mutually reinforcing relation between metaphysics and the basic assumptions behind mechanics. It is in fleshing out the last of these that Kant parts company with Euler and Wolff in the *Kritik*, for there he argues that metaphysical argument has clearly definable limits. Beyond these limits lie fundamental assumptions, for example about the nature of space and time, but also about the existence of God, as well as questions of morality.

The route that Kant describes in the *Kritik* identified three stages in thought: dogmatism, scepticism, and criticism. 'Dogmatism', a term of art in Kant, is the view that, on the basis of pure reason, one can attain knowledge of the existence of God, of the existence of freedom in a world otherwise governed by necessity, and of the existence and immortality of the soul.[111] Plato, Spinoza, and Mendelssohn are singled out at one time or another for criticism in this respect, but Kant's prime target is Wolff. Yet Kant does not want to abandon what he sees as the strengths of Wolffian procedures. In particular, his own proposed critical method is

> not opposed to the *dogmatic procedure* of reason in its pure cognition as science (for science must always be dogmatic, i.e., it must prove its conclusion strictly a priori from secure principles); rather, it is opposed only to dogmatism, i.e., to the presumption of getting on solely with pure cognition from (philosophical) concepts according to principles which reason has been using for a long time, without first inquiring in what way and by what right it has obtained them.[112]

Scepticism is identified with Hume, and is presented as a necessary step beyond dogmatism, one that reveals its limits, and Kant describes it as 'a resting place for human reason, which can reflect upon its dogmatic peregrinations and make a survey of the region in which it finds itself, in order to be able to choose its path in the future with greater certainty, but it is not a dwelling-place for permanent residence'.[113]

attack on the Wolffian defence of monads at the 1746 Berlin Akademie prize essay competition. Euler was the Akademie's most prestigious member, and the prize was awarded to Johann Heinrich Gottlob Justi, whose anti-monad argument was very close to that which had been elaborated by Euler himself. See Roberto Palaia, 'Berlino 1747: Il dibattito occasione del concorso dell'Academia della scienze', *Nouvelles de la République des Lettres* (1993), 91–119; Thomas Broman, 'Metaphysics for an Enlightened Public: The Controversy over Monads in Germany, 1746–1748', *Isis* 103 (2012), 1–23.

111 *Kritik der reinen Vernunft*, A5/B8. 112 Ibid., Bxxxv.

113 Ibid., A761/B789. See his *Prologomena zu einer jeden künftigen Metaphyik als Wissenschaft wird auftreten können* (Prologomena to any Future Metaphysics): 'Since the Essays of *Locke* and *Leibniz*, or rather since the rise of metaphysics as far as the history of it reaches, no event has occurred that could have been more decisive with respect to the fate of this science than the attack made upon it by *David Hume*. He brought no light to this kind of knowledge, but he certainly struck a spark

Hume, on Kant's reading, rejected any 'pretensions of reason in general to go beyond the empirical'.[114]

The connection between dogmatism, scepticism, and criticism is then set out in a general statement of his strategy in the *Kritik*:

> One can call a procedure of this [Humean] sort, subjecting the *facta* of reason to examination and when necessary to blame, the *censorship* of reason. It is beyond doubt that this censorship inevitably leads to *doubt* about all transcendent use of principles. But this is only the second step, which is far from completing the work. The first step in matters of pure reason, which characterizes its childhood, is *dogmatic*. The just mentioned second step is *sceptical*, and gives evidence of the caution of the power of judgement sharpened by experience. Now, however, a third step is still necessary, which pertains only to the mature and adult power of judgement, which has at its basis firm maxims of proven universality, that, namely, which subjects to evaluation not the *facta* of reason but reason itself, as concerns its entire capacity and suitability for pure a priori cognitions; this is not the censorship but the *critique* of pure reason, whereby not merely *limits* but rather the determinate *boundaries* of it—not merely ignorance in one part or another but ignorance in regard to all possible questions of a certain sort—are not merely suspected but are proved from principles.[115]

In other words, the task is not just to register the limits of reason/metaphysics but to provide a general theory about how these limits arise and how we are to determine the standing of those things that fall outside the limits. Among the novel ingredients in Kant's strategy in the *Kritik*, two are of particular interest. First, he generalizes Hume's scepticism about the ability of metaphysics to establish basic truths, which Kant takes to be focused on causation, to the whole of metaphysics, arguing that for any metaphysical thesis (he has Wolff's metaphysics particularly in mind), it can be shown that its contrary can be defended with as much legitimacy. He looks at four basic types of case, termed 'antinomies of pure reason'.[116] The first places the thesis that the world has a beginning in time and is finite in extent, against the thesis that the world has no beginning and is infinite in extent. The second opposes the thesis that everything that exists is either simple or composed of simples, against the thesis that nothing that exists is simple. The third opposes the thesis that as well as causation ruled by the laws of nature we also need to allow causation that arises due to freedom, to the thesis that all causation must be in accord with the laws of nature alone. The fourth opposes the thesis that there is a necessary being, to the thesis that there is no necessary being. In each case, Kant shows that the two opposing theses are each

from which a light could well have been kindled, if it had hit some welcoming tinder whose glow had then been carefully kept going and made to grow.' *Werke*, iv. 257; *Theoretical Philosophy after 1781*, (Cambridge, 2002), 54–5.

114 *Kritik der reinen Vernunft*, A760/B788.

115 Ibid., A760–1/B788–9.

116 Ibid., A405–567/B432–595.

equally compelling. The argument seriously undermines the claims of metaphysics to establish fundamental truths.

Kant's second crucial move is to argue that, included in the range of things that we take to be true but cannot establish by reason/metaphysics, are not just the kinds of things that he has included in the antinomies, but basic truths about the nature of space, time, and causation. In other words, at a general level there is nothing special about (what were considered to be) the truths of religion and morality: fundamental truths about the nature of the physical world were also at stake, and for exactly the same kind of reason. Religion and morality did not present local problems, but were rather manifestations of a wholly general problem. This wholly general problem concerned the nature of a priori truths, and Kant proposed to resolve it by arguing that while the truths of metaphysics are analytic a priori (that is, they are regulated solely by the principle of contradiction), there are also a priori truths which fall outside the purview of metaphysics, which are synthetic. The difference between the two lies in the fact that analytic judgements, in predicating something of a subject, simply reveal something that had been contained in the subject all along, even if that had not been evident, whereas a synthetic judgement 'adds to the concept of the subject a predicate which had not been in any way in it, and which no analysis could possibly extract from it'.[117] Kant's argument is that an account of synthetic a priori judgements provides us with a way of understanding, and establishing the legitimacy of, those truths that we recognize as truths, but which reason is unable to establish.

He begins his account with something that would generally have been recognized to be secure, namely our understanding of space, more specifically of geometrical extension. In 1768 he had published a short tract on the differentiation of regions in space,[118] where he presents a distinctive argument rejecting the Leibnizian conception of relative space and defending the idea of a privileged frame of reference in space. In purely geometrical terms, a right-hand glove and a left-hand glove are identical: they have the same parts standing in the same geometrical relations to one another. Yet we know—if only in trying the gloves on—that the one is in fact different from the other: they are 'incongruous counterparts' in Kantian terminology. Since geometry cannot distinguish these, we must conclude that there is something about spatial relations that goes beyond geometry, and that this extra ingredient can only be captured intuitively, not in terms of some further extension of geometry.

In the *Kritik*, Kant elaborates in detail on the geometrical case, and adds arithmetic. The general question is how geometrical and arithmetical judgements can be both necessarily true and yet informative. It would seem that they can be one (analytic a priori) or the other (synthetic a posteriori), but not both. Kant's

[117] Ibid., A7/B11.

[118] *Von dem ersten Grunde des Unterschiedes der Gegenden im Raume*: *Gesammelte Schriften*, ii. 377–83; *Theoretical Philosophy, 1755–1770*, 365–72.

argument is that the truths of arithmetic and geometry are not analytic, as Leibniz and Wolff had claimed, but they are nevertheless necessary and they are independent of experience, contrary to abstractionist theories such as that of Locke. That space is not something empirical is revealed, Kant argues, when we reflect on the difference between genuine empirical things, namely bodies, and space. We can imagine bodies being removed from the universe, leaving us with empty space, but we cannot imagine a universe from which space has been removed, or a part of the universe which is empty of space. Space is not something empirical, and we do not come by our concept of space by abstracting it from particular spatially extended bodies or regions. Rather, it forms, along with time and causation (and, in later works, an ether) a condition of possibility of physical experience. It is not something that we recognize as a content of our experience: the notion of space that we work with in (Euclidean) geometry is infinite and there is nothing in experience of this kind. Such a condition of possibility of experience has the form of a synthetic a priori judgement, Kant argues. The world is accessible only insofar as we are able to impose structure upon it, and not all of this structure is conceptual, in the sense of comprising general notions which we have derived from experience or which we have formulated for ourselves in some other way. Some of the structuring—which Kant refers to in terms of 'intuition'—simply comes as part of the package, as it were, if we are to have physical experiences of the kind that we have. We must bring both intuitions and concepts to our experience and judgement, and these must be combined in some way to produce the kinds of experience that we have.

Kant's 'critical' philosophy takes the contrast between propositional and non-propositional forms of understanding to a new level. Developing the Humean idea of the limits of metaphysics, he does not dispense with non-propositional forms of understanding from philosophy, but offers a comprehensive unified picture of the whole of human understanding, one in which practical reason plays as important a role as theoretical reason, and is equally subject to philosophical exploration. Even though the kind of understanding yielded is very different in the two cases, he makes it clear that both are ultimately to be incorporated into the one single system.[119] What we have in effect is a taming, as it were, of non-propositional forms of knowledge, not via a reduction of the non-propositional to the propositional along the lines of Condorcet, but through a transcendental treatment that unites the two, incorporating the non-propositional into a comprehensive systematic account of our understanding of the world. This proved to be crucial to how Kant conceived of a philosophical anthropology, as we shall see in Chapter 4. It lends an attractiveness to Kantian philosophy which is best illustrated by its reception in Britain in the last decade of the eighteenth century. The first book-length exposition of Kant's philosophy in

[119] For example, *Kritik der reinen Vernunft*, A840/B868.

English was the 1796 work by Friedrich Nitsch, whose title draws attention to the religious apologetic dimension of Kantian thought: *A General and Introductory View of Kant's Principles concerning Man, the World, and the Deity*.[120] But it was the radical Thomas Beddoes, physician and chemist, and follower of Priestley, who in his 1793 *On the Nature of Demonstrative Evidence* had first introduced Kant to British audiences and, despite some Lockean-inspired antipathy towards Kant, Beddoes was the earliest source of the new German philosophy for English readers.[121] Coleridge, a member of Beddoes' circle, became a committed Kantian, as, under his influence, did Humphry Davy, and the attraction of Kant for them was that Kantianism and idealism more generally allowed British thinkers from the end of the eighteenth century 'to navigate the slender channel between the rocks of French revolution and British reaction', as one commentator has put it, 'and pursue a route to progress that was tainted by neither divisive ideology. Germany was a powerhouse of research and progress, but its novelties, unlike those of France, were seen as purely intellectual and unconnected with any political programme.'[122]

Kant's critical philosophy relied on a range of assumptions, however, and particularly on a model of understanding taken from a particular area of physics—rational mechanics—which experimentalists such as Beddoes and Davy could hardly have accepted. While elements of the critical philosophy can be appreciated or defended on independent grounds, this model ultimately lies at the core of the critical philosophy, and it stands or falls on its appropriateness and legitimacy. The problem was that many of the assumptions that underlay rational mechanics had come into question. It was not only far from the cutting edge of science in the late decades of the eighteenth century, but had become very unproductive in physical terms. Most problematic of all for those who were still taking it as a model of scientific certainty, however, was the fact that the late eighteenth-century revival of the fortunes of celestial mechanics—of all the mechanical disciplines, the one that was held up, from Newton onwards, as the epitome of mathematical certainty and predictive power—was due to a recognition of uncertainty and the introduction of new techniques for dealing with it.

THE LIMITS OF CERTAINTY

One of the features that marked out the non-mechanical physical disciplines from the mechanical ones was the aspiration to certainty in the mechanical disciplines.

[120] Friedrich A. Nitsch, *A General and Introductory View of Kant's Principles concerning Man, the World, and the Deity* (London, 1796).

[121] Mike Jay, *The Atmosphere of Heaven: The Unnatural Experiments of Dr Beddoes and His Sons of Genius* (New Haven, 2009), 61–2.

[122] Ibid., 193.

The model was the celestial mechanics set out in Newton's *Principia*. But celestial mechanics relied not only on the mathematical techniques of geometry, algebraic analysis, and calculus, but also on observation. From the middle of the eighteenth century, there were a number of attempts to explore the relation between the two.[123] Buffon, whose piece written for admission to the Académie des Sciences was on the application of probability theory to games of chance, set out to apply probability theory to physical facts and made this a key part of his method in natural history. In the 'Première Discours' (1749) of the *Histoire Naturelle*, he raises the question of how one 'synthesizes' mathematical and physical truths:

These general effects are for us the true laws of nature: all the phenomena that we recognize as obeying these laws and depending on them are as much facts explained as truths understood; those that we are unable to relate to one another are simple facts that need to be reserved to one side, awaiting a greater number of observations over a longer time so that we might discover the physical cause, that is, the general effect from which these particular effects derive. It is here that the union of the two sciences of mathematics and physics can be advantageous. The one tells us the 'how many' of things, the other the 'how' of them; and since it is a question here of combining and calculating probabilities in order to judge whether an effect depends more on one cause than on another, when you have used physics to imagine the 'how', that is, when you have observed that such and such an effect might well depend on such and such a cause, you then apply mathematics in order to assure yourself as to how often this effect happens in conjunction with its cause. And if you find that the result accords with the observations, the probability that you have estimated correctly is so increased that it becomes a certainty.[124]

Nineteen years later, Condorcet was arguing that comparison of mathematical laws, which yielded necessary truths, and observational laws, which were imperfectly known, was the prime concern of those working in the sciences.[125] Observational laws did not yield necessary truths because they were the product of the free will of an intelligent being who simply willed the world to be as it is rather than in any other way. God could have willed the world to be regulated by a completely different set of laws.[126] Necessary laws were a manifestation of human rationality, whereas contingent ones were a manifestation of divine choice. Condorcet does not see the two as being qualitatively and unbridgeably different however, and he seeks a connection in hypothetical terms:

Perhaps these laws differ only because of the present relation between things and us, which means that we require different degrees of knowledge to grasp them, so that if the law of

[123] The issues were tied up with the questions of whether God was required for anything more than creating the world, and to what extent the laws governing it could have been other than they are. See, for example, Charles Batteux, *Histoire des causes premières: ou, Exposition sommaire des pensées des philosophes sur les principes des êtres* (Paris, 1769).

[124] Buffon, *Œuvres complètes*, i. 111–12.

[125] Condorcet, *Le Marquis de Condorcet à Mr. D'Alembert sur le Système du Monde et sur le Calcul Intégral* (Paris, 1768), 4.

[126] Cf. Jean Baptiste d'Alembert, *Traité de Dynamique* (Paris, 1743), xxiv–xxv.

continuity were to hold in the universe, one could regard any instant to be the consequence of the initial arrangement of matter into a particular order and then left to itself. An Intelligence that knew the states of all phenomena at a given instant, the laws to which matter is subject, and their consequences at any given time from then, would have perfect knowledge of the *System of the World*. This knowledge is beyond our capacities, but it is the goal towards which all the efforts of philosophical geometers are directed, and which they always gradually approach, without ever hoping to reach it.[127]

Five years later, Laplace offered what was to become the canonical statement of this view, writing that: 'The present state of the system of Nature is manifestly a consequence of its state in the preceding moment, and if we conceive of an intelligence who, at a given instant, grasps all the relations between beings in this universe, it would be able to determine, at whatever time in the past or the future the respective position, motion and, generally, the properties of these beings.'[128] In following up this claim, he goes beyond Condorcet's suggestion, opening up uncertainty to a measure of quantitative treatment:

Physical Astronomy, that subject which, of everything we know, does the greatest honour to the human mind, offers us the idea, imperfect as it is, of what such an intelligence would be like. The simplicity of the laws describing what makes celestial bodies move, and the relationship between their masses and their distances, allows us, using Analysis, to follow their motions up to a certain point; and to determine the state of the system of these large bodies in past or future centuries it is enough that observations provide the mathematician with their position and speed at a given moment. Man owes this capacity to the power of the instrument he employs, and to the small number of relations involved. But ignorance of the different causes that are involved in the occurrence of these events, and their complexities, together with the imperfection of Analysis, prevent him from pronouncing with the same certainty on the greatest number of phenomena. For him, therefore, there are things that are uncertain, and there are some that are more probable and others that are less probable. Given the impossibility of complete knowledge, he has sought to compensate by determining their different degrees of likelihood; so that we owe to the frailty of the human mind one of the most delicate and ingenious [forms of] Mathematics, namely the science of chance or probabilities.[129]

The way in which the role of probability theory is conceived here is particularly interesting. Leibniz had thought of the use of algebraic analysis and infinitesimal algorithms (both of which came under 'analysis' as Laplace uses the term) as ways of extending our cognitive faculties, just as we might use a telescope as an extension of our sensory faculties.[130] These do not 'improve' our reasoning or eyesight, bringing them back to some prelapsarian state, and they certainly do not approximate to divine modes of reasoning and seeing. Rather, they provide a way

[127] Condorcet, *Système du Monde*, 4–5.
[128] Pierre-Simon Laplace, *Oeuvres complètes* (14 vols, Paris, 1878–1912), viii. 144.
[129] Ibid., 144–5.
[130] See my *The Collapse of Mechanism and the Rise of Sensibility*, 139–41.

of circumventing certain natural inadequacies. To take an example from Leibniz,[131] quadrature of the circle is impossible for us, because we cannot express the ratio between a square and a circle by a single number, but we can express the ratio by an infinite series of numbers whose way of continuing we know: $\pi = 4(1/1 - 1/3 + 1/5 - 1/7 \ldots)$. Probability theory can serve the same role, providing an indirect means of overcoming a difficulty.

The introduction of probability theory into celestial mechanics raises some fundamental questions about models of understanding which bear directly on the relation between the natural and the moral sciences. We have seen that the mechanical disciplines traditionally aspired to certainty, an aspiration associated with the ability to formulate theories in a rigorously mathematical form. The mathematical techniques involved here are geometry, algebraic analysis, and calculus. By contrast, the mathematics that we find in chemistry, for example, is arithmetic, often a simple manipulation of proportions. But there was another form of mathematics, probability theory, first developed in the correspondence between Pascal and Fermat in the 1650s, and shaped into a mathematically powerful discipline by Jacob Bernoulli in the early eighteenth century. In his *Ars Conjectandi* (1713), Bernoulli writes of this discipline:

> To *conjecture* is to measure the probability of something: and the *Art of Conjecture*, or the *Stochastic Art*, is for this reason defined as the art of measuring as exactly as possible the probabilities of things, so that in our judgements and actions we can always choose or follow the path that it would be better, safer, more preferable, and more considered to take; all the wisdom of the Philosopher and the prudence of the Statesman consists in this alone.[132]

This conception of probability locates it firmly in the moral sciences, and at first sight it may seem that probability theory could have little relevance to a mechanical discipline such as astronomy, which, from Newton's *Principia* onwards, was treated by many as a model of geometrical and analytical precision. But Laplace's resort to probability theory in astronomy turned out to pave the way for very significant advances. The transformation was achieved by taking astronomy out of the realm of absolute certainty that was so dear to those who took mechanics as the paradigm form of scientific activity. It did not accomplish this by making astronomy less mathematical, but by making it more mathematical, adding powerful new techniques to its resources without depriving it in any way of those resources it already relied upon. The new mathematical techniques explored and accounted for uncertainty, rather than treating uncertainty simply as an obstacle to progress and something to be overcome. Such an approach to

[131] See ibid., 140.

[132] Jacob Bernoulli, *Ars Conjectandi, opus posthumum* (Basel, 1713), 213. For an assessment of the importance of Beroulli's contribution, see Ian Hacking, 'Jacques Bernoulli's *Art of Conjecturing*', *British Journal for the Philosophy of Science* 22 (1971), 209–29.

uncertainty had direct consequences for a model of scientific enquiry that promoted certainty as the hallmark of successful enquiry. This, together with the fact that statistics, a crucial area of mathematical enquiry, was developed in the moral sciences and then subsequently taken up successfully in the physical sciences, rather than the other way around, means that these developments are of significance for our understanding of the projects pursuing investigation of the human realm via naturalization.

Probability theory is a natural tool of statistical thinking, which was originally the concern of 'statists', that is, reformers and bureaucrats concerned with various features of populations. The use of statistics, in the form of birth and death rates, was pioneered in a social context as early as the 1660s in Britain by John Graunt and William Petty by means of what was referred to as 'political arithmetic', a procedure originally designed to estimate each individual's contribution to the public good by measuring what he or she contributed to, or withdrew from, the national stock. Graunt, for example, in his *Natural and Political Observations*, applies what he terms his 'Shop-Arithmetique' to London bills of mortality, construing these as Baconian-style tables of discovery which might be made to yield a natural history of man, and initiating the investigation of death as a function of age.[133] Populations comprise large numbers of inherently variable objects, and the statists came to realize that systems consisting of autonomous individuals could be studied at a level where many particular individual variations could be ignored. They realized that such systems could, as Theodore Porter puts it, 'be presumed to generate large-scale order and regularity that would be virtually unaffected by the caprice that seemed to prevail in the actions of individuals'.[134] The crucial point was that changes to the state of a large-scale system could only be consequences of large-scale causes, so that one was now able to work with relative frequencies rather than individual actions. It was not so surprising that births and deaths, despite the huge variations in individual causes of death for example, would follow regular patterns. But it subsequently turned out that there were large-scale regularities in the incidence of marriages, crime, and, most surprisingly of all given the very individual nature of the act, suicides. Kant, who assumed that what was at issue here was something that fell outside science proper, described the discovery of statistical regularities in these terms:

> Thus marriages, the births that come from them, and deaths seem—since the free will of human beings has so great an influence on them—to be subject to no rule in accordance with which their number could be determined in advance through calculation; and yet the annual tables of them in large countries prove that they happen just as much in accordance

[133] John Graunt, *Natural and Political Observations mentioned in a following Index, and made upon the Bills of Mortality* (London, 1662), 7. See also William Petty, *An Essay Concerning the Multiplication of Mankind together with another Essay in Political Arithmetick* (London, 1683). See Lorraine Daston, *Classical Probability in the Enlightenment* (Princeton, 1988), 127–9.

[134] Theodore M. Porter, *The Rise of Statistical Thinking 1820–1900* (Princeton, 1986), 5.

with constant laws of nature as do weather conditions, which are so inconstant and whose individual occurrence one cannot determine in advance, but which on the whole do not fail to sustain the growth of plants, the course of streams, and other natural arrangements in a uniform uninterrupted course.[135]

These properties of a population fall under the more general rubric of collective properties of human beings, and pursuing enquiry into human behaviour in these terms forms an important independent stream in the naturalization of the human, which we shall be examining in Chapter 6.

From the late eighteenth century, there is a rise in the deployment of statistical and probabilistic techniques in dealing with questions that do not lend themselves to traditional means of establishing certainty in results. One crucial resource was error theory. In the early eighteenth century it became evident to those studying the distribution of errors in repeated measurements that observations of a large number of different measurements often tended to show a similar form of distribution. In his *Ars Conjectandi* of 1713, Jacob Bernoulli showed how to calculate what degree of approximation was appropriate for given forms of enquiry, enabling one to fix the range of acceptable variance in advance and calculate its effect.[136] In 1733 the mathematician Abraham de Moivre formulated an equation for the distribution of errors,[137] which took a bell shape (later called a normal or Gaussian distribution). De Moivre's equation, which, for the first time, allowed one to apply probability theory to an indefinite number of independent events, was nevertheless completely neglected at the time. Laplace took up de Moivre's work in the 1770s, however, and he combined it with a new technique, using exponential approximation, whereby a knowledge of past events could be used to predict distributions of events in the future.[138] On the questions of the ratio of male to female birth rates, for example, he dealt with Buffon's report that communes in Burgundy had showed a significantly higher birth rate for girls over a five-year period, by showing that these were anomalies due to chance which could be identified as such by taking figures for the century. He then went on to show that the surplus of male over female births in Paris in every year was not a discrepancy and not something requiring providential intercession: given the general average male to female ratio and the total number of births, the probability that female births would exceed male ones was 1/259, which made such an event extremely unlikely. But the most immediately fruitful development

[135] Kant, *Werke* viii. 17; *Anthropology, History, and Education*, 108.

[136] See the discussion of Bernoulli in Daston, *Classical Probability in the Enlightenment*, 230–53.

[137] Abraham de Moivre, *Approximatio ad summam terminotum binomii* $(a+b)^2$ *in seriem expansi* (1733): facsimile reprint in Raymond C. Archibald, 'A Rare Pamphlet of De Moivre and Some of His Discoveries', *Isis* 8 (1926), 671–84. Reprinted in the 2nd edition of De Moivre's *The Doctrine of Chances* (London, 1738).

[138] See Porter, *The Rise of Statistical Thinking*, ch. 4, to which I am indebted in this paragraph. See also Stephen M. Stigler, *The History of Statistics: The Measurement of Uncertainity before 1900* (Cambridge, Mass., 1986), 109–13.

was that of error analysis. Laplace had applied error analysis in a number of cases, an early one being changes in atmospheric pressure at different times of the day. The problem in determining whether diurnal variation was a cause of atmospheric pressure variation, was that there were so many irregular causes of variation that they exceeded the magnitude of the effect itself. The solution was to compile results for a large number of variations, and calculate the probability with which the effect was indicated. Laplace was subsequently able to devise a procedure for determining altitude from barometric pressure, a procedure that Humboldt was to use to great effect in his travels, starting in 1799 with his demonstration that central Spain was a plateau, not a valley as had been thought up to that time.

Another especially fruitful application of error theory was in astronomy, where there was a pressing need to reduce a large number of observations to a single value or a single curve, as well as a need for some way of estimating the accuracy of the final result from an account of the occurrence of single errors. In 1805 Legendre, in an account of how one might determine the orbits of comets, published an ingenious general method for reducing multiple observations of any object.[139] This was 'the method of least squares', and it consisted in choosing that curve which best minimized the sums of the squares of the deviations of individual measurements from it. It resulted in the concept of the 'standard deviation' of a set of observations: the square root of the arithmetic mean of the squares of the differences from the mean. Application of this method transformed the way in which astronomy was pursued in the nineteenth century, allowing one to range observations by statistically determined reliability.[140]

Two things are distinctive about these statistical and probabilistic procedures for our purposes. First, not only do they not originate in developments in highly quantified areas of the sciences, such as mechanics, but they cover all disciplines—natural and moral sciences—without distinction, by contrast with geometry and analysis for example. Second, they are models of rigorous quantitative enquiry. Yet, whether in celestial mechanics or population studies, their value consists in showing how to determine reliable approximations rather than establishing certainty. This is a development of some consequence in considering models of successful scientific practice, something crucial to naturalization projects. It is an important consideration when we encounter, for example, Kantian rejections of naturalistic forms of enquiry in anthropology on the grounds that they cannot achieve the kind of certainty characteristic of mechanics. The idea that mechanics

[139] Adrien Marie Legendre, *Nouvelles méthodes pour la détermination des orbites des comètes* (Paris, 1805). Gauss may have discovered the procedure as early as 1795, but did not publish his results until after the appearance of Legendre's book.

[140] See Oscar B. Sheynin, 'On the Mathematical Treatment of Astronomical Observations', *Archive for History of Exact Sciences* 11 (1973), 17–126; and 'Laplace's Theory of Error', *Archive for History of Exact Sciences* 17 (1977), 1–61.

offers canons of certainty to which other disciplines can aspire comes now, in the last three decades of the eighteenth century, to be undermined by developments in mathematics itself.

THE ANATOMY OF REASON

Changes in conceptions of the role and standing of reason are fundamental to the development of eighteenth-century understanding of the factors that shape human behaviour. In attempting to capture what was at issue in these changes, I have pursued a comparative strategy in this chapter, setting out a number of contrasts not just with reason itself but with its cognates: systematic understanding, propositional understanding, and the search for certainty. The aim has been to identify the main issues that were at stake in the rethinking of the role of reason. None of these correspond to that standard trope of Enlightenment studies, reason versus religion. This is not because questions of the role of reason do not impinge directly on the standing of religion, for they do. It is just that, with some isolated exceptions, there is not just one issue that is at stake, namely that in which religion and reason are mutually exclusive alternatives. Nor can this question be taken as a model or starting point for the discussion of the other conflicts over the role of reason. These conflicts hinge rather on a range of questions, including that of how reason and religion can be reconciled, raising the issues of whether the truths of Christian dogma should be treated in terms of rational demonstration, and what kind of historical reconstruction is appropriate in understanding how religious beliefs were formed and what their effects were. There were of course some who argued that reason was not an appropriate basis on which to establish Christian teaching, but most were concerned to reconcile Christianity and reason, not to oppose them to one another.

It is crucial to realize, if one is to take these questions seriously, that Christianity, when properly contextualized, is not something that simply stands or falls on a set of dogmas, but is rather something that has an inner complexity. We shall be exploring some of these questions in Chapter 7. But, for our concerns in Part II, it is more important to appreciate the extent to which reason too has an inner complexity. It also has a historical dimension: specifically, what is meant by reason in the various different debates with which we are concerned is not something uniform and ahistorical. Quite apart from the question of whether reason is analysed as something individual or collective, or whether it is construed as something essentially embodied or as something essentially disembodied, one thing that is clear is that there were significant changes in conceptions of what reason can and cannot do. In Hume for example, there is an influential rejection of the idea that morality is the product of reason, or that it can be regulated by reason. Kant is more radical. There was nothing unusual about the way his conception of reason hinged on absolute commitment to propositional understanding, systematic structure, a priori deductive certainty, and a sharp demarcation between reason and sensibility. The four assumptions do naturally go together and are mutually reinforcing, and Leibniz and Wolff had linked them. The difference is that no one before Kant had drawn the consequences of putting them together in such a thorough fashion. What he set out to show was that putting them together yields a very different conception of reason from that with which metaphysicians had started. Kant argued that a defence of 'theoretical' reason, as he termed it, was possible only by restricting it to a fraction of what his predecessors would have considered to be its proper domain. Not only are religion and morality excluded from the realm of theoretical reason, but so too are those sciences—chemistry, electricity, and the life sciences—that fail to match up to his conception of reason.

In the debates that we shall be concerned with in Part II, what is at issue is rarely reason in the form of all four assumptions taken together, but rather particular themes: a refusal to separate propositional and non-propositional forms of understanding; a defence of the view that non-propositional understanding, in the form of affective states, underlies all cognitive states; a commitment to forms of empirical enquiry that are antithetical to the notion that understanding consists ultimately in deductive certainty; and so on. My aim in this chapter has been to identify and examine, largely in the abstract, the basic dichotomies that underlie worries over the role of reason, so that we are better equipped, as we come to the details of the highly contextualized debates, to grasp what is at issue. But crucial as it is to be clear about the different questions that are at stake in disputes about reason, our enquiry is more specific than just changing conceptions of reason and the way in which they bear on changing conceptions of the human being. It is with how these conceptions change as a result of projects of naturalization. Naturalization is an attempt to draw on the successes of empirical disciplines by posing in empirical terms questions that had not been posed in this way before. It might be thought that this is simply a matter of putting these questions into a scientific form, but this is not the case. For one thing, one of the main sources of naturalization was the opening up of questions that had been treated as purely conceptual and ahistorical to historical investigation. Second, there is no one 'scientific form'. In particular, there are empirical sciences and there are highly mathematical non-empirical sciences. Everything depends on which one takes as one's model. Rational mechanics, as pursued by d'Alembert, Euler, and others, had an axiomatic deductive structure and sought absolute certainty, but had become an almost purely mathematical a priori discipline by the mid-eighteenth century. It provided no resources for naturalization, and Kant, for example, who took it as a model for enquiry, rejected the naturalization of the human in large part because he rejected naturalization: the core of science lay not in empirical enquiry but rather in deduction of results from absolutely certain a priori principles, a procedure Kant associated with 'reason'. In the light of this, before we can begin to look at the naturalization programmes in detail, we need to consider the merits of the various approaches to the natural sciences, to understand the advantages of choosing one model over another, and above all to grasp how the choice of naturalizing resources might shape one's understanding of human behaviour.

2

Rethinking the Nature of Matter

The naturalization of the human and the humanization of nature of necessity depended on some basic beliefs about nature. These typically involved an understanding of matter: what it was and how it acted. Matter theory did not by any means exhaust the understanding of the natural realm, but it had traditionally played a fundamental role in accounting for physical processes, and as a result was something in which disciplines across the spectrum had a stake, from metaphysics to the physical and life sciences. The period between the early decades of the eighteenth century and the early decades of the nineteenth was one in which a unified conception of matter that had predominated in the seventeenth century collapsed, with basic conceptions of matter diverging along new disciplinary lines. Mechanism had provided a unifying basis for the understanding of physical properties, and, with its decline, relatively autonomous disciplines—such as chemistry, electricity, and the life sciences—became the domains that were looked to for an account of matter and its properties. The problem was that they were motivated to a large extent by the need to make sense of experimental results, results that tended to be at best discipline-specific and at worst idiosyncratic, and consequently they remained largely autonomous.

Micro-corpuscularianism had been such a successful form of matter theory because it had provided a common basis for the reduction of macroscopic physical properties. Crucial to its success was its ability to unify the whole of the natural realm. From the early to middle decades of the eighteenth century onwards, in the wake of the decline of this form of matter theory, there was no single account of matter to replace it. Freed from the constraints mechanism had imposed, not only the life sciences and adjunct disciplines such as physiology, but also physical disciplines such as chemistry and electricity, became autonomous fields of enquiry and made significant progress as a result. But the idea that there was some intrinsic principle of unification of the natural realm, and the idea that such a principle would lie in an account of the nature of matter, were not wholly abandoned. Kant and others were to offer metaphysical accounts of the nature of matter in terms of quasi-mechanical forces, and electricity and the life sciences both worked with the idea of fundamental forces, albeit of different kinds.

By the end of the century, there were moves to reintroduce some unity and coherence into the understanding of matter, one that eschewed the dominance of post-Lavoisian chemistry for example, and sought a single overriding conception that could be used to underpin not just the inert substances studied in chemistry, but also the material effects of electrical currents, and the properties of living tissue. This was motivated in part by the fact that various connections were beginning to be made between chemical, electrical, and vital phenomena at the level of force. As this idea was developed, force began to replace micro-corpuscular notions of matter as the underlying basis of all natural processes. This was particularly attractive in the case of living things, which could now be accounted for in a way that avoided the reductionism of mechanist accounts.

THE REACTION TO MECHANISM

It had been the defining assumption of mechanism that, to account for the behaviour of macroscopic bodies, all one ultimately needed to understand was the behaviour of the microscopic corpuscles of which they were composed. Everything that occurred at the macroscopic level was simply an effect of activity at the microscopic level, and this microscopic activity was characterizable wholly in terms of mechanical interactions between corpuscles of matter differentiated only by their position, shape, size, speed, and direction of motion. This matter was inert, so there could be no distinctions between mineral, vegetable, and animal matter: there was only one kind of matter (material extension, on the dominant Cartesian view) and it was completely homogeneous. While matter behaved in various different ways at a macroscopic level—flames and solids behaved differently in virtually every respect for example—it was argued that these qualitative differences were superficial manifestations of quantitative microscopic differences. The microscopic level was one of common causation: every physical state and process could be traced back to causes at this level, so that the traditional classifications were rendered redundant.

Two key questions that arose here were how one identified microscopic constituents, and how one connected postulated microscopic behaviour with observed macroscopic behaviour. The first was a wholly a priori question, as might be expected. The tiny homogeneous particles of matter had not been detected with the use of microscopes, or inferred from the results of microscopic observation: such observation, which was almost exclusively of organic matter, displayed a micro-world of great complexity,[1] and pointed in a different direction entirely from that expected by those looking for simpler constituents at a microscopic level. Nor was micro-corpuscularianism associated with any experimental tradition: there was no experimental procedure or form of controlled observation by which the postulated atoms could be identified. In many respects, the second

[1] See, for example, the illustrations in Robert Hooke, *Micrographia: or some Physiological Descriptions of Minute Bodies made by Magnifying Glasses* (London, 1665); and Anton von Leeuwenhoek, *Arcana Naturae* (Leiden, 1695).

question also was largely a priori, with the bulk of the mechanist project outside biomechanics devoted to absolving natural philosophy of responsibility for a wide range of phenomena whose explanation had traditionally been thought to fall centrally within its purview. Virtually every aspect of physical bodies other than shape, mass, and state of motion were now deemed additions of the perceiving mind, and hence no longer part of natural philosophy proper. Clearly, the postulated atoms or corpuscles could not have been chosen to play a fundamental role because it was believed that they explained a vast range of phenomena in a satisfactory way. Of the available kinds of fundamental explanation, if they had advantages over other kinds, it was not because they explained more but because of the transparency of the explanations to which they aspired. But this transparency came at an immense cost. The trouble was not that they explained so little, but rather that they explained nothing at all. They were completely speculative, and there was simply no experimental or observational way of connecting them to macroscopic behaviour. Even worse, they made many parts of physical enquiry, from chemical reactions and electrical behaviour to physiology and medicine, wholly inexplicable.

The point is worth stressing, as there has been a great deal of misunderstanding on the subsequent importance of seventeenth-century atomism. It has often been assumed that the rise of mechanism marked a permanent shift—in many ways taken to be part of the core of the 'Scientific Revolution'—from basic characterizations of nature in terms of four elements, for example, to characterizations in terms of the micro-corpuscular constituents of bodies.[2] But the actual situation is quite different. As Klein and Lefèvre point out in their comprehensive study of eighteenth-century chemistry:

> As to eighteenth-century chemists' ways of identifying and classifying substances and their writing histories of substances, atomism and corpuscularianism were totally irrelevant. The many ways of identification and classification of substances we have studied in the different fields of eighteenth-century chemistry never included a classificatory order according to composition from atoms or atomic structure. In the eighteenth century, identification and classification of substances according to chemical composition always meant classification according to substance components rather than number, size, figure, or any other parameter of atoms and atomic structure.[3]

If one traces a genealogy of modern atomic thinking—of the kind first developed by Boltzmann, Thompson, Perrin, and others in the late nineteenth and early twentieth centuries—back to the developments that inspired it, the micro-corpuscularianism that was postulated by seventeenth-century mechanists plays no role at all. Even the mass points of mechanics, which had initially been

[2] The classic account is E. J. Dijksterhuis, *The Mechanization of the World Picture* (Oxford, 1961).

[3] Ursula Klein and Wolfgang Lefèvre, *Materials in Eighteenth-Century Science: A Historical Ontology* (Cambridge, Mass., 2007), 62.

correlated with micro-corpuscles (for example in the first half of Book II of Newton's *Principia*) had by now become part of an almost exclusively mathematical discourse, rather than a physical (observational or experimental) one, so that any idea that, in investigating the properties of matter, one was investigating the behaviour of the ultimate microscopic material constituents of bodies was effectively abandoned even at a theoretical level. Micro-corpuscularianism was a dead end, as ancient atomism had been, and any resemblance to modern atomism, in physics or in chemistry, is superficial.[4]

Attempts to mechanize chemistry, for example, never provided any remotely satisfactory account of the outstanding problems facing explanations of chemical behaviour: the persistent identity of certain substances, how they can have completely different properties from their constituents, why they cannot be separated by mechanical means into their original constituents, why they combine in definite simple proportions, or even how they can combine in the first place. The failure of a mechanical approach to these questions had been diagnosed as early as 1706 by Stahl as lying in its inability to penetrate beyond the surface of bodies.[5] Mechanists had effectively assumed that, because they were dealing with ultimate constituents, it was a question of the behaviour of smaller bodies explaining the behaviour of larger ones, but the new turn in chemistry completely marginalized such considerations, making the issues hinge instead on internal and external properties, where questions of size are irrelevant to the contrast between interior and surface. Moreover, it was a premise of the kind of micro-corpuscularianism with which mechanism worked that matter was ultimately homogeneous, but Stahl insisted that at a sensory level every indication was that matter was heterogeneous, and, while accepting that matter may indeed be composed of micro-corpuscles, he advocated a basic number of larger-scale elements—made up from aggregates of micro-corpuscles in some way that remained to be explained—to account for chemical properties, since the micro-corpuscles themselves had no explanatory value as far as chemistry was concerned. Stahl was adamant in his insistence that chemistry was a discipline quite independent of any micro-corpuscular forms of enquiry, and that the two were concerned with different kinds of phenomena: it was distinctive of chemical compounds, for example, that their properties were quite different from any of their constituents. This approach became enshrined in the long entry on chemistry in the *Encyclopédie*.[6] Here a sharp contrast was drawn between the uncertain

[4] See, for example, Alan Chalmers, *The Scientist's Atom and the Philosopher's Stone: How Science Succeeded and Philosophy Failed to Gain Knowledge of Atoms* (New York, 2011).

[5] Georg Ernst Stahl, *Disquisitio de mechanismi et organismi diversitate* (Halle, 1706). See the discussion in Hélène Metzger, *Newton, Stahl, Boerhaave et la doctrine chimique* (Paris, 1930), 93–188.

[6] 'Chimie', *Encyclopédie*, viii. 12–63. Stahl appeared hopelessly out of date in the first half of the eighteenth century, but there was a significant and rapid turn around in his fortunes mid-century: see Peter Hanns Reill, *Vitalizing Nature in the Enlightenment* (Berkeley, 2005), 123–8.

and speculative nature of basic 'physics' (i.e. mechanics) and the careful experimental and observational results of chemistry.

The abandoning of micro-corpuscularianism provoked a fundamental rethinking of matter theory, and by the 1740s the question of what the activity of matter consists in was no longer one for mechanics, but rather one on which chemistry, electricity, and the life sciences each had a claim. In the case of chemistry for example, in order to secure this fundamental role for itself, it needed to mark out a domain of enquiry fully autonomous from that of mechanics. But the battle for autonomy inevitably brought with it the question of which of the disciplines had priority with respect to the other. Up to this point, there had been a general assumption that it was mechanics that was prior because it was more fundamental than chemistry, in that it was mechanics that characterized the constituents of the bodies that chemistry studies, and it was taken as given that it was the behaviour of these mechanically characterized microscopic constituents that ultimately determined the chemical behaviour of the larger bodies that chemistry investigated. From the 1740s and 1750s onwards, we witness a reversal of fortunes, as chemistry comes to play the role of the dominant partner. In 1777, for example, the chemist and naturalist Torbern Bergman writes:

> The Science of Nature seems to have three degrees. The first fixes our attention to the outsides, and teaches us to collect external characters, in order to enable us to distinguish various natural bodies; and that this is the proper object of NATURAL HISTORY. If we penetrate still deeper by our contemplation, and examine the *general qualities* of matter (its extension, impenetrability and *vis inertiæ*) in regard to its peculiar relations; it is that which is commonly called NATURAL PHILOSOPHY (*Physica*). But CHEMISTRY is the *innermost part*, since it examines the material elements, their mixtures, and proportions to one another. The first teaches us the elementary rudiments, the alphabet of the great book of nature; the second instructs us in spelling; and the third, to read distinctly. The two first therefore are no more than subsidiary sciences, which conduct us to the last, as the proper great subject.[7]

The mathematician Wenceslaus Johann Karsten notes in 1786 that, prior to the mid-eighteenth century, discussion of the basic qualities of matter had turned on its divisibility, but these are 'useless' questions, and no one concerns themselves with them any more. Instead, they concern themselves with 'the decomposition of physical matter into non-identical elemental matter'.[8] Johann Samuel Gehler

[7] Carl Wilhelm Scheele, *Chemical Observations and Experiments on Air and Fire . . . with a Prefatory Introduction by Torbern Bergman, translated from the German by J. R. Forster* (London, 1780), xv–xvi. Originally published as *Chemische Abhandlung von der Luft und dem Feuer* (Uppsala and Leipzig, 1777). The same classification appears in a number of places: see, for example, Johann Friedrich Gmelin, *Grundriß der allgemeinen Chemie zum Gebrauch bei Vorlesungen* (2 vols, Göttingen, 1789), i. 2–3.

[8] Wenceslaus Johann Gustav Renger Karsten, *Physische-chemische Abhandlung, durch neuere Schriften von hermetischen Arbeiten und andere neue Untersuchungen veranlaßet* (2 vols, Halle, 1786–7), ii. 69. Quoted in Reill, *Vitalizing Nature*, 72.

makes the same point in the context of the theory of heat in 1789, noting that mechanical explanations are inadequate, and that if one wants 'to correctly order these phenomena and bring them under specific laws, one had to speak the language of the chemists'.[9]

At the same time, the life sciences were also claiming autonomy, and it was being argued that they had a fundamental standing. Matter as studied in mechanics and chemistry, it was maintained, was simply matter that lacked living properties. As Diderot put it in his *Pensées sur l'interpretation de nature* of 1753: 'Could so-called living matter not simply be matter which moves by itself? And could so-called dead matter not be one type of matter moved by another'?[10] Far from being primary forms of matter, what mechanists and chemists were studying was dead matter, an empty shell. It was in the living realm that matter manifested its full range of qualities, and consequently this must now be the paradigm domain of investigation for the study of matter.

In the wake of the demise of mechanism as an explanatory strategy, what was at issue in understanding the broad features of the natural realm were questions of classification, and they overlap in distinctive ways. They are quite different from anything that had been envisaged in the seventeenth century, and in many respects from anything that had been envisaged earlier. We can take our bearings from two fundamental forms of identification and classification of the basic kinds of physical things that were common in the eighteenth century. The first was a traditional division of general natural kinds of things that was renewed in Linnaeus' division of the 'empire of nature' into three 'kingdoms': mineral, vegetable, and animal. The second, which had come down from antiquity and persisted throughout medieval and renaissance discussions, was the division into the four elements: earth, air, fire, and water.

In the seventeenth century, mechanists had done away with both these classifications, and they were abandoned in favour of accounts that focused instead on hypothesized differences in the behaviour of the postulated atomic constituents of things. The first classification in many respects mirrored, and perhaps had found a rationale in, the traditional Aristotelian fourfold division into: inert things; those things having a vegetative 'soul' (that is, capable of nutrition and growth); those also possessing a sensitive 'soul' (that is, capable of motion and sensation); and those also possessing a rational 'soul' (that is, those possessing rational faculties). Descartes for one had considered this kind of classification useless, and had pointed out, with justification, that the Aristotelian 'souls' explained nothing, merely labelling differences with no account of how these differences came about, or why they emerged where they did. His own solution, which was to dominate

[9] Johann Samuel Gehler, *Physikalisches Wörterbuch* (5 vols, Leipzig, 1787–92), iv. 549. Quoted in Reill, *Vitalizing Nature*, 72.

[10] Denis Diderot, *Œuvres complètes*, ed. J. Assézat and M. Tourneux (20 vols, Paris, 1875–9), ii. 59.

thinking about matter theory into the eighteenth century, was that there is only inert matter and rational souls: nothing else is needed to account for the phenomena. At the same time, and on the same general basis, the four elements were deemed redundant since on the mechanist account everything was made of the one single homogeneous kind of matter, and Descartes for example devoted a great deal of attention to demonstrating how various natural phenomena—from rainbows to flames, static electricity, and magnetism—could be accounted for in micro-corpuscularian terms. There were some areas, physiology being the most important, where mechanism had seemed, at least in some respects, to work well, and biomechanics was definitively abandoned only in the later decades of the eighteenth century. By the middle of the eighteenth century, however, the explanatory ambitions of mechanism had fallen apart in those disciplines at the cutting edge of physical investigation such as chemistry and electricity,[11] and the numbers of adherents of mechanist physiology had dwindled significantly. At the same time, the question of the nature of life was becoming pressing, and the answers given increasingly contradicted the mechanist assumption of the inertness of matter.

The rejection of the atomist idea that everything, living and non-living, is made up of, and is to be explained in terms of, microscopic pieces of solid homogeneous matter, reopened the question of how the natural realm should be divided up between disciplines. There were some areas, such as mineralogy, botany, and what became zoology, that had remained untouched by mechanism. There were other areas, such as chemistry, in which attempts at mechanization turned out to be so fruitless that practitioners came to ignore them, and to continue to work with procedures that assumed the traditional natural classifications, albeit in increasingly reworked forms. There was also the investigation of areas such as electricity, where initial attempts at mechanization failed and where traditional distinctions seemed irrelevant, and where as a result it remained unclear just what kind of phenomenon one was dealing with.

There were two main strategies in response to these issues, and it was possible to employ them in tandem. The first relied on the powerful new tradition of experimental natural philosophy, which we looked at in Chapter 1, and which was developed at much the same time as mechanism but which ran along a different trajectory. Rather than seeking explanations in deeper underlying structures, it bracketed off questions of ultimate constituents in favour of explanatory strategies that worked on a macroscopic phenomenal level, relying on experimental manipulation. The experimental philosophy tradition was crucial in shaping conceptions of the natural realm in the eighteenth century, displacing the kind of speculative natural philosophy that had played this role from classical antiquity up to the seventeenth century. The development of chemistry in the early decades of

[11] See Gaukroger, *The Collapse of Mechanism and the Rise of Sensibility*, chs 8, 9, and 10.

the eighteenth century, for example, is in many ways a triumph of experimental natural philosophy over micro-corpuscularianism. It offered a model for a precise form of experimental enquiry, one that was quantitative, but not in the way in which those paradigms of seventeenth-century natural philosophy—mechanics, optics, and astronomy—had been quantitative. The resources it deployed did not come from the highly technical field of geometry, nor did they come from the newer, even more technical fields of algebraic analysis and calculus, but from the very basic subject of arithmetic, and even then it was usually no more than a matter of establishing ratios.[12]

The second strategy was to return to earlier natural classifications where possible, classifications which had in any case never quite died out in practice for many disciplines. In chemistry, for example, not only did French chemistry textbooks organize their material around the division into mineral, vegetable, and animal,[13] but, more importantly, the traditional categories of earth, air, fire, and water still played a role, even if these were transformed in fundamental respects. The theory of the elements, as deployed in the eighteenth century, cuts across more familiar distinctions of physics and chemistry, but it helps us capture concerns in the matter theory of the time that we might otherwise overlook. In particular, the assumption that any comprehensive account of the natural world had to account for the four basic types of substance was a fundamental one. When, towards the end of the eighteenth century, this assumption began to be challenged, there nevertheless remained four basic types, even though they were no longer considered types of substance, and they remained on a par. The key late eighteenth-century development was the gradual move away from considering the traditional elements as components of matter responsible for the properties of bodies, to considering them rather as vehicles or instruments of chemical change. That is to say, they begin to be conceived not as different substances but as different states of substances. But the different states of substances—solids, liquids, 'airs' or gases,[14] and combustion—stood in one-to-one correspondence with the now superseded elements: earth, water, air, and fire.

[12] Grasp of arithmetic had been very elementary in the seventeenth century, with knowledge of multiplication tables being beyond the scope of most educated gentlemen. There was an increase in numeracy in the eighteenth century with the rise of government bureaucracies, commerce, mining, etc. See John Heilbron, 'The Measure of Enlightenment', in T. Frängsmyr, J. Heilbron, and R. Rider, eds, *The Quantifying Spirit in the Eighteenth Century* (Berkeley, 1990), 207–42: 211.

[13] See, for example, Nicolas Lemery, *Cours de Chimie* (Paris, 1675); Guillaume François Rouelle, *Cours d'experience chymique* (Paris, 1759); Pierre Joseph Macquer, *Elemens de chymie-practique, contenant les descriptions des opérations fondamentales de la chymie, avec des explications et des remarques sur chaque opération* (2 vols, Paris, 1751); Antoine Baumé, *Manuel de Chymie, ou exposé des opérations et des produits d'un cours de chymie* (Paris, 1763). Note however that they did not apply organizational principles beyond this basic division, and in any case the division between mineral and vegetable, for example in the case of acids, was not always a strictly functional one. See Klein and Lefèvre, *Materials in Eighteenth-Century Science*, 11–13.

[14] 'Air' was the English word. Lavoisier was the first to use the term 'gaz', and the English version of the word was used thereafter.

CHEMICAL COMPOSITION

Chemistry held the key to a number of fundamental developments in thinking about matter from the middle decades of the eighteenth century onwards. In crucial respects it provided the vehicle by means of which the natural realm was explored. For example, there was deep concern over how the constituents of the natural realm were to be characterized, and how they were related to one another. Although they could never offer the kind of substantive unity that mechanism had purported to provide, the Linnaean 'kingdoms' provided a standard unifying account of these questions. Classificatory schema were the same in each case, and minerals were ordered like plants and animals, into classes, orders, families, genera, species, and varieties.[15] And as with botany and zoology, classification in mineralogy proceeded on the basis of diagnostic characters: in this case such features as crystalline form, hardness, colour, and cleavage (splitting along structural planes). The classifications themselves took their model largely from botany, which was Linnaeus' starting point. In his comprehensive botanical taxonomies, which began to appear in the 1730s, he had set out to group together and order all plants[16] in a single system, with spectacular success.[17] The classifications were intended to yield affinities between various kinds of things, and botany began to provide general models for classification. Botanical models for the classification of diseases, for example, existed from at least the 1730s,[18] and Linnaeus' botanical classification had an impact on these, with nosology taking a particularly Linnaean turn in subsequent decades: 1763 saw the publication not only of Sauvages' Linnaean *Nosologia methodica* but also Linnaeus' own *Genera morborum.*[19]

Botany also provided a model for chemistry. A fundamental problem for those engaged in studying chemical processes, from the sixteenth to the eighteenth centuries, was the apparently limitless and heterogeneous character of the subject matter. The substances recorded in the work of eighteenth-century chemists, for example, included entire plants, roots, leaves, flowers, bones, hair, nails, balsams, resins, gums, oils, fats, blood, ceramics, porcelain, and glass, as well as processed

[15] See, for example, Jean-Claude de La Méthrie, 'Discours préliminaire, contenant un précis des nouvelles découvertes', *Observations sur la physique, l'histoire naturelle et les arts* 28 (1787), 1–53: 11–12.

[16] Linnaeus did not include cultivated plants, only those found in nature.

[17] Carl Linnaeus, *Systema Naturæ per Regna Tria Naturæ, secundum Classes, Ordines, Genera, Species* (Leiden, 1735).

[18] See for example François Boissier de Sauvages de la Croix, *Nouvelles classes de maladies, qui dans un ordre semblable à celui des botanistes, comprennent les genres et les especes de toutes les maladies, avec leurs signes et leurs indications* (Avignon, 1737).

[19] François Boissier de Sauvages de la Croix, *Nosologia methodica sistens morborum classes, genera et Species juxta Sydenhami mentem et botanicorum ordinem* (2 vols, Amsterdam, 1763); Linnaeus, *Genera morborum, in auditorum usum* (Uppsala, 1763).

chemical substances such as metals, mineral acids, alkalis, and salts.[20] Bringing order to this diverse range had been an aim of more systematic thinkers engaged in matter theory, and Bacon for example had devoted a great deal of effort to trying to impose some order on natural substances, but his attempts led nowhere.[21] Chemistry remained more like natural history, which, from Pliny's *Historia Naturalis* onwards, had always had a somewhat haphazard appearance in its choice of materials, making connections freely and avoiding overarching systems.[22] Nevertheless, I suggest the most fruitful way to think about chemistry in the eighteenth century is in terms of breaking down substances into distinct species, along the lines of mineralogy, botany, and zoology. As the chemist and great reformer of chemistry William Cullen put it in 1753: 'it may be often useful to refer the productions of chemistry to their proper classes, orders, genera, species, and varieties as is commonly done with respect to the objects of natural history. This will lead us to distinguish between the general and particular properties of bodies.'[23] Twenty-six years later, Lavoisier would adopt the same natural history nomenclature:

> As for those bodies which are formed by the union of several simple substances, we give them new names, composed as the nature of the substances directs; but, as the number of double combinations is already very considerable, the only method by which we could avoid confusion was to divide them into classes. In the natural order of ideas, the name of the class or genus is that which expresses a quality common to a great number of individuals. The name of the species, on the contrary, expresses a quality peculiar to certain individuals only.[24]

The importance of chemistry in the eighteenth century lay in the fact that the procedures used for analysis (liquid reagents or intense heat), whether in pharmaceutical botany or mineralogy, were those of chemistry. In providing so many of the tools of analysis, chemistry offered a crucial key to understanding how natural phenomena were to be accounted for, and provided an entry point into conceptions of natural processes. Chemists believed that it was important for the development of chemistry, having freed itself from micro-corpuscularianism and flourished as a result, that it secure its standing as an autonomous discipline. If it was to lay plausible claim to being an autonomous form of enquiry, it was crucial that it be considered as a something with an identity in its own right. Seeking

[20] I take the list from Klein and Lefèvre, *Materials in Eighteenth-Century Science*, 11.

[21] See Stephen Gaukroger, *Francis Bacon and the Transformation of Early-Modern Philosophy* (Cambridge, 2001), chs 5 and 6.

[22] On the uses to which Pliny's *Historia Naturalis* was put by French Enlightenment thinkers, for example, see Aude Doody, *Pliny's Encyclopedia: The Reception of the* Natural History (Cambridge, 2010), ch. 2.

[23] Leonard Dobbin, 'A Cullen Chemical Manuscript of 1753', *Annals of Science* 1 (1936), 138–56: 146.

[24] Antoine-Laurent de Lavoisier, *Traité Élémentaire de Chimie, présenté dans un ordere nouveau et d'Après les Découvertes Modernes* (Paris, 1789), xix–xx.

some form of chemically guided differentiating features within the open-ended range of things that potentially fell under the purview of chemical enquiry was clearly not going to work. As we trace the development of chemistry in the middle decades of the eighteenth century, however, three features emerge which did provide it with an identity. The first was a shift from thinking of chemistry as a study of the properties of substances to a study of the composition of substances. Micro-corpuscularianism had worked in terms of the composition of substances, but with their physical composition, not their chemical composition. The idea of a distinctively *chemical* composition made little sense in micro-corpuscularian terms, since the ultimate constituents of things were materially identical. Moreover, it had proceeded in the opposite direction to eighteenth-century chemistry. Mechanists had set out the properties of what they considered to be the ultimate constituents of any material substance and then, not so much moved up to their manifestation at the macroscopic level, since this proved impossible, but instead used these to specify the ultimate and only legitimate resources that could be used in accounting for macroscopic events. Eighteenth-century chemists moved in the opposite direction, taking chemically characterized substances and asking how they could be treated, for example through heating or by being dissolved, so as to yield constituents.

The second feature is the thinking through of chemical operations and their products in instrumental terms. How chemical enquiry is directed is largely a matter of the contingencies that constrain human engagement with the natural world. What instrumentation is available (e.g. the quality of glassware) is crucial[25] but so too is what we can characterize as the conversion of chemical substances into a canonical form as a prerequisite to analysis. The chemist's practice of working with paradigm forms of substances is not something peculiar to the seventeenth and eighteenth centuries. Rather, it is a permanent feature of analytical chemistry. The current paradigm form is the crystal: if one wants to find out about the internal structure of a substance, one attempts to get it into a crystalline form. This is driven not by any intrinsic interest that crystals have, but by the spectacular success of crystallography in producing individualized diffraction patterns when crystal samples are targeted by X-ray (or neutron or electron) beams, patterns that reveal internal structure. Crystals typically allow for a complete specification of structure.[26] The point here is that developments outside

[25] Glassware, crucibles, and furnaces had not changed between the end of the sixteenth century and the middle of the eighteenth, when significant improvements began to be introduced: see Frederic Lawrence Holmes, *Eighteenth-Century Chemistry as an Investigative Enterprise* (Berkeley, 1989), 17–20. By contrast, later in the century Lavoisier was using the services of over seventy carefully selected manufacturers of laboratory instruments: see Marco Beretta, 'Between the Workshop and the Laboratory: Lavoisier's Network of Instrument Makers', *Osiris* 29 (2014), 197–214.

[26] I say 'typically' because there are problem cases. Chemical reactions will often depend crucially on the surfaces of the crystals for example, whereas crystallography only tells us about their internal

molecular chemistry, developments that had no theoretical connection with it, have quite fortuitously provided it with a powerful tool that dictates a paradigmatic form of chemically characterized substances for analytical purposes.

In the eighteenth century, the paradigmatic form of substances was not of course the crystal. In fact in the early decades of the century it shifted from solids to liquids, and this is a revealing development. The contrast with mechanics is instructive here, for mechanics and chemistry operate with different paradigm forms of material substances. Mechanics was ultimately about solids, whereas chemistry was ultimately about fluids or liquids.[27] In seventeenth-century mechanics, for example, it was collisions between microscopic hard bodies that were considered to underlie macroscopic physical processes. In the eighteenth century, the mechanics of mass points was developed with great mathematical sophistication in rational mechanics, and there was an elaborate programme of constructing flexible solids and fluids out of mass points. The direction of construction—from mass points, to rigid bodies, to flexible bodies, to elastic bodies, to fluids—meant that solids were primitive with respect to fluids, in that one had to understand solids in order to understand the more complex case of fluids. In chemistry, by contrast, the operational path whereby the properties of matter were identified meant that the paradigm form of material substances was thought of in terms of the action of fluids. The language of chemical analysis was that of solvents and distillation. To discover the constituents of something, one did not physically break it down into its homogeneous micro-corpuscles, but rather began by liquefying it, dissolving it in an acid or alkali: its properties could be revealed only once it was in a liquid form. Alternatively, one could heat the substance to be analysed, breaking it down in this way, but for this to be instructive from the point of view of analysis the nature of heating had to be understood. The models that were developed to understand the action of heating in chemistry, by contrast with the primitive mechanist understanding in terms of increased motion, worked in terms of fluids. There were various possibilities: heat was conceived along the lines of a fluid; or heating was thought of in terms of matter being energized by an active subtle fluid; or it was thought of as flowing between bodies due to the pull of attractive forces; and so on. All the options substantialized heat in one way or another, and the substantialization was in every case reduction to a fluid.

The third feature shaping the identity of eighteenth-century chemistry is the operational partition of the subject matter of chemistry into two. The analysis of organic matter was pursued via destructive distillation, whereby the 'essence' of

structure. Moreover, there are forms of substances (such as liquids) that are not crystalline, for which X-ray analysis has very limited value, and others (such as glass) for which it has no value at all.

27 I use the terms 'liquid' and 'fluid' interchangeably here. In modern usage, by contrast, fluids include not just liquids but also gases and plasmas. Flames are plasmas, made up predominantly of free electrons and positive ions. Consequently, the traditional division into solids, liquids, gases, and flame reduces, at the most general level, to that between solids and fluids.

the substance (or, as we might now say, its active ingredients) could be isolated. It was in the Paris Jardin du Roi, established in 1635 for the supplemental education of medical students, that these techniques were developed,[28] and given the pharmaceutical importance of recovering the essence of a substance in the production of medicines, what mattered was the end product. The process was destructive in that the original material was lost in the process, but given the aim of the exercise, this was irrelevant. Analysis via solvents was different. It was developed for extracting minerals in metallurgy, but it was not a question of extracting the essence and simply discarding everything else, but rather one of dissolving substances and the recovering of substances from solution. For this, reversible processes were of paramount importance. Substances were treated through heating or by being dissolved so as to yield constituents. These same processes were then used to resynthesize the original compound from the separated components.[29]

A division emerges here that separates the chemistry of plants, roots, flowers, bones, hair, resins, fats, and blood, for example, from the chemistry of metals, mineral acids, alkalis, and salts. This corresponds to the division between the pharmaceutical and mineralogical traditions, not just in techniques of analysis or even in kinds of product, but more importantly in what kind of information one wants from analysis. The crucial developments in eighteenth-century chemistry occur with the elaboration of an account of reversible chemical operations, and the key early figure in these developments is Geoffroy. The kind of investigation in which Geoffroy engages, which deals with those processes in which the original reagents can be recovered by chemical means, necessarily restricts it to a narrow range of reagents. The purity of the substances examined is crucial here, in a way that it is not when one is dealing with organic materials: while it makes perfectly good sense to talk about pure nitric acid or pure lead, for example, there is no analogue in the case of nails, or resin, or plant roots. A very specific kind of laboratory 'chemical' emerges, quite different from those substances that make up the subject matter of chemistry more generally construed. This radical restriction of the substances open to investigation changes the nature of chemical enquiry.

Within this laboratory or mineral chemistry, there were two main competing understandings of how chemical explanations worked. Competing, but not wholly exclusive, for both were thoroughly experimental in approach, and there were areas of overlap. The first was one that sought to discover the composition of chemical substances, and among the principal exponents of this kind of approach were Geoffroy and Lavoisier. The second of these understandings was one that sought to discover the chemical 'principles' that underlay chemical processes, and

[28] See Jean-Paul Contant, *Enseignment de la chimie au Jardin Royal des Plantes de Paris* (Cahors, 1952).

[29] This is one of the central arguments of Klein and Lefèvre, *Materials in Eighteenth-Century Science.*

among the principal exponents of this kind of approach were Stahl and Priestley. Matters became complicated in the last decades of the eighteenth century. Up to the middle of the century, chemists confined their attention to solids and liquids, but this changed with the discovery that there was not a single 'element', air, but rather chemically distinct gases. This, combined with the discovery that solids, liquids, and gases were simply different states of substances and that the transformation from one state to the next was effected through heat, opened up the possibility of a general account of chemistry, and at the same time sharpened the conflict between the two ways of thinking about chemical explanation. What resulted was a rethinking of the nature of matter, a rethinking that was further motivated with the discovery that electrolysis had chemical effects, and by the realization that merely specifying ratios of chemical components—or just specifying chemical components—was of very limited use for those working with organic substances.

It was in Geoffroy's work that an experimentally driven discipline successfully emerged which eschewed considerations of the micro-corpuscularian constitution of chemical substances. Instead, he offers a classification of various chemical substances in terms of their affinities, that is, in terms of the degree to which they are found to combine with one another.[30] He organizes the results of his experimental enquiries in a tabular form (Fig. 2.1), listing chemical reagents against one another by order of affinity, that is, in terms of the ability of the substance to displace those lower down the list, but not themselves displaceable by those lower down the list. So, for example, the first column lists, from the top, acid spirits, fixed alkali salt, volatile alkali salt, absorbent earth, and metallic substances, and it records the information gleaned from experiments that fixed alkali salts react most favourably with acid spirits and will displace all the substances listed below it from their existing combination with acid spirits, but the reverse never occurs. The aim of the table was both to set out what happened in various chemical reactions, and to predict what would result from various combinations.

Note that this classification is a purely empirical or phenomenological one: it is an economical compendium of a body of experimental results. Geoffroy is especially keen to avoid importing any theoretical assumptions into the presentation, and what is so important about his table is that, by classifying substances in this way by degree of affinity, the table pushes the question of chemical composition to the fore. Chemical composition was not one of the traditional questions posed by matter theory, either in an Aristotelian element version or in a corpuscularian version.[31] Moreover, the substances themselves are taken as elemental,

[30] Étienne François Geoffroy, 'Tables des différens rapports observés en chymie entre différentes substances', *Mémoires de l'Académie royale des sciences* (1718), 202–12.

[31] See, in particular, Ursula Klein, 'E. F. Geoffroy's Table of Different "Rapports" Observed Between Different Chemical Substances—A Reinterpretation', *Ambix* 42 (1995), 251–87.

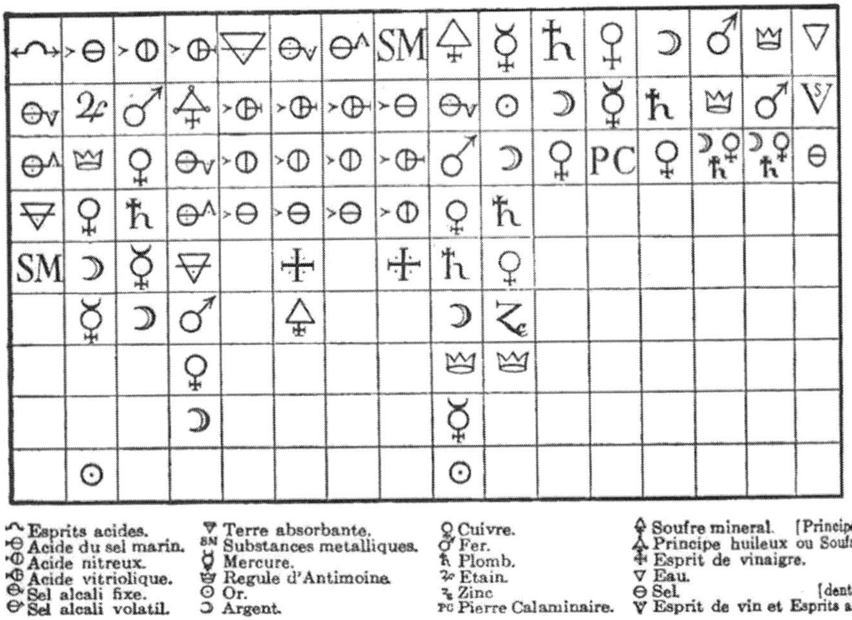

Fig. 2.1 Geoffroy's Table of Affinities, from his 'Tables des différens rapports observés en chymie', *Mémoires de l'Académie royale des sciences* (1718).

not in the sense that they correspond to anything posited by mechanism as primitive, but because they are the products of distillation procedures whereby more complex compounds are broken down into their constituents. The idea of a basic substance becomes transformed here from postulated ultimate constituents of bodies to isolable components of bodies that have resisted all attempts at further analysis.

Geoffroy encountered a problem soon after the first version of the table appeared, however. Stahl had observed that if one mixed vitriolic acid with salt of tartar, there was a way of treating the product such that the acid could be separated out again. The problem was that vitriolic acid was the strongest acid and salt of tartar the strongest alkali, so no other acid or alkali should be able to break their union.[32] Experimenting on this process, Geoffroy melted salt of tartar saturated by vitriolic acid on a crucible with some salt of tartar and some inflammable material. The contents violently burst into flames and produced copious fumes. Taking the residue, a saline sulphurous material, he dissolved it in water, filtered it, poured it over a distilled weak acid spirit (vinegar), and obtained

[32] See Mi Gyung Kim, *Affinity, That Elusive Dream: A Genealogy of the Chemical Revolution* (Cambridge, Mass., 2003), 146–51 for a full account of the experiments.

sulphur. He identified the active agent as phlogiston, the 'principle of inflammability' which he placed just below vitriolic acid in a revised version of his table. Phlogiston figures centrally in Stahlian chemistry, but it has no natural place in, and no rationale in, the compositional chemistry that yields Geoffroy's table of affinities. A 'principle' is not a component of substances in the sense that lead is a component in various lead salts. Rather, it is a special kind of substance that, when combined with other substances, imparts a characteristic quality to them. Phlogiston imparts combustibility to substances, and when the substance is burned it releases its phlogiston, which is manifested in the flame that results. A chemistry that works in terms of principles and one that works in terms of the composition of substances have largely different aims or problem fields.[33] But the resources of compositional chemistry are insufficient for what Geoffroy needs, and phlogiston is imported from a set of resources different from, and in some respects alien to, that with which he is working. It occupies a crucial place in subsequent tables of affinities right up to Bergman's comprehensive table in his 1785 chemistry textbook, the last such comprehensive list before Lavoisier's reforms.

Geoffroy argues that what has happened in the vitriolic acid case is that phlogiston, contained in the inflammable material, has been rarefied and set in motion by the element of fire, and having more affinity with vitriolic acid than with the fixed alkali salt, it has united with the former, which has become detached from the alkali salt. On this reading there are three ingredients in the combustion process: the principle of combustion, namely phlogiston; the combustible substance; and the substance with which it combines. Phlogiston is thought of as forming the food or nourishment for combustion. It leaves the combustible substance and enters the other until the other is full, and the rate and violence with which this is effected is a function of the degree of affinity between the two substances.

Geoffroy's tables describe a kind of ranking of reagents, but he does not and could not describe the chemical reactions that produce the displacements that he catalogues, because the only substances identified are solids and liquids, whereas, to put the issue in traditional terms, 'airs' and combustion also need to be taken into account for a full characterization of what occurs in the chemical processes underlying the transformations he identifies. Once these other two 'elements' were joined to Geoffroy's account of the composition of substances, a more general chemical understanding was possible. It is his approach to the composition of substances that anchors the enterprise, however, and without it the other developments could not have made the contribution they did to a more general chemical theory of the composition of material substances. But these other developments are nevertheless crucial, and the key lay in the discovery of chemically distinct 'airs'.

[33] See Hasok Chang, *Is Water H_2O? Evidence, Realism and Pluralism* (Dordrecht, 2012), 19–22.

Of the problems bequeathed by Harvey to his successors in the mid-seventeenth century—bodily heat, respiration, and the nature of blood—the most pressing was respiration, and it became closely associated with the more general issue of the 'spring of the air', which had preoccupied a number of mid-seventeenth-century natural philosophers, most notably Boyle. Boylean pneumatics was concerned with the properties of various kinds of 'airs', that is, compressible, dispersed, and—as Boyle showed—elastic types of matter. Aristotle had in fact postulated a fifth kind of matter, the aether, to account for what he considered to be the distinctive properties of celestial bodies (such as the fact that their natural motion was circular rather than rectilinear, and unlike rectilinear motion did not culminate in rest), and something called an aether had been postulated in Newton. But Newton had proposed it in a wide range of differing roles, not least that of transmitting gravitational attraction between separated bodies, and, as the person who had definitively unified celestial and terrestrial mechanics, he could hardly have associated it with something celestial as opposed to terrestrial. Moreover, since the properties of the aether, such as elasticity, that were of interest were also those that were being investigated in studies of the spring of the air, concern with the physical properties of the aether fell under pneumatics, rather than under a discipline concerned with some purportedly completely separate form of matter.

It is in this context that we need to address the work of Stephen Hales, which initiated the fundamental changes in thinking about pneumatology that occurred in the second half of the eighteenth century. In the experiments, dating from the early 1720s, described in his *Haemastaticks*,[34] Hales carried out a number of detailed measurements which had the effect of undermining the views of physiologists such as Borelli, Keill, Pitcairne, and Freind on the role of pressure: he shows that perspiration is not a function of increase in blood pressure for example, that the pressure of blood in the veins and arteries of horses, dogs, sheep, oxen, and deer could not possibly explain muscular motion, that health could not be a direct function of hydraulic pressure, that blood pressure could not cause uterine vessels to burst, and that dilation of the lungs is insufficient to cause the passage of blood through the lungs into the heart.

It is, however, in the *Vegetable Staticks* (1727), that the most important innovations come.[35] There, Hales undertakes 'an analysis of air', that is to say, an experimentally guided account of its internal constitution.[36] The starting point

[34] Stephen Hales, *Haemastatics*, which appeared as volume 2 of his *Statical Essays: Containing Haemastatics; or An Account of Some Hydraulic and Hydrostatical Experiments Made on the Blood and Blood-vessels of Animals* (2 vols, London, 1740). On Hales, see the discussion in Robert E. Schofield, *Mechanism and Materialism: British Natural Philosophy in an Age of Reason* (Princeton, 1970), ch. 4.

[35] Stephen Hales, *Vegetable Staticks: Or, An Account of Some Statical Experiments on the Sap of Vegetables: being an Essay towards a Natural History of Vegetation. Also, a Specimen of an Attempt to Analyse the Air, By a great Variety of Chymio-Statical Experiments* (London, 1727). References are to the modern edition: Stephen Hales, *Vegetable Staticks* (London, 1969).

[36] Hales, *Vegetable Staticks*, ch. 6 (89–180).

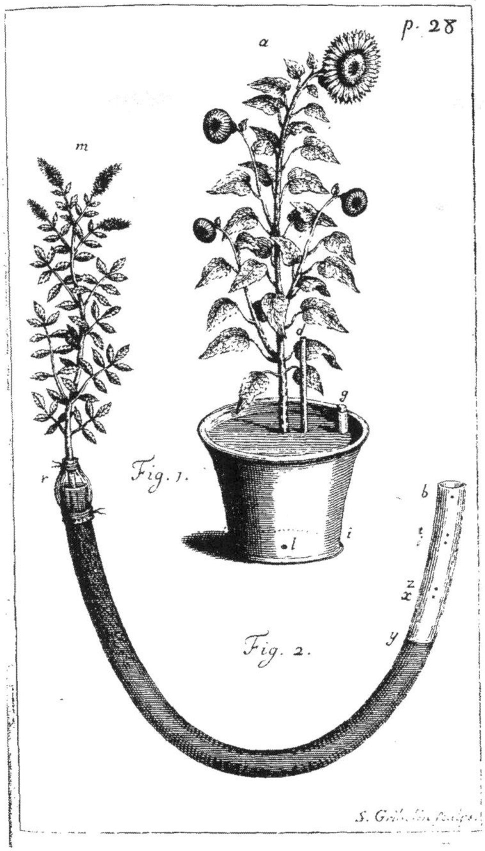

Fig. 2.2 Instrument to measure the rise of sap in plants. Hales, *Vegetable Staticks* (1727).

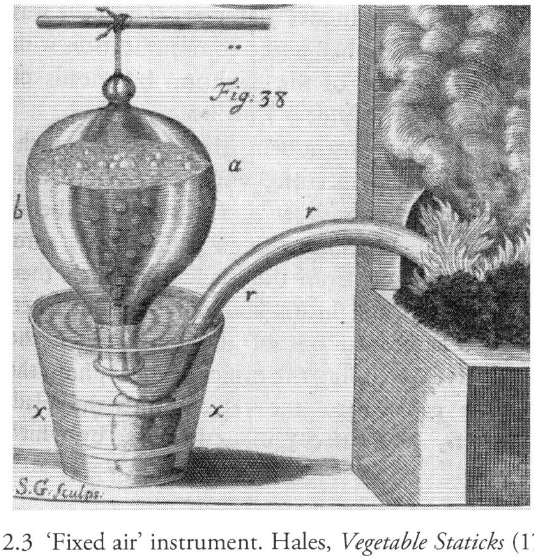

Fig. 2.3 'Fixed air' instrument. Hales, *Vegetable Staticks* (1727).

is a series of experiments on the rise of sap in plants, and he set out to measure how much water and 'air' the plant absorbed, and then how much 'air' it released, under different conditions (see Fig. 2.2). Taking his bearings from Newton's account of the attractive and repulsive relations between particles, Hales argues that air differs from other forms of matter in that its particles exhibit elastic repulsive powers.[37] However, these repulsive powers can be overcome by the attractive force between particles of material bodies. In this latter form they constitute 'fixed air', and are responsible for the coherence of material bodies. This fixed air can be released, the particles of air regaining their elasticity, and it can be captured from mineral substances, for example, in a device based on the same principles as the experiments on plants, namely a water-filled vessel held upside down in a container of water, allowing the air to rise and be collected (Fig. 2.3). As Schofield notes, the whole of eighteenth-century pneumatic chemistry takes its starting point from Hales' chapter on the analysis of air, with Black, Cavendish, and Lavoisier all taking their bearings from it.[38]

The significance of Hales' experiments is twofold. First, although he himself did not question whether 'fixed air' was in fact a single substance, his successors, particularly in Britain, were gradually able to distinguish different airs, and, on the model of the chemical analysis of solids and liquids, began to ask about the chemical composition of these airs. It was the heating of solid or liquid substances that produced the airs, or gases as they were subsequently known. These gases came to be considered no longer as varieties of air but as a state that all bodies could attain, given sufficient heat. Second, the heating of solid and liquid substances could result not just in the release of gases but also in combustion, with flame one further state of substances. This approach points to the need for a general theory of analogues of the four traditional elements—solids, liquids, gases, and flame.

To start with the first set of questions, identifying 'airs' or 'vapours' as states of material substances was a prerequisite to raising the question of chemical composition. The first to find a new 'air' different from atmospheric air was Joseph Black, in the work that led to his Glasgow dissertation of 1754. He set out to discover a substance that would dissolve kidney stones,[39] but instead of focusing on limewater (calcium hydroxide[40]) as his contemporaries were doing, he

[37] Ibid., 165–8. [38] Schofield, *Mechanism and Materialism*, 75.

[39] In the eighteenth century, the roles of physicians and chemists were combined to a significant extent, many of the most notable chemists of the period having originally been trained as physicians. This was particularly true of those working on gases, which were of interest to physiologists and physicians concerned with respiratory diseases as well as to chemists.

[40] I give modern chemical names because the nomenclature of the time is variable, and in any case will mean nothing to most modern readers. Little hinges on the anachronism for our purposes, except in the case of oxygen, which I initially refer to as 'dephlogisticated air' so as not to pre-empt Lavoisier's innovations. In contrast, I do not give any modern formulas, for the introduction of these is a novel and especially significant nineteenth-century development.

examined another alkaline substance, magnesia alba (magnesium carbonate).[41] When heated, magnesia alba yielded a product, magnesia usta (magnesium oxide), which unlike quicklime was neither caustic nor soluble in water. When magnesia usta was treated with acid, it yielded the same product (the magnesium salt of the acid used) as did magnesia alba when it was heated, but the vigorous effervescence that accompanied the reaction in the first case was entirely lacking in the second. What was striking was that heating magnesia alba and treating magnesia usta with acid resulted in identical weight losses. Black concluded that the weight loss was caused in both cases by fixed air becoming freed: in the one case by the action of acid (causing effervescence), in the other by the action of heat. In carrying out comparable experiments on chalk, he received similar results. What was unexpected, however, was that this fixed air had different properties to atmospheric air: it had a peculiar odour and extinguished flames.

The significance of the discovery that there was a form of fixed air that was chemically different from atmospheric air would have been negligible for Black's contemporaries in Britain, just as Hales' results from thirty years earlier had achieved no resonance in the chemical community. The exception was William Cullen, Black's teacher at the University of Glasgow, who played a crucial role by reorganizing the teaching of chemistry.[42] Cullen was convinced that the chaotic state of chemistry was due to its subservience in Britain to a corpuscularian reduction, generally associated (though not by Cullen himself) with Newtonianism. As he puts it, 'chemistry not only considers a set of properties different from these which the mechanical philosophy considers, but a set of properties also that may depend on different circumstances'.[43] A reductionist approach was not only unproductive but robbed chemistry of the autonomy that was crucial if it was to generate results.[44]

Working on the assumption of the autonomy of chemistry means looking for different things than those sought on a reductionist model. There is a parallel here with Newton's early investigation of spectral colours. Newton was intrigued by a feature of the spectral image that just looked wrong: it was lozenge-shaped rather than circular, and there was no explanation for this. For his contemporaries, the shape of the image was a distraction, and could have nothing to do with how spectral colours were produced, since this was a question of micro-corpuscular

[41] Joseph Black, 'Experiments Upon Magnesia Alba, Quicklime and some other Alcaline Substances', *Essays and Observations, Physical and Literary, read before a Society in Edinburgh* 2 (1756), 157–225.

[42] See Arthur L. Donovan, 'Pneumatic Chemistry and Newtonian Natural Philosophy in the Eighteenth Century: William Cullen and Joseph Black, *Isis* 67 (1976), 217–28: 218. More generally, see idem, *Philosophical Chemistry in the Scottish Enlightenment: The Doctrines and Discoveries of William Cullen and Joseph Black* (Edinburgh, 1975), Part I.

[43] Glasgow University Library, Cullen MS, n. 34: cited in Donovan, 'Pneumatic Chemistry and Newtonian Natural Philosophy in the Eighteenth Century', 220.

[44] Cullen was particularly appreciative of Geoffrey's work. See Donovan, 'Pneumatic Chemistry and Newtonian Natural Philosophy in the Eighteenth Century', 221 n. 21.

Fig. 2.4 Priestley's improved version of Hales' pneumatic trough. *Experiments and Observations on Different Kinds of Air* (1774).

activity, not macroscopic effects. Yet it was precisely reflection on the lozenge shape that led Newton to establish the heterogeneity of white light.[45] In the present case, Black set up an experiment in which a vacuum pump is used to extract the air dissolved in water, and he was able to show that there were two kinds of fixed air in water, and two kinds of elastic air in the atmosphere, contrary to the mechanical idea, first developed by Boyle, that air is a homogeneous elastic substance. The fact that there might be more than one chemically distinct air is not something that had any relevance for those preoccupied with the micro-corpuscular basis for chemical properties. It proved to be crucial however, although its significance was not grasped immediately.

While Black and Cullen saw these results as an important achievement, for example, neither attached any great theoretical significance to them. This changed over the next two decades. In the 1760s, Henry Cavendish made very precise measurements of the densities of several 'factitious airs', showing that 'atmospheric air' was chemically simple. In 1766 he published the results of his analysis of the properties of 'inflammable air' (hydrogen), showing that it was distinct from other combustible gases by careful measurement of its density.[46] In the early 1770s, Joseph Priestley, temporarily putting to one side his numerous political,

[45] See Gaukroger, *The Emergence of a Scientific Culture*, 379–99.

[46] Henry Cavendish, 'Three Papers, containing Experiments on Factitious Air', *Philosophical Transactions* 56 (1766), 141–84. There is a comprehensive account of Cavendish's experiments in

religious, and educational projects, set up a sophisticated laboratory in his house, with an improved version of Hales' pneumatic trough (Fig. 2.4). Between 1774 and 1777 Priestley published the three volumes of his *Experiments and Observations on Different Kinds of Air*, the first volume of which identifies 'nitrous air' (nitric oxide), 'vapour of spirit of salt' (anhydrous hydrochloric acid), 'alkaline air' (ammonia), 'dephlogisticated nitrous air' (nitrous oxide), and 'dephlogisticated air'. Now, for the first time, there was a documented range of chemically distinct gases.

CHEMICAL 'ELEMENTS'

The discovery of gases resulted in some spectacular practical applications, not least the medical use of nitrous oxide, and the use of hydrogen in balloon flights, but it also had significant theoretical consequences. In particular, the question could now be raised whether a general characterization of what occurs in chemical processes can be provided, one that complements the account of solids and liquids with an account of gases and combustion. As far as most chemists were concerned, solids, gases, and liquids were no longer different substances but different states of substances, which could be transformed from one into the other. The role of heat in the transition between states held the key, and studies took their initial bearings from Boerhaave's 1732 chemistry textbook, *Elementa Chemiae*.[47] Heat had generally been conceived, from the second quarter of the seventeenth century onwards, as a form of motion in the micro-corpuscles making up matter. Such an account was unable to explain not only such pressing questions as why some chemical reactions produce cold and some heat, but also to account for phenomena that Boerhaave was studying in the 1730s, such as heat transfer, change of state, and heat retention. Heat, he concluded, could not just be a general property of matter, but must rather be associated with a particular kind of matter, something that was so subtle that it could move through the spaces between corpuscles, causing friction. This substance was, he believed, manifest in those phenomena such as heat transfer and retention that the corpuscular model could not account for. Boerhaave's account incorporated what had been the separate domains of heat and fire into the one phenomenon, and in doing so he made heat an integral part of the chemical constitution of substances.[48] He identified expansion—rarefaction—as the reliable indicator of the action of heat, and in a number of experiments he tried to match thermometric measurements and

Christa Jungnickel and Russell McCormach, *Cavendish: The Experimental Life* (Lewisburg, 1999), 191–217.

[47] Hermann Boerhaave, *Elementa chemiae, quae anniversario labore docuit in publicis privatisque scholis* (2 vols, Leyden, 1732).

[48] See Kim, *Affinity*, 184–5.

conceptions of heat to the chemical notion of fire, showing expansion of solid bodies to be proportional to the amount of fire and inversely proportional to the density of the body.[49]

It was Black who, in his lectures at the University of Glasgow from 1762 onwards, went furthest in resolving the problems surrounding heat.[50] He rejects both a mechanical conception of parts rubbing against one another, as well as the idea that it consisted in the agitation of a subtle elastic fluid, but he also rejects Boerhaave's model of a fluid contained in the interstices between the parts of hot bodies. Nevertheless, he assumes that heat is a material fluid of some kind, and he takes his starting point from Boerhaave in a number of respects.[51] He accepts that expansion as measured by a thermometer is the only reliable indicator of heat, but he rejects Boerhaave's assumption that this is a measure of the quantity of heat: rather, it is a measure of the intensity of the heat. Boerhaave had assumed that, at thermal equilibrium, equal areas contained the same amount of heat because they registered equal temperatures, but Black was able to show that temperature and heat did not correlate in this way: equal amounts of mercury and water, when exposed to the same amount of heat, reached equilibrium temperature at different rates. The slower rate at which water reached equilibrium temperature meant, as Black put it, that it had a greater capacity for heat, which was a function of the substance itself. This capacity for heat was not limited to substances that retained their physical state, however, but extended in a striking way to changes of state, where heat is required to change a substance from one state to another, say ice to water, without any change in temperature.

We saw earlier that Geoffroy's tables of affinities only identified solids and liquids. By the later decades of the century, it became very clear that gases and combustion also needed to be taken into account for a full characterization of what occurs in chemical processes. But there were conceptual as well as practical difficulties here. Air and fire do not fit into the new chemical understanding of 'elements', accepted by Stahl and Geoffroy, as those substances so simple that there is no known method of decomposing them.[52] Air and fire were treated not as chemical constituents but as instruments effecting particular kinds of activity. Geoffroy had initially managed to keep such questions out of his investigation of chemical affinities, but he discovered he needed to introduce phlogiston to make sense of the reaction involving vitriolic acid and salt of tartar. This duality was

[49] Ibid., 185–6. See also Hasok Chang, *Inventing Temperature: Measurement and Scientific Progress* (New York, 2004), 57–8.

[50] Black's lectures on chemistry were not published in his lifetime, but were edited by John Robinson for publication after his death as *Lectures on the Elements of Chemistry* (Edinburgh, 1803).

[51] See Schofield, *Mechanism and Materialism*, 185–90, to which I am indebted here.

[52] This definition holds for the whole of the period with which we are concerned. Cf., for example, Humphry Davy's remark in his *Consolations in Travels*, published in 1830 after his death: 'I cannot demonstrate to you what are the true elements of things: but I can exhibit to you those substances, which, as we cannot decompose them, are elementary for us.' Humphry Davy, *The Collected Works* (9 vols, London, 1839–40), ix. 385.

subsequently rationalized in the lectures of Guillaume François Rouelle at the Jardin du Roi.[53] These established Stahl's conception of chemistry in France, and Rouelle dealt with the dichotomy between instrumental and compositional elements by giving every element a compositional and an instrumental feature. It was this conception that Lavoisier focused on, and he removed the instrumental features of chemical elements, transferring them to a single substance, caloric. Caloric thereby took on a range of quite disparate roles. Despite ceasing to play any part in chemical change, it was the vehicle of repulsion, the cause of fluidity and elasticity, and the producer of light, and the cause in the change in size of bodies when they were heated.[54] Nevertheless, it helped significantly to unify the field of enquiry of chemistry, since it allowed different gases to be treated as resulting from a temporary combination with caloric, so that the 'element' itself did not change nature between a solid or liquid form and a gaseous one.[55] As Geoffroy had found, a purely compositional chemistry did not allow one to make sense of chemical processes: just as Geoffroy had to introduce phlogiston, so Lavoisier had to introduce caloric, the 'matter' of heat, although he tells us that we do not have to treat it as something material ('une matière réelle').[56]

Phlogiston and caloric are very similar, and even Fourcroy, Lavoisier's collaborator, maintained that 'even that numerous train of experiments which have been of late made on combustible bodies and combustion, have not fully demonstrated that there is no such principle as fixed fire in material bodies: its existence seems to be allowed, while its name is changed; and instead of the phlogistic, it is called the caloric, principle'.[57] In spite of manifest similarities, however, phlogiston and caloric play different chemical roles. On the Stahlian picture, phlogiston is emitted during combustion and in the calcination of metals. The metal was composed of calx (the true elemental substance left when a metal has been calcinated, i.e. lost its phlogiston) and phlogiston, and burning resulted from the loss of phlogiston. Lavoisier's caloric, by contrast, could be thought of along the lines of a substantialized version of Black's latent heat. In vaporization, caloric combined with the substance being vaporized, depending on the chemical affinity between this substance and the caloric.

In the early 1770s, Lavoisier had begun to experiment on the related processes of combustion and calcination, and he was particularly interested in the

53 Rouelle published a few papers on salts but little else. His influence was considerable however: see Kim, *Affinity*, 188–201. The influential article 'Chymie' in the *Encyclopédie* (viii. 12–63), by Gabriel François Venel, treats Rouelle's work as being at the cutting edge in chemistry.

54 See Schofield, *Mechanism and Materialism*, 228.

55 Ree Robert J. Morris, 'Lavoisier and the Caloric Theory', *British Journal for the History of Science* 6 (1972), 1–38; Robert Siegfried, 'Lavoisier's View of the Gaseous State and Its Early Application to Pneumatic Chemistry', *Isis* 63 (1972), 59–78. More generally see Arthur Donovan, *Antoine Lavoisier: Science, Administration and Revolution* (Cambridge, 1993), ch. 6.

56 Lavoisier, *Traité Élémentaire de Chimie*, 5.

57 Antoine François de Fourcroy, *Elements of Natural History and Chemistry* (3rd edn, 3 vols, London, 1790), i. 107.

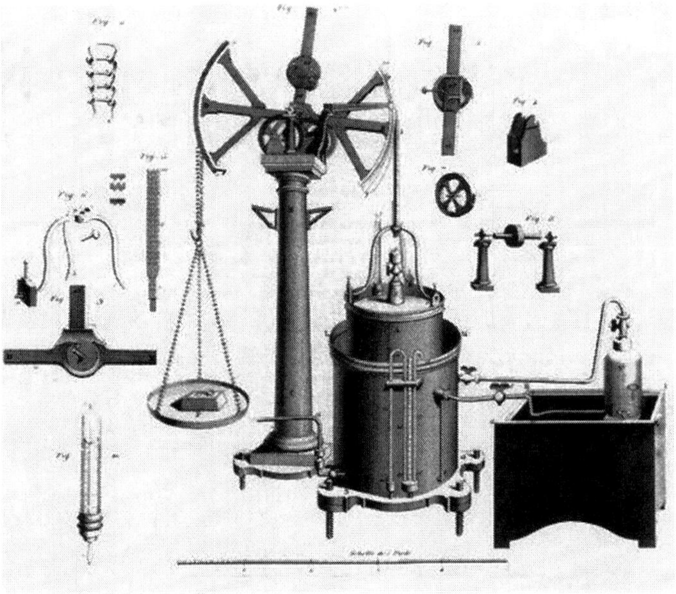

Fig. 2.5 The gasometer of Lavoisier and Meusnier. Lavoisier, *Traité élémentaire de chimie* (Paris, 1789).

experiments of Louis Bernard Guyton de Morveau, which showed that metals gained weight as they underwent calcination. This was explained by phlogiston theorists in terms of phlogiston having a negative weight. But Lavoisier looked for a different explanation. He began to start taking losses and gains in weight as the key to what had occurred in chemical reactions, and to this end gradually built up very precise results for weight differences, devising elaborate apparatuses for weighing (Fig. 2.5). He was able to show that, in all cases of combustion where an increase in weight was observed, air was absorbed, and conversely when a calx was burned with charcoal, air was liberated. The guiding principle was conservation of weight in chemical processes, and it is such a conservation principle that gives the quantitative results the significance they have. This enabled him to formulate a comprehensive alternative to phlogiston theory, one which included the role of oxygen (as he now referred to dephlogisticated air) in combustion, calcination (henceforth referred to as oxidation[58]), and the action of acids. In

[58] In contemporary chemistry the process of 'oxidation' is explained in terms of a process of removing electrons from a compound or element. Lavoisier believed that oxygen alone had the power to oxidize, and he treated what we now know to be chlorine, which has oxidizing powers, as an oxide of hydrochloric acid, to explain its oxidizing power. This was shown to be mistaken by Davy, and in 1811 he identified chlorine as an 'element' (in the sense of being irreducible). In 1961 it was discovered that oxygen itself can be 'oxidized', by the newly synthesized platinum hexafluoride, such a powerful oxidizer that it will even react with the noble gases.

combustion, for example, oxygen combined with a combustible substance, and the caloric present in the oxygen was released as a result. Lavoisier believed that this process could be investigated quantitatively, and the release of caloric was measured by a calorimeter, just as temperature was measured by a thermometer.

Lavoisier and his supporters took the synthesis of water from hydrogen and oxygen to be a crucial blow to phlogiston theory, whose proponents, such as Cavendish and Priestley, had construed water as elemental (in the sense of substances not capable of further decomposition by any known means). It was in fact Cavendish and Priestley who had first performed this synthesis, and Lavoisier had learned the procedure from them.[59] They had at first supposed that 'inflammable air', hydrogen, was pure phlogiston, but later refined their account and construed hydrogen now as 'phlogisticated water', water containing an excess of phlogiston, and oxygen or dephlogisticated air as 'dephlogisticated water', so that when the two are combined the phlogiston content balances out and we get plain water. Lavoisier, by contrast, construed what was happening as hydrogen combining with oxygen to produce water. This explanation was to win widespread support, especially when, in 1800, six years after Lavoisier's death, water was decomposed via electrolysis into hydrogen and oxygen by Nicholson and Carlisle.[60]

Combined with fine discriminations in terms of weight, Lavoisier built on the work of the chemist Pierre-Joseph Macquer, whose textbooks *Elémens de chymie-théorique* (1749) and *Elémens de chimie-pratique* (1751), played a formative role in French chemistry. Macquer took affinity tables far more seriously than Rouelle, and they were central to his enterprise,[61] as they were for Lavoisier. Moreover in his *Dictionnaire de chymie* (1766), he tackled the vexed question of chemical nomenclature, something brought to fruition in Lavoisier, who, working with Berthollet and Fourcroy, provides each of his 'elements' with a fixed name, and provides names for other substances in a way that reflects the elements in their composition, where this was known. Lavoisier's famous 'table of elements' (Fig. 2.6) separates four groups. The first, 'simple substances belonging to the three realms and which can be regarded as the elements of bodies', lists five elastic fluids: light, caloric, oxygen, nitrogen, and hydrogen. The second, 'simple substances that are oxidizable and acidifiable non-metallic elements', lists sulphur, phosphorous, carbon, hydrochloric acid, hydroflouric acid, and boric acid. The third, 'simple substances that are metallic, oxidizable, and able to neutralize an acid to form a salt', lists silver, bismuth, cobalt, copper, tin, iron, manganese,

[59] See Chang, *Is Water H_2O?*, 6–8.

[60] William Nicholson, 'Account of the New Electrical or Galvanic Apparatus of Sig. Alessandro Volta, and Experiments Performed with the Same', *Journal of Chemistry, Natural Philosophy, and the Arts* 4 (1800), 179–87. Actually, phlogiston theorists could account for this—see Chang, *Is Water H_2O?*, ch. 2—but by this time Lavoisier's chemistry had taken centre-field, with his table of elements now providing the basis for chemical practice.

[61] See Kim, *Affinity*, 208.

	Noms nouveaux.	Noms anciens correſpondans.
Subſtances ſimples qui appartiennent aux trois règnes, & qu'on peut regarder comme les élémens des corps.	Lumière	Lumière.
	Calorique.	Chaleur.
		Principe de la chaleur.
		Fluide igné.
		Feu.
		Matière du feu & de la chaleur.
	Oxygène	Air déphlogiſtiqué.
		Air empiréal.
		Air vital.
		Baſe de l'air vital.
	Azote	Gaz phlogiſtiqué.
		Mofète.
		Baſe de la mofète.
	Hydrogène.	Gaz inflammable.
		Baſe du gaz inflammable.
Subſtances ſimples non métalliques oxidables & acidifiables.	Soufre	Soufre.
	Phoſphore	Phoſphore.
	Carbone	Charbon pur.
	Radical muriatique .	Inconnu.
	Radical fluorique. . .	Inconnu.
	Radical boracique. .	Inconnu.
Subſtances ſimples métalliques oxidables & acidifiables.	Antimoine	Antimoine.
	Argent	Argent.
	Arſenic.	Arſenic.
	Biſmuth	Biſmuth.
	Cobalt	Cobalt.
	Cuivre.	Cuivre.
	Etain	Etain.
	Fer.	Fer.
	Manganèſe.	Manganèſe.
	Mercure	Mercure.
	Molybdène	Molybdène.
	Nickel.	Nickel.
	Or.	Or.
	Platine	Platine.
	Plomb	Plomb.
	Tungſtène.	Tungſtène.
	Zinc	Zinc.
Subſtances ſimples ſalifiables terreuſes.	Chaux.	Terre calcaire, chaux.
	Magnéſie	Magnéſie, baſe du ſel d'epſom.
	Baryte	Barote, terre peſante.
	Alumine	Argile, terre de l'alun, baſe de l'alun.
	Silice	Terre ſiliceuſe, terre vitrifiable.

Fig. 2.6 Lavoisier's table of 'elements'. Lavoisier, *Traité élémentaire de chimie* (1789).

mercury, molybdenum, nickel, gold, platinum, lead, tungsten, and zinc. Finally, the fourth, 'simple substances that are earthy and able to form salts', lists lime, magnesia (magnesium oxide), 'baryte' (barium oxides), alumina (aluminium oxide), and silica (silicon dioxide).

The mixed nature of this list reinforces the point that these are substances that cannot be further decomposed by known methods. They are not elements in the atomic sense. This is particularly evident in the way in which the compound substances that are made up from these are treated. If there is an instructive model here, it is that of recipes: it is a question of mixing substances in certain proportions. Discovering in what proportions already compounded substances are mixed is a delicate matter which involves knowing the weight of the elemental substances. It was all too easy to get these wrong and Lavoisier, for example, concluded that water contained equal parts of hydrogen and oxygen. Greater precision entered the picture when Dalton introduced atomic considerations. His work on gases led him in the direction of atomism, and he came to the conclusion that each element is made up of identical atoms.[62] Working on this basis, he argued that each of the least parts of compounds of elementary substances consists of characteristic combinations of small numbers of these elementary substances, so that he was able to devise laws of proportion that reflected their relative weights. Dalton's theory that elementary substances are composed of atoms of equal weight, and that compounds containing those atoms must always be multiples of that weight, led him to conclude that a compound always contains a particular number of atoms characteristic of that compound. The proportions of the constituents of the compound must always remain the same, so that relative atomic weights can be assigned for each elementary substance. The law of proportions was a breakthrough, but the atomic hypothesis itself did not actually do any work in chemistry. Dalton's representations of atoms as little spheres which when combined took on new properties added nothing to the understanding of chemical processes: how the new properties were supposed to be generated, for example, remained a complete mystery. It is also worth noting that he was resolutely opposed to the structural formulae of Berzelius, which transformed chemistry by allowing an understanding of how elements might replace one another in compounds, something Dalton's representations cannot do.[63] Nevertheless, it should be said that there was no urgency in providing structural formulae in mineral chemistry. Recipes that aimed to identify proportions of ingredients, of the kind that Dalton and others were providing, served well enough, even though they occasionally got the proportions wrong. The chemistry of organic materials, carbon chemistry, was a completely different matter. Such carbon compounds typically have very few elemental constituents, but can nevertheless be extremely complex because of the way in which large numbers

[62] John Dalton, *A New System of Chemical Philosophy* (3 vols, Manchester, 1808–27).
[63] See Chalmers, *The Scientist's Atom*, 190–3.

of the constituents are arranged. Because of this, it was essential that a means of investigating and representing structure be devised. This was the achievement of Berzelius in the second decade of the nineteenth century, brought to fruition in the 1830s, transforming tables of elements from recipe guides into something that held the key to the structure of chemical composition.[64] But this is not to underestimate the achievements of mineral ('inorganic') chemistry. What was important about the development of mineral chemistry was the way in which it opened up a new understanding of matter, one that was thoroughly experimental. This led to the abandonment of the idea of inert micro-corpuscles being the ultimate constituents of the physical world, in favour of a search for constituents that were characterized in a wholly chemical way.

Mineral chemistry was a quantitative, autonomous, experimentally grounded discipline. Pursuing its explanatory goals in a way that could not be further from that of mechanics, by the end of the eighteenth century it stood at the forefront of the physical sciences. It might be thought that its success in carving out a free-standing area of enquiry would have acted as a model for those working in the life sciences and elsewhere. But in the wake of its reform by Lavoisier, chemistry faced some significant criticisms.[65] On the one hand, there were those who believed that Lavoisier had destroyed the autonomy of chemistry, and that this autonomy had to be restored. On the other, there were those who thought that his work had established its autonomy to the extent that it had assumed a position of such dominance that it was blocking progress in other disciplines, offering something that was as useless and counterproductive for these other disciplines as mechanism had been a century earlier.

On the first set of criticisms, I shall be brief, and we can restrict our attention to the questions they raise about the identity of chemistry as an experimental discipline. Metallurgy, for example, which had always been a crucial part of chemistry, was reformed in the 1770s from a somewhat chaotic set of practices into a methodologically sound experimental discipline. Abraham Gottlob Werner, director of the leading college of mines, the Bergakademie in Freiberg, and author of a series of textbooks on descriptive mineralogy that completely transformed the field,[66] based his influential system of mineral classification on colour, hardness, texture, smell, and taste. These were the kinds of criteria that those critics of Lavoisier who worried about the identity of chemistry were concerned were being ignored. What was basically at issue for them was the relative standing of precise quantitative assessments of chemical products, on the one

[64] See Ursula Klein, *Experiments, Models, Paper Tools: Cultures of Organic Chemistry in the Nineteenth Century* (Stanford, 2003).

[65] This was particularly true of the German reaction to Lavoisier: see Karl Hufbauer, *The Formation of the German Chemical Community (1720–1795)* (Berkeley, 1982), ch. 7.

[66] Abraham Gottlob Werner, *Von den äusserlichen Kennzeichen der Fossilien* (Leipzig, 1774); idem, *Kurze Classification und Beschreibung der Gebirgsarten* (Dresden, 1787); idem, *Neue Theorie über Entstehung der Gänge* (Freiberg, 1791).

hand, and the chemist's skill in recognizing substances and processes in terms of a highly cultivated responsiveness to such phenomena as vapours, colours, and smells.[67] Priestley, for example, insists that the exposure of iron to heat in a confined space results in a strong smell which cannot be ignored, but which Lavoisier and his followers take no notice of, being concerned solely with weight differences.[68] The German pharmacist Johann Christian Wiegleb likewise denies arguments against the existence of phlogiston on the grounds that in calcination, we can pick up 'a particular matter, sensitive to the sense of smell'.[69] Similarly, Richard Kirwan argues that the colours of the vapours released cannot be ignored, as they are specific signs of particular chemical substances.[70] Botanza and Gal sum up nicely what is at issue, when they write that these critics of Lavoisier's chemistry were concerned to investigate

> the world of matter in all its complexity. Its practitioners came to view this complexity as essential to matter: matter comprised particular substances, each carrying unique properties, related to each other in a variety of ways. This complexity embedded uniformities and regularities, but could not be reduced to the universal bulk and force which the new chemistry imported from physics. Proper chemical inquiry had to pay full attention to the various causal interactions and their manifestations; it could not be reduced to the measurement and comparison of weight.[71]

The early successes of eighteenth-century chemistry, effectively beginning with Geoffroy, depended crucially on a resolutely experimental approach, something deemed central to the establishment of its autonomy. In other words, it was not just in the content of particular chemical theories that its new identity lay, but above all in its ability to devise a range of experimental and observational practices that rid it of a reliance on micro-corpuscularianism and provided it with the means to think through the transformations of substances in purely chemical terms. Priestley and others believed that this feature of chemistry was being lost: it had become an arithmetic of weights that did not take into account the authentic and productive chemical procedures developed from the beginning of the century.

At the same time there emerged a different kind of criticism that focused on the inability of the new chemistry to take into account the relation between chemistry and other disciplines, disciplines that were part of the eighteenth-century move

[67] See Victor Boantza and Ofer Gal, 'The "Absolute Existence" of Phlogiston: The Losing Party's Point of View', *British Journal for the History of Science* 44 (2011), 317–42: 324–30.

[68] Joseph Priestley, *Experiments and Observations Relating to the Analysis of Atmospherical Air* (London, 1796), 42.

[69] Johann Christian Wiegleb, 'Doctrine de Stahl sur le Phlogistique, rectifiée et appuyée par des preuves, en opposition au nouveau Systême chimique des François, dont on cherche en même tems à démontrer le peu de solidité, Extrait des Annales de Chimie de Crell', *Observations sur la physique* 41 (1792), 81–5: 84.

[70] See, for example, Richard Kirwan, *An Essay on Phlogiston, and the Constitution of Acids* (London, 1789), 128–9

[71] Botanza and Gal, 'The "Absolute Existence" of Phlogiston', 341–2.

towards experimental and observational investigation. I noted earlier Cullen's advice of 1753 that chemistry be modelled on natural history. This idea became more pronounced as the century progressed. Part of the classificatory problem lay in the realization that Geoffroy's simple law of affinities failed to capture many chemical processes. In particular, Bergman, in compiling his more comprehensive table of affinities, was aware that various kinds of factors could affect affinities. He set out in detail the factors which had an effect on affinities: temperature, solubility, the joining of three or more substances, excessive amounts of one substance, graduated changes of substances producing different types of attraction, and double elective affinities creating intermediary relations.[72] In short, as Reill has pointed out, almost all of the environmental conditions in which a chemical reaction, as characterized in terms of affinities, takes place can alter and direct the specific form of the product. Other chemists were aware of this, and Lavoisier, for example, knew that each affinity relation could proceed in steps of saturation levels, each step producing different qualities.[73]

It was becoming clear that what reactions occurred, and when they occurred, was a matter of the conditions under which the substances interacted, and it was Berthollet who gave the question of the conditions under which reactions took place a central role in understanding chemical processes. Noting that salt and limestone produce sodium carbonate in the shores of Lake Natron (a salt lake in present-day Tanzania), something that never happened under normal circumstances because the chemical affinities between the two substances were too weak, he enumerated the factors he believed facilitated the reaction: the great heat of the region, the great preponderance of limestone, and the trickling of salt solution through the lime. In 1801 and 1803, he published two studies[74] that started from Bergman's list of six variables affecting chemical combinations, adding powers of cohesion, electricity, effervescence, tendency towards crystallization, and effect of the solvent. As Reill notes:

> Berthollet's approach virtually transformed chemistry into a form of natural history, whereby the chemist had to deal with each reaction both as a type, illustrating a general law, and as a specific event in which individual conditions play a major role. Not only was the chemical universe made up of many unique individuals whose number was being multiplied enormously through chemical research, but the individuals themselves were influenced directly by the conditions under which they were formed and existed. Thus, as in natural history, chemical genera, species, and varieties could be grasped only through a procedure combining diachronic and synchronic understanding.[75]

[72] Torbern Bergman, *Traité des affinités chimiques ou attractions électives* (Paris, 1788), 10–56.

[73] See Reill, *Vitalizing Nature*, 87, and more generally 86–9, to which I am indebted here.

[74] Claude Louis Berthollet, *Recherches sur les Lois de l'Affinité* (Paris, 1801); idem, *Essai de statique chimique* (2 vols, Paris, 1803).

[75] Reill, *Vitalizing Nature*, 87. Cf. Kim, *Affinity*, 439–55.

Berthollet's approach meant that the possibilities for potential chemical combinations were limitless, and raised the question of whether there was some other type of unifying factor or factors lying behind natural processes. In what follows, I want to explore two kinds of approach to bringing chemistry under more general organizational principles. The first, which questions whether Lavoisier's chemical elements are the fundamental constituents of all physical substances, is an attempt to make chemistry come to terms with the life sciences. There is no significant direct connection between the two (if we exclude medicine for these purposes), but there is an indirect connection, via electricity. Second, there is the search for a general understanding of the processes underlying all forms of physical activity, where there is an attempt to treat chemistry as part of a project—in mechanics, electricity, and the life sciences—to go beyond matter as traditionally understood and to think of it in terms of variations in properties of fundamental forces, or even of a single fundamental force. The speculative aspirations of such a view were summed up nicely in Richard Lovett's *The Electrical Philosopher* of 1774, when he writes that:

> the electrical Fluid is in form of an extremely elastic Air, and . . . appears to be a universal Agent in the strictest Sense, so as the occupy and fill all Space, not only all free and open Spaces, but the *minutest vacuities and interstices or pores of the most compact and solid Bodies*, agreeable to the Prediction of Sir *Isaac Newton* as we find it in the concluding Paragraph of his Principia. . . . Were I to declare my own private Sentiments to you, I shou'd affirm, without any hesitation, that this grand Principle is the primary or first, and indeed, sole Cause of all other secondary causes whatsoever.[76]

VITAL FORCES

In his *Florae Fribergensis* of 1793, Alexander von Humboldt writes that the 'internal fire which breaks the bonds of chemical affinity and prevents the elements from being joined to one another at random, we call the vital force'.[77] The life sciences cannot be subordinated to chemistry, for processes that, left to their own devices as it were, would be subject to chemistry, follow a wholly new direction as vital forces come to regulate them. The surgeon Richard Saumarez summed up what was at issue when he wrote that life was an active power 'by the energy of which various species of matter are converted into one kind, under one system, so that the matter thus converted possesses the power of resisting the operation of external causes and of preserving itself from putrefaction and

[76] Richard Lovett, *The Electrical Philosopher. Containing a new System of Physics founded upon the Principle of an Universal Plenum of Elementary Fire* (Worcester, 1774), 18–19.

[77] Alexander von Humboldt, *Florae Fribergensis specimen plantas cryptogamicas praesertim subterraneas exhibens* (Berlin, 1793), 135.

decay',[78] and when the vital powers have stopped acting, the substances of which the organism is composed return to the regulation of 'the laws of physics in general'.[79] In the last decades of the eighteenth century the question of the relation between chemistry, in particular, and the life sciences had become pressing,[80] with the distinguished Scottish surgeon John Hunter complaining in his 1786 lectures that physiologists had been so concerned with mechanics and chemistry that they had 'entirely lost sight of life'.[81]

In his *Recherches sur les causes des principaux faits physiques*, published in 1794, Lamarck complained that chemistry had in effect cut itself loose from other branches of science. Results in other disciplines that bore on processes discussed in chemistry were either ignored or accommodated by means of ad hoc hypotheses. What Lamarck is seeking is some more general system that connects chemistry, physics, and the life sciences. Explaining the project that he devotes himself to in the *Recherches*, he writes:

> As for the usefulness of this work, can it be doubted for an instant? Can it be called a mass of futile hypotheses when it is known that the principles that are developed are connected throughout with one another, and are all truly dependent on one another, which would not be the case if these principles were wholly without foundation? Can it be said that these same principles are sterile as far as results are concerned, when it is seen that they offer simple and natural explanations not only of the facts of the ordinary course of life, passing daily in front of our eyes, and whose true causes must interest us, but also of the main *organic* facts, which it is so important to understand properly, when one is aware that these principles lead us to grasp the true cause of the development of our organs and the growth of the parts of our bodies, and to the cause that puts an end to these and consequently to our lives.[82]

Part of the problem, Lamarck argues, lies the proliferation of chemical elements in Lavoisian chemistry, which instead of introducing order into the discipline has lead to a lack of clarity:

> Since the beginning of the examination of gases by many contemporary physicians and chemists, one can say that the sciences that they have cultivated have proved in large part to be a kind of revolution, which will perhaps not be as favourable to true progress of these

[78] Richard Saumarez, 'On Generation and the Principle of Life', *Medical and Physical Journal* 2 (1799), 242–7, 321–6: 325.

[79] Richard Saumarez, *A New System of Physiology, comprehending the laws by which animated beings in general, and the human species in particular, are governed, in their several states of health and of disease* (2nd edn, 2 vols, London, 1799), ii. 529.

[80] On the debates in Germany, which were particularly intense, see Joan Steigerwald, 'Rethinking Organic Vitality in Germany at the Turn of the Nineteenth Century', in S. Normandin and C. Wolfe, eds, *Vitalism and the Scientific Image in Post-Enlightenment Life Science, 1800–2010* (Dordrecht, 2013), 51–75.

[81] John Hunter, 'Lectures on the Principles of Surgery', in *The Works of John Hunter*, ed. J. F. Palmer (4 vols, London, 1835–7), ii. 216.

[82] Jean-Baptiste-Pierre-Antoine de Monet de Lamarck, *Recherches sur les causes des principaux faits physiques* (2 vols, Paris, 1794), i. xiii.

sciences as they think. It appears in fact that in this period their principles . . . have become daily more complicated, less clear, and far less general.[83]

Lamarck denies that chemical substances form naturally through affinities or through crystallization. Rather they formed by the degeneration and decay of organic substances. This is a view common to those, ranging from Maupertuis to Bordeu, who considered that matter must contain vital powers. The matter of chemists was not matter per se, any more than that of mechanists had been, but simply dead matter: that is, matter that had lost its active and vital properties.[84] On such a view, the kind of investigation of the medicinal and psychological effects of gases that Priestley and Beddoes had pursued might have been considered of more general value, but that of Lavoisier was inevitably going to appear as if it had taken a fruitless path. Nevertheless, inorganic matter is the natural state of matter for Lamarck: irritability and sensibility were, as one commentator has put it, 'momentary interruptions of nature's ordinary course towards death and destruction'.[85] In contrast to Lavoisier's elements, however, for Lamarck inorganic matter consists of only the four traditional elements. These are combined in various ways by plants into the compounds that sustain life, and are in turn absorbed by animals and processed into even more complex substances.[86] But there is no combination of nature's powers with any set of possible circumstances in which these powers could produce 'a being endowed with organic movement, capable of reproducing itself and subject to death'.[87]

Lamarck's response to what he saw as the insulation of chemistry from other pursuits was to restrict the number of basic elements with which it worked so that more general resources might be provided to capture other natural processes. It is difficult to envisage just how a return to a theory of the four elements could satisfy his demand for increased clarity, on the one hand, and provide a material basis for both organic and inorganic forms on the other. Yet it was clear that chemistry and

83 Ibid., i. 7.

84 See Gaukroger, *The Collapse of Mechanism and the Rise of Sensibility*, ch. 10. It is widely thought that Wöhler's 1828 synthesis of the organic compound urea from an inorganic substance (ammonium cyanate) dealt a severe blow to the idea of living matter, to the idea that inorganic substances were just the detritus of living ones, and above all to the idea of vitalism generally. But in fact it had very little impact. Ursula Klein labels it a 'persistent myth': see her discussion in *Experiments, Models, Paper Tools*, 212–13.

85 Guido Giglioni, 'Jean-Baptiste Lamarck and the Place of Irritability in the History of Life and Death', in S. Normandin and C. Wolfe, eds, *Vitalism and the Scientific Image in Post-Enlightenment Life Science, 1800–2010* (Dordrecht, 2012), 19–49.

86 Lamarck, *Recherches sur les causes des principaux faits physiques*, ii. 285–7. See the discussion in Pietro Corsi, *The Age of Lamarck: Evolutionary Theories in France 1790–1830* (Berkeley, 1988), 47–55.

87 What drives this process Lamarck believes we will never know: 'I believe that it is as impossible for man to know the physical cause of the first member of each species, as it is to give a similarly physical specification of the existence of matter or of the whole universe.' *Recherches sur les causes des principaux faits physiques*, ii. 214. Note that the *Recherches* was composed well before Lamarck had developed his distinctive theory of transformation.

natural history (mineralogy and the life sciences) could not be completely unconnected, and the basic divisions in nature were mirrored in what were in effect the distinct disciplines of mineral (inorganic) chemistry and organic chemistry. Indeed, part of the problem may have been that mineral chemistry was simply being extrapolated into the realm of organic chemistry, an impossible task on Lamarck's view. But there was one area where there did indeed seem to be a deep connection between chemistry and the life sciences: electricity. Connections between electricity and chemistry became evident when it was discovered that electrical charges could promote chemical synthesis, as in the synthesis of water from hydrogen and oxygen, and even chemical decomposition, as in the decomposition of water into hydrogen and oxygen. And at the same time that these processes were discovered, Galvani was not only showing that an electric current could stimulate nervous and muscular activity, but was also offering a theory of animal electricity.[88]

There had been a number of experiments on the use of electricity to stimulate growth in plants since mid-century,[89] and John Feke in 1746 was explaining the behaviour of the 'sensitive' plant, *Mimosa pudica*, in terms of a surplus of electrical fluid.[90] More generally, there are examples of earlier writers identifying electricity with a life force. Benjamin Rackstrow writes in 1748 that 'This Æthereal or Electrical Fire I take to be the vivifying spirit that resides in Air, which, taken into the lungs of an animal, is separated from the heterogeneous particles'.[91] William Stukeley, in 1752, writes that: 'All life is owing to the subtle, electric fire pervading the macrocosm; & operating in divers proportions.'[92] And Marmaduke Berdoe, in his 1771 *Enquiry into the Influence of the Electric Fluid*, erected an elaborate account of the development of the embryo and the development of

[88] There had been several suggestions of a connection between electricity and muscular activity earlier in the eighteenth century, but such associations remained in the realm of suggestive speculations. Stephen Hales postulated a 'vibrating electrical virtue' acting through the muscles on the nerves, and he had tried to generate electrical sparks by shaking blood (*Statical Essays*, ii. 59, 93–7). Bryan Robinson had speculated that electricity vaporized liquids in the blood causing motion in muscles and nerves: *A Dissertation on the Aether of Sir Isaac Newton* (Dublin, 1746), 141. I am particularly indebted to the identification and discussion of primary sources on electricity and vital forces in Philip C. Ritterbush, *Overtures to Biology: The Speculations of Eighteenth-Century Naturalists* (New Haven, 1964). An invaluable bibliographic sourcebook is Francis Ronalds, *Catalogue of Books and Papers Relating to Electricity, Magnetism, the Electric Telegraph, Etc.* (Cambridge, 1880).

[89] See, for example, Karl Heinrich Koestlin, *Dissertatio Physica Experimentalis de Effectibus Electricitatis in Quaedam Corpora Organica* (Tübingen, 1775); Pierre Bertholon, *De l'électricité des végétaux* (Paris, 1782); Abée d'Ormoy, 'De l'influence de l'électricité sur la végétation, prouvée par de nouvelles expériences', *Observations sur la Physique, sur l'histoire naturelle et sur les arts* 35 (1789), 161–76. See the discussion in Ritterbush, *Overtures to Biology*, 28–35, from which I have drawn references.

[90] John Feke, *An Essay to Shew the Cause of Electricity* (London, 1746), 40. The idea that sensitive plants operate by means of electricity was decisively refuted in 1795 in Christoph Heinrich Pfaff, *Über thierische Elektricität und Reizbarkeit. Ein Beytrag zu den neuesten Entdeckungen über diese Gegenstände* (Leipzig, 1795), 118.

[91] Benjamin Rackstrow, *Miscellaneous Observations, together with a Collection of Experiments on Electricity* (London, 1748), 25.

[92] MS. Letter, 13 May 1752, to Peter Collinson: cited in Ritterbush, *Overtures to Biology*, 33.

species in terms of a gradual divesting of electrical fluid.[93] But it was only with Galvani's work that the possibility began to be raised more seriously that electricity formed a bridge between the chemistry of inert substances and the living realm.

Electricity was of equal importance with chemistry in the reform of the understanding of nature in the eighteenth century,[94] and its standing was transformed in 1749, when Franklin demonstrated that lightening was an electrical discharge. Henceforth, it was clear that electricity was a natural phenomenon rather than a special effect that needed to be artificially induced. As a natural phenomenon, its role in matter theory was no longer marginal, and it became one of its core concerns, especially once the role of electrolysis in decomposing chemical compounds was discovered. But, even more significantly for the developments with which we are concerned, it also had a role as a natural phenomenon par excellence when, as a result of the work of Galvani in the 1780s, it began to be intimately connected with the nature of life.

In 1781 Galvani, professor of anatomy and obstetrics at the Institute of Arts and Sciences in Bologna, noted, in the course of his research on the role of nerves in muscular contractions, the contraction of a detached frog's leg when an exposed nerve was touched by one investigator while another drew a spark from an electrostatic machine some distance away.[95] In 1786 he observed that the leg could kick in the absence of external electrical stimulation. The kick of the leg was explained in terms of the completion of an electrical circuit through the crural nerve and the leg muscle, caused by a nervous electrical fluid contained in the muscle, which, he surmised, acted like an organic Leyden jar. It was known that a Leyden jar could be discharged by a heterogeneous conductor, so Galvani and his associates set up experiments to show how contraction of the frog's leg could be produced in the same way, using connecting arcs of different materials. Animal electricity had been known for some time, and the torpedo fish had been used by physicians to treat patients suffering from epilepsy and heart failure,[96] but Galvani

[93] Marmaduke Berdoe, *An Enquiry into the Influence of the Electric Fluid in the Structure and Formation of Animated Beings* (Bath, 1771).

[94] The standard account is John L. Heilbron, *Electricity in the Seventeenth and Eighteenth Centuries* (Mineola, NY, 1999). See also the comprehensive introductory essay in Roderick W. Home, *Aepinus's Essay on the Theory of Electricity and Magnetism* (Princeton, 1979). Developments in electricity before mid-century are covered in Gaukroger, *The Collapse of Mechanism*, 196–206; 336–50.

[95] Luigi Galvani, *De viribus electricitatis in motu musculari* (Bologna, 1791). The contemporary notes that Galvani kept from 6 November 1780 to 25 April 1787 are to be found in *Memorie ed esperimenti di Luigi Galvani con la iconografia di lui e un saggio di bibliografia degli scritti* (Bologna, 1937), 233–411.

[96] Cavendish undertook an analysis of how the effects in the torpedo fish were generated, establishing that they were compatible with what was known about electricity: Henry Cavendish, 'An Account of Some Attempts to Imitate the Effects of the Torpedo by Electricity,' *Philosophical Transactions* 66 (1776), 196–225. The first definitive demonstration that the effects of the torpedo fish were electrical was that of Frans van der Lott, 'Kort bericht van den Conger-all, ofte drilvisch;

went well beyond this. What Galvani was arguing was that all animals—not just torpedo fish, electric eels, and such—were able to store electricity in their muscles. The conclusion he drew was that electricity held the key to life, for it was electricity that animated living matter.

Galvani's work marked a major step in the understanding of life for many of his contemporaries. George Cadogan Morgan wrote in his influential *Lectures on Electricity* of 1794 that: 'The knowledge of the electrical fluid is leading us rapidly into a full investigation of the nervous influence and of all the mysteries connected with the sensations and motions of animal life.'[97] At a more popular level, Galvani became widely known through the efforts of his nephew, Giovanni Aldini, who was professor of physics at the University of Bologna, and who devised some spectacular public demonstrations of Galvani's theory of electricity and animal movement. Dispensing with frogs' legs, he attached electrodes to a severed sheeps' and cows' heads (Fig. 2.7), and on connecting the current the animal's eyes moved, its tongue shot out, and its vital force seemed to have been revived. The demonstrations were followed by the electrical stimulation of the exposed brains of the severed heads of executed criminals, causing them to grimace, their jaws to move, and their eyes to open and shut.[98] These demonstrations were a great popular success, and news of them travelled rapidly throughout Europe: some of the details were incorporated into Mary Shelley's *Frankenstein*, for which Aldini himself was the model.

The questions of the relation between electricity and life raised here should be seen against the background of general concern with the nature of life that had been prompted by the discovery by the Swiss naturalist and microscopist Abraham Trembley, in 1740, of the reproductive behaviour of the freshwater polyp, or hydra.[99] Although in all other respects the polyp displayed the signs of animal behaviour[100]—it could move and feed for example—it regenerated by budding,

getrokken uit eenen brief van Frans van der Lott, gedateerd Rio Essequebo den 7 Juny 1761', *Verhandelingen uitgegeeven door de Hollandsche Maatschappye der Weetenschappen, te Haarlem, 6* (1762), part ii, supp. 87–95.

[97] George Cadogan Morgan, *Lectures on Electricity* (2 vols, Norwich, 1794), i. pp.xxiiii–iv.

[98] On these experiments, see André Parent, 'Giovannio Aldini: From Animal Electricity to Human Brain Stimulation', *Canadian Journal of Neurological Sciences* 31 (2004), 576–84.

[99] See Abraham Trembley, *Mémoires pour servir à l'histoire d'un genre de polypes d'eau douce* (Leiden, 1744). On the background to the discovery, see Marc J. Ratcliffe, *The Quest for the Invisible: Microscopy in the Enlightenment* (Farnham, 2009), ch. 5. Regeneration on a more limited scale was not unknown and Réaumur had reported on regeneration of crayfish appendages earlier in the century: René Réaumur, 'Sur la reproduction de quelques parties des écrevisses', *Histoire de l'Académie des sciences* (1712), 35–7. The phenomenon was accorded no general significance however, at least compared with the polyps case. See Charles W. Bodemer, 'Regeneration and the Decline of Preformationism in Eighteenth Century Embryology', *Bulletin of the History of Medicine* 38 (1962), 21–31; and Catherine Abou-Nemeh, 'Réaumur's Crayfish Experiments in Hartsoeker's *Système*: Regeneration and the Limits of Mechanism', in O. Nachtomy and J. E. H. Smith, eds, *The Life Sciences in Early Modern Philosophy* (Oxford, 2014), 157–80.

[100] Despite its brilliant green colour, naturalists had routinely treated it as an animal since Leeuwenhoek discovered it in 1703. Leeuwenhoek himself described it as an animalcule.

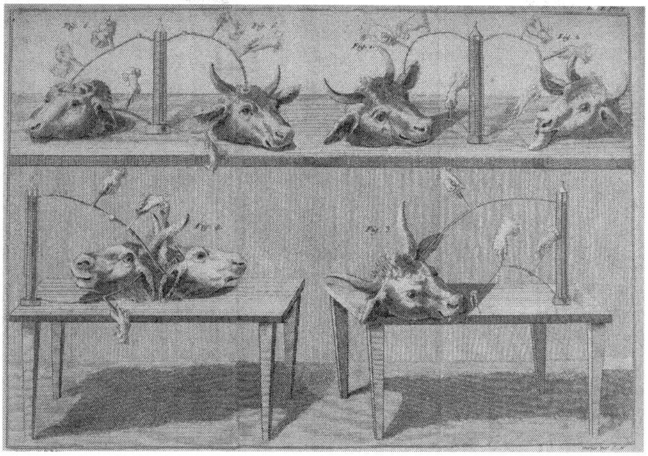

Fig. 2.7 Galvanic experiment on cows' heads. Aldini, *Essai théorique et expérimental sur le galvanisme* (1804).

like a plant. Even more remarkably, when cut into multiple parts, each part would regenerate, with all the features of the whole polyp—mouth, arms, legs, stomach—appearing, no matter from what segments of the original the cut part came (see Fig. 2.8). This was the most disconcerting discovery of the eighteenth-century life sciences.[101] When he was first informed of it, the great naturalist Bonnet wrote: 'everything I knew in natural history seemed like almost nothing compared to the polyp. It overthrew all my ideas and put my head into combustion as it were.'[102] From the point of view of natural history, the issue was the distinction between animals and plants, and more generally the problem of zoophytes, but it also had profound implications for matter theory. It suggested that life was not a property of the whole animal but of the matter from which it was constituted, for what else was there in the slices of the polyp? If matter could regenerate into a living organism, the conclusion drawn was that matter itself could harbour life. Growth was in any case not restricted to living things: Buffon had studied the growth of salt crystals, and Maupertuis had drawn attention to the 'Tree of Diana', produced when one mixes silver and spirits of nitre with mercury and water, with the result that after some time the 'particles of these substances

[101] As a sign of its importance, it is worth noting that its discovery prompted an anonymous satirical pamphlet in which it is discovered that gold coins can multiply by being cut in half: Henry Fielding, *Some papers proper to be read before the R____L SOCIETY, concerning the terrestrial chrysipus, golden-foot or guinea; an insect, or vegetable, resembling the polypus, which hath this surprising property, that being cut into several pieces, each piece becomes a perfect animal, or vegetable, as complete as that of which it was originally only a part* (London, 1743).

[102] Charles Bonnet, *Mémoires autobiographiques de Charles Bonnet de Genève*, ed. Raymond Savioz (Paris, 1948), 65.

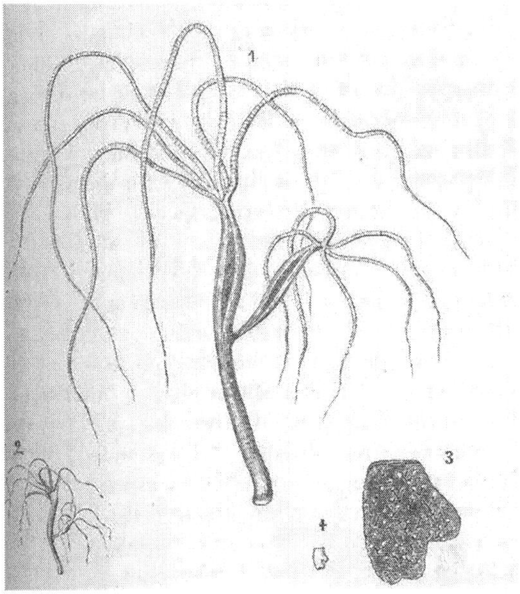

Fig. 2.8 The freshwater polyp. Trembley, *Mémoires pour servir à l'histoire d'un genre de polypes d'eau douce* (1744).

come together themselves to form a growth so similar to a tree that it is impossible to refuse it the name'. Moreover, Maupertuis notes, since the discovery of this kind of growth, various others have been found,[103] and, in a vain attempt to reconnect living things with the physical sciences, he suggests that the regeneration of limbs in the polyp should be seen on the model of attraction as described in physics and chemistry.

It was reflection on the behaviour of the polyp that led Johann Friedrich Blumenbach, one of the most celebrated anatomists of his time, to argue that there must be an inborn active drive (*Trieb*) in all living things, which dictates their form, preserves it, and where possible repairs the organism accordingly if it is damaged. This active force is quite different from other forces and he argues that reproduction, nourishment, and restoration of parts are all just modifications of one and the same force, which he calls *Bildungstrieb*, a formative or life force.[104] The issues at stake in embryology between preformationists (for whom the

[103] *Vénus physique*, ch. 17: *Œuvres de Mr de Maupertuis, nouvelle edition* (4 vols, Lyon, 1756), ii. 85–6.

[104] Johann Friedrich Blumenbach, *Über den Bildungstrieb und das Zeugungsgeschäfte* (Göttingen, 1781). See Peter McLaughlin, 'Blumenbach und der Bildunstrieb: Zum verhältnis von epigenetischer Embryologie und typologischem Artbegriff', *Medizinhistorisches Journal* 17 (1982), 357–72; and more generally Stefano Bertoletti, *Impulso, formazione e organismo. Per una storia del concetto di Bildungstrieb nella cultura tedesca* (Florence, 1990).

'germs' of every living thing were formed at creation) and epigenecists (for whom living things are formed from unorganized matter) are an important dimension in Blumenbach's general argument. Haller, a preformationist, had interpreted the apparent spreading of blood in the yolk of a fertilized egg, for example, as the pumping of blood into preformed vessels, whereas Kaspar Fredrick Wolff, an epigenecist, had argued that the adult structure resulted from a differential solidification of an originally homogeneous fluid.[105] To account for what drove this process—what controlled the formation of blood vessels for example—Wolff postulated the existence of a vital force, *vis essentialis*, which organized matter into a living form. Blumenbach developed the epigenecist position further, taking regeneration, as in the case of the polyp, as decisive.[106] Preformationists had to posit the existence of 'germs' distributed throughout the adult body, but Blumenbach notes that not only is there 'not a trace, not a shadow . . . of a preformed germ, of such a developed fibre as a miniature out of which a bud should be formed',[107] but also that the new parts are always much smaller than those they replace, which indicates, not accretion of material to a 'germ', but rather reorganization of living matter at the site of the regeneration. Blumenbach replaces Wolff's *vis essentialis*, which is primarily a mechanism of nutrition,[108] with his more ambitious notion of *Bildungstrieb*, the most basic vital power.[109] *Bildungstrieb* regulates the structure of the organism by mediating between an external stimulus and a capacity to activate the corresponding organs.[110] Once it has played this role successfully, however, purely mechanical processes take over the specialized functional tasks. *Bildungstrieb* is not a vital force simply supervening on material processes as it has no existence independent of material processes, but nor can it be reduced to these processes.

Both Kant and Herder were indebted to the idea of *Bildungstrieb*, even though they interpreted it differently. Kant focused on mechanical causes within the organism taking organization—in the form of *Bildungstrieb*—as given, whereas Herder focused on the active agency itself.[111] Just what *Bildungstrieb* consisted in was an open question, but it was not unlike animal electricity in a number of

[105] See Shirley A. Roe, Matter, *Life, and Generation: Eighteenth-Century Embryology and the Haller–Wolff Debate* (Cambridge, 1981).

[106] Earlier in the century it has been assumed that regeneration was something restricted to primitive life forms, but in 1768 Spallanzani demonstrated regeneration of the legs and tails of larval frogs, i.e. in vertebrates: Lazzaro Spallanzani, *Prodromo di un opera da imprimersi sopra le riproduzioni animali* (Modena, 1768).

[107] *Über den Bildungstrieb*, 53.

[108] It was a standard move in epigenecist accounts from Descartes onwards to provide a general account of growth in terms of nutrition, and then to argue that the development of the foetus is just a special case of nutrients being turned into growing material: see Stephen Gaukroger, *Descartes' System of Natural Philosophy* (Cambridge, 2002), 190–6.

[109] See Blumenbach, *Über den Bildungstrieb*.

[110] Johann Friedrich Blumenbach, *Handbuch der Naturgeschichte* (Göttingen, 1797), 12.

[111] See Robert J. Richards, *The Romantic Conception of Life: Science and Philosophy in the Age of Goethe* (Chicago, 2002), 216–37.

respects. If so, it was possible that animal electricity held the key to life, in that it was this that animated matter. But the connection was not made. The problem was that different approaches to electricity worked within different disciplines, and looked for different things. Galvani's project was one very much within the life sciences, rather than physics, and a key aim was to establish that animals possess an electrical fluid analogous to a physical electrical fluid.[112] Volta, who was motivated by physical considerations (although his interest was initially stimulated by the torpedo fish[113]), offered a very different explanation. Galvani had hung his frogs outside by brass or iron hooks attached to an iron railing, and Volta argued that it was the combination of two different metallic conductors that produced the electromotive force in the frog, not animal electricity. This required a new understanding of electricity because the metals were no longer passive conductors. He reasoned that the two different metals provoked a disequilibrium between their electrical fluids, and that this was balanced by fluid flowing from the metal with an excess to the one with a deficit. Galvani countered by showing how the contraction could be caused by directly connecting the nerve and the muscle of the frog, without the intermediary of a metallic link. Volta in turn responded by setting up an experiment in which electrical fluid passed between two different metals separated only by piece of wet cloth, showing that organic material was not needed. The stacking of such metal and cloth combinations on top of one another formed what came to be termed a Voltaic pile. In the pile, surplus and deficit alternated in such a way that what resulted was the circulation of the electrical fluid.

Galvani had had the ambitious project of establishing a connection between electricity and vital forces. Volta's electric pile, the first electrical battery, converted one kind of force to another, chemical to electrical. Nicholson and Carlisle's electrolysis broke chemical bonds electrically, decomposing water into hydrogen and oxygen. The consequences for matter theory were enormous. Whether electricity could be identified with a vital force in nature was certainly a pressing question, but it was not the only one, and the various forms of interaction between chemical and electrical phenomena complicated the question of just what forces there were in nature.

THE UNITY OF FORCE

Much advocacy of the unity of force was speculative and metaphorical. John Lyon, writing in 1780, for example, announces that: 'Experience and observation,

[112] See Marcello Pera, *The Ambiguous Frog: The Galvani–Volta Controversy on Animal Electricity* (Princeton, 1992).

[113] See Giuliano Pancaldi, *Volta: Science and Culture in the Age of Enlightenment* (Princeton, 2003), ch. 6.

the two great guides to human knowledge, have convinced us, that there is diffused within the substance, and on the surface of all bodies, whether animal, vegetable, or mineral, a certain fluid, which some have called æther, some phlogiston, others the sulphureous principle, the inflammable substance, the electric fluid, or the elementary fire.'[114] If chemical forces could be converted to electrical ones, and electrical ones into chemical ones, then it was clear that no understanding of matter was complete without an understanding of electricity. But this required a new theory of matter. Above all, the interconvertibility of electrical and chemical forces tended to suggest the notion of the unity of forces. Even Newtonian conceptions of force, which rejected the idea of the unity of forces, leant some support to this. In 1785 Coulomb, using a very precise torsion balance, had shown how electrical attraction and repulsion followed the inverse square law that Newton had established for gravitational attraction. In his first report, he establishes that the electrical repulsion between two like charged bodies is inversely proportional to the square of the distance between them.[115] In a second report, he establishes that the electrical attraction between two oppositely charged bodies is proportional to the quantities of charge that they carry, and inversely proportional to the square of the distance between them.[116] In other words, at the formal level, the action of electricity (and magnetism) mirrors that of gravitation: the equations governing them take the same form.

But what about at a more substantive level? There was a non-Newtonian tradition, in its continental forms taking inspiration from Leibniz, which was committed to the continuity of matter and forces—as opposed to the conception of discrete matter and discrete forces of the Newtonian tradition—as well as to the unity of physical forces. In the course of the eighteenth century, there were a number of attempts, starting with Benjamin Worster in the 1720s, to construe all material activity in terms of attractive and repulsive forces.[117] In 1754, Gowin Knight, who had devised a procedure for strongly magnetizing steel and had been awarded the Copely medal of the Royal Society, set out an ambitious programme of reduction of light, magnetism, and other forces in nature to a basic pair.[118]

The most developed theories along these lines are those of Boscovich, in his *Theoria philosophiae naturalis* of 1758, and Kant, in his *Metaphysiche*

[114] John Lyon, *Experiments and Observations Made with a View to Point Out the Errors of the Present Received Theory of Electricity* (London, 1780), 195.

[115] Charles-Augustin Coulomb, 'Sur l'éléctricité et le magnetisme. [I] Construction et usage d'une balance éléctrique', *Académie Royale des Sciences: Mémoires de mathematique et de physique, présentés . . . par divers savans* (1785), 569–77: 574.

[116] Charles-Augustin Coulomb, 'Sur l'éléctricité et le magnétisme. [II] Ou l'on détermine suivant quelles lois le fluide magnétique ainsi que le fluide éléctrique agissent', *Académie Royale des Sciences: Mémoires de mathematique et de physique, présentés . . . par divers savans* (1785), 578–611: 579.

[117] Newton himself had speculated on this idea in his short manuscript *De gravitatione*, but he did not follow it up, and the work remained unknown in the eighteenth and nineteenth centuries.

[118] Gowin Knight, *An Attempt To Demonstrate, That all the Phenomena in Nature, May be Explained by Two Simple Active Principles, Attraction and Repulsion* (London, 1754).

Anfangsgründe der Naturwissenschaft of 1786.[119] Boscovich tried to account for all physical phenomena, including chemical ones, in terms of two primitive forces, but he found himself limited to showing how one might assimilate experimental results already reached by others, and on the basis of a more traditional understanding, to his underlying conception: they were made to fit the two-force model, a model that did no independent work.[120] Kant—unlike Newton and Euler, who had treated impenetrability as an absolute force—thinks of it as a repulsive force in the immediate environment of the centre of body,[121] so that the boundary formed by this is not fixed but, at least theoretically, can be compressed to infinity.[122] There is also a second force, which is attractive—gravitation—and it pervades the whole of space.

The problem was that there was no basis in any empirical study of matter for such a two-force approach. Boscovich had developed the two-force account in most detail, trying to explain all physical phenomena, including chemical ones, in terms of two primitive forces. But this was just an empty exercise in assimilation to already known results. These were regimented into a two-force model, but this was a purely formal exercise and offered nothing at all that was new. Boscovich's approach was unable to generate anything beyond what was already known. Kant faced the same problem, but there was another more serious difficulty for Kant. His model was Euler's rational mechanics, for this provided the archetypical a priori deductive structure that he considered to be constitutive of science. But in Euler's scheme there was no way of connecting mechanical forces with material properties: the two exercises remained distinct. Rational mechanics could not have provided a basis for Kant's conception of matter. His own solution simply foists matter theory on to mechanics in the hope that this would produce a unified physical theory, but the hybrid that results not only has no empirical basis but, having deviated so radically from rational mechanics, can no longer claim any share in what he considered to be the real strength of rational mechanics, its a priori deductive structure. It is worth noting that, in Euler, moving from one kind of phenomenon to another, from mass points to rigid bodies, or from flexible bodies to elastic bodies for example, is a complex and demanding business, often involving the elaboration of novel mathematical resources. Boscovich and Kant face no such tasks: a two-force (or single attractive/repulsive force) model simply stipulates that physical patterns of attraction and repulsion, seemingly varying

[119] Kant, *Gesammelte Schriften*, iv. 467–565; *Theoretical Philosophy after 1781*, 183–270. The text is discussed in detail in Friedman, *Kant and the Exact Sciences*.

[120] This is in contrast to his work on measuring the length of the meridian arc near Rome, where he deals with inconsistencies between different arc measurements using innovative statistical procedures: Roger Joseph Boscovich and Christopher Maire, *De Litteraria Expeditione per Pontificiam ditionem ad dimetiendas duas Meridiani gradus* (Rome, 1755). See the discussion in Stigler, *The History of Statistics*, 39–50.

[121] Kant, *Gesammelte Schriften*, iv. 516; *Theoretical Philosophy after 1781*, 227.

[122] Kant, *Gesammelte Schriften*, iv. 501; *Theoretical Philosophy after 1781*, 213.

only with distance from the source, somehow correspond, in some unspecifiable way, to various material qualities. The centres of force that are invoked are not like mass points, in which case Kant's assumption that they can be considered in terms of the a priori deductive discipline of mechanics loses all credibility. Above all, it is no longer clear what the attraction and repulsion that Kant uses to characterize fundamental properties are. He is forced to abandon the rigour of rational mechanics, failing to distinguish resisting forces (such as the force of inertia) and repulsive forces (such as magnetic repulsion), for example, simply treating them as 'negative' forces. Yet these are supposed to be the basic principles from which all material phenomena can be deduced.

Nevertheless, what is being proposed is more than simply an empirically equivalent variant of rational mechanics. The advocacy of a continuity of matter and force in effect assimilates the one to the other, opening up the possibility that the advances made in mechanics can be transferred to the understanding of material properties. Were it successful, this move would transform mechanics into a truly general theory of physical-material properties, one that inherited the a priori deductive structure of rational mechanics, extending it to a range of questions in matter theory that even the most ambitious practitioners of rational mechanics had not proposed. The trouble is that there is nothing in rational mechanics as such to support such an extension, and if one considers the disciplines that made the greatest advances in matter theory in the eighteenth century, chemistry and electricity, then it is clear that the assertion of such continuity cannot be guided by empirical considerations.

The kinds of material transformations studied in chemistry, for example, are autonomous with respect to mechanics. Yet both Boscovich and Kant maintained that chemistry could be subsumed under their two-force model. Kant believed that we should be able to deduce chemical affinities from 'the approach or withdrawal of the parts of matter (as, say, in proportion to their densities and suchlike)'.[123] This was of course in the realm of speculation, and at least in Kant's case, from someone who had only a very superficial familiarity with the kinds of developments in chemistry that we have been discussing. Boscovich and Kant were in effect attempting to revive a foundationalist mechanical model of physical processes. But the failure of such an attempt did not mean the failure of the idea of the reducibility of natural processes to fundamental forces, or of that of fundamental substances differing in the kind of force they exert. There were other moves along these lines that, while still speculative, took their starting point from concrete problems in chemistry, for example.

A number of chemists, including Berthollet, had considered that a more economical account of basic substances was needed than that currently available in theories of chemical elements, and they suggested that polar properties must lie

[123] Kant, *Gesammelte Schriften*, iv. 471; *Theoretical Philosophy after 1781*, 186.

at the basis of chemical reactions. It was Jakob Joseph Winterl, a professor of chemistry in Budapest, who developed this line of reasoning furthest. Returning to the notion of chemical principles, he offers an account in which, at the most basic level, there are only acidic or alkaline principles. In this way, Lavoisier's chemical elements differ from one another to the extent to which they manifest a preponderance of one of these.[124] Unlike the mechanical models, what Winterl is offering is something that does some independent work: Lavoisier had linked oxygen and acidity but Winterl argues that acids need not contain oxygen; and he recognized the existence of substances, such as many metal oxides, that exhibit both acidic and alkaline properties.[125] Developments in electro-chemistry play a key role in Winterl's characterization, for the acidic principle turns out to be identical with positive electricity, and the alkaline principle with negative electricity. Positive electricity acts on water to produce oxygen, for example, whereas negative electricity acts on it to produce hydrogen. Conversely, when hydrogen and oxygen combine, there is an exact balance between the principles in the water that results.

Winterl's work was taken up by a number of chemists, who tried in vain to detect his two principles experimentally.[126] But in the hands of one of his most devoted followers, Johann Wilhelm Ritter, there was some progress. Ritter was associated with the Romantic movement, notably Goethe, Herder, and Schelling, and it was Ritter who introduced Winterl's idea of chemical polarities not just into Romantic thought about nature, but also to Ørsted, who worked with him in 1801.[127] Ritter's interest in chemistry was initially stimulated by Galvani–Volta controversy. In 1799, Alexander von Humboldt—who two years earlier had called for a new branch of science, 'vital chemistry', investigating the chemical changes that take place in matter during the exercise of vital functions[128]—employed him to devise experiments with a view to supporting Galvani's theory of animal electricity.[129] He concluded that the flow of the electrical fluid, galvanism, was neither organic nor simply a question of contact between metals, but rather the outcome of two processes taking place at the same time: an electrical process that results from a disequilibrium between two metals, and a

[124] Jakob Joseph Winterl, *Accessiones novae ad Prolusiones suam primam et secundam* (Buda, 1803). Cf. his earlier *Prolusiones ad chemiam saeculi decimi noni* (Buda, 1800).

[125] See H. A. M. Snelders, 'The Influence of the Dualistic System of Jakob Joseph Winterl (1732–1809) on the German Romantic Era', *Isis* 61 (1970), 231–40.

[126] See ibid., 234–5.

[127] See Olaf Breidbach, 'The Culture of Science and Experiments in Jena Around 1800', in R. M. Brain, R. S. Cohen, and O. Knudsen, eds, *Hans Christian Ørsted and the Romantic Legacy in Science* (Dordrecht, 2007), 177–216; and Alexander Gottfried Friedrich Gode-von Aesch, *Natural Science in German Romanticism* (New York, 1941).

[128] Alexander von Humboldt, *Versuche über der Gereizte Muskel- und Nervenfaser* (2 vols, Posen and Berlin, 1797), ii. 41–2.

[129] See Dan C. Christensen, *Hans Christian Ørsted: Reading Nature's Mind* (Oxford, 2013), ch. 10. Ritter's work was published as *Beweis, dass ein beständiger Galvanismus den Lebensprozess in Therreich begleite* (Weimar, 1798).

chemical one that involved a reaction between metals and a dilute acid. In short, Ritter's experiments supported neither Galvani nor Volta, at least as their theories had developed up to that time: Volta's electric pile subsequently vindicated the idea of electricity being produced by a chemical reaction. Ritter concluded something rather different from the experiments, however, namely that at the most fundamental level, physical processes were regulated by polar forces: light, heat, electricity, and magnetism he concluded were all manifestations of one force. As well as magnetic poles, he postulated that the earth had electric ones, and he was convinced that the electrolysis of water could be produced by a series of magnets.[130]

The theme of the unity of forces was to become a dominant theme in *fin-de-siècle* natural philosophy. The idea of primitive forces being at the basis of physical behaviour encouraged the idea that physical and vital forces were, if not manifestations of the same single force, at least part of the same general picture.[131] Indeed, even some followers of Kant's theory of matter saw no obstacles in extending it to cover living forces.[132] The conception of nature that resulted, *Naturphilosophie*, was, like rational mechanics, a priori. Schelling spells out the programme in characteristically unambiguous terms:

> The assertion is, that all phenomena are correlated in one absolute and necessary law, from which they can all be deduced; in short, that in natural science all that we know, we know absolutely a priori. Now, that experiment never leads to such a knowing, is plainly manifest, from the fact that it can never get beyond the forces of Nature, of which itself makes use as means. . . . The assertion that natural science must be able to deduce all its principles a priori, is in a measure understood to mean that natural science must dispense with all experience, and, without any intervention of experience, be able to spin all its principles out of itself.[133]

The mineralogist Henrich Steffens became an ardent follower of Schelling on meeting him in 1798, and in 1801 he published a general account of the 'inner natural history of the earth' that combines a number of the themes that characterize the *naturphilosophische* picture of the natural realm as an interplay between

[130] Roberto de Andrade Martins, 'Ørsted, Ritter and Magnetochemistry', in R. M. Brain, R. S. Cohen, and O. Knudsen, eds, *Hans Christian Ørsted and the Romantic Legacy in Science: Ideas, Disciplines, Practices* (New York, 2007), 339–85.

[131] Just confining our attention to German works in the last five years of the century, there are: Johann Christian Reill, 'Von der Lebenskraft', *Archiv für die Physiologie* 1 (1795), 8–162; Joachim Dietrich Brandis, *Versuch über die Lebenscraft* (Hannover, 1795); Christoph Wilhelm Hufeland, *Ideen über Pathologie und Einfluß der Lebenskraft auf Entstehung und Form der Krankheit* (Jena, 1795); Jacob Fidelis Ackermann, *Versuch einer physikalischen Darstellung der Lebenskräfte organischer Körper* (2 vols Frankfurt-on-Main, 1797–1800).

[132] See, for example, the discussion of the Dutch Kantian physicist Jan Rudolph Deiman in Henricus Snelders, 'Romanticism and Dutch Scientists', in S. Poggi and M. Bossi, eds, *Romanticism in Science* (Dordrecht, 1994), 175–88: 176.

[133] Friedrich W. J. Schelling, *Einleitung zu dem Entwurf eines Systems der Naturphilosophie. Oder über den Begriff der Speculativen Physik und die innere Organisation eines Systems dieser Wissenschaft* (Jena & Leipzig, 1799), §4. Translated as 'Introduction to the Outlines of a System of Natural Philosophy; or, on the Idea of Speculative Physics and the Internal Organization of a System of this Science', *The Journal of Speculative Philosophy* 1 (1867), 193–220: 196.

polarities.[134] The work begins with an account of the formation of rocks via chemical processes, which he argues express two fundamental vital powers: a carbon-based power of vegetation and a nitrogen-based power of animation. The vegetative power is responsible for primitive siliceous rocks, which contain vegetable fossils, and the animating power for calcareous rocks, which contain animal fossils. Having established this basic connection between rocks, plants, and animals, Steffens proceeds to a classification of metals into two series: one identified with nitrogen, which is 'fluid' and contains a series from arsenic to mercury, and another identified with carbon, which is 'cohesive' and contains the series from antinomy to gold. The question now is how the existence of these two series are to be accounted for, and it is here that Steffens, following Schelling (who was attempting to replace what he considered to be a Newtonian corpuscular theory with the idea of a dynamic equilibrium between attractive and repulsive forces[135]), introduces a theory of the 'double polarity' of the Earth. There is a North–South polarity which is that of positive and negative magnetism and of nitrogen and carbon, associated with the opposing poles. The East–West polarity is manifested in positive and negative electricity, which are associated with hydrogen and oxygen. He deduces that we would expect a preponderance of the cohesive metals at the Earth's poles, and fluid metals in equatorial regions. In moving to the plants and animals formed from these materials, Steffens follows Kielmeyer's 'law of the succession of organic forces',[136] whereby a different type of force predominates in each level: reproductive in plants and lower animals, irritability in insects, sensibility in higher animals. In a later work on the 'foundations of philosophical natural science',[137] he expands his earlier account in a number of respects, and sets out a level of reflection in which human beings are able to achieve a hitherto lost union with nature, which is now mirrored in scientific understanding. For Steffens the ultimate aim is to grasp the unity of spiritual processes in nature and ourselves: human self-understanding approached through chemistry, geology, and the life sciences. The key theme here is the unity of all knowledge, something central to Schelling's programme for the reform of university teaching,[138] which was one of the founding documents of the new university in Berlin in 1809.[139]

[134] Henrich Steffens, *Beyträge zur inneren Naturgeschichte der Erde* (Freiberg, 1801). In my account of Steffens, I draw on Nicholas Jardine, '*Naturphilosophie* and the Kingdoms of Nature', in N. Jardine, J. A. Secord, and E. C. Spary, eds, *Cultures of Natural History* (Cambridge, 1996), 230–45.

[135] See in particular Part II of his *Ideen zu einer Philosophie der Natur als Einleitung in das Studium dieser Wissenschaft* (Jena & Leipzig, 1797).

[136] Carl Friedrich Kielmeyer, *Über die Verhältnisse der organischen Kräfte* (Stuttgart, 1793).

[137] Henrich Steffens, *Grundzüge der philosophischen Naturwissenschaft* (Berlin, 1806).

[138] Friedrich Wilhelm Joseph Schelling, *Vorlesungen über die Methode des Akademischen Studiums* (Leipzig, 1803).

[139] See Elinor S. Shaffer, 'Romantic Philosophy and the Organization of the Disciplines: The Founding of the Humboldt University of Berlin', in A. Cunningham and N. Jardine, eds, *Romanticism and the Sciences* (Cambridge, 1990), 38–54.

NATURALIZATION

I have set out in this chapter to describe a number of the key developments in the natural and life sciences, and to highlight the move from a reductionist account, along mechanist lines, to the attempts to establish autonomous disciples, notably in chemistry and the life sciences, to, finally, attempts to provide a new unifying synthesis in which the life sciences play a central role. It will be clear from this that, to the extent to which the natural sciences could have served as a general model of understanding, mechanics was an unlikely exemplar. Kant's advocacy of a mechanical model was almost unique in this respect, and the lessons that were drawn from the natural sciences were typically drawn from chemistry, or the study of electricity, or from the life sciences. These were resolutely empirical, non-reductionist forms of enquiry in which autonomy, rather than claims to foundational standing, was the pressing issue.

The emergence of the 'moral' or 'human' sciences, to which we are about to turn, increasingly forced open the question of how one might consider human behaviour in empirical terms. The natural sciences occasionally provided a model. In the 'Preface' to the first edition of his *Das Wesen des Christentums* (1841), Feuerbach wrote about his rejection of Christianity in favour of a form of humanism: 'The method that the author follows here is thoroughly objective. It is the method of analytical chemistry.'[140] But the natural sciences were not the only forms of naturalizing discourse. They did not provide the only models for naturalization, nor were they the only naturalizing resources drawn upon. Our concern will be with the formulation in empirical terms of questions about the human realm that had up to that point taken a non-empirical form, and had typically been deemed to be the concern of theologians and metaphysicians. This raises deep questions about the legitimacy of such an approach in particular cases, and to appreciate the significance of these, there are some general issues concerning naturalization that need to be clarified before we can begin examining the specific question of the naturalization of the human.

Naturalization is the translation of questions that had previously been taken as exclusively conceptual or a priori matters, and had been treated accordingly, into a form in which empirical evidence becomes appropriate to answering these questions. The reformulation in empirical terms of questions that had up to that point taken a non-empirical form is motivated in large part, in the eighteenth and succeeding centuries, by the success of empirical projects. This was a success which, it was considered, could be emulated if, through naturalization, one could transform what were seen by their critics as stagnant discourses into fruitful ones. But whatever benefits naturalization was able to secure in particular cases, we

[140] Ludwig Feuerbach, *Das Wesen des Christentums*, ed. W. Schuffenhauer, (2 vols, Berlin, 1956), i. 4.

cannot assume that it would be an appropriate or fruitful strategy in every case. In our own time, there are many examples where naturalization is widely argued to be a problematic and fruitless approach—from attempts to reduce mathematics to empirical psychology, to attempts to account for the existence of morality in terms of evolutionary advantages—and cognate anti-naturalization arguments were directed against the eighteenth-century naturalizing strategies that we shall be examining. The situation is complicated by the fact that naturalization inevitably changes the original question, because it is typically part of the argument that the question had originally been posed in the wrong way, with the result that it could not be resolved until it was rethought in more fruitful terms, ones that identified the appropriate kinds of reasoning and evidence. Nevertheless, however much the idea of making questions amenable to empirical treatment might seem an unqualified good, naturalization may in fact be misguided in many cases, and it will be a matter of serious dispute in the cases that we shall be looking at.

Naturalization is primarily an eighteenth-century phenomenon rather than a seventeenth-century one. In the seventeenth century, naturalization had been eclipsed by quantification, that is, the formulation in mathematical terms of questions that had up to that point taken a qualitative form. Quantification of course continues to play a significant role in the eighteenth century, but most of the kinds of issues that we shall be considering turned rather on naturalization. Although intimately connected from the seventeenth century onwards, naturalization and quantification are independent goals. It is obvious that one can have the former without the latter: it is difficult to conceive how quantification could plausibly be relevant to, or issue from, the naturalization of religious discourse (such as an empirically motivated reduction of religion to documented primitive fears and superstitions), for example. Nevertheless, it is tempting to think of naturalization as a necessary prelude to quantification, that is, to think of posing questions in an empirical form as a *sine qua non* of quantification, should quantification be appropriate. Matters are not so straightforward, however, for while naturalization followed by quantification is a typical route—one which goes back to the founding moment of early-modern mechanics, namely Galileo's establishment of his law of falling bodies[141]—it is not the only one.

If we think of naturalization in terms of opening up questions to empirical enquiry, then we can see reduction and assimilation as being just two strategies among others. The way in which Galileo, in his treatment of falling bodies, was able to show how we can conceive of a case which we never experience, namely bodies falling in a void, in empirical terms (e.g. by comparing rates of fall of bodies of different densities in media of different densities and treating fall in a void as a limiting case), opened up questions to empirical investigation, and as such it is a form of naturalization, as I have defined it. By contrast, the strategy for

[141] See Gaukroger, *The Emergence of a Scientific Culture*, 403–13.

quantification deployed by proponents of rational mechanics in the eighteenth century, for example, was predominantly a matter of assimilating mechanical questions to mathematical ones,[142] and in the hands of its greatest exponents—such as d'Alembert, Euler, and Lagrange—it was a move away from empirical questions towards an attempt to pursue physical enquiry in an a priori fashion, via deduction from what, in Euler for instance, were regarded as conceptual truths about the physical world. Here, one discourse, mechanics, is in effect being assimilated to another, mathematics. Rather than opening up questions to new forms of empirical investigation, it abandons a number of avenues of empirical enquiry in favour of mathematical development of theories. Quantification and naturalization here pull in different directions.[143]

To the extent to which naturalization has been associated with reduction to physics, there are two general points worth noting. The first is that, even if one thought of physics as being in some way the fundamental discipline in accounting for the natural world, it would still be mistaken to treat naturalization as a reduction of a non-physical discipline to a physical one. Questions of naturalization can be raised *within* physics: Galileo's discussion of the kinematics of fall, without which Newton's treatment of dynamics would not have been possible, is a matter of showing how one can transform conceptual questions into empirical ones.[144] But one does not go outside mechanics in doing this. Transforming conceptual questions into empirical ones has been a central part of physical enquiry since Galileo. Far from being considered as essentially a procedure for connecting non-physical and physical disciplines, naturalization can be internal to physical disciplines. Second, just as naturalization can connect one physical discipline with another, so too can it be a way of connecting one non-physical discipline to another. The modern preoccupation with naturalization via reduction to physical sciences completely misses the point of naturalization as a means of rendering questions amenable to empirical treatment. History, for example, acted as a naturalizing source from the sixteenth century onwards, and was one of the chief forms of naturalization from the early eighteenth to the mid-nineteenth century, providing a rich evidential source for questions in the 'moral' or human sciences. The physical sciences were irrelevant here.

The range of resources that naturalizing strategies might draw upon are varied, and they are not necessarily consistent with one another. In particular, there is no such thing as naturalization per se. Naturalization was not a single programme. It

[142] For details, see Clifford Truesdell, 'Rational Fluid Mechanics, 1687–1765', *Leonhardi Euleri opera omnia* series 2, vol. xii, section 1 (Zurich, 1954); idem, 'Rational Fluid Mechanics, 1765–1788', *Leonhardi Euleri opera omnia* series 2, vol. xiii (Lausanne, 1955); idem, 'The Rational Mechanics of Flexible or Elastic Bodies, 1638–1788', *Leonhardi Euleri opera omnia* series 2, vol. xi, section 2 (Zurich, 1960); idem, 'A Program Towards Rediscovering the Rational Mechanics of the Age of Reason', *Archive for History of Exact Sciences* 1 (1960), 3–36.

[143] See Gaukroger, *The Collapse of Mechanism and the Rise of Sensibility*, 145–9, 304–17.

[144] See Gaukroger, *The Emergence of a Scientific Culture*, ch. 11.

was not effected through a single procedure, or even by a unified set of procedures. There is—by contrast with physicalism for example—no single pre-given realm of evidence to which any form of enquiry can appeal. Quite the contrary, the various naturalistic strategies that we shall be examining were often treated by their proponents as alternatives, and were often at odds with one another. Decisions about how to formulate issues so that certain kinds of evidence can be identified and assessed means that the evidence is directed to particular kinds of questions, and is to a large extent shaped around the concerns that those questions bring to the enquiry. This is nowhere more evident than in the changing fortunes of anthropological medicine, to which we now turn.

PART II

3
Anthropological Medicine

Each of the developments that we shall be looking at in the next four chapters offers a distinctive form of naturalization in explaining core aspects of human nature and human behaviour. In this chapter and in Chapter 4, we shall be concerned with naturalizing strategies that focus explicitly on the competing demands of reason and sensibility. Anthropological medicine, our concern in this chapter, focuses on nervous sensibility as the route to understanding human behaviour, and by medicalizing this notion seeks to establish medicine as the key to an ambitious naturalistic account of what might be termed the 'human condition', one that sets the agenda not just for questions of health, but also for moral and political questions. Philosophical anthropology, which will be our concern in Chapter 4, focuses on the cognitive aspects of sensibility and, in one of its versions—that pursued by Herder—seeks to naturalize these by extending the discipline of empirical psychology, and then using this as a model for the naturalization of reason.

We may begin by noting that one of the distinctive features of sensibility is the extent to which it was becoming naturalized by the middle of the eighteenth century. This took many forms, from the attempts of moral philosophers such as Hume and others to treat political, social, and economic questions in terms that went beyond both traditional Christian thinking and civic humanism, to the attempts of those working within the life sciences, most notably Buffon, to extend their account of zoology into the human and social realms. Its characteristic feature is the replacement of speculative by empirical forms of enquiry. This has consequences for what we consider the role of reason to be, and how we think of reason. The key question is that of how our beliefs and values are shaped. Whereas philosophers had traditionally assumed that reason and the passions played the crucial role here, typically (if not universally) thinking of the passions as an obstacle to reason, we are led in a different direction if we approach reason from a consideration of sensibility.

THE IDEA OF AN ANTHROPOLOGICAL MEDICINE

Anthropological medicine, like much else in the eighteenth century, found its initial stimulus in Locke. Locke approached epistemology in the *Essay* not through the philosophical texts of his predecessors and contemporaries but through travel books, and he was acutely sensitive to both the historical and the regional variation in moral and other precepts. Moreover, the medical origins of his interest in questions of epistemology should not be forgotten, for among other things this motivates a rejection of the idea that a search for underlying causes is the only worthwhile form of explanatory strategy: such a search would abstract from the easily overlooked empirical connections that Locke considers to be of interest in their own right.[1] Intimately connected with this, and of even more importance in the present context, is Locke's exploration of just what follows from making sensation into a primary source of knowledge. Once sensation takes over as the sole source of knowledge, in the radicalized form of Lockeanism initiated by Condillac, this question becomes even more pressing, and the incorporation of sensation into the general notion of sensibility transforms the questions so that what, in Locke, had an implicit moral and political dimension, now explicitly comes to the core of moral and political thinking.

The connection between sensibility and moral and political programmes took two rather different forms in France: a direct form, and one mediated by a new broad conception of medicine. In his 'Mémoire sur la faculté de penser' (1796), Destutt de Tracy sets out the distinction as one between 'physiological ideology' and 'rational ideology'. The former, he writes, requires vast learning, and in its present state can do little more than help us identify a few truths, albeit scattered and uncoordinated, and isolate a few errors. This is the project in which medicine is centrally involved. Rational ideology, by contrast, is action-oriented and requires far less knowledge: it can get by with far fewer facts, which makes their connections easier to establish, and its chief advantage is that it is directly applicable and 'forms a complete system' in its own right.[2]

Two examples of 'rational ideology' will suffice to indicate the kinds of things at issue.[3] The first is the political programme of Condorcet. Reflecting on the question of moral diversity, a question brought to a head in Locke and in Bayle in the last decades of the seventeenth century,[4] and one that had plagued eighteenth-century attempts to establish a universal set of moral precepts, Condorcet maintains that the solution lies not in the discovery of some special moral sense, but rather in the realization that it is a combination of sensation and reflection that provides moral guidance. We seek pleasure, but this does not mean that it is a matter of self-interest, for our spontaneous sympathy for others

[1] See Gaukroger, *The Collapse of Mechanism and the Rise of Sensibility*, ch. 4.

[2] Antoine-Louis-Claude Destutt de Tracy, 'Mémoire sur la faculté de penser', *Mémoires de l'Institut national des sciences et arts. Classe de sciences morales et politiques*, 1 (1798), 283–451: 344–5. On Destutt de Tracy, see Sergio Moravia, *Il pensiero degli idéologues: scienza e filosofia in Francia (1780–1815)* (Florence, 1974), 319–65.

[3] I take the examples from Martin S. Staum, *Cabanis: Enlightenment and Medical Philosophy in the French Revolution* (Princeton, 1980), 168–70.

[4] See Gaukroger, *The Collapse of Mechanism and the Rise of Sensibility*, 402–9.

is also a source of pleasure, and this is what guides our behaviour.[5] The ideas of justice and rights are formed in all beings as a result of their common sensibility, and are therefore universal and uniform.[6] From these Condorcet derives our 'natural rights', which comprise the rights of personal liberty and security, freedom to enjoy property and trade, equality before a properly codified law, and a participatory politics.[7] A second example is Volney, and the full title of his 1793 work effectively summarizes the whole programme: *La Loi naturelle, ou Catéchisme du citoyen français, ou Principes physiques et de la morale déduits de l'organisation de l'homme et de l'univers*. The argument is that self-preservation logically requires cooperation, so cultivating the natural faculty of sympathy is necessary for us, and leads to much the same kinds of natural rights advocated by Condorcet. Liberty, equality, and property were inherent in the natural attributes of sensible man. These are the kind of projects that Destutt de Tracy praises for the direct connections that they are able to establish, and this is the programme he recommends and pursues, while making it clear that the path of 'physiological ideology' is the only truly secure route.

'Physiological ideology' is a development of the work of the *médecins philosophes*, those physicians who explored the physiological basis of sensibility and used this to expand the remit of medicine into the social, political, and moral realms.[8] What the *médecins philosophes* offered was a form of naturalization, in that questions that would earlier have been dealt with in theological or metaphysical terms now come to be thought through in ways that appeal to empirical evidence. In order to do this, they were removed from theology and metaphysics and anchored in another field, medicine, because this had by far the strongest claim to be the most appropriate discipline to deal with sensibility in empirical terms.

Earlier forms of medical naturalization, which began in a systematic way with Galen, were concerned above all with the relation between the spiritual and the corporeal, and the extent to which the capacities deemed specific to the spiritual realm should in fact be included within the corporeal. In the sixteenth century—in the thinking of the philosophically inclined Aristotelians of the medical faculties of the North Italian universities—and in the seventeenth century (for example in the Cartesian biomechanics that came to dominate physiological thinking), what was at issue was the relation between differently motivated

[5] 'Fragment de l'histoire de la Xe époque', *Œuvres de Condorcet*, vi. 542–9. The origins of the idea that selfish needs must have been transformed in the transition from a simple world into one with a significant degree of interdependency can be traced back to Samuel Pufendorf, *De Jure Naturae et Gentium* (London, 1672), which provided the first source for Enlightenment anthropology: see Robert Wokler, 'Anthropology and Conjectural History in the Enlightenment', in C. Fox, R. Porter, and Robert Wokler, eds, *Inventing Human Science: Eighteenth-Century Domains* (Berkeley, 1995), 31–52: 37–8.

[6] 'Propositions morales', *Œuvres de Condorcet*, iv. 439.

[7] 'De l'Influence de la Révolution', *Œuvres de Condorcet*, viii. 5–6.

[8] In the rare cases where they were not actually qualified physicians, it seems to have been assumed that they were. This was the case with Diderot, for example: see John H. Zammito, '*Médecin-philosophe*: *Persona* for Radical Enlightenment', *Intellectual History Review* 18 (2008), 427–40.

accounts of cognitive functions. The North Italian Aristotelians, most notably Pomponazzi, stressed the irreconcilability of two understandings of mental functions. The first was that offered by the Aristotelian natural philosophy that underlay the understanding of the soul or mind, which, at least in its most developed version, was a functionalist understanding: the soul/mind was the organizing principle of the body. On this account, the soul/mind was not something that had any identity, or perhaps even existence, apart from the body. The second understanding was one in which the soul/mind continued to exist with its own identity after the death and corruption of the body. This was the understanding underpinning the Christian doctrine of personal immortality: a core doctrine, of equal standing in Christian theology with those of the Trinity and the Incarnation. What worried thinkers such as Pomponazzi was that these accounts seemed fundamentally irreconcilable, even though Aristotelian natural philosophy had been treated since the thirteenth century as the natural partner for Christian theology.

Cartesian biomechanics, by contrast, had reconstructed animal physiology in a fundamental respect, treating it as comprising nothing but mechanically characterizable processes. Although Descartes set out to explain as much as possible in these terms—and was ingenious in showing how various processes that one might consider resisted such forms of explanation could be accommodated[9]—the minimal explanatory resources deployed by mechanism meant that, unlike the Aristotelian naturalist accounts of the soul/mind of Pomponazzi and others, biomechanics simply did not even aspire to account for the mind/soul. Descartes, at least when he is arguing at a general metaphysical level, treats the realms of the spiritual and the corporeal as wholly separate. But at the level of areas of overlap—how acts of will can affect corporeal bodies, how corporeal bodies can affect the mind, or how a disembodied soul could have memories, sensations, or even an identity in its own right, as well as on the question of the 'passions of the soul'—the sharp separation caused immense difficulties. For both Cartesians and those who reacted critically to Descartes, these difficulties were treated as being primarily metaphysical ones. Nevertheless, the 'mental' included not just intellectual faculties and the will, which is what metaphysical accounts largely confined themselves to, but also a range of other phenomena, not least those included under the rubric of sensibility. In *Les Passions de l'âme* (1649), Descartes had provided an account of affective states generally, not in metaphysical terms but rather in terms of a mechanist physiology combined with a more traditional understanding of the passions that took its bearings from classical and humanist authors.

Subsequent developments, particularly those originating with the physicians of the medical university of Montpellier[10] in the 1730s, offered a far more integrated

[9] Examples include such apparently goal-directed processes as the development of the foetus and perceptual cognition: see Gaukroger, *The Emergence of a Scientific Culture*, 337–46.

[10] The Montpellier Faculty of Medicine was, uniquely, an autonomous university of medicine, at least until it was downgraded by Revolutionary-era legislation.

account of cognitive and affective states, and proceeded in a rather different fashion. One can think of sixteenth- and seventeenth-century concerns in terms of a two-stage process, in which one first accounts for the natural realm, and then proceeds to explore our place in this realm; or accounts for the natural realm and then proceeds to try to reconcile this with an independent account (supplied for example by Christian teaching) of our place in this realm. But anthropological medicine does not follow this sequence: it takes the question of our relation to the natural realm as something on a par with our attempt to understand this realm.[11] The two must be taken as part of a whole. Far from being autonomous forms of enquiry, they must form an integrated whole, deriving from the same basic principles.[12]

Because of the way in which it expanded the range and responsibilities of medicine, anthropological medicine postulated an intimate connection between maladies of the body and maladies of the soul. It sought a general account of the interconnections between the realms of the physical, the mental/intellectual, and the emotions. Anything external to the body that affected these three realms was included, for it had an effect on the well-being of the person. These concerns should be contextualized in terms of what one commentator has identified as 'four nodes of reference' that characterize anthropological medicine: holism, an insistence on an intimate physical-moral relation, the construction of medicine as a necessarily social enterprise, and a disposition to typologize human beings according to categories discernible only to medical observers.[13]

MATTER AND SENSITIVITY

The idea that medicine might take over ambitions traditionally fostered in metaphysics and theology—that Hippocrates might replace Socrates, and perhaps even Jesus, as the model for understanding human behaviour—was something

[11] On earlier understandings of natural philosophy, and the point of pursuing it, which share a sense of a connection between development of the self and an understanding of nature, see Sorana Corneanu, *Regimens of the Mind: Boyle, Locke, and the Early Modern* Cultura Animi *Tradition* (Chicago, 2012).

[12] Christian theology from the Church Fathers onwards had taken our relation to the natural realm as something at least on a par with our attempt to understand this realm. It is consequently perhaps not so surprising that this was also the view of some Catholic thinkers in the nineteenth century, when there emerged a form of Christianized anthropological medicine in which the unifying factor was a 'common end' deriving from God. But this was quite a radical move, both for Catholic orthodoxy, which held that medicine was confined to the body while Christian teaching dealt with the soul, as much as for anthropological medicine, which operated with the traditional secularized conception of medicine. See particularly Philippe Buchez, *Essai d'un traité complet de philosophie du point de vue du catholicisme et du progrès* (3 vols, Paris, 1838–40). On Buchez see François-André Isambert, *Politique, religion et science de l'homme chez Philippe Buchez (1796–1865)* (Paris, 1967).

[13] Elizabeth A. Williams, *The Physical and the Moral: Anthropology, Physiology, and Philosophical Medicine in France, 1750–1850* (Cambridge, 1994), 22–3.

novel and unprecedented. There were, however, two earlier developments that enabled it to rise to prominence in the late eighteenth century. The first was the widely perceived failure of mechanical conceptions of nature to account for the properties of matter, and the subsequent shift to the life sciences as an area in which investigation of matter might proceed more fruitfully. This was a view shared by proponents of a philosophical anthropology and the natural history of man, which we shall be looking at in Chapters 4 and 5, but it also turned out to be particularly important in establishing an expanded role for medicine. The second development was something more specific to anthropological medicine. It was the increasing awareness of the place of sensibility in our emotional and cognitive lives, and the belief that medicine might be better placed than philosophy or more traditional forms of physiology (such as biomechanics) to deal with questions of sensibility. As the Montpellier vitalist Théophile de Bordeu put it, sensibility 'is most suitable as a basis on which to explain all the phenomena of life, whether in a state of health or of sickness'. As a result, 'this is the way of considering the living body that has been adopted by those who, among modern thinkers, have pursued their speculations beyond practical medicine and the received systems of the schools at the beginning of the century. Such is the scope that philosophical medicine has assumed concerning the purely material functions of the body.'[14]

We looked at the question of the failure of mechanical conceptions of nature in Chapter 2, and we saw that the discovery of the reproductive behaviour of the hydra, for example, prompted many to conclude that matter must contain vital powers. Although some medical theorists continued to be committed to elements of iatromechanics, this was exceptional and more a pragmatic continuation of a physiological tradition. Above all, it was certainly not accompanied by a defence of the kind of matter theory that underlay iatromechanism. It is worth noting, in this respect, that even materialists such as La Mettrie had no interest in reducing vital and mental phenomena to micro-corpuscular interactions. Like their contemporaries, the materialists conceived of matter as intrinsically active, and their reductive programme was not a reduction to a physics of inert matter, but more like a reduction to forms of activity described by medicine, by chemistry, and by the newly emerging physical disciplines such as electricity that were premised on the idea that activity was an inherent feature of matter.

In short, if one wanted an account of the full range of properties of matter, then the investigation of matter in its impoverished inert form, of the kind that mechanics postulated, was not the way to proceed. Rather, one needed to look at it in a richer form, as in chemistry and the study of electricity, and in the richest form of all, the living matter described in the life sciences. Since matter theory had been the core of how 'natural philosophy' was understood from antiquity onwards, the displacement of the understanding of what matter was and how it

[14] Théophile de Bordeu, *Œuvres complètes* ed. A. Richerand (2 vols, Paris, 1818), ii. 668.

behaved from the quantitative physical disciplines to the life sciences was momentous. And while it was by no means universally accepted that the life sciences held the key, it was generally recognized that they harboured problems which matter theory had to face. Consequently the life sciences came to occupy centre stage in thinking about natural processes, from around the middle of the eighteenth century. Given the close connection between the life sciences and medicine, the latter gained a theoretical significance from the newly elevated theoretical standing of the former.

On the mechanist conception, such phenomena as nervous sensitivity had been reduced to biomechanics: basically, inert matter moving under pressure, with health being conceived in terms of the unobstructed flow of life-sustaining fluids. This kind of account bore no relation to how physicians were coming to think of sensitivity and sensibility more generally. In particular, a number of crucial developments had taken place in the wake of Haller's path-breaking work on irritability, developments which located sensitivity within matter itself, and brought sophisticated observational and experimental resources to bear on the phenomenon.[15] Haller was explicitly concerned to distinguish the living from non-living, and to subject the former to its own kind of empirical investigation: physiology.[16] His initial interests were in involuntary and semivoluntary animal motions, such as the beating of the heart and circulation of the blood, and he established that the cerebellum was not, as had been thought, the primary regulatory mechanism for heart activity and respiration. Rather, regulatory mechanisms were decentralized, and he concerned himself specifically with local regulation, and with specific functions of muscle and nerve fibres, correlating defined functions with particular structures. In seeking the cause of cardiac activity in the structure of the heart,[17] he developed an account which located this cause in muscle irritability, that is, contractibility of muscle fibres. Every animal muscle fibre, he showed, contracts upon stimulation,[18] and there is a scale of degrees of irritability, depending on how strong the stimulus needs to be to provoke a response. Irritability is a completely different phenomenon from sensibility/sensitivity on Haller's account.[19] The latter is a property of tissues

[15] In what follows, I summarize the discussion in Gaukroger, *The Collapse of Mechanism and the Rise of Sensibility*, 394–402.

[16] Note, however, that the term itself was not widely used until the end of the century.

[17] This work was presented in his comprehensive notes on Boerhaave's lectures: Albrecht von Haller, *Hermanni Boerhaave Praelectiones academicae in proprias institutiones rei medicae edidit*... (7 vols, Göttingen, 1739–44), ii. 13–100.

[18] One problem that was to concern later naturalists was in what respects irritability in animal muscles was different from the behaviour of 'sensitive' plants, particularly *Mimosa pudica*, which respond to touch by contracting. See, for example, Jean-Baptiste-Pierre-Antoine Lamarck, *Mémoires de physique et d'histoire naturelle, établis sur des bases de raisonnement indépendantes de toute théorie* (Paris, 1797), 288; and Jean-Claude Delamétherie, 'Discours préliminaire', *Journal de physique* 48 (1798), 3–134: 38–9.

[19] Albrecht von Haller, *A Dissertation on the Sensible and Irritable Parts of Animals* (London, 1755), 25.

imbued with nerves, whereas the nerves play no role in irritability. Moreover, the typical reaction is different: irritability manifests itself in contraction caused by a stimulus, whereas in the case of sensibility, the stimulus causes pain.[20]

Haller's firm distinction between sensibility and irritability was rejected by those who believed that irritability was just a form of sensibility, as well as by those who argued that sensibility was just a form of irritability. The staunchest advocate of the universal prevalence of sensibility was Whytt, who, in his *Essay on the Vital and other Involuntary Motions of Animals*, offered a general theory of sensibility, arguing that the capacity of the muscles to contract depended on the nerves, and that sensibility was widespread throughout the body. Whytt's rejection of irritability resulted from his refusal to accept that there can be genuinely local powers in organs, on the grounds that to envisage such powers would be to ascribe intrinsic active powers—sentience—to matter, but matter is incapable of any kind of sensation, so local action of the kind envisaged by Haller is impossible.[21] Haller responds by presenting experimental findings to show that a number of parts of the body that exhibited irritability were insensitive, in that animals showed no pain when these parts were stimulated, and there was no evidence of nerves.[22] In response, Whytt disputed the lack of sensitivity in some of the parts listed by Haller. What was basically at stake for Whytt was the belief that the power of contraction could only be awakened by the sentient activity of the soul in the nerves, whereas for Haller the nerves are not involved. Sensitivity and irritability are marked out from one another on Haller's view in that, for the case to be one of sensibility, the stimulation results in a conscious sensation in the mind, whereas irritability results in no conscious sensation. Whytt, by contrast, invoked unconscious sensations as well as conscious ones, the former playing an important role in his rejection of Stahl's view that the sentient cause of life must be a consciously reasoning soul.[23] But there are cases of responses to stimulation which cannot be characterized as unconscious sensations. As Haller points out, one can detect irritability in amputated arms and surgically removed intestines, where, although nerves are present, there can be no question of their being attached to a centralized agency such as the soul.[24]

Whereas Haller opted for sensibility and irritability, and Whytt for sensibility alone, La Mettrie opted for irritability alone. In his *L'Homme-machine* (1747), dedicated to Haller, La Mettrie dispenses with the notions of soul and substantial forms: all that is required to account for the internal principle by which living matter is regulated is muscular irritability. The combination of muscular irritability and reflex is, La Mettrie believes, all one needs for a self-moving organism. In

[20] Ibid., 8–9.

[21] Robert Whytt, *An Essay on the Vital and other Involuntary Motions of Animals* (Edinburgh, 1751), 241–2.

[22] Albrecht von Haller, *De partibus corporis humani sensibilibus et irritabilibus* (Göttingen, 1752).

[23] Whytt, *An Essay*, 268.

[24] Albrecht von Haller, *Ad Roberti Whyttii nuperum scriptum Apologia* (Göttingen, 1764), 20.

his *L'Homme plante* (1748), he draws attention to a number of analogies between plants and humans, from matters of anatomical detail to reactions to climate. There is what he refers to as an 'imperceptibly graduated ladder' between the two,[25] and a progression in the faculties of feeling and movement as we ascend from the one to the other, these being provided by nature to the extent required in each species.[26]

La Mettrie's materialism met with widespread rejection, and against La Mettrie, Haller argued that, among other things, the fact that irritability in organs persists after death shows that it is independent of the nerves and its action cannot be assimilated to that of the soul.[27] Haller's own view was a traditional one, namely that the powers that organs have which are manifested in irritability derive directly from God, and require no intermediary soul. But Haller's approach does nevertheless have radical consequences. Organs manifest something that can be characterized as a life force in Haller's view, by contrast both with a view of matter as inert, and that of being acted upon from outside.[28] The local, intrinsic regulation of the operations of organs prompts a rethinking of the idea of an organism. Centralized, extrinsic regulation of organs, of the kind advocated by Whytt, provided a source of unification for the organism whose organs these are. Local, intrinsic regulation of the kind advocated by Haller, where the macroscopic hydraulic vessels of the iatromechanists are replaced by microscopic muscle fibres, provokes the question of what the unity of the organism consists in. If all there were to the organism were localized centres of irritability, then there could be no unity. But it is not as if a centralized sensibility could simply be added to a localized irritability to provide a unifying principle for the organism, for Haller's point was that organs exhibiting irritability do so independently of whether or not nerves are present, and if they are not present, then such organs can hardly be connected through sensibility, which requires nerves.

In the light of this kind of difficulty, and in the light of the fact that he provides no details about the nature of sensibility, it is not surprising that Haller's sharp distinction between sensibility and irritability was gradually blurred and then effectively abandoned by those who saw themselves as continuing his work.[29] Whereas Haller had confined his attention to fibres, Bonnet went beyond this, extending the account of 'sensible fibres' to mental acts, arguing that they too depend on the arrangement of fibres. In doing this, he not only made the action of

[25] Julien Offray de La Mettrie, *Oeuvres philosophiques* (3 vols, Berlin, 1796), ii. 69.
[26] Ibid., 66–8. [27] Haller, *A Dissertation*, 45–6.
[28] A direct connection between irritability and life was subsequently made in a 'treatise on irritability as the principle of life in organized nature': Christoph Girtanner, 'Abhandlungen über die Irritabilität', *Journal der Physik* (1791) 3.2: 317–51, 3.3: 507–37 (originally published in French in 1790).
[29] Bordeu is exceptional here. As Anne Vila points out, unlike most of his contemporaries, he 'does not so much conflate the properties of irritability and sensibility as assign them different roles in the general mechanism of local organic activity': *Enlightenment and Pathology*, 68.

all types of fibres the same, but in his assimilation of the action of brain fibres to those in other parts of the body, he undermined the distinction between irritability and sensibility.

A crucial issue underlying this extension of physiological investigation into the realm of the psychological, and thereby into that of sensation in its fullest sense, is that of the relations between parts and wholes. This had significant consequences not only for how phenomena involving sensibility were examined, but in marking out sensibility from reason and indeed unifying the epistemological question of sensation and physiological question of sensitivity.

With the publication of Condillac's *Essai sur l'origine des connaissances humaines* (1746) and *Traité des sensations* (1754), there emerged a very influential general account of the whole/parts relation which stood in direct contrast to accounts that assumed the primacy of reason in perceptual cognition, typically through the postulation of innate ideas. Condillac's model was developed to bolster a sensationalist account of perceptual cognition, one in which sensation, rather than reason, was the key to understanding how cognition was possible. Whereas reason had always been construed very much in terms of a single source of rules and procedures regulating our cognitive life, in Condillac's account of sensation we find the inverse of this: a thoroughly decentralized account of our cognitive life, as reflected in language. Whereas reason encourages a top-down model in which a grasp of the whole is necessary if we are to establish, deductively, a necessary order in the parts,[30] sensibility on this view works in the opposite direction, starting with the parts and building up from these a picture of the whole.

When Condillac took over a Lockean model of perceptual cognition, he radicalized it, rejecting Locke's theory that sensation and reflection were the origins of our ideas, and arguing instead that sensation was the sole source of ideas.[31] Even more significantly, to support this, he transformed the discussion of the nature of knowledge by arguing that the only way to understand our ideas was to trace their genesis, identifying those primitive ideas with which we start and exploring how other ideas are generated from these. In the penultimate chapter of the *Essai sur l'origine des connaissances humaines*, on 'the method that must be followed in the search for truth', he writes that the key question is to discover how

[30] I say 'encourages', rather than 'dictates', because Kant, for example, who was a staunch defender of 'reason', was criticized by Reinhold—in the 1790s, when he becomes a critic of Kant—for using an 'analytic' method, one that starts from the parts, instead of a 'synthetic' one, which begins with the whole and then deduces the order of the parts through a priori deduction: see Karl Leonhard Reinhold, *Beyträge zur Berichtung bisheriger Missverständnisse der Philosophen* (2 vols, Jena, 1790–4), i. 263–4; and idem, *Ueber das Fundament des philosophischen Wissens* (Jena, 1791), 72–5. We can think of Reinhold's work at this stage as an early stirring of German idealism, in which there is a very sharp line drawn between a top-down holism and everything else. See Paul W. Franks, *All or Nothing: Systematicity, Transcendental Arguments, and Skepticism in German Idealism* (Cambridge Mass., 2005).

[31] Condillac's sensationalism is far less qualified in the *Traité des sensations* than it is eight years later in the *Essai sur l'origine des connaissances humaines*.

ideas are connected with one another, and that to achieve this we must examine the way in which, and the order in which, ideas are generated. He gives the example of the Scholastics and the Cartesians, who 'know neither the origin nor the generation of our knowledge; and this is because the principle of innate ideas and the vague notion of understanding with which they start' means that they are unaware that all knowledge starts with sensation. Locke, Condillac tells us, fared better because he started with the senses. But he was unable to follow up this insight because he did not explore the first stages of the soul's operations. Therefore Condillac himself has set out, he tells us, 'to do what this philosopher forgot to do: I have gone back to the first operation of the soul, and I believe that I have not only given a complete analysis of the understanding, but have also discovered the absolute necessity of signs and the principle of the connection of ideas'.[32]

The concern with origins pervades Condillac's work in two respects.[33] First, Condillac's contemporaries devoted much attention to tracing the origins of language, society, and the arts and sciences, and there was a widespread assumption that the original state could be identified with something authentic and truthful. Condillac's own sense of origins mirrors this assessment: one can only reach true understanding by going back to origins. Second, this return to origins is manifested in a variety of ways in Condillac's writings. In the *Traité des sensations*, he starts with origins and draws out the implications of the sensory origins of ideas. In the *Traité des sistêmes* (1749), he sets out a detailed account of the need to start from origins. In the *Essai sur l'origine des connaissances humaines*, he works his way back to origins.

The tool that Condillac uses in his project of tracing origins is what he terms 'analysis', but he distinguishes this from the kind of procedure employed in logic or geometry, which he construes as demonstrating that the conclusion is contained in the premisses. On his understanding of analysis, it is not a deductive procedure of this kind, but rather one by which we

> compose and decompose our ideas in order to make different comparisons of them, and in this way to discover what the relations between them are, and what new ideas they can produce. This analysis is the true secret of discovery, for it always takes us back to the origin of things. It has the advantage of only ever offering a few ideas at the same time, and always in the simplest ordering. It is the enemy of vague principles, and of everything that can be contrary to exactness and precision. It does not seek the truth by means of general propositions, but always by a kind of calculation, that is, by composing and decomposing ideas in order to compare them in the way most conducive to the discoveries that one

[32] *Œuvres de Condillac, revues, corrigées par l'auteur* (23 vols, Paris, 1798), i. 502–3. As we shall see in Chapter 4, the importance of signs for Condillac lies in mediating the relation between ideas and things.

[33] See John C. O'Neal, *The Authority of Experience: Sensationalist Theory in the French Enlightenment* (University Park, Penn., 1996), 24–7.

seeks. It is no longer a matter of definitions, which usually only serve to multiply disputes, but rather one of explaining the generation of each idea.[34]

Condillac's concern here is with sensation. Nevertheless, as I have indicated, sensation is drawn under the general notion of sensibility for French writers of this era, very explicitly in the accounts of *sensibilité* in the *Encyclopédie* for example. Indeed, one can detect a general shift in discussion of sensation to discussion of sensibility in the 1750s and 1760s. A striking example is Antoine Le Camus' very popular *La Médicine de l'esprit*, first published in 1753, and followed by an expanded second edition in 1769. Whereas the term 'sensation' appears regularly in the first edition, in the second edition 'sensation' has disappeared and the discussion now proceeds in terms of 'sensibilité'. It is largely a question of substituting the one term for the other, although sensibility, while including sensation, has a wider remit, being responsible for all human activities from the development of the foetus to cognition.

In the physiological writings of Bonnet, we can see how Condillac's account of sensation is extended to another form of sensibility, namely sensitivity, thereby providing something more like a general bottom-up 'analytic' model for sensibility which marks it out from the top-down 'synthetic' model for reason. Like Condillac, Bonnet is committed to analysis as the only method of discovery, telling us that 'we must anatomize each fact, decompose it into its smallest parts, and separately examine all these parts. We must look for the relations linking these things to one another and to analogous things, and find results that can become principles'.[35] Moreover, just after Condillac had published his account of the 'human statue', Bonnet, unaware of Condillac's work, was independently developing the idea of a statue which we transform into a human being by progressively endowing it with sensory faculties.[36] His account is primarily concerned with smell, however, investigating the shift from experience of a single sensation to the comparison of two successive sensations. The aim is to explore the physiological and psychological dimension of sensory experience. His focus, he tells us in the *Essai analytique*, is not on the soul, about which we know little, but on its physical manifestation in the body:

Man is a mixed being: he has ideas only through the intervention of the senses, and even his most abstract notions derive from the senses. It is on his body and through his body that the soul acts. It is therefore always necessary to return to the physical, as the original

[34] *Œuvres de Condillac*, i. 109–10.

[35] Charles Bonnet, *Essai analytique sur les facultés de l'ame* (Copenhagen, 1760), iv.

[36] It is impossible to say whether Bonnet's interest in these questions may have been partially motivated personally. He had to be removed from school during his adolescence because of hearing loss, and, having been an outstanding microscopist in his early twenties, he began to suffer loss of eyesight from the age of twenty-three. Although the condition later stabilized, he was unable to continue his microscopy studies and he began to move to more philosophical questions, which he had earlier distained. On Bonnet's chequered career, see Lorin Anderson, *Charles Bonnet and the Order of the Known* (Dordrecht, 1982).

source of everything that the soul experiences. We no more know what an idea in the soul is than we know what the soul itself is. But we know that ideas are tied to the action of certain fibres. Thus we are able to reason about these fibres by means of what we see of the fibres. We can devote a little study their movements, the resultants of their movements, and the relations that they enter into.[37]

Of some importance in Bonnet's account is his view of what distinguishes the soul and the body. In the earlier *Essai de psychologie* (1755), he claims ignorance about the nature of the soul, and there is no discussion of it.[38] But in the *Essai analytique*, he adopts a view that goes back to Plato and Aristotle and permeates the philosophical tradition of thinking about the nature of the soul, namely that the soul is simple and indivisible whereas matter is divisible. 'It is to satisfy this sense of the *self*, always *one*, always *simple*, always *indivisible*, that we have recourse to the existence of this *immaterial* substance that we call the *soul*'.[39] These properties define the soul for Bonnet. The only reason he attributes 'a soul to man', he tells us, is because matter alone is insufficient to account for 'all the phenomena of man'.[40] If we take the central contrast here to be that between indivisible soul and divisible matter, we can identify distinct forms of investigation or explanation, namely a top-down account in terms of reason and a bottom-up 'analysis', which starts from autonomous parts and builds up an account of the relations between these parts.

The physiology of sensation was taken further in the mid-eighteenth century by the Montpellier physicians, who expanded it into a medicine of sensibility. In the process, sensation and sensitivity come to be joined up more intimately with the question of moral sensibility, as one's mental life comes to be comprehensively medicalized.[41] We can separate out three issues in this thinking through of questions of sensibility. The first is that of the priority of the parts with respect to the whole. In his *Encyclopédie* entry on 'observation', the Montpellier-trained physician Jean Jacques Ménuret de Chambaud[42] provides an account of one of the core ideas behind this conception: and that he finds this core idea in medicine—'observation' is classified as a medical entry—is not unexpected. Observation, he writes,

[37] Bonnet, *Essai analytique*, xii.

[38] Charles Bonnet, *Essai de psychologie; ou considerations sur les operations de l'ame, sur l'habitude et sur l'education* (London, 1755), 1–5. On the changes in his account of the soul between the *Essai de psychologie* and the *Essai analytique*, see O'Neal, *The Authority of Experience*, 61–6.

[39] Bonnet, *Essai analytique*, 297. [40] Ibid., xxiv.

[41] As we shall see in Chapter 4, the connection between sensation and aesthetic sensibility was being drawn at the same time by Baumgarten. In this way, the four forms of sensibility treated in the *Encyclopédie* under the same rubric—sensation, sensitivity, moral sensibility, and aesthetic sensibility—come together.

[42] Ménuret played an important part in promoting the doctrines of the Montpellier school. See Roselyne Rey, *Naissance et développement du vitalisme en France de la deuxième moitié du 18ᵉ siècle à la fin du Premier Empire* (Oxford, 2000), chs 3 and 6.

is the primary foundation of all the sciences, the most reliable way of reaching one's goal, the principal means of extending the boundaries of scientific knowledge and of illuminating all its points. The facts, whatever they are, constitute the true wealth of the philosopher and the subject of observation: the historian collects them, the theoretical physicist combines them, and the experimenter verifies the results of their synthesis. Several facts taken separately appear dry, sterile, and unfruitful. The moment we compare them, they acquire a certain power, assume a vitality that everywhere results from the mutual harmony, from the reciprocal support, and from a chain that binds them together.[43]

The commitment to building up from parts goes beyond methodological questions, however. On the kind of account offered by the leading Montpellier figure, Bordeu, the whole is as it is because the parts are as they are: in particular, the organism is a living thing because the fibres and organs that make it up are living. The question of the relation between the whole and the parts is now central to the understanding of the nature of life.

The second issue is the anatomical question of what organs one can explore in terms of irritability and sensibility, and what one can learn from this. Once anatomical attention was no longer confined to muscles, the autonomous nature of the behaviour of other organs, particularly the way in which they react to stimulation, becomes evident, and the idea of life and sensibility consisting in the central organization of otherwise inert matter now becomes questionable on empirical grounds. A good deal here depended on what kinds of organ one investigated with respect to irritability. Haller himself had at one stage thought of exploring irritability in the case of 'arteries, glands, secretions, and fevers',[44] but decided to confine himself to muscles, partially at least on the grounds that such further investigation would be beyond the bounds of anatomical investigation.[45] Bordeu's early work, by contrast, focused precisely on the action of the glands, a subject which enabled him to question iatromechanical accounts whereby glandular activity was the result of compression of the glands by surrounding bones and muscles. The biomechanical resort to pressure, he argued, made no anatomical sense. As he shows in his *Recherches anatomiques sur la position des glands et leur action* (1752), minute anatomical observation shows that the excretion of liquids from glands where they have been stored cannot be the result of pressure, because the space around the gland cannot be altered in size: they are lodged in such a way that no muscles can exert any pressure on them. A water-soaked sponge inserted into the parotid gland cavity, for example, cannot be made to yield any water no matter how strenuous the movement of the jaw. The same absence of susceptibility to pressure can also be found in all the other glands, and he records his

43 'Observation', *Encyclopédie*, xxiii. 294.

44 Albrecht von Haller, *Mémoires sur la nature sensible et irritable des parties du corps animal* (4 vols, Lausanne, 1756–60), iv. 34–5.

45 Ibid., 96.

anatomical investigations of the lachrymal glands, the pituitary, the thymus, the thyroid, the pancreas, and the suprarenals. It was not just mechanism that was at issue however. His anatomical investigation of glands also enabled him to question elements of the anti-mechanist Stahlian doctrines that had, since the 1730s, been crucial in shaping a distinctive identity for the Montpellier physicians.[46] Stahl had insisted on a sharp distinction between the living and the non-living, where, in the case of the former, all bodily functions were characterized by the direct intervention of the soul. Bordeu rejected the doctrine of direct intervention, and the study of the action of the glands proved a testing ground for combining anatomy and medicine to establish the autonomy of glandular action.

The third issue is that of the nature of life. The theory that what is 'living' about an organism is its parts, rather than the whole, was a radical one. But as I have indicated, reflection on the reproductive behaviour of the hydra suggested that life was a property of matter, or at least organic matter—something not only smaller than, but far more primitive than, organs. As might be expected, there was considerable dispute and uncertainty about just what the minimal living entities—often referred to as 'molecules'—were, but it was now evident that organs at least were living in their own right. This contradicted not only Cartesian biomechanics but also Aristotelian theories whereby life was due to a form suffused throughout the material body.[47]

The questions of sensibility, life, and the ultimate units of the living clearly become intimately connected here, and it is in this context that we should understand Bordeu's account of the living. Bordeu attributes sensibility to all organs, where this sensibility has now become localized: each organ leads a life of its own,[48] and the lives of organs contribute to, indeed constitute, the collective life of the organism. 'Life is only feeling and movement',[49] he claims. It consists of the 'ability of the animal fibre to feel and to move itself', something that is 'inherent in the primary elements of the living body such as gravity, attraction, and the mobility of various bodies'.[50] In the context of physiology what is at issue here is irritability and sensibility. The structure of fibres making up the organs of the body is the core issue for Bordeu. These he regards as extensions of the nerves, embedded in a spongy, mucous, cellular substance,[51] which serves both to provide nourishment for the fibres that it encases, and, as it stretches from organ to organ, to connect these different organs.[52] This mucous cellular substance

[46] See François Duchesneau, *La physiologie des lumières: Empirisme, modèles et théories* (The Hague, 1982).

[47] See, for example, Dennis Des Chene, 'Life and Health in Cartesian Natural Philosophy', in S. Gaukroger, J. Schuster, and J. Sutton, eds, *Descartes' Natural Philosophy* (London, 2000), 723–35.

[48] Théophile de Bordeu, *Recherches sur les maladies chroniques* (Paris, 1775); in Bordeu, *Œuvres*, ii. 829.

[49] Bordeu, *Œuvres*, ii. 831. [50] Ibid., ii. 924. [51] Ibid., ii. 829.

[52] Théophile de Bordeu, *Recherches sur le tissu muqueux* (Paris, 1767); in Bordeu, *Œuvres*, ii. 738–9.

orders the agitations of the fibres, making them act harmoniously and producing the visible functions of the body. In general terms, one might say that it is because the parts are living that the whole is living, and it is because the living parts are connected in the way they are that the whole is the way it is.

The distributive account of life and sensibility proposed here is one in which these qualities exist at a level that is smaller than that of the organism, and in which the smaller organs (or, in the limiting case, simply pieces of organic matter) are autonomous,[53] in that they have these qualities in their own right, not to the extent to which they play some functional role in a hierarchical organization dictated by the needs of the organism. In particular, note that sensibility is a *sine qua non* not only of the autonomy of the organs, but more importantly of the unifying connections that they are able to form.

The distributed nature of sensibility, and hence life, is crucial here, because a healthy body is one in which the autonomous parts are in harmony with one another:

In order that we might grasp the particular action of each of its parts better, we compare the living body to a swarm of bees, which gather together in a cluster and hang from a tree like a bunch of grapes. We do not find it amiss that one ancient author has said of one of the viscera of the abdomen that it was an '*animal in animali*': each part is, so to say, not an animal, certainly, but a kind of separate machine, which contributes, each in its own way, *to the general life* of the body. To continue with the comparison with the swarm of bees, it is a *whole* glued to the branch of a tree by the action of a good many of the bees which need to act in unison in order to hold on; there are those that attach themselves first, and those that attach themselves to these; they combine to form a very solid body yet each has its own separate action; were one to give way or act too vigorously the whole mass would be disturbed; when they conspire to huddle together and to clasp one another in the required order and proportions, they compose a whole which persists until it is disturbed. Applying [this metaphor] is straightforward: the bodily organs are joined together; they each have their own area and action. The relation between these actions, and the harmony that results, is what constitutes well-being. If this harmony is disturbed, whether it be by some part relaxing or because it dislodges an antagonist, if the actions are upset, if they no longer follow the natural order, these changes will constitute more or less serious disorders.[54]

In his entry on 'observation' in the *Encyclopédie*, Ménuret spells out the nature of the connections even more explicitly:

Following these authors, one could compare man to a flock of cranes, which fly together in a particular order, without any mutual assistance or dependence on one another. The physicians or philosophers who have studied and carefully observed man have noticed this sympathy in all animal motions: this constant and necessary agreement in the interaction of the various parts, however disparate or distant from one another. They have also noticed

[53] Bordeu identifed three centres of sensitivity: one primary one, the brain, and two subsidiary ones, the heart and the epigastric region: *Recherches anatomiques sur la position des glands et leur action* (Paris, 1752); in Bordeu, *Œuvres*, i. 193n.

[54] Bordeu, *Œuvres*, i. 187.

the disturbance that results in the whole from the sensible disagreement of a single part. A famous physician (Bordeu) and an illustrious physicist (Maupertuis) likewise compared man, from an illuminating and philosophical point of view, to a swarm of bees that strive together to hang to a branch of a tree. One can see them pressing together and holding one another in place, forming a kind of whole in which each living part contributes in its way, by the correspondence between and direction of its movements, to sustain this kind of life of the whole body, if we may refer in this way to a mere combination of actions.[55]

Ménuret identifies the key innovation in these developments as having been made by Louis de La Caze (Bordeu's cousin and patron of sorts), who developed Bordeu's ideas in a distinctive way, albeit not one that Bordeu himself always acknowledged. Following up the idea of origins, so prominent—as we have seen—in the understanding of 'analysis' as defined by Condillac and Bonnet, La Caze seeks what Ménuret calls 'the primitive function that preceded all others and gave rise to them'. What La Caze does is to trace sensibility back to its origins, through the embryo to a seminal fluid which he claims can be found in both male and female.[56] Rejecting Haller's view that the heart is the first thing to develop in the embryo, he maintains that the nerves must be the first things to form since they are the primary bearers of sensibility.[57] Although these claims are in fact wholly speculative, they result from an attempt to remodel medicine along methodological lines with a view to increasing its scientific standing. What is distinctive about La Caze's approach is the combination of the methodological commitment to analysis and the concomitant search for origins, with an attempt to flesh this out as a return to Hippocrates. Like many of his contemporaries, he wanted to pit Hippocratism against some grand medical theory, but he rejected the idea that medicine was just a collection of successful therapeutic techniques. La Caze saw medicine as a science of organized living beings whose basic principles were sensibility and movement, and the method by which this science proceeds is by breaking phenomena down, and careful observation and recording, in order to establish relations of cause and effect, something that other advocates of the Hippocratic method had neglected.[58]

A distributive view of life provided a popular model of bottom-up unification for other areas in the 1760s and 1770s. In his *Rêve de d'Alembert*, written in 1769, and in the *Éléments de physiologie*, published in 1778, Diderot used the model of the unity of the organism to show how there can be a unity of the subject of thought without postulating an immaterial soul.[59] And Rousseau moved

[55] *Encyclopédie*, xxiii. 301–2. See Charles T. Wolfe and Motoichi Terada, 'The Animal Economy as Object and Program in Montpellier Vitalism', *Science in Context* 21 (2008), 537–80.

[56] Louis de La Caze, *Idée de l'homme physique et moral, pour servir d'introduction à un traité de médecine* (Paris, 1755), 74–5.

[57] Ibid., 100–1: 'primary' bearers because sensibility is possible without nerve endings.

[58] Ibid., 66.

[59] For the discussion in the *Rêve de d'Alembert*, see Diderot, *Œuvres complètes*, ii. 126–30. See also Timo Kaitaro, *Diderot's Holism* (Frankfurt am Main, 1997), ch. 3; and François Duchesneau, 'Diderot et la physiologie de la sensibilité', *Dix-Huitième Siècle* 31 (1999), 195–216.

effortlessly from the biological version of the question to a political one, asking in *Du Contrat Social* how an aggregate of individual wills can be transformed into a collective will, and identifying the discovery of a form of government that achieves this as the fundamental problem to which the social contract is a solution.[60] Nevertheless, it was not to remain a central doctrine of Montpellier physicians. With the publication of Barthez's *Nouveaux éléments de la science de l'homme* in 1778,[61] it was replaced by the idea of a single principle of life, which he termed the 'vital principle' and which was to become the defining feature of Montpellier medical theory from that point onwards. As the title of his book indicates, his primary concern is with human beings, and here a distributive view of life had always been most controversial, if only because questions about identity arise—by contrast with the case of plants for example—that are especially pressing if one makes the kind of naturalist assumptions about the mind and affective states that Montpellier physicians do: that is, if one cannot simply displace such questions into the realm of an immaterial soul. Barthez's 'vital principle' looked to his critics to be something as mysterious and metaphysical as the 'soul', but in response, he tells us that he

> has never employed the term 'vital principle' to *explain* any of the phenomena of life, but simply in order to make evident and certain the formulation of new findings regarding these phenomena, which I attempt to combine in a way that is simpler, more general, and consequently more useful than anyone before me has done.[62]

The idea of making things evident by combining elements here is clearly designed to indicate a commitment to 'analysis', but there is an added methodological ingredient. Barthez takes up a form of empiricist explanation rationalized by Locke and generalized by Hume, which insists on the completeness and fruitfulness of purely phenomenal descriptions, eschewing the idea that explanation only comes to an end when one has demonstrated the phenomena in terms of underlying causes. The model is gravitation, and Barthez makes it clear that his vital principle accounts for all the known facts even though its nature is unknown.[63]

MÉDECINS PHILOSOPHES

What is at stake here is not just the question of deciding whether understanding a living thing is a matter of understanding its living parts and establishing how they

[60] Jean-Jacques Rousseau, *Du Contrat Social, ou Principes du Droit Politique* (Amsterdam [i.e. Germany], 1762), book i, chs 6 and 7; idem, *Oeuvres Complètes* (4 vols, Paris, 1883), i. 644–6.

[61] Paul-Joseph Barthez, *Nouveaux éléments de la science de l'homme* (Montpellier, 1778). This subsequently appeared in a much expanded edition with extensive additional notes, which are separately paginated: *Nouveaux éléments de la science de l'homme, seconde édition* (2 vols, Paris, 1806). This is the edition that I shall cite. Barthez had published two earlier works on the same topic: *Oratio academica de principio vitali hominis* (Montpellier, 1773); and *Nova doctrina de functionibus naturae humanae* (Montpellier, 1774).

[62] Barthez, *Nouveaux éléments*, i. notes 4.

[63] Ibid., i. 7–8.

are connected, or even of understanding how these parts partake in a shared vital principle. There is also the question of what it is that we have to know in order to comprehend the cognitive and affective states of organisms so constituted. In the case of everything except human beings, there is a clear assumption that there is nothing further that we need to understand to have a complete grasp. But what about human beings? Bonnet, as I have indicated, believed it necessary to introduce a soul in the case of human beings, but the introduction of the soul would seem to add nothing to our understanding of human cognitive and affective states. If we have successfully constructed our fully functioning statue-man, Bonnet asks, what would happen if we then introduce a soul into this automaton? The answer is that all the feelings, and all the faculties, such as memory and imagination, would remain as they were, for these are all a function of the behaviour of the fibres in the brain, which are 'independent of the soul'. Moreover, if we placed the soul of a native American Huron in Montesquieu's brain, the Huron soul would have the same sentiments, perceptions, and abstract ideas as Montesquieu, as well as speaking French and writing *De l'esprit des lois*.[64]

This is a development with radical consequences. As Anne Vila has pointed out, there is a move here from 'sensibility as a power to be observed to a vision of sensibility as a power to be harnessed and redirected', for 'if the development and successful conduct of the intellect could be shown to be determined entirely by the action of fibers in the common sensorium, then the soul could be completely marginalized from discussions of human nature. It was now the body that, through its inherent reactive properties, seemed to offer the most effective site of intervention in the moral, intellectual, and physical constitution of human beings.'[65] What we act upon when we proceed in this way is sensibility, not reason. Bordeu emphasizes this new standing of sensibility:

> The reign of feeling [*sentiment*] or sensibility [*sensibilité*] is among the most extensive; feeling is involved in all the functions; it directs them. It dominates over illness; it guides the action of remedies; it sometimes becomes so dependent upon the soul that the soul's passions take the upper hand over all the changes of the body; it varies and modifies itself differently in almost all the parts.[66]

This is instructive, for in its new expanded domain, sensibility no longer looks like something for which physiology or psychology alone could account. Rather it is medicine—a form of medicine in which control of sensibility holds the key—that now becomes the primary tool of investigation. In his *Encyclopédie* entry on 'prognostic', for example, Ménuret argues that the medical technique of prognosis enables the physician to lift 'the thick veil that obscures future events', and, 'guided by the brilliant torch of repeated and well-judged observation, he sees with a confident eye both objects that exist now and those that will emerge'. The

[64] Bonnet, *Essai analytique*, 370.
[65] Vila, *Enlightenment and Pathology*, 38.
[66] Bordeu, *Œuvres*, ii. 669.

method is so successful that Ménuret wonders whether it should be confined to medicine, or can perhaps be extended, so that 'through a considered examination of moral man, there could be a body of science that concerned itself with the means of knowing and foreseeing the actions of men'.[67]

The principal significance of La Caze's work, and the feature that secured the prominence of his *Idée de l'homme physique et moral* in the decades to come, was, as the title intimates, his explicitness in drawing out the moral and social consequences of this newly conceived medical science. Although he considered himself to be developing Bordeu's work, he went beyond anything that Bordeu had advocated (or in many cases was prepared to accept).[68] His starting point is a revival of the doctrine of 'non-naturals', whose first systematic formulation can be traced back to Galen's *Ars medica*. Factors relating to health are divided there into the naturals, the non-naturals, and the contra-naturals. The naturals were structural and functional elements innate in each body such as the temperaments, humours, parts of the body, faculties, and functions. The non-naturals were those factors that determined the state of the body without being controlled by the natural functioning of the body: ambient air, food and drink, movement or exercise and rest, sleep and waking, excretion and retention, and the passions of the soul. The contra-naturals comprised diseases, and these could result from an internal imbalance in the naturals or from an imbalance between naturals and non-naturals.[69] Health on this account was the result of a proper ordering of the naturals and a proper regimen of the non-naturals, brought under the general notion of 'hygiene'. Since one cannot avoid the effects of the non-naturals, in the Galenic tradition they were included in the curriculum of the medical schools, and physicians were required to learn about them not only to treat diseases but to preserve health and to prevent diseases.

This Galenic conception was revived in eighteenth-century medicine in France. Popular medical manuals of the period, such as the *Conservateur de la santé* (1763) of Achille Le Bègue de Presle, were written directly for a bourgeois reading public, and stressed the perils to health of misuse of the non-naturals, particularly for those with a sedentary lifestyle. The theme is pursued with vigour in the *Encyclopédie*, above all in the articles by the Montpellier-trained physician Arnulfe d'Aumont,[70] whose contributions included the entries on 'non-naturelles,

[67] 'Prognostic', *Encyclopédie*, xxvii. 525. A similar view can be found in Haller's one-time German student, Johann Georg Zimmermann: 'Moral and medical observations require the same spirit of observation . . . . A true physician determines the illnesses of the body by means of immediately and correctly observed signs, just as a true moralist does the dispositions of minds.' *Von der Erfahrung in der Arzneykunst* (2 vols, Zürich, 1763–4), i. 169–70.

[68] See Williams, *The Physical and the Moral*, 41–5.

[69] See Peter H. Niebyl, 'The Non-Naturals', *Bulletin of the History of Medicine* 45 (1971), 486–92; and Chester R. Burns, 'Nonnaturals: A Paradox in the Western Concept of Health', *Journal of Medicine and Philosophy* 1 (1976), 202–11.

[70] D'Aumont was not a vitalist, however, and was associated with the iatromechanist faction at Montpellier, led by Antoine Fizes.

choses', 'santé', and 'régime', and most likely the unsigned entry on 'hygiène'. The non-naturals are identified as preserving health, so long as they are used properly, and they are seen in terms of the healing power of nature, for 'the role of the medical art in the curing of diseases is in fact less than is commonly believed.... In acute diseases the cure is usually the work of nature.... Chronic diseases, especially those that are intractable, are almost always beyond all the cures of medicine'.[71] There is a shift in the direction of thinking about the role of medicine here. As one commentator has pointed out, d'Aumont 'exhibited the fact that the basic medical concern in the popular expression of the doctrine of the non-naturals was care, not cure, and that the physician's primary role had become that of pedagogue and advocate'.[72] It is in such a context that we should consider Diderot's famous remark that he enjoyed reading about medicine and talking to physicians more than anything else, but only when he was feeling well.[73]

One non-natural that was of particular interest to many physicians was the sixth, the passions of the soul.[74] D'Aumont, whose main interest was in the role of ambient air,[75] does not go into detail on the passions of the soul,[76] but other writers took them more seriously. La Caze in particular focused on them. Armed with the broadened notion of sensibility which he had inherited from Bordeu, he set out to establish connections between physiologically characterized sensibility and affective states in terms of a general notion of health as a harmonious 'animal economy'.[77] The crucial thing about this animal economy was that sensations were not just physiological in nature but were pleasurable or painful, with an intensity corresponding to the needs of the animal economy. Our moral habits are determined by this animal economy, in terms of the pleasure or pain with which particular behaviour is associated. But although this means that there is no direct social shaping of behaviour for example, there is nevertheless an indirect influence, because the 'constitution' of the body, and hence its animal economy, are themselves shaped by the 'constitution' of society, which La Caze argues is variable: regionally (Northern versus Southern Europe), in terms of the social distribution of tasks, and in terms of whether the society is in a savage state, where muscular action is at a premium, or civilized, where sensibility may become too

[71] 'Hygiène', *Encyclopédie*, xvii. 919.

[72] William Coleman, 'Health and Hygiene in the *Encyclopédie*: A Medical Doctrine for the Bourgeoisie', *Journal of the History of Medicine and Allied Sciences* 29 (1974), 399–421: 406.

[73] Denis Diderot, *Eléments de physiologie*, ed. Jean Mayer (Paris, 1964), 299.

[74] There is a good discussion in Rey, *Naissance et développement du vitalisme*, ch. 3.

[75] The importance of the role of ambient air, and climate more generally, was crucial to writers in the field of what might be termed comparative geography: see, in particular, John Arbuthnot, *Essay Concerning the Effects of Air on Human Bodies* (London, 1733) and, the most famous argument for the importance of climate, Montesquieu, *De l'esprit des lois* (Geneva, 1748).

[76] 'Non-naturelles', *Encyclopédie*, xxiii. 113.

[77] On the notion of the animal economy in French medical and biological thought at this time, see Rey, *Naissance et développement du vitalisme*, 123–39.

refined.[78] As Elizabeth Williams notes, this line of argument 'culminated in an extension of Bordeu's definition of health as harmony and balance to social practice: moderation was the key to health because any kind of excess disturbed bodily harmony, promoted "disorder", and put the all-important connective tissue into a state of "vicious sensibility"'.[79]

The issues that are at stake here form the core of disputes in the second half of the eighteenth century over the extent to which human faculties and behaviour can and should be shaped by social and medical intervention. Beginning in the 1750s, there was a burgeoning stream of literature advocating the use of medicine in these matters, duly accompanied by a number of warnings about the over-stimulation of sensibility. It is in this context that the *médecins philosophes* thrived, for their interests transcended those of traditional medicine and projected it firmly into the moral or human sciences.

The key early figure in this development was not French, however, but Dutch. It was the *De regimine mentis* (1747) of Jerome Gaub,[80] professor of medicine and chemistry at the University of Leiden, that transformed—and indeed naturalized—the philosophical question of the mind–body relation so that it turned on the relationship between the mind and the body in health and in disease.[81] Gaub argues that there are two sets of circumstances in which the physician should intervene: where a condition of the body causes a mental disturbance, or, where this is not the case, it can nevertheless best be dealt with by treating the body; and where the mind is the cause of the disease, or, where this is not the case, can nevertheless best be dealt with by treating the mind. The comparison of healthy and unhealthy states plays a crucial role in his discussion of mind–body interactions. But so too does the more radical thesis that the 'physician's duty is to care for the whole man; mind and body are abstractions'.[82] The idea that the 'whole man' is prior to the abstractions of mind and body, has some significant consequences for treatment, for the physician is now encouraged to treat the body to deal with maladies of the mind, and to treat the mind by means of agents acting on the body. *De regimine mentis* opens with a characterization of health in terms of harmony between the mind and the body, and a malady is anything that disturbs this harmony. Consequently, it is not just the form of treatment advocated that is radical, but the range of issues with which the physician is now called upon to deal.

In his *La Médecine de l'esprit* (1753), Antoine Le Camus sets out to develop the idea of the medically conceived connection between the mind and the body,

[78] La Caze, *Idée de l'homme*, 363–7. [79] Williams, *The Physical and the Moral*, 45.

[80] Jerome Gaub, *Sermo academicus de regimine mentis quod medicorum est* (Leiden, 1747). English translation in L. J. Rather, *Mind and Body in Eighteenth Century Medicine: A Study Based on Jerome Gaub's De regimine mentis* (London, 1965).

[81] Note, however, that there are precedents in Locke's *Essay*: see Paul B. Wood, 'The Natural History of Man in the Scottish Enlightenment', *History of Science* 27 (1989), 89–123: 96.

[82] Title of section XI of *De regimine mentis*.

pursuing Gaub's mind–body holism in terms of an investigation of sensibility, which he treats as the key to understanding the connection between mind and body. He begins by arguing that the sensationalist tradition is, as it stands, insufficiently developed to provide the kind of understanding needed. In particular, he urges that sensation not simply be taken as primitive, as those in the Lockean tradition do, but must be traced back to sensibility, of which it is just a particular manifestation. He then takes up the argument, which we have already seen in La Caze, that sensibility must be traced back to its origins.[83] In the second Book of *La Médecine*, he identifies the causes affecting our mind and will: generation and heredity, sex, climate and seasons, regimen, age, and degree of health. These factors affect the blood and thereby have an effect on the size of vessels and the elasticity of the fibres, which in turn affect mental powers and character. The key lies in the cultivation of sensibility, which, like health more generally, is a matter of harmonious balance, and overcultivation can be injurious. The balance is secured through such factors as diet, hygiene, and moderation. These are the subject of the third book, where detailed instructions are given as to how to adjust climate, education, and regimen so as to remedy excesses or deficiencies in sensibility. Yet while personal regimen is one kind of answer to the question, another is selective breeding, a solution associated above all with the physician Charles Augustin Vandermonde, who, in his *Essai sur la manière de perfectionner l'espèce humaine* (1756), moved from individual self-improvement to the self-improvement of the species through what is in effect an early form of eugenics.

The idea of a harmonious state was generally perceived as a 'natural' state, both in the case of the correct balance of the naturals, and in the case of the balance of the naturals and the non-naturals. The task was that of guiding individuals to the achievement of a natural state. Personal regimen remained the preferred option, and cultivation of sensibility was what this personal regimen involved.[84] But just what the cultivation amounted to was contested. Above all, questions began to be raised about 'the perfectibility of man'.

With the publication of Rousseau's *Émile, ou de l'éducation* in 1762, this question had been thrust into the forefront of the Republic of Letters.[85] Rousseau, in the imaginary educational experiment to which he subjects Émile in the novel, sets out to show how we might cultivate the kind of sensibility that enables us to

[83] Antoine Le Camus, *La Médecine de l'esprit* (2nd edn, 3 vols, Paris, 1769), i. 8–9.

[84] In fact cultivation of sensibility was not restricted to individual self improvement, and even projects to improve the fitness of the race deployed these means: see Jacques Domenech, *L'éthique des Lumières. Les fondements de la morale dans la philosophie française du XVIIIe siècle* (Paris, 1989), 57–98.

[85] Rousseau, *Émile, ou de l'éducation*. The literature on this question is vast, but see: Nicholas Dent, *Rousseau* (London, 2005); and Jonathan Marks, *Perfection and Disharmony in the Thought of Jean-Jacques Rousseau* (Cambridge, 2005); Annette Pierdziwol, 'Extending Nature: Rousseau on the Cultivation of Moral Sensibility', in M. Lenz and A. Waldow, eds, *Contemporary Perspectives on Early Modern Philosophy: Nature and Norms in Thought* (Dordrecht, 2013), 135–58.

act morally. The idea that moral capacities were not equal and were something that could be cultivated was not something new, and indeed it was common ground for generations of thinkers from the late sixteenth century onwards. Descartes is a case in point. When Descartes dealt with ethical questions in the last decade of his life, in the letters to Princess Elizabeth or in *Les Passions de l'âme*, the context was not that of determining which actions were moral and which were not, or whether there was some general principle (such as universalizability) that we were compelled to follow in genuinely moral judgements. Rather, it was a question of what might be termed a theory of psychological preparation for moral agency: how to reshape one's life so that the anxieties, melancholia, and uncontrolled passions that prevent one taking full control of one's life can be overcome. It is to the degree that one overcomes these, and as a consequence is able to act freely, that one can be said to be a morally responsible agent.[86] Descartes attempted to provide Princess Elizabeth with advice on what to do to transform herself from someone subject to dispersed sensations and affections, with a reduced sense of self, into a unified locus of subjectivity—one that could exercise genuine agency, above all genuine moral agency.

What thinkers such as Diderot and Rousseau rejected in this account was the individualistic solution, as if one could simply reshape one's own psychological resources. In its place they offered a developmental account in which the social and cultural context, especially throughout childhood and adolescence, is crucial for the shaping of sensibility. As Anne Vila puts it, 'Rousseau's *morale sensitive* would provide the key for making people virtuous not by training their minds, but rather, by carefully controlling the impressions made upon their bodies—or more precisely, upon their sensitive systems'.[87] This is of particular importance in Rousseau's account in *Émile*, because it is the excessive stimulation, the constant excess of impressions, that is the cause of a distorted mode of experiencing and understanding our relations with others. Consequently, one needs to moderate one's exposure to this excess of impressions, by choosing with care one's society, occupations, and pleasures. The sensationalism that drives much of the programme, in Rousseau for example, enables one to include, not just the traditional elements of formal education here, but also the whole social and cultural milieu in which one lives. In short, the cultivation of what French Enlightenment thinkers would refer to as sensibility is quite different from Descartes' cultivation of a sense of the self, in that the source of the problems is located primarily in the social and cultural context in which one is raised, and the impact of excessive stimulation on one's sensibility.

It differs from it also in a second respect. The discussions of Rousseau, Diderot, and others are in terms of sensibility. Descartes had not talked in these terms and even though, in *Les Passions de l'âme*, he had offered an account of

[86] See Stephen Gaukroger, *Descartes, An Intellectual Biography* (Oxford, 1995), ch. 10.
[87] Vila, *Enlightenment and Pathology*, 183.

psycho-physiology, it was hopelessly underdeveloped compared to the theories of nervous sensibility that emerged in the wake of Haller's work. The new understanding of sensibility was, as we have seen, one that was thoroughly integrated with questions of nervous sensitivity and muscular irritability. Rousseau is in fact exceptional in pursuing the questions he does without—or at least without very much—attention to these physiological questions, even though he is adamant that excess stimuli work through the whole body. His criticism that modern urban sensibility is out of alignment with nature is pursued in terms of comparison with a hypothetical state of nature. But such lack of alignment can also be thought of in physiological terms.[88]

There were a number of physicians who combined a Rousseau-like worry about excessive stimulation with a concern with nervous sensitivity. The Scottish physician John Gregory, for example, laments the 'softness and effeminacy' of manners in modern urban living and draws attention to a 'debility and morbid sensibility of the nervous System, which lays the foundation of most of our diseases, and deprives us at the same time of the spirit and resolution to support them'.[89] An even more influential example is the Swiss physician Samuel August Tissot, author of one of the best-selling medical advice books of the eighteenth century, *Avis au peuple sur sa santé* (1767).[90] In 1768, in his *De la santé des gens de lettres*, and in 1770 in his *Essai sur les maladies des gens du monde*, Tissot launched an attack on the overcultivation of sensibility. Accepting that intellectual processes are generated and regulated by the activity of sensible fibres, he rejected the prevalent view that one could perfect one's sensibility by engaging in the arts and sciences. Tissot argues that, on the contrary, such activity had deleterious effects on health, producing delicate oversensitized individuals who are unable to deal with daily life in a satisfactory way. The route to health for sickly scholars is not more reading in the arts and sciences but more attention to the advice of their physician.

There are two questions of particular significance that were prompted by these kinds of considerations. The first was whether the problems of sensibility were common to all human bodies, or whether the sensibility of men and women, for

[88] Rousseau's appeal to a state of nature was rejected by contemporaries as diverse as Ferguson and Herder as based on idle speculation rather than observed facts: see David Carrithers, 'The Enlightenment Science of Society', in C. Fox, R. Porter, and R. Wokler, eds, *Inventing Human Science: Eighteenth-Century Domains* (Berkeley, 1995), 232–70: 247–9. In these circumstances, the appeal of a medical approach is increased considerably.

[89] John Gregory, *A Comparative View of the State and Faculties of Man with those of the Animal World* (London, 1765), 69–70.

[90] Samuel August Tissot, *Avis au peuple sur sa santé* (Lausanne, 1767 with multiple reprints). See Patrick Singy, 'The Popularization of Medicine in the Eighteenth Century: Writing, Reading, and Rewriting Samuel Auguste Tissot's *Avis au peuple sur sa santé*', *Journal of Modern History* 82 (2010), 769–80: 776. See also Séverine Pilloud, 'Mettre les maux en mots, médiations dans la consultation épistolaire au XVIIIe siècle: Les malades du Dr Tissot (1728–1797)', *Bulletin Canadien d'Histoire de la Médecine* 16 (1999), 215–45; and Antoinette Emch-Dériaz, *Tissot: Physician of the Enlightenment* (New York, 1992).

example, might differ. There was broad agreement that they did in fact differ.[91] The *locus classicus* was Pierre Roussel's *Système physique et moral de la femme* (1775), a bestseller from the time of its publication, and going through seven editions by 1820.[92] But there was a significant literature before Roussel,[93] much of which focused on the particular susceptibility of women to what was termed 'vapours', that is, various nervous afflictions including melancholia, hypochondria, and hysteria, thought to issue from a nervous system oversensitized by pampered urban living.[94] The 'vapours' primarily affect women, whose organs have an especially fragile equilibrium with one another, but they can affect men as well, making them effeminate.[95]

The second issue that arose was whether it was simply a matter of increasing one's level of sensibility, or whether one's sensibility could be harmfully overdeveloped. For many of Tissot's contemporaries, the cultivation of sensibility was above all a matter of increasing one's degree of sensibility. Diderot, in the *Lettre sur les Aveugles* (1749), maintains that it is beyond doubt that 'the state of our organs and senses has a significant influence on our metaphysics and morals', and that virtues depend on our manner of sensing and 'the degree to which external things affect us'.[96] He concludes from this that the sensory deficiency of the blind man leads to a moral deficiency in him. Developing one's sensibility to the fullest degree is crucial to one's formation as a healthy, well-formed, balanced, morally responsive individual. Moreover, since this sensibility is a function of the sensory

[91] See Jocelyne Livi, *Vapeurs de femmes: Essai historique sur quelques fantasmes médicaux et les pertes de sang* (Dijon, 1984); Londa Schiebinger, *The Mind Has No Sex? Women in the Origins of Modern Science* (Cambridge, Mass., 1989); Lindsay Wilson, *Women and Medicine in the French Enlightenment: The Debate over 'Maladies des Femmes'* (Baltimore, 1993). More generally, see Sylvana Tomaselli, 'Reflections on the History of Science of Woman', *History of Science* 29 (1991), 185–205.

[92] Pierre Roussel, *Système physique et moral de la Femme, ou tableau philosophique de la Constitution, de l'Etat organique, du Tempérament, des Mœurs, et des Fonctions propres au Sexe* (Paris, 1775).

[93] See Vila, *Enlightenment and Pathology*, ch. 7, to which I am indebted here.

[94] Pierre Hunard, *Dissertation sur les vapeurs et les pertes de sang* (Paris, 1765); Joseph Raunlin, *Traité des affections vaporeuses du sexe* (Paris, 1758); Pierre Pomme, *Traité des affections vaporeuses des deux sexes* (Lyon, 1760); Jean-Baptiste Pressavin, *Nouveau traité des vapeurs* (Lyon, 1770); Edmé-Pierre Chauvot de Beauchêne, *De l'Influence des affections de l'âme dans les maladies nerveuses des femmes, avec le traitement qui convient à ces maladies* (Montpellier and Paris, 1781); Joseph Bressy, *Recherches sur les vapeurs* (London and Paris, 1789). There was an analogous concern in Britain with George Cheyne, in his *The English Malady; or, A Treatise of Nervous Diseases of All Kinds, as Spleen, Vapours, Lowness of Spirits, Hypochondriacal and Hysterical Distempers* (Dublin, 1733), identifying these diseases as a peculiarly English problem. In 1807, Thomas Trotter, in his *A View of the Nervous Temperament: being a Practical Treatise on Nervous, Bilious, Stomach, and Liver Complaints* (London, 1807), announced that nervous diseases had grown dramatically in the intervening period. On climatic pathology in Britain, see Jan Golinski, *British Weather and the Climate of Enlightenment* (Chicago, 2007), ch. 5.

[95] Raunlin, *Traité des affections vaporeuses*, 42. On women and vapours see Paul Hoffmann, *La Femme dans la Pensée des Lumières* (Paris, 1977).

[96] Denis Diderot, *Oeuvres philosophiques et dramatiques* (6 vols, 'Amsterdam' [Paris], 1772), 240, 242.

sources of experience generally, it should in theory be quite possible to control the social and pedagogic environment in such as way as to foster a high degree of sensibility. Of course, not all forms of sensory exposure will have the same effect, and may have different effects in different people.[97] Nevertheless, it was widely considered that many sensory experiences did in fact have a uniform effect: experience of weakness and misery, for example, produced certain affective and emotional states in people.

The problem with this kind of account is its assumption of routinely beneficent effects of increasing levels of sensibility. This is what Tissot rejects. He documents a number of cases where patients have overtaxed themselves in attempting to cultivate sensibility with a view to intellectual and moral development, only to find themselves sick and incapacitated, and constantly agitated.[98] Instead of first making sure to preserve their sensible constitution, they have engaged in a wholly unnatural exertion.[99] Sensible fibres are liable to wear and tear just like any other organic bodies, and become hardened and less sensitive through overuse.[100] What results is a form of desensitization, an effect compounded in the case of exposure to intense cognitive and affective stimulation, and this affects not just the local fibres, but the whole animal economy.[101] There are very few, he concludes, who have a sensible constitution sufficiently robust to follow a demanding moral and intellectual trajectory.[102]

SOCIAL MEDICINE

As we have already seen, for the *médecins philosophes*, the subject matter of medicine included every aspect of human experience, and anything external to the body that affected physical, mental, intellectual, or emotional states was included within the scope of medicine, for it had an effect on the well-being of the person. The traditional Christian idea that medicine should confine itself to the body while the Church dealt with the soul was no longer viable. In this way, the objectives of anthropological medicine became far more ambitious than anything to which physicians had aspired up to this point.

[97] Note also that, for Diderot, the delicate ways in which nervous activity and qualities of character are associated means that there are questions concerning the structure of the nerves that affect sensibility, although one looks in vain for any subtlety in his account. In *La Rêve de d'Alembert*, for example, he remarks confidently that if the trunk of the network of nerves is too vigorous in relation to its branches, what results are poets, artists, and madmen, whereas if the branches and threads are disproportionately active, the result would be mediocrity. If the whole network is slack, what we get are imbeciles.

[98] Samuel August Tissot, *Essai sur les maladies des gens du monde* (Lausanne, 1770), 38. My thinking on these questions was helped along greatly by discussion with Annette Pierdziwol.

[99] Samuel August Tissot, *De la santé des gens de lettres* (Lausanne, 1768), 134.

[100] Ibid., 52–6. [101] Ibid., 18. [102] Ibid., 58.

In the late 1770s, the leading edge in French medicine shifted from Montpellier to Paris, and, particularly in the wake of the Revolution, became significantly politicized. The crucial pre-Revolutionary development was the foundation of the commission in 1776 by the Controller-General, Turgot, to investigate the cattle plague of 1770–6, which was causing economic hardship because of rises in the prices of meat, milk, and livestock. Its terms of reference included not only the investigation of the cattle plague, but human and animal epidemics more generally, as well as sanitary conditions, endemic and occupational diseases, and a number of surgical and anatomical questions. In 1778 it was formed into the Société Royale de Médecine, with the aim of compiling a medical topography. The anatomist Félix Vicq d'Azyr became permanent secretary, and he focused on the broader questions, especially environmental factors affecting health, including water quality (the Société Royale was granted exclusive rights to inspect and distribute mineral water) and manner of life.[103] This required the establishment of a network of national correspondents who reported on climatic and meteorological factors as well as social and domestic ones. In these respects, it formed a possible model for a reform of the medical system, dominated as this was by the Paris Medical Faculty, which was considered by its critics to be producing ill-trained physicians with inadequate skills.[104] In the course of the 1790s, French education came to be extensively debated, and, in particular, the faults of the medical teaching of the *ancien régime*, as identified by Vicq d'Azyr, were taken as posing pressing problems. With the abolition of the universities in 1793, their medical faculties were disbanded.[105] They were replaced—in Paris, Montpellier, and Strasbourg—in 1795 by new *écoles de santé*, which were staffed by a new generation of physicians. As Elizabeth Williams has pointed out, the new professors who took up appointments in the Paris École—Pinel, Chaussier, and Cabanis—were figures who either had never been considered for appointment or had been deliberately excluded from the old Faculty of Medicine.[106] The Paris École very quickly evolved into groups or societies that not only instituted new clinical ideals, but pursued medically directed social and political programmes: in 1796 the Société de Santé de Paris was formed, with an explicit agenda of

[103] See Caroline Hannaway, 'The Société Royale de Médecine and Epidemics in the Ancien Régime', *Bulletin of the History of Medicine* 46 (1972), 257–73. On Vicq d'Azyr, see idem, 'Vicq d'Azyr, Anatomy and a Vision of Medicine', in Ann La Berge and Mordechai Feingold, eds, *French Medical Culture in the Nineteenth Century* (Amsterdam, 1994), 280–95.

[104] See, for example, Vicq d'Azyr's own criticisms in his 'Nouveau Plan de constitution pour la médecine en France', *Histoire et mémoires de la Société royale de médecine, 1787–1788*, ix (Paris, 1790), 1–201: 3. In fact, the traditional Paris Faculties were not as inadequate as their critics claimed: see Laurence W. B. Brockliss, 'Before the Clinic: French Medical Teaching in the Eighteenth Century', in Caroline Hannaway and Ann La Berge, eds, *Constructing Paris Medicine* (Amsterdam, 1998), 71–116. Cf. Staum, *Cabanis*, 96–7 on the quite stringent requirements that Cabanis had to meet as a student at the Paris Faculty of Medicine.

[105] See Robert R. Palmer, *The Improvement of Humanity: Education and the French Revolution* (Princeton, 1985).

[106] Williams, *The Physical and the Moral*, 71.

pursuing 'a science intimately tied to the well-being of humanity', and this was quickly followed by the establishment of the Société Médicale d'Émulation, by a younger group of physicians, which fostered similar aims.[107]

Central to both these groups and societies was the idea that physical, mental, and moral questions formed an intimately linked whole, and could not be considered in isolation from one another. Jean-Louis Alibert, secretary to the Société Médicale d'Émulation from 1796 to 1802, played a crucial role in promoting the centrality of medicine with respect to the other sciences and to moral and social questions. In his 'Discours sur les rapports de la médecine avec les sciences physiques et morales' (1798),[108] he argued that medicine was not only autonomous with respect to other sciences but oversaw them by directing their discoveries along useful paths. Just as attraction and affinity were the subject matter of those sciences dealing with inanimate things, so medicine dealt with sympathy and sociability.[109] Moreover, no moral question was foreign to medicine, and Alibert envisaged an intimate alliance between medicine and politics.[110]

It is not Alibert, however, but Cabanis who represents this conception of medicine as a form of political anthropology to its fullest degree.[111] In his 1804 *Coup d'oeil sur les Révolutions et sur la réforme de la médecine*, Cabanis refers to the integration of physical and mental questions—more specifically the synthesis of physiology, analysis of ideas, and moral philosophy—which medicine seeks to achieve by a new discipline in terms of what he says the Germans called 'anthropology or the science of man'.[112] Like Alibert, Cabanis sees medicine as a master discipline to which subjects such as anatomy and chemistry, and procedures such as laboratory experimentation, are strictly subordinate.

In 1802, Cabanis published a systematic account of the medical foundations of the human sciences, *Rapports du physique et du moral de l'homme*. The two volumes of the *Rapports*, over a thousand pages, deal in turn with the general nature of the relation between one's physical state and one's intellectual and moral faculties; the physiological account of sensations; the impact on one's ideas and one's moral 'affections' of age, sex, temperament, illness, regimen, and climate. The book then turns to an analysis of the nature of life in animals, the first determinants of sensibility, instinct, sympathy, sleep, and delirium. Finally, there are two chapters on the influence of the moral on the physical, and on 'acquired' temperaments.

107 Ibid., 72.

108 Jean-Louis Alibert, 'Discours sur les rapports de la médecine avec les sciences physiques et morales', *Mémoires de la Société Médicale d'Émulation* 2 (1798), i–cxii.

109 Ibid., lxxxi. 110 Ibid., xcv.

111 On Cabanis, see Staum, *Cabanis*; and Moravia, *Il pensiero degli idéologues*, Part I.

112 *Oeuvres philosophiques de Cabanis*, ed. C. Lehec and J. Cazeneuve (2 vols, Paris, 1956), ii. 77. On the German term, it is likely that Cabanis is referring to Ernst Platner, *Anthropologie für Ärzte und Weltweise, erster Theil* (Leipzig, 1772).

This last chapter contains the crux of the reform programme, because it is here that the question is raised of how one might acquire those temperaments that are socially desirable. What was needed was a combination of two things. The first was what Cabanis refers to as 'climate', which is something more comprehensive than the term might suggest. Each latitude and climate has its characteristic features, he writes, but we must also take into account the relations between water and soil, between the air and sunlight, river flows, mountains, and vapours which affect the atmosphere; correlatively, there are the ways in which plants imitate the qualities of the soil, water, and the state of the air. Human beings are particularly susceptible to these factors, and he cites Hippocrates' observations on the great variation in human beings and races, depending on such factors.[113] The second factor is regimen: diet, sleep, physical exercise, occupation. Vegetarianism and monastic asceticism are condemned, but so too is manual labour. While no specific recommendations are offered which a legislator might put into effect, various areas of endeavour—hygiene in the case of bodily improvement, public festivals in the case of patriotic sensibility, pedagogic method in the case of education—are identified. Nevertheless, the changes in sensibility that one wishes to effect are achieved largely by simply inculcating new habits: 'If then the causes of particular impressions act frequently enough, or for a long enough time, on the system, they may change it habits and those of its organs; consequently they can introduce the accidental dispositions, or the new temperaments, that these habits constitute. This is the true source of the acquired temperaments.'[114]

If one compares Cabanis with other defenders of a programme for the medically based 'science of man', then there is really no area in which he goes beyond writers such as Gaub, or Le Camus, or Tissot in ambition or sophistication. Although his version of the programme is more detailed, it is far from clear that there is anything that is significantly new. What is distinctive about Cabanis is his ability actually to undertake the proposed reforms. He was a medical reformer before he developed a general theory, active not only in hospital reform, but also in the setting of standards for the medical profession, as well as being involved in the movement for universal primary education, and in national public assistance. On hospital reform, there was trenchant criticism of existing hospitals in the report compiled, after twenty years' work, by Jacques Tenon in 1788, identifying

[113] Pierre Jean Georges Cabanis, *Rapports du physique et du moral de l'homme* (2 vols, Paris, 1802), ii. 236–41. The theory about the influence of climate was in fact subject to many basic objections, which its supporters never took the trouble to answer. As Helvétius points out: 'If the different temperature of climates had such an influence on the mind and its capacities, how comes it that the Romans so magnanimous, so brave under a republican government, are now so dastardly and effeminate? How comes it that the Greeks and Romans, who were formerly so worthy of esteem for their wit and virtue, and were the admiration of the earth, are now dwindled into contempt?' Claude-Adrien Helvétius, *De l'esprit: or, Essays on the Mind, and its several faculties* (London, 1759), 222. Hume was equally caustic in his essay 'Of National Characters': David Hume, *Essays and Treatises on Several Subjects* (2 vols, Edinburgh, 1793) i. 198–213.

[114] Cabanis, *Rapports*, ii. 596.

conditions that were archaic, unhygienic, overcrowded, and mixed foundlings, the aged homeless, the sick, the insane, and the contagious.[115] Various reforms were made in response, although not everything was to Cabanis' satisfaction.[116] In particular, he was sympathetic to the recommendation of Du Pont advocating small hospitals which could cultivate the 'natural' atmosphere of one's home,[117] but the commission decided instead in favour of four new 1,200-bed hospitals on the outskirts of Paris. In his *Observations sur les hôpitaux* (1790), Cabanis sought a compromise on size, and, in connection with this, linked the needs of medical research and teaching to medical and hospital reform, arguing that medical research required small hospitals. Cabanis had attended to Mirabeau during his final illness, and after Mirabeau's death in 1791, he was quickly catapulted into senior administrative positions, starting with appointment to a five-man hospitals committee to assume administration of the Paris hospitals.

Cabanis was a central figure in the reforms that culminated in what is generally referred to as the Paris School. There is some controversy over whether the developments that marked out this 'school' were original, and the conclusion that recent historians of medicine have come to is that the sharpness of the break with the earlier teaching of the Paris Faculty of Medicine has been exaggerated, and also that it was reliant on work from other countries, most notably the British Hunterian tradition.[118] But no one doubts the significance of the way in which the parts were put together to form a novel coherent model for medicine. The model has five generally accepted features: first, the rise of pathological anatomy, that is, the systematic correlation of external manifestations of disease with the lesions found in the organs at the time of autopsy; second, the move from observing hundreds of cases to observing thousands; third, the shift beyond listening to the patient's account to active physical examination using such new procedures as percussion, auscultation, with extensive use of the stethoscope; fourth, the transformation of the hospital into the locus of medical activity and research; and fifth, the use of medical statistics in the analysis of case histories and the evaluation of therapy.[119]

115 Jacques Tenon, *Mémoires sur les hôpitaux de Paris* (Paris, 1788).

116 See Straum, *Cabanis*, chs 4 to 6, on which I rely in what follows.

117 Pierre-Samuel Du Pont de Nemours, *Idées sur les secours à donner aux pauvres dans une grande ville* (Paris, 1786).

118 Two important studies from the 1960s—Michel Foucault, *La Naissance de Clinique* (Paris, 1963), and Erwin Ackernecht, *Medicine at the Paris Hospital, 1794–1848* (Baltimore, 1967)—stressed the novelty of the Paris School. However, its clean break with earlier developments in Paris has been questioned in Brockliss, 'Before the Clinic'. Another criticism of the idea that the developments in the Paris School were *sui generis* is set out in Othmar Keel, 'Was Anatomical and Tissue Pathology a Product of the Paris Clinical School or Not?', in Caroline Hannaway and Ann La Berge, eds, *Constructing Paris Medicine* (Amsterdam, 1998), 117–84, in which its indebtedness to the Hunterian tradition is set out. On these questions, see Ann La Berge and Caroline Hannaway, 'Paris Medicine: Perspectives Past and Present', in Hannaway and La Berge, *Constructing Paris Medicine*, 1–70.

119 I take these features from La Berge and Hannaway, 'Paris Medicine', 4. Pinel, whose concern was with the hospitalized insane, will rebel against this last development in particular, advocating

What is striking about these reforms is that, while they are the product of the kind of broadened conception of medicine that Cabanis and his contemporaries had sought to establish, at the same time they undermine the whole idea of medical anthropology—that is, a project in which exclusively medical resources are used to build a 'science of man'—and indeed mark the beginning of its decline. We can identify three major reasons for this decline, and I want to look first at the way that its increasingly politicized nature began to undermine anthropological medicine as the political situation in France changed between the 1780s and the first decade of the next century. We shall then turn to two different kinds of factors explaining the demise of anthropological medicine: the discovery that sensibility turned out to be as inadequate as reason as a means of explaining the connection between minds and the physical world, and the inability of anthropological medicine to come to terms adequately with competing forms of naturalization.

Anthropological medicine developed an explicit socio-political programme in its Parisian (i.e. post-Montpellier) phase. While medical advice on a range of matters that fell outside physical remedies for physical maladies was certainly sought from a significant section of the reading public, and while there is likewise no doubt that this was accompanied by a new interest in medical matters on the part of this same public, the changed standing of medical expertise could simply have placed it more favourably within a range of possible sources of advice on moral and social questions. The emergence of the *médecin philosophe* could just have been the emergence of yet one more group claiming social and moral authority. But in fact what happened was rather more significant. In the Revolutionary period, the centralized organization that had been characteristic of French culture disappeared.[120] By 1793, the medical structure of pre-Revolutionary France was dismantled, and for a decade the state exercised no supervision of medical practice: not only was the corporatist organization, with its central system of supervision and licensing, abolished, but so too were the traditional distinctions between physicians, surgeons, and apothecaries. During this period, anthropological medicine had to reshape an identity for itself, in the context of deep ideological conflicts between clerical and anti-clerical, and conservative and radical, as well as deep divisions over such questions as the relation

close personal relationships with patients. See his three formative papers: Philippe Pinel, 'Mémoire sur la manie périodique ou intermittente', 'Recherches et observations sur le traitement moral des aliénés', and 'Observations sur les aliénés et leur division en espèces distinctes', in *Mémoires de la Société médicale d'émulation* 1 (1796–7), 94–119; 2 (1797–8), 215–55; and 3 (1798–9), 1–26 respectively.

[120] See Elizabeth A. Williams, 'The French Revolution, Anthropological Medicine, and the Creation of Medical Authority', in Bryant T. Ragan Jnr and Elizabeth A. Williams, eds, *Re-Creating Authority in Revolutionary France* (Westport, Conn., 1992), 79–92, which I summarize here. See also Moravia, *Il pensiero degli idéologues*; and Georges Gusdorf, *La Conscience révolutionnaire, les idéologues* (Paris, 1978).

between the material and the spiritual. In a period when the authority of church and state was collapsing, physicians found themselves as authorities in a range of intellectual and social domains, as anthropological medicine was transformed by Cabanis, Alibert, Pinel, Mesmer, Gall, and others into psychological, social, and educational programmes.

As a result, it was inevitably caught up in political developments, and these started to work to its detriment in the last decades of the century. Of particular importance here is the fact that dissatisfaction with scientific authorities from the 1780s tended to take a political form. Mesmerism became the vehicle for this dissatisfaction, so that it came to be treated very much as a threat to political stability.[121] From the late 1770s, the established scientific and medical authorities in Paris were subjected to a number of attacks from those whose unorthodox medical and scientific theories, and whose medical practice, had been condemned by these bodies. At one level, there was nothing new in such challenges. Attacks on scientific authority were not unknown in the seventeenth century, for example, although it was predominantly mathematical questions that were at issue then. There was no shortage, for example, of those who claimed to have a technique for squaring the circle, and the proponents of such techniques composed strident attacks on establishment mathematicians who (understandably) failed to take any notice of them.[122] But there is a new element in the late eighteenth-century assaults on scientific and medical authority. It was not just a question of scientists or physicians asserting their authority over recalcitrant cranks and quacks who refused to accept that what they were proposing was misguided. There is a new, political element at issue, absent from earlier disputes. With the cultural standing of science elevated in the way it was, the stakes were considerably higher. To be excluded from serious science was now to be excluded from the Enlightenment project, and the groups so excluded included those who believed they had the most to offer to this project. The progress of civilization had been tied to the progress of science from d'Alembert's preliminary Discours to the *Encyclopédie* onwards, and was to receive a definitive statement in Condorcet's *Esquisse*, as we have seen. At the same time, civilization had been strongly associated with free speech: to the extent to which one can talk of any general notion of a transnational Enlightenment, this is its most distinctive feature.[123]

[121] The detailed case is made in Robert Darnton, *Mesmerism and the End of the Enlightenment in France* (Cambridge, Mass., 1968).

[122] See, for example, Jeff Loveland, 'Panckouck and the Circle Squarers', *Eighteenth-Century Studies* 37 (2004), 215–36.

[123] The idea of trans-national Enlightenment has been a subject of much dispute for some decades. There is a good summary of the issues in Charles W. J. Withers, *Placing the Enlightenment: Thinking Geographically about the Ages of Reason* (Chicago, 2007), ch. 2. For a revealing study of the commonality in the nevertheless contextualized nature of debates in different centres, see Robertson, *The Case for Enlightenment*, ch. 1.

Here we have a situation more complex than might at first be apparent, for the question arises not only as to what the relation is between the pursuit of science and freedom of expression, but also whether any body has the power to dictate what counts as a scientific matter falling under its jurisdiction. These questions became entangled: the latter was, after all, a presumed power reminiscent of the traditional far-reaching claims of the Church to decide on whether any matter fell under its jurisdiction. In the politically charged atmosphere of late eighteenth-century Paris, the questioning of scientific authority was something that motivated a move to revolutionary politics,[124] and a number of those associated with unorthodox scientific and medical systems, such as Jean Louis Carra,[125] Jean-Paul Marat,[126] and Marat's supporter Jacques-Pierre Brissot de Warville, became prominent revolutionaries, with one of their chief aims being the dismantling of the 'autocratic' organization of scientific enquiry. The Académie des Sciences, with the support of the government (which had set up the Académie in the first place), was accused of trying to circumscribe science so as to make it entirely a matter for professionals. The only way to fight this was to appeal to the public. Brissot de Warville put the argument in these terms:

> The ancients never imagined any institution so bizarre as our academies. Neither in Athens nor in Rome does one find those who have certification as philosophers. To enter the school of Chrysippus or that of Epicurus, it was not necessary to take a ballot or to have a vote.... The empire of the sciences should have nothing to do with despots, aristocrats, or electors. Its image is that of a perfect republic, in which the most useful talent is the sole entitlement to honour. To admit a despot, or aristocrats, or electors officially empowered to set the seal of certification on works of genius, is to violate the natural order and the liberty of the human mind. It is a crime against public opinion, which alone has the right to crown genius. It introduces an outrageous despotism, making each elector a tyrant and turning all other savants into slaves. In debasing the nobility of minds, it holds back the perfection of the sciences and prejudices the search for truth.[127]

The vehicle for the appeal to the public was mesmerism. Carra, Marat, and Brissot de Warville may each have had their own sources of discontent, but they united around the common cause of mesmerism, which had a public standing that none of their own projects could ever have achieved. Mesmerism in France began with

[124] See Darnton, *Mesmerism and the End of the Enlightenment in France*, ch. 4.

[125] Carra's writings on scientific topics include *Nouveaux principes de physique ornés des planches* (2 vols, Paris, 1781); *Systême de la raison, ou le Prophete philosophe* (Paris, 1782); and *Examen physique du magnétisme animal* (Paris, 1785).

[126] For an example of his writings on scientific subjects, which cover heat, electricity, and optics, see his *Recherches physiques sur le feu* (Paris, 1780). For his polemics against the Académie, see *Les Charlatans modernes; ou, Lettres sur le charlatanisme académique publiées par M, Marat, l'Ami du peuple* (Paris, 1793).

[127] Jacques-Pierre Brissot de Warville, *De la verité, ou méditations sur les moyens de parvenir à la verité dans toutes les conaissances humaines* (Neufchâtel, 1782), 165–6.

Franz Mesmer's arrival in Paris in 1778,[128] when he announced the discovery of an extremely subtle fluid that surrounded and penetrated all bodies, acting as a medium for gravitation and sensation alike.[129] He was not the first to have countenanced such a fluid—in his later years Newton had postulated the existence of an aether that had these characteristics[130] and such an aether seemed to a number of his successors to be a suitable medium for the transmission of gravitational and magnetic attraction—but Mesmer was the first to put it at the forefront of medicine, in the form of a theory of animal magnetism. Sickness, he argued, resulted from obstacles to the flow of this fluid, and its cure consists in the skilled redirection of the flow, by seeking out the small magnetic centres that exist throughout the body by aligning them with stable magnets of the mesmerist physician's body, namely the fingers and the nose, which were directed at the hypochondria, on the sides of the upper abdomen, the body's 'equator' and—for Mesmer—its 'common sensorium', that is, the place where the sensations from the various sense organs are unified (see Fig. 3.1).[131] In the case of violent convulsives, patients were placed in tubs filled with bottles of iron filings and mesmerized water oriented towards their body.

The set-up was not completely unprecedented, for in some ways it mirrored the Leyden jar, which stored electrical 'fluid' and radiated it out in various ways. It was, as in electrical experiments, a matter of focusing the power to produce the requisite effect. Medical uses of electricity had been fostered since mid-century, with a special focus on the treatment of paralysis.[132] In 1746, the English physician John Feke had advised that cures of illnesses could be effected by replacing lost electrical energy (which he identified with a life force, *vis vitae*),[133] and ten years later Richard Lovett was maintaining that since electricity and the aether were 'the very same identical substance', then 'electrical ether must necessarily, as you observe, discover a natural tendency to heal'.[134] In 1785,

[128] The reception of mesmerism in France, where it was used by radical elements to promote their programmes, was different from that in Germany, where it was attacked by radicals as indicative of superstition, and it was associated with the popular exorcisms of Johann Joseph Gassner: see for example, Carl von Knoblauch, *Taschenbuch für Aufklärer und Nichtaufklärer auf das Jahr 1791* (Berlin, 1790), 44; idem, *Euclides Anti-Thaumaturgicus, oder demonstrantiver Beweis von der Unmöglichkeit hyperphysischer Begebenheiten* ('Germany' [i.e. Berlin], 1791), 39, 41. See Jonathan Israel, *Democratic Enlightenment: Philosophy, Revolution, and Human Rights, 1750–1790* (Oxford, 2011), 686–7.

[129] See Franz Mesmer, *Mémoire sur la découverte du magnétisme animal* (Geneva, 1779). Cf. the work of one of Mesmer's prime supporters in the Paris medical community: Charles Deslon, *Observations sur les deux Rapports de MM. les Commissaires nommés par Sa Majesté pour l'Examen du Magnétisme animal* (Paris, 1784).

[130] See Betty Jo Teeter Dobbs, *The Janus Faces of Genius: The Role of Alchemy in Newton's Thought* (Cambridge, 1991), ch. 7.

[131] See Darnton, *Mesmerism*, ch. 1, for a full discussion.

[132] See Abbé de Mangin, *Histoire générale et particulière sur l'électricité* (Paris, 1782), 31.

[133] Feke, *An Essay to Shew the Cause of Electricity*, 43–4.

[134] Richard Lovett, *The Subtil Medium prov'd: or, That Wonderful Power of Nature, so Long ago Conjectur'd by the most Ancient and Remarkable Philosophers, which they call'd sometimes Aether, but oftener Elementary Fire, verify'd* (London, 1756), 73.

Fig. 3.1 Mesmerism. Franz Mesmer, *Mémoire sur la découverte du magnétisme animal* (Paris, 1779).

George Adams was explaining disease as a disturbance in electrical equilibrium, an equilibrium which it was the task of the physician to restore.[135] Jean Jallabert in Geneva, François Boissier Sauvages de la Croix and Jean-Etienne Deshais in Montpellier, Giuseppi Verati in Bologna, Gianfrancesco Pivati in Venice, and Giovanbattista Bianchi in Sardinia, all reported successful electrical therapies,[136] and in 1780 the Montpellier physician Pierre Bertholon published an influential treatise lauding the success of such remedies.[137] By the late 1770s, two schools of

135 George Adams, *An Essay on Electricity* (2nd edn, London, 1785), 312.

136 See Pera, *The Ambiguous Frog*, 18–25.

137 Pierre Bertholon, *De l'éléctricité du corps humain dans l'état de santé et de maladie* (2 vols, Paris, 1780). Cf. Tiberius Cavallo, *An Essay on the Theory and Practice of Medical Electricity*, (London, 1780), and Giovanni Aldini, *General Views on the Application of Galvanism to Medical Purposes; Principally in Cases of Suspended Animation* (London, 1819).

electrotherapy had emerged in Paris.[138] The first was represented by Bertholon and Pierre Jean Claude Mauduyt de la Varenne,[139] who construed health in terms of a balance between vital forces in the body and atmospheric electricity. The second, which was primarily represented in the work of the conjuror and illusionist Nicolas-Philippe Ledru, argued that one cause of illness was a contortion of the nerves, and that the subtlety of electrical fluid enabled it to clean the nerve canals, thereby unblocking them and allowing the course of vital fluids.[140] Mesmerist therapies, which attracted significant hostility from the medical authorities, were not very different from the electrical therapies of Ledru and Mauduyt de la Varenne, for example, yet these had the full support of the Société Royale de Médecine.[141]

The medical establishment may have found fault with Mesmer, but the members of fashionable Paris society did not. Mesmerist cures developed a very significant and devoted public following in the 1780s. It attracted converts not only from the general population, but from the medical and even from the religious community, and with the discovery of induced hypnosis by the Chastenet de Puységur brothers there were some spectacular demonstrations of putting people into a strange sleep in which they followed the hypnotist's orders to the letter.[142] These were not unauthenticated reports, and many of these mesmerist phenomena were carefully documented and notarized. Mesmerism produced quite unexpected effects and quite unexpected responses in patients, and it was these, more than anything else, that accounted for its great popularity.

Matters became sufficiently serious that in 1784 the government appointed a royal commission to investigate mesmerism, comprising four doctors from the medical Faculty of the University, and five members of the Académie des Sciences. At the same time, they appointed another commission from the Faculty's rival, the Société Royale de Médecine. There are two features of the response to mesmerism that are of particular interest in this respect.

First, many of the criticisms that were made of bodies like the Académie des Sciences, in response to their rejection of the new theories of medicine, magnetism, optics, etc., recapitulated criticisms going back to the seventeenth century,

[138] See Pera, *The Ambiguous Frog*, 23–5.

[139] Pierre Jean Claude Mauduyt de la Varenne, *Mémoire sur les différentes manières d'administrer l'électricité: et observations sur les effets quélles ont produits* (Paris, 1784). See M. U. Zelbstein, 'Mauduyt de la Varenne (1733–1792), un médecin au siècle des "Lumières"', *Histoire des sciences médicales* 18 (1984), 295–306.

[140] Nicolas-Philippe Ledru, *Rapport de MM. Cosnier, Maloet, Darcet, [et al] sur les avantages reconnus de la nouvelle méthode d'administrer l'électricité dans les maladies nerveuses, particuliérement dans l'épilepsie, & dans la catalepsie, par M. Ledru, connu sous le nom de Comus* (Paris, 1783). It is true that Ledru's credentials do not inspire confidence, but this kind of account was also supported by Tiberius Cavallo, who was a serious scholar: Cavallo, *An Essay on the Theory and Practice of Medical Electricity*, 11–13.

[141] See Geoffrey Sutton, 'Electric Medicine and Mesmerism', *Isis* 72 (1981), 375–92.

[142] Armand Marie-Jacques de Chastenet, Marquis de Puységur, *Mémoires pour servir à l'histoire et à l'establissement du magnétisme animal* (London 1786). See Darnton, *Mesmerism*, 58.

which we looked at earlier. Here commitment to systematic bodies of theory is pitted against raw experiment, the former associated with an inability to respond to the evidence, the latter with openness to new ideas and a refusal to accept anything on authority. There is typically both a political and an anti-clerical edge in these criticisms in their eighteenth-century forms. The battleground was shaped in 1783 in a legal case concerning the erection of a lightning rod, by Charles Dominique de Vissery de Bois-Valé, a retired lawyer and enthusiast for science, on top of his house three years earlier. A dispute over property ensued (the tail of the lightning rod ran down his neighbour's wall) and the case would have remained one of local interest, were it not for the fact that, by the time it had come to court, one of the lawyers defending Vissery, the young Maximilien Robespierre, made the issue one of the claims of scientific and legal authority.[143] Denying that officials could actually have the general knowledge to which they pretended, that they could offer interpretations that legitimately moved from the facts presented to general theories, Robespierre in effect denied expertise, by insisting that one consider only the raw facts. The prosecutor's arguments were merely 'reasonings of theory', whereas 'against experience . . . what good are all the reasons?' As Riskin points out, for Robespierre the 'natural tendency' of 'electrical matter towards metal bars in preference to tile roofs was a fact. It represented all the explanations that physics required and superceded any questions about its mechanism.'[144] In other words, observation was pitted against expertise and authority, and truth was on the side of the former.

If the committees set up by the government to investigate mesmerism had simply responded by showing how the effects claimed by Mesmer and others were impossible, this would have been not only unconvincing but counterproductive. But they did not respond in this way, and in fact the Académie des Sciences included two of the greatest experimenters of the eighteenth century, Franklin and Lavoisier. Their examination was not only thoroughly experimental, and took great care to examine a significant number of cases, but was also highly innovative in the way it conducted its experiments: the reports of patients subjected to mesmeric procedures were tested against reports of those who were told they were being subjected to these procedures, but in fact weren't. Here, for the first time, a placebo comparison was carried out. In short, this was exactly the kind of response needed to neutralize the claims that the scientific authorities were putting preconceived ideas before empirical enquiry.

Second, there was the question of democracy. Why should the Académie des Sciences, the Sorbonne Faculty of Medicine, and the Société Royale de Médecine be entitled to sit in judgement on those who were not members of those bodies and had sought no authentication from them for their scientific theories and therapeutic practices? Was this not a violation of basic democratic principles?

[143] See the discussion of the case in Riskin, *Science in the Age of Sensibility*, ch. 5.
[144] Ibid., 183.

Such criticism clearly goes beyond anything that can be resolved by experiment, and it was Condorcet who was to respond to these criticisms at the time, then, ten years later, to formulate his responses in more general terms, ones that held both for monarchy and republic, in his *Mémoires sur l'instruction publique* (1791–2).[145] The Académies, he argued, were recognized by monarchical authority but were not created by it. In fact, this is not literally true, for the monarchy had established and funded them (and very much for its own purposes),[146] but it remained the case that they were not established *ex nihilo*, for their membership had been formed out of earlier private societies. The Académie des Sciences had been nurtured by monarchical authority, Condorcet concedes, but it was now ready for independence.[147] Reversing the anti-authoritarian arguments of the Mesmer generation, he argues that, with this independence, the scientist becomes the natural enemy of power, for 'any man who makes a profession of seeking and announcing the truth, will always be odious to those who exercise authority'.[148] Baker sums up the Condorcet view nicely, pointing out that science for him is public knowledge, being 'openly discussed and publicly validated by the consensus of the republic of science. Acknowledging no other source than the free and rational consensus of scientists, the authority of science exemplifies the public authority itself.'[149]

Given the highly charged political construal of scientific authority to which Condorcet was responding, we can appreciate the delicate situation in which any programme of social reform pursued by means of medicine found itself, even one whose proponents were antithetical to mesmerism. This situation shifted from delicate to dire in the course of the 1790s, when the political landscape changed significantly, to the detriment of anthropological understanding generally. I have indicated that the emergence of the *médecin philosophe* could just have been the emergence of yet one more group claiming social and moral authority, had it not been for the fact that, in the wake of the Revolution, there was an authority vacuum, as it were, due to the collapse of church and state authority, and this created a situation in which physicians found themselves as authorities in a range of intellectual and social domains. Crucial here is the kind of intimate connection that Destutt de Tracy made between the broader claims of anthropological medicine, which he referred to as physiological ideology, and a revolutionary political programme deriving its principles from what were argued to be basic natural rights, which he termed rational ideology. As we saw earlier, he was adamant that physiological ideology provided the ultimate rationale for

[145] A related response, which stresses the importance of science as a public resource, is to be found Jean-Baptiste Biot, *Essai sur l'histoire générale des sciences pendant la révolution française* (Paris, 1803). Biot's response was important because after the fall of Robespierre there was a rebuilding of the education system in which it became inextricably linked to the goals of the state: see Dorinda Outram, *Georges Cuvier: Vocation, Science and Authority in Post-Revolutionary France* (Manchester, 1984), ch. 4.

[146] See Gaukroger, *The Collapse of Mechanism and the Rise of Sensibility*, 232–47.

[147] Condorcet, *Œuvres*, vii. 305. [148] Ibid., 420. [149] Baker, *Condorcet*, 79.

rational ideology. This presumed connection now became problematic. With the arrest and execution of Robespierre on 27 July 1794, and the institution of a constitutional republic—the Directoire—there was an immediate rounding up of Jacobins, and rejection of the kind of radical politics that Robespierre had pursued. With the coup of 18 Brumaire 1799, and the institution of the Consulate under Napoleon, there was a counter-revolutionary shift to the right. This autocratic regime, one in which the Catholic Church regained much of its standing in the wake of Napoleon's signing of a concordat with the Vatican in 1801, fostered an environment not conducive to the kind of naturalizing claims that characterized anthropological medicine, claims now inextricably connected to a revolutionary programme, even though they were conceived quite independently of it. Support for the kinds of very strong claims to which anthropological medicine was committed declined sharply during and after the Consulate, especially in cases where they had been associated explicitly with a now-discredited revolutionary programme.

THE DEMISE OF SENSIBILITY

The problems with mesmerism went beyond those arising from the political agendas of some of its supporters. Far more significant in the long term was the nature of the behaviour induced by mesmerists in their patients. Mesmer and his followers assumed a very holist view of the patient and her environment, one that did away with any internal/external distinction. This was something to which the *médecins philosophes* of late eighteenth-century Paris could hardly have taken exception, and certainly the idea of subjecting the body to a non-invasive regimen would have been commonplace by this time. Even the control of the ubiquitous animal spirits through movements of the mesmerist's hands may not have seemed that bizarre: the dissenter from the report of the Société Royale de Médecine, the botanist Antoine-Laurent de Jussieu, postulated 'animal heat' streaming from the fingers of the mesmerist, so the idea of fingers being used to effect a cure was not one confined to mesmerists. Nor was the postulation of an all-encompassing fluid—animal spirits—in itself so unusual, as this was not that dissimilar to the various aethers that had been postulated from Newton onwards to account for a range of phenomena from gravitation to sensation. Animal magnetism was not only a sixth sense for Mesmer and his followers, but it underlay the other five senses, in that it was the material basis of sensibility in general, and specifically of sensation. In an account reminiscent of ancient philosophical theories of perception, it was maintained that sensory perception—whether vision, hearing, smell, touch, or taste—was effected when animal spirits passed from the objects sensed to the sense organs and thence to the brain. These animal 'spirits' were actually a flow of a physical substance. It was a physical substance that could not be detected directly, only through its effects, but that in itself was not a problem since it was generally accepted that gravity acted in this way, as did 'mineral' magnetism, on which animal magnetism was clearly modelled in this respect. In cases such as heat, mineral magnetism, and electricity, various fluids had been posited as the carriers of the effects, and these fluids had a distinctive feature: they were 'imponderable' (weightless) so could not be detected directly, only through their effects. Finally, as I have indicated, the procedures used by mesmerists were not that different from the electrical therapies that received support from the medical community at the same time. In short, at the level of theoretical understanding, and to some extent practical application, what Mesmer was proposing fell within the purview of the sensibilist assumptions of anthropological medicine.

The issues for anthropological medicine arose rather as a result of the reports of the investigating commissions.[150] The mesmerists did not claim effects that they could not demonstrate, for the commissioners were able to reproduce them. The problem was that they were able to reproduce them at will, whether or not the patients were being mesmerized: what mattered was that they thought they were being mesmerized. The patients' behaviour, unusual as it was, was the work of their imaginations, brought about by the dramatic, highly charged circumstances, often with a theatrical element, that accompanied mesmerist cures. While the commissioners were able to discredit the mesmerist explanations for their results without difficulty, what they discovered in the course of tests was that the imaginations of patients could be excited in an often violent way, even when they were misled into believing that they were being mesmerized. This was not a result that made any sense in traditional sensibilist terms.

In pushing sensibility in a particular direction, Mesmer revealed a dimension of sensibility that turned out to undermine the rationale behind explaining cognitive and affective states in terms of sensibility rather than in terms of reason. This rationale, at its most basic level, was that sensibility put us in touch with the world in a way that reason did not. Reason could not even establish a relation between our cognitive states and a physical world, let alone between our affective states and the physical environment in which we lived. The claim was that sensibility not only could do these, but, starting from nervous sensitivity, it could offer an account in which empirical evidence played a deciding role, by contrast with reason, where everything remained at a speculative level. The explanatory and therapeutic value of anthropological medicine depended crucially on the idea of sensibility as a faculty that establishes a connection between the mind and the physical world, by contrast with reason, which was condemned because of its internal self-reflective nature. What the investigation of mesmerism showed was that sensations could be induced that had no relation to the physical world. They could be self-contained, like reason, in which case one of the basic assumptions

[150] See Riskin, *Science in the Age of Sensibility*, ch. 6.

motivating the preference for sensibility over reason in explaining our experience of the world, and the cognitive and affective relation that we have to it, has to be abandoned.

As well as problems with sensibility, there were also difficulties with the attempt to naturalize it. There is a fundamental problem with the project of naturalization employed in anthropological medicine: not with the fact of naturalization itself, which is of course what Catholic and other critics focused on, but with competition between naturalizing strategies. For the *médecins philosophes*, sensibility was the vehicle of naturalization. The legitimacy of their claims to offer a new and fruitful account of human beings that bypassed the dead ends of metaphysical and theological speculation rested crucially on naturalization, that is, the appeal to empirical evidence in the treatment of questions which had hitherto been dealt with in a priori or purely conceptual terms. But naturalization is not any one single procedure: it rests on strategies of naturalization that privilege certain forms of naturalization and marginalize or exclude others. This can be overlooked if one thinks of naturalization in monolithic reductionist terms: as micro-reduction for example, or some other form of reduction to physics. But naturalization, as I have defined it, is about what kinds of resources one deploys and what kind of evidence one appeals to in an investigation. There is no reason to think that this must take a reductive form, and in the late eighteenth and early nineteenth centuries one crucial kind of naturalization takes the form of opening questions up to historical evidence.

In the present context, it is a question of competing naturalizing strategies on the issue of where exactly the empirical dimension of one's enquiry lies and what kind of empirical evidence one should be appealing to. There was a widespread rejection among those coming from the *médecin philosophe* tradition of experimental work in medicine, which had taken its cue from Morgagni's 1761 textbook of pathological anatomy,[151] which had exercised a great influence on French anatomists such as Antoine Portal, Joseph Lieutaud, and above all Philippe Pinel.[152] Not only pathological anatomy, but the relevance of any experimental enquiry was questioned. Barthez, for example, in the second (1806) edition of his *Nouveaux éléments* was adamant in his opposition to the new chemical theory of respiration, preferring vitalistic explanations of animal heat. And as Staum notes, leading medical figures such as La Caze, Fouquet, and de Sèze castigated 'cruel animal experiments', and denigrated the 'optical illusions' of microscopy as an aid to pathology.[153] Bichat was no exception, arguing that 'when people observe in

[151] Giovanni B. Morgagni, *De sedibus et causis morborum per anatomen indagatis libri V, dissectiones et animadversiones nunc primum editas, complectuntur* (2 vols, Venice, 1761).

[152] See Rey, *Naissance et développement du vitalisme*, 10.

[153] Staum, *Cabanis*, 84–5. Before the introduction of achromatic lenses into Paris in 1827, the quality of microscopic images was poor, and subject to criticism by many anatomists and physiologists. See for example the detailed criticisms in the standard anatomy textbook of Ernst Heinrich Weber, *Allgemeine Anatomie des menschlichen Körpers* (Brunswick, 1830).

conditions of obscurity, each sees in his own way and according to the way in which he is affected. It is, therefore, observation of vital properties that must guide us, rather than the blurred images provided by the best of microscopes.'[154]

Cabanis brings out what is at stake in a revealing way. While warning that medicine must rid itself of unwarranted intrusions from the other sciences, at the same time he insists that the time was ripe for Hippocratic reform 'to place medicine in harmony with the other sciences and to determine precisely their mutual relationships'.[155] These scientific truths, he argues, will eventually form a chain, and their points of contact will enable one to deduce new truths from basic principles. Yet the physician had to know these foundations of his discipline without subverting its identity or autonomy. Defending Stahl's view of the uselessness of chemistry for medicine, he warns that the physician must guard against 'specious deductions' from chemistry.[156] Medicine is, in true Hippocratic fashion, a matter of observation of the patient, and cannot be replaced by laboratory observation which employs animate and inanimate bodies. The only certain results in medicine are those that come from observing animate bodies, for to the extent to which they concern inanimate bodies, physical and chemical hypotheses could provide no insight into the nature of living things, by contrast with sound Hippocratic medical practice, that is, detailed observations at the sickbed and in infirmaries. The crucial point, however, is not that physiological research and experimental medicine were rejected as being of no value to medicine: rather they were rejected as being of no value to medicine *in their own right*. Medicine needed independently to establish its own truths in its own way before relating them to chemical discoveries.[157]

If we compare this approach with developments in England, particularly in the therapeutic use of gases, in the last decades of the eighteenth century, where there was a very different kind of concern to pursue the social aspects of medicine, we can begin to appreciate just how Cabanis' strictures on the relations between medicine and the other sciences were something peculiar to French culture at the turn of the century.[158] The key developments in England were initiated by Joseph

[154] Xavier Bichat, *Anatomie générale appliquée à la physiologie et à la médecine* (Paris, 1801), 51. The key figure in eighteenth-century physiology, Haller, had built up his project on an experimental basis, and Bichat himself had carried out significant experimental work, as his *Anatomie générale* makes clear. On Bichat's ambivalent position on the question of experimentation, see William Randall Albury, 'Experiment and Explanation in the Physiology of Bichat and Magendie', *Studies in History of Biology* 1 (1977), 47–131.

[155] See *Oeuvres philosophiques de Cabanis*, ii. 70–5.

[156] The Montpellier physicians had offered a sustained criticism of chemistry from around 1750 onwards: see Rey, *Naissance et développement du vitalisme*, 92–3.

[157] *Oeuvres philosophiques de Cabanis*, ii. 169. See Staum, *Cabanis*, 153–5; and Sergio Moravia, 'Philosophie et médecine en France à la fin du XVIIIe siècle', *Studies on Voltaire and the Eighteenth Century* 89 (1972), 1089–151.

[158] It is interesting in this respect that opponents of Lavoisier's proposed reforms of chemistry singled out 'pneumatic chemistry' as their target in the *Journal de physique*, a forum for opposition to the new chemistry when it was revived in 1798.

Priestley, who in the mid-1770s reported various new kinds of gas, including nitrous and nitric oxides, ammonia, and oxygen. By the last decades of the eighteenth century, pneumatics was having a significant social impact: gases that Priestley had identified began to be used for a variety of medical purposes and in order to aid more general well-being. In 1772 he devised a process for producing artificially carbonated water,[159] the carbon dioxide being generated by a mixture of chalk and acid, and bubbled through a container of water. Within four years this sparkling water had replaced distilled water on ships undertaking long voyages, as the source of unadulterated water.[160] By the end of the century, soda water had become a very popular worldwide commodity, with significant health benefits being claimed for it.

The impact of nitrous oxide was no less spectacular. Priestley had vigorously promoted the medicinal use of the gases he had produced, but in the wake of the burning of his house and laboratory by a Tory mob in 1791, he found it difficult to pursue further scientific work, and in a climate where dissenters were being charged with seditious libel, he emigrated to Philadelphia in 1794.[161] His medicinal work on gases was continued after he left, however, by Thomas Beddoes. Beddoes was also a sympathizer with the French Revolution, and had had to leave his Oxford readership in chemistry in 1792. Later in the 1790s Beddoes worked to set up an institution for the treatment of diseases by the inhalation of different gases. He was aided in his laboratory work from 1798 by his young assistant, Humphry Davy, as well as by his friend James Watt, whose engineering skills proved useful in the manufacture of gases. His 'Pneumatic Institution' finally opened in Bristol in 1799 with Davy taking over its running.[162] Beddoes held back at first from applying what, in the circumstances, were pneumatic remedies that might have been considered politically sensitive, until his patients and others had built up confidence in the medical skills of those at the Institute, although he experimented continuously, working on gases, electricity, and natural phosphorescence. The main focus in the clinic was on the treatment of patients with tuberculosis, but ulcers and paralytics were also of special interest. Once pneumatic procedures had been reintroduced, inhaled oxygen and hydrogen were the principal means of treatment, although results on consumptives proved disappointing. At the same time, Davy was also experimenting on the

[159] Joseph Priestley, *Experiments and Observations on Different Kinds of Air* (3 vols, London 1774–7), i. 25–43.

[160] There was subsequent improvement and commercialization of the production process by others, and in 1783 the Schweppes company was founded in Geneva, manufacturing soda water using pressure to introduce the gas into the water.

[161] See Robert E. Schofield, *The Enlightened Joseph Priestley: A Study of His Life and Work from 1773 to 1804* (University Park, Penn., 2004).

[162] The experiments and treatments carried out are described in Thomas Beddoes, *Notice of Some Observations Made at the Medical Pneumatic Institution* (Bristol, 1799). On Beddoes and Davy see Jan Golinski, *Science and Public Culture: Chemistry and Enlightenment in Britain, 1760–1820* (Cambridge, 1992).

effects of nitrous oxide, experiments that he had started two years earlier but had not been able to follow up at the time.[163] He was able to produce a pure form of nitrous oxide, and on pinching his nose and inhaling it through a tube he experienced 'a sensation analogous to gentle pressure on all muscles, attended by a highly pleasurable thrilling in the chest and extremities'.[164] The objects around him became 'dazzling and my hearing more acute', and he ended the experiment in a frenzied state. Beddoes and Davy began to try small doses of the gas on palsy patients, since 'a propensity to action' was one of the effects that Davy had recorded from his own experiences with the gas, and this was exactly what palsy sufferers lacked. The inhaling of the gas seemed to have the requisite effects in the few cases in which it was administered, and as one commentator has noted, after years of administering foul-tasting medicines and painful treatments to victims of palsy, Beddoes had finally discovered a medicine whose only side effect appeared to be pleasure.[165]

The early positive results in the treatment of palsy were encouraging, but were not able to be reproduced in treatment of other conditions. Davy suggested a number of possible medical uses of nitrous oxide, focusing on its use for resuscitating unconscious or debilitated patients, adding in passing that since it relieved pain, it could perhaps be used in certain types of surgical operations,[166] although this suggestion of an anaesthetic use for the gas was not followed up.[167] By the turn of the century, however, nitrous oxide was increasingly being used as an aid to understanding perception and consciousness by Beddoes and his circle.[168]

The contrast with the kind of programme that Cabanis and his colleagues were promoting could not be starker. For them, medicine had to be in the driving seat: not chemistry, not anatomy, not pathology, not laboratory tests. These areas could be pursued, but only if they could be assimilated to the naturalized sensibility that forms the basis for medicine with its newly acquired anthropological aspirations. To appreciate what is at issue here, we need to consider the achievements of anthropological medicine. Above all, its claims were that the reductionist aspirations of biomechanics had finally been overcome, and that the existence of a vital principle, which only medicine can capture, had been established. At the turn of the century, its successes were significant: from the wholesale reform of hospitals to the new understanding of nervous diseases. This was potentially undermined if the autonomy of medicine was threatened by

[163] See Humphry Davy, *Researches Chemical and Philosophical, Chiefly concerning Nitrous Oxide, or Dephlogisticated Nitrous Air, and its Respiration* (London, 1800).

[164] Ibid., 458. [165] Jay, *The Atmosphere of Heaven*, 174.

[166] Humphry Davy, *Researches Chemical and Philosophical*, 556.

[167] The use of nitrous oxide as an anaesthetic had to be rediscovered in the 1840s, and it was not until 1868 that the Royal College of Surgeons approved its use in this role.

[168] By contrast with the daily medicinal use of gases, in the evenings they experimented with the effect of nitrous oxide on their perceptual and psychological states and, despite significant variation in the effects of the gas, it was to be portrayed in the anti-Jacobin press in terms of wild nitrous oxide parties.

anything that suggested a return to something that ignored or did not respect life as the core issue.

The key to anthropological medicine is the naturalization of sensibility, not naturalization for its own sake, because the naturalization of sensibility led to a fruitful understanding of how something whose sole source of knowledge was sensation could connect with the world in the way in which we know human beings do, an understanding that put knowledge of the human condition on a new and fertile path. Physiological research and experimental medicine that in effect attempted to bypass sensibility were tangential to this project and could not usurp the controlling position of medicine. The stakes were high, and Cabanis for one was clear on the importance of securing medicine as the guiding discipline for scientific enterprise in general. This is strikingly evident in his answer to the question whether one should explain every kind of attraction—such as chemical affinities for example—in nature in terms of sensitivity, or whether it is by means of (or on the model of) gravity that such forms of attraction operate. His answer is that gravitation itself must be explained in terms of sensitivity, for even in ordinary inert matter, which is subject to gravitation alone, one senses an inherent tendency to harmonious, orderly motion guided by a single principle.[169]

Cabanis' view on the priority of sensibility over gravity is not one that he defended in public, and we do not know what position other *médecins philosophes* would have taken. What we do know is that, even leaving mesmerism to one side, gravitation was widely taken as a model for how life forces might act: its action at a distance, lawlikeness, and non-reducibility to mechanistic conceptions of matter had been attractive to those thinking about the nature of life and sensation from Maupertuis onwards. Barthez, as I have indicated, in advocating a single life principle instead of the distributed life forces of Bordeu, explicitly took Newton's account of gravitation as his model. Not only did one need a single principle that could account for a range of physiological responses as coherently as Newtonian gravitation accounted for physical and astronomical ones, but his vital principle, like Newton's gravitation, should be an experimentally established principle: it is not a question of postulating metaphysical causes.[170] But as Cuvier pointed out in a 1806 paper,[171] Barthez's vital principle was quite unlike Newtonian gravitation in the relevant respects. Whereas gravity was precisely defined, the vital principle was described only in vague terms. Its status was wholly mysterious because it was neither material nor immaterial, neither mechanical nor intelligent. To say that everything from muscular contraction to the formation of the foetus and the reproduction of species was due to a vital principle had no explanatory power at

169 *Oeuvres philosophiques de Cabanis*, ii. 195–7. See Staum, *Cabanis*, 153–5.

170 See Elizabeth Haigh, *Xavier Bichat and the Medical Theory of the Eighteenth Century* (London, 1984), ch. 2, to which I am indebted here.

171 Georges Cuvier, 'Histoire de la classe des sciences mathématiques et physiques', *Mémoires de l'Institut national des sciences et arts. Sciences mathématiques et physiques* 7 (1806), 1–79.

all, by contrast with Newton's claim that the orbit of the planets around the sun and the fall of bodies to the ground were each a function of gravitational attraction. In short, to the extent to which anthropological medicine was tied in to the vitalist programme, its naturalist credentials were now called into question.

In sum, the problem now was how to hold together a programme which had extolled the virtues of naturalization against metaphysical and theological speculation, but which at the same time resisted forms of naturalization in physiological research and experimental medicine that were the result of the increasing use of microscopy and chemistry. Anthropological medicine had no solution to this problem, and it went into decline in the early decades of the nineteenth century.[172] The new medical works that appeared at the time of the Restoration of the Bourbon Monarchy, such as Magendie's 1816 *Précis élémentaire de physiologie* and Laënnec's 1819 *Traité de l'auscultation médiate*,[173] offered experimentally oriented accounts quite different in style from that of anthropological medicine. Magendie did not take his starting point from particular organs but from particular functions, such as respiration, which involved many organs. As a result, sensibility and irritability became effects rather than causes, thereby losing their central explanatory role.[174] At the same time, the emergence of cell biology redirected the focus of the life sciences. In the 1820s, Dutrochet started to consider cells as physiological entities, as the basic units of metabolic exchange.[175] Developments in microscopy in the 1830s lead to the investigation of fine structures, and in 1837 Schleiden and Schwann, comparing their respective investigations of plant and animal cells, realized that all cells had nuclei, and performed experiments on a wide variety of organic substances to establish that all living things are composed of cells and cell products.[176] The combination of the work of Dutrochet and Raspail on cell chemistry,[177] and Schleiden and

[172] See, for example, the revealing discussion of the decline of vitalism in August Comte, *Cours de Philosophie Positive* (6 vols, Paris, 1830–42), iii. 272–4 (Quarantième Leçon).

[173] François Magendie, *Précis élémentaire de physiologie* (2 vols, Paris, 1816–17); René-Théophile-Hyacinthe Laënnec, *Traité de l'auscultation médiate ou Traité diagnostic des maladies des poumons et du cœur, fondé principalement sur ce nouveau moyen d'exploration* (2 vols, Paris, 1819).

[174] See Haigh, *Xavier Bichat*, ch. 8.

[175] René Joachim Henri Dutrochet, *Mémoires pour servir à l'histoire anatomique et physiologique des végétaux et des animaux* (Paris, 1824).

[176] Theodor Schwann, *Mikroskopische Untersuchungen über die Uebereinstimmung in der Struktur und den Wachstum der Thiere und Pflanzen* (Berlin, 1839). The 'Preface' provides a clear statement of the project: 'The object of the present treatise is to prove the most intimate connexion of the two kingdoms of organic nature, from the similarity in the laws of development of the elementary parts of animals and plants. The principal result of this investigation is, that one common principle of development forms the basis for every separate particle of all organized bodies, just as all crystals, notwithstanding the diversity of their figures, are formed according to similar laws.' Theodor Schwann, *Microscopical Researches into the Accordance in the Structure and Growth of Animals and Plants* (London, 1847), ix.

[177] Dutrochet, *Mémoires*; François-Vincent Raspail, *Nouveau système de physiologie végétale et de botanique fondé sur les méthodes d'observation, qui ont été dévelopées dans le nouveau système de chimie organique* (Paris, 1833).

Schwann's account of cell structure, meant that tissues, which had played such a crucial role in the physiology underpinning developments in anthropological medicine, gave way to cells as the ultimate units of life.[178] Experimental physiology, now bypassing sensibility, took over the task of providing an empirical understanding of life, and anthropological medicine was finally doomed beyond redemption.

[178] See Henry Harris, *The Birth of the Cell* (New Haven, Conn., 1999).

4
Philosophical Anthropology

In an essay from 1765, Herder, the pivotal figure in the elaboration of a philosophical anthropology in the eighteenth century, wrote: 'If philosophy's viewpoint gets changed in the manner in which the Copernican system developed from the Ptolemaic system, what new fruitful developments must not occur here, if our whole philosophy becomes anthropology.'[1] Unlike anthropological medicine, which pits reason against nervous sensibility, which is then broadened to sensibility in general, philosophical anthropology pits reason against sensation, which is then broadened to sensibility in general.[2] What is important about both sensation and nervous sensibility, and what distinguishes them from reason as traditionally conceived, is that they are amenable to empirical enquiry. This is the key to the strategies of naturalization.

The strategy is nowhere more evident than in the work of Herder, and we can think of his project in terms of a threefold naturalization. He approaches reason by understanding it in terms of thought; he approaches thought via what the Wolffian tradition characterized as empirical psychology; and the tool that he uses in this empirical psychology is language, which he treats as an empirically analysable manifestation of thought.[3] This empirically analysable manifestation has a historical dimension: the history of languages—'history' in the eighteenth-century sense, as something that has both a diachronic and a synchronic dimension—is a history of thought, which in turn is a history of reason, albeit one of a far more subtle form than the 'triumph of the present' type history of thought offered, for example, by Diderot and d'Alembert in the preliminary 'Discours' to the *Encyclopédie*, or by Condorcet in his *Esquisse*. At the same time, it is far more plausible and fruitful than the materialist reductions of the mind offered by La Mettrie and Helvétius. In many respects, Herder's is a model

[1] Johann Gottfried Herder, 'Wie die Philosophie zum Besten des Volks allgemeiner und nützlicher werden kann', *Werke*, i. 134; idem, *Philosophical Writings*, ed. and trans. Michael N. Forster (Cambridge, 2002), 29.

[2] We shall be concerned with philosophical anthropology as a movement in Germany in the second half of the eighteenth century. On earlier developments which can be described as anthropology, see Simone De Angelis, *Anthropologien: Genese und Konfiguration einer 'Wissenschaft vom Menschen' in der Frühen Neuzeit* (Berlin, 2010).

[3] Actually, as we shall see, it is richer than merely a manifestation of thought, but this characterization will do for our immediate purposes.

of successful naturalization, one in which reductionist strategies play a far less significant role.

Nevertheless, Herder's anthropology was just one of the two main forms of philosophical anthropology proposed in the second half of the eighteenth century. The other, that of Kant, rejected the kind of understanding that Herder sought as impossible, and rejected naturalizing projects generally on the basis of an understanding of scientific enquiry which took rational mechanics as its model, and was very different not only from that of Herder, but also from that of their contemporaries, for whom chemical, electrical, and biological models had displaced mechanical ones. Kant thought there was an autonomous discipline of anthropology, and indeed accorded it great significance, but argued that it could not issue in scientific understanding, and indeed could not aspire to the standing of 'knowledge' as he construed this. Nevertheless, he assumes that it produces a distinctive and legitimate form of understanding. At stake here is the question of the relation between anthropology and naturalization, at its most fundamental level the question of what kind of understanding we should be seeking of human nature.

METAPHYSICS AND PSYCHOLOGY

From the Renaissance onwards, study of the sources of human behaviour had been claimed by three different kinds of enquiry: Christian theology, metaphysics, and civic humanism. Much of this literature had focused on moral questions, although civic humanism added a social and political dimension to the discussion. It was assumed in this literature that there were innate, universal moral precepts, and for almost all the writers in these traditions these precepts were those that had been set out in Christian teaching. But from the beginning of the sixteenth century, travel books about remote regions had started to appear, and these raised the spectre of moral diversity, casting doubt on the innate nature of morality. Closely tied in with this question was that of the universality of religion, which had been taken as a *sine qua non* of morality. This too had been thrown into question by the travel literature. The problem of moral diversity had come to a head at the end of the seventeenth century in the writings of Locke and, in a more polemical form, those of Bayle. The conclusions that were drawn were that: morality was not innate, and there was significant moral diversity, with incest, cannibalism, and infanticide, for example, sanctioned in some societies; there were societies without religion, such as some of those in the Caribbean, where there appeared to be no understanding at all of the notion of a god among the indigenous population; there existed moral and socially harmonious large-scale societies, such as China, which were deemed atheist; and the religious tenets to which citizens in Christian countries were expected to subscribe could change radically over the course of one's life, for example in early-modern England, to doctrines that blatantly contradicted those to which one had been earlier expected to subscribe, with the result that the notion of religious authority (and by implication moral authority) was undermined.[4] There were various attempts to circumvent these problems: by redefining atheism, for example, or by insisting that societies without religion or a recognizable morality were savage and as such irrelevant to the discussion of human affairs, or, more ambitiously, by seeking to establish that there was a 'natural' morality to which we naturally responded without any need for innate ideas,[5] and so on. But none of these were successful.[6] As Hume puts it in the 'Dialogue' appended to the *Inquiry concerning the Principles of Morals*, in which the question of moral diversity is presented in the sharpest terms: 'What wide difference, therefore, in the sentiments of morals, must be found between civilized nations and Barbarians, or between nations whose characters have little in common? How shall we pretend to fix a standard for judgments of this nature?'[7]

In the course of the eighteenth century, the scope of these questions extended beyond those of morality and religion, to what might be termed mentality. None of the traditional treatments of human behaviour were able to deal with questions of variations in mentality. Metaphysics dealt with questions of human behaviour in a priori terms, as purely conceptual issues, effectively assuming that there could not be any significant difference in mentalities. Theology, when it wasn't being pursued via metaphysics, of the kind we find in Aquinas for example, was concerned with the interpretation of sacred texts in a fashion that read these as habouring timeless truths which it was the aim of interpretation to uncover. While biblical hermeneutics was coming under pressure by the eighteenth century,[8] this was not in a way that bore directly on the study of human behaviour. Civic humanism by contrast relied significantly on historical study, and hence had an empirical element. It was the one traditional discipline that had some degree of continuity with the philosophical anthropology of the eighteenth century, although it was not itself part of a naturalizing project.

The recognition of the variability of mentalities in an anthropological context does not seem to have been an isolated phenomenon in later eighteenth-century thought, but part of a more general rejection of the imposition of pre-given

[4] This was a genuine source of worry for religious and secular authorities. In England in the 1640s and 1650s, for example, a number of radical Protestant sects had advocated complete sexual freedom: see Christopher Hill, *The World Turned Upside Down: Radical Politics During the English Revolution* (London, 1972), 247–60.

[5] Sometimes it was not even a question of a natural morality but more of a natural necessity. John Graunt, for example, argues that the greater number of male over female births means that polygamy, i.e. sexual licentiousness, is unnatural, in a way that it would not be (though it would still be immoral) if women outnumbered men: *Natural and Political Observations*, 64–71.

[6] See Gaukroger, *The Collapse of Mechanism and the Rise of Sensibility*, 40–54.

[7] Hume, *Essays and Treatises*, ii. 388.

[8] See, for example, Hans W. Frei, *The Eclipse of Biblical Narrative: A Study in Eighteenth and Nineteenth Century Hermeneutics* (New Haven, 1974).

systems. Buffon, for example, in rejecting Linnaean classifications for animals, treats them as a kind of parochialism in which one extrapolates from what one knows, as if the world were the same everywhere. In their article on 'animal' in the *Encyclopédie*, Diderot and Buffon's collaborator Daubenton begin by discussing Buffon's Lockean definition of the concept of an animal, as a general idea formed from particular ideas, where these particular ideas are taken from our ideas of dogs, horses, etc.—animals with which we are familiar and which are regarded as paradigm exemplars of the concept. But the further we get from such everyday cases, they point out, the less the things we identify as particular animals on this basis seem to manifest the paradigm qualities, and the more conventional the concept appears.[9]

Philosophical anthropology marks a departure from the kind of enquiry in which the basic categories are set out in advance and material is then assimilated to them, for it is directed against a fundamental assumption of metaphysical treatments, that there is a fixed human nature—implicitly typified in early-modern Christians—which can be examined in conceptual or a priori terms. On the metaphysical view, this fixed human nature may be manifested in different ways in different cultures and different historical periods, but this is treated as a superficial feature of human behaviour, dependent on contingent factors such as religion, climate, and upbringing: it does not impinge on human nature itself.[10] Yet while this is the kind of picture that Herder rejects, insisting that the exploration of human nature must be empirical, he nevertheless takes his starting point from Wolffian metaphysics, which insisted on a fixed human nature. He does not naturalize metaphysics so much as take a part of a metaphysics that its proponents had recognized to be an empirical matter, if one that was a means to an end and of minor significance in its own right, and shows how this can be built up to form the core of a new account of the mind.

This core is what Wolff had called 'empirical psychology', distinguishing it from 'rational psychology'.[11] It proceeded primarily by means of an analysis of our own psychological states, but also through an analysis of the behaviour of others, and Wolff writes that it 'is really a *history* of the soul and can be known without

[9] *Encyclopédie*, ii. 673.

[10] See for example Du Marsais: 'The different paths that different peoples have taken to express themselves are subject to the two sovereign rules of uniformity and variety; there is uniformity in the essential nature of thought, and variety in the path followed and in the expression.' *Exposition d'une méthode raisonnée pour apprendre la langue latine* (1722), in César Chesneau, sieur du Marsais, *Oeuvres* (7 vols, Paris, 1797) i. 23.

[11] Wolf devoted two short tracts to these questions: *Psychologia empirica, methodo scientifica pertractata, qua ea, quae de anima humana indubia experientiae fide constant, continentur* (2nd edn, Frankfurt and Leipzig, 1732); and *Psychologia rationalis, methodo scientifica pertractata, qua ea, quae de anima humana indubia experientiae fide innotescunt, per essentiam et naturam animae explicantur* (2nd edn, Frankfurt and Leipzig, 1740). The Prologomena to the two works are translated in Robert J. Richards, 'Christian Wolff's Prolegomena to Empirical and Rational Psychology: Translation and Commentary', *Proceedings of the American Philosophical Society*, 124/3 (1980), 227–39.

any other discipline'.[12] Empirical psychology is designed to act as an ingredient in scientific understanding, namely the deduction of propositions from secure or certain premises, but it also has the related task of providing empirical support for conclusions that have been arrived at via deduction. It deals with perception, and Wolff's elaborate understanding of what is involved in perception is crucial here. Its three components are the represented content, the mind's act of representing, and the consciousness by which the perception itself becomes present to the mind ('apperception').[13] It is apperception—the consciousness, which accompanies perceiving something, of being affected in a certain way by one's perception—that enables one to transform vague and obscure perceptions into clear and distinct ones. Because apperception is involved in every perception, careful attention to our perceptions can yield a veridical experience of the mind's activity.

Wolff defines rational psychology as the science that provides reasons for what occurs in the soul, or what can occur in it,[14] and it takes its subject matter from 'ontology, cosmology, and empirical psychology',[15] because only these have the requisite degree of certainty. The function of empirical psychology was to provide the raw materials for rational psychology, and the task of rational psychology was then to derive the results of empirical psychology from a priori first principles. As he puts it in the *Psychologia rationis*:

> When you rest content with knowledge of the soul a posteriori, you remain satisfied with empirical psychology . . . ; you do not advance to rational psychology, but leave it untouched. You thus fail to complete all parts of philosophy, whose aim is to provide the reason for whatever exists. Therefore, those of us who intend to do philosophy ought to try, as much as we can, to exhibit the connecting links in demonstrations of those matters we wish to consider, so as to fill in any gaps we might detect. The noblest part of philosophy works toward this goal.[16]

The procedure by which we achieve this consists in 'deriving a priori what we establish a posteriori in empirical psychology'.[17] Wolff compares this to the procedure of physics, where we use experiments to test the theories of 'dogmatic physics'. Rational psychology enlarges empirical psychology and borrows from it, but 'it returns with interest what it has borrowed'.[18]

[12] Christian Wolff, *Discursus praeliminaris de philosophia in genere. Einleitende Adhandlung über Philosophie im allgemeinen; Historische-kritische Ausgabe*, ed. Gunther Gawlick and Lothar Kreimendahl (Stuttgart, 1996), §34. On the sources of Wolff's psychology, see Hans-Jürgen Engfer, 'Von der Leibnizischen Monadologie zur empirischen Psychologie Wolffs', in S. Carbocin and L. Madonna, eds, *Nuovi studi sul pensiero di Christian Wolff* (Hildesheim, 1992), 193–215.

[13] *Psychologia empirica*, §§24–5. On apperception see Udo Thiel, *The Early Modern Subject: Self-Consciousness and Personal Identity from Descartes to Hume* (Oxford, 2011), 295–311.

[14] *Psychologia rationalis*, §4.

[15] Ibid., §3.

[16] Ibid., §2; trans. in Richards, 'Christian Wolff's Prolegomena to Empirical and Rational Psychology', 234.

[17] *Psychologia rationalis*, §5. [18] Ibid.

Wolff's distinction between empirical psychology and rational psychology mirrors his distinction between 'historical' knowledge, which concerns itself with empirical matters, and 'philosophical' knowledge, which works from basic principles.[19] In the context of his account of empirical and rational psychology, the task is to convert merely historical knowledge into philosophical knowledge. Note, however, the way in which epistemological questions about the distinction between historical and philosophical knowledge are treated here as psychological questions about thought and cognition. To a significant extent, epistemology is being pursued via psychology. There are conceptual questions, such as the standing of the laws of logic and the principles of mathematics, that Wolff treats in a priori terms, but even logic, for him, has a psychological function in that its aim is to allow us to achieve clarity and distinctness in our ideas. Nevertheless, for Wolff himself, the focus on thought rather than reason has no naturalizing consequences. Empirical psychology deals with thought in the form of sensibility, and rational psychology with thought in the form of reason, and he makes it clear that the former is to be subsumed under the latter. Sensibility in effect has no autonomy, any more than have experiments, as he conceives them: they are a means to an end, and simply provide the raw materials for reason to work on. Indeed, pure sensibility is characteristic of animals, who, not having language and reason, have very limited powers of representation. As a consequence, they lack those characteristic features of human beings: will, morality, personality, and immortality of the soul.[20] Because of this, 'the more the human being clings to his senses and to his imagination, and the more he is as a consequence subject to slavery, the closer he comes to the beast or the brute'.[21]

The most influential development of this approach comes not in Wolff but in the work of his pupil Alexander Baumgarten, the prime representative of what its critics deemed the 'rationalist' view.[22] Baumgarten was committed to the Wolffian programme, and his textbook on metaphysics—the state-sanctioned recommended text in metaphysics courses in German universities in the second half of the eighteenth century—was an attempt to revise and update this programme. Baumgarten went beyond Wolff in crucial respects, most notably in giving sensation a significant role as a source of knowledge. Wolff's metaphysics had accorded sensation no significant role: he construed it simply as a form of

[19] There is a comprehensive account of Wolff's psychology in Thomas Sturm, *Kant und die Wissenschaften vom Menschen* (Paderborn, 2009), ch. 2.

[20] Wolff, *Vernünfftige Gedancken von Gott, der Welt und der Seele des Menschen, auch allen Dinge überhaupt* (Halle, 1751), §872.

[21] Ibid., §892.

[22] There is even some question, for example, as to whether Kant had acquired his knowledge of Wolff from reading Wolff, or from reading Baumgarten: see Jean École, 'De la connaissance qu'avait Kant de la métaphysique wolffienne, ou Kant avait-il lu les ouvrages métaphysiques de Wolff?', *Archiv für Geschichte der Philosophie* 73 (1991), 161–76. Cf. Soo Bae Kim, *Die Entstehung der Kantischen Anthropologie und ihre Beziehung zur empirischen Psychologie der Wolffschen Schule* (Frankfurt am Main, 1995), ch. 1.

cognition, namely sensory perception, which produces only obscure ideas. Such a view of sensation was a highly problematic one by the middle decades of the century, and the most significant respect in which Baumgarten went beyond Wolff was on this question. He had already noted in his early *Meditationes* that, in promising to improve our knowledge, logic actually promises more than it can deliver, for logic deals with intellectual cognition, whereas human cognition generally also involves sensory cognition, so if we rely on logic alone we will leave a whole realm in its original unanalysed state.[23] In particular, Baumgarten argued, if we allow the intellect to rule the lower cognitive faculties, this will have detrimental consequences for art.[24] Wolff had argued that philosophy should aspire to become a science, and that it can only achieve this if it starts from self-evident principles and follows a mathematical method. In the case of architecture, the one art that he writes about in any detail, Wolff applies this procedure, construing the subject in mathematical terms.[25] But for Baumgarten, to proceed in this way risks producing something frigid and scholastic, for we find a richness and vividness in sensory perception that is lacking in purely intellectual cognition.[26]

A basic criterion for recognizing the truth of an idea, one to which the metaphysical tradition from Descartes onwards had adhered, was that of clarity and distinctness. This is where Baumgarten's innovation comes. The criterion had played a crucial role in Wolff, where a clear idea was one that enabled us to discriminate an object in its own right, whereas a distinct idea was one that picked out things in terms of their properties. This latter required comparison of properties, which in turn required an analysis at some level of abstraction. Because of this, it was the work of reason, encapsulated, for Wolff, in the form of logic. The role of logic on this conception was to make confused concepts distinct through analysis. Baumgarten set out to show that a clear but confused idea can be of value in its own right. Indeed, abstraction can be a problem, because it causes one to lose sight of the original unity of the object: there is nothing in the properties of a thing that accounts for how they combine to characterize a single entity. The unity that is thereby lost is most evident in the case of art, where perfection is equated with beauty, and it provides a basis on which to study clarity in its own right.

Baumgarten terms the general study of sense perception, including art, 'aesthetics', and he argues that it is the 'younger sister' of logic: whereas logic deals with intellectual cognition, aesthetics deals with the 'lower cognitive powers',

[23] Alexander Gottlieb Baumgarten, *Meditationes philosophicae de nonnullis ad poema pertinentibus* (Halle, 1735), §115.

[24] Baumgarten, *Aesthetica* (2 vols, Frankfurt an der Oder, 1750–8), §§105, 560, 619.

[25] Wolff, *Elementa Matheseos Universae Tomvs IV, Qvi, Geographiam cum Hydrographia, Chronologiam, Gnomonicam, Pyrotechnicam, Architectvram Militarem atqve Civilem Complectitur* (Magdeburg, 1738), 319–488. On Wolff's aesthetics, see Frederick C. Beiser, *Diotoma's Children: German Aesthetic Rationalism from Leibniz to Lessing* (Oxford, 2009), ch. 2.

[26] Baumgarten, *Aesthetica*, §§105, 560, 619.

namely sensory cognition (*cognitio sensitiva*). As a means to achieving clarity, as opposed to distinctness, his aim is to perfect our powers of perception,[27] and to this end, he sets out to establish that perfected perception is equivalent to beauty, so that his 'science of sensory cognition' is also a 'critique of taste'.[28] That is to say, he makes a direct link between sense perception and aesthetic sensibility. Moreover, there is a set of rules by which the perfectibility of the two can be achieved. There is one important difference between intellect and sensory faculties, however: whereas the former operates by means of demonstrative inference, the confused elements of the latter are related through association, and Baumgarten identifies 'the law of the imagination', namely, the principle that 'the partial perception of an idea conjures up (*recurrit*) the whole idea', with the law of association of ideas.[29] Just as there is a 'natural logic', that is, a grasp of inferential relations prior to any education in logic, so too is there a 'natural aesthetic' which children exercise in their sensory exploration of the world, and above all in playing games.[30] And just as natural logic can be perfected through training in rules developed through an analysis of inference, so too can natural aesthetics be improved in an analogous way.

Nevertheless, in crucial respects the model for analysing the sensory realm is still drawn from that devised for analysing the intellectual realm, and aesthetics falls under a general comprehensive metaphysics for Baumgarten. In particular, the fact that it works via association of ideas rather than deductive inferences between ideas does not mean that aesthetics qua the study of sensation is a genuinely autonomous discipline, and it does not mean that it is not ultimately subservient to reason for Baumgarten. But matters are complicated by the fact that Baumgarten, following Wolff, also posits a continuum between the higher and lower faculties, between understanding and sensibility.[31] This opens up the possibility of another form of enquiry, one in which the aim is to explore the relation between the faculties as placed along the continuum. In itself, it would not rule out the project of subsuming sensibility under reason, but if one had independent objections to such a project—objections which were plentiful from the 1750s—then it would provide a potentially promising route. What, in Wolffian terms, were the lower faculties would now be reassessed, being called upon to provide the model for analysis of the higher ones. It was a matter of using

[27] Ibid., §§1–13. Cf. Baumgarten, *Metaphysica* (Halle, 1739), §§519–33. There are a number of developments in Baumgarten's thinking between the *Metaphysica* and the *Aesthetica*, but the differences need not concern us here, and in any case are slighter than was once thought: see Ursula Franke, *Kunst als Erkenntnis: die Rolle d. Sinnlichkeit in d. Ästhetik d. Alexander Gottlieb Baumgarten* (Wiebaden, 1972).

[28] Baumgarten, *Metaphysica*, §§521, 622.

[29] Baumgarten, *Aesthetica*, §561.

[30] Ibid., §§54–5.

[31] See the discussion of these questions in Zammito, *Kant, Herder, and the Birth of Anthropology*, ch. 2.

the resources employed to explore sensibility, resources that were highly naturalized, to explore the understanding.

This is the route that Herder would take, but it is important to distinguish it from some other ways in which a naturalizing programme that took its bearings from empirical psychology could be pursued. Some of these were in conflict with Wolff's programme, and a good example of this is Bucher's attempt to model perceptual cognition on the *camera obscura*. For Wolff, the distinction between empirical psychology and rational psychology was crucial in the struggle against materialism, and this is nowhere clearer than in the 1724 *Anmerckungen* to the '*German Metaphysics*', where Wolff takes issue with the use of the model of the *camera obscura* for understanding the workings of the mind.[32] The *camera obscura* had been used a model for vision since Kepler's *Paralipomena* (1604), which established that the optics of vision is the same as that of the *camera obscura*.[33] As an account of how vision works, it was generally accepted from that point onwards, revolutionizing the understanding of visual perception. In 1690, in the *Essay Concerning Human Understanding*, Locke, in comparing the operation of visual cognition to that of a *camera obscura*, extended it beyond the external action of light on the eye to the internal 'understanding', that is, to the process whereby ideas come into the brain and are retained as impressions in the form of memories:

> For, methinks, the *understanding* is not much unlike a closet wholly shut from light, with only some little opening left, to let in external visible resemblances, or *ideas* of things without; would the pictures coming into such a dark room but stay there, and lie so orderly as to be found upon occasion, it would very much resemble the understanding of a man, in reference to all objects of sight, and of the ideas of them.[34]

Although Locke does not argue that such a process models thought as such, only the activity of the brain, the idea that thought itself can be explained on the model of the *camera obscura* comes in 1713 with the physician Urban Gottfried Bucher.[35] Bucher makes the claim—distinctive of a radicalized form of Lockeanism—that all mental acts, including those of the intellect and the will, are in fact grounded in sensation, and he moves from this to the argument that, since sensation can be accounted for purely in mechanical terms, thought can be accounted for in a similar fashion.

[32] Christian Wolff, *Anmerckungen über die Vernünfftige Gedancken von Gott, der Welt und der Seele des Menschen, auch alle Dinge überhaupt, zu besserem Verstande und bequemerem Gebrauch derselben* (Frankfurt am Main, 1724). See Michael J. Olson, 'The Camera Obscura and the Nature of the Soul: On the Tension between the Mechanics of Sensation and the Metaphysics of the Soul', Intellectual History Review 25 (2015), 279–92.

[33] Johannes Kepler, *Ad vitellionem paralipomena, quibus Astronomiae pars optica traditvr* (Frankfurt, 1605).

[34] John Locke, *An Essay concerning Humane Understanding* (London, 1690), Book II, ch. x.

[35] [Urban Gottfried Bucher and Johann Baptist Röschel], *Zweyer Guten Freunde vertrauter Brief-Wechsel vom Wesen der Seelen* (The Hague, 1713).

This is Wolff's target, and he sets out to show that 'no body or subtle matter can think, as the materialists claim' by demonstrating that what is distinctive about the soul/mind is that it 'is conscious of itself and what comes before it'.[36] The crucial difference is that if a machine, working for example on the principle of the *camera obscura*, were able to represent an external body, this representation would not be experienced as external, whereas the soul or the mind represents everything as external to itself. We are back here with the question that motivated French Lockeans from Condillac onwards, namely how it is possible for our cognition of the world to present it as something external to ourselves. Their answer, as we have seen, was that this was in fact impossible simply on the basis of a cognitive relation to the world, and that it was only sensibility that could establish the independence of the world from us. By contrast, Wolff clearly believes that cognition alone can do this, but only on condition of an immaterial soul, for purely material processes are not sufficient. It is not possible, he tells us, for a machine to compare the motions that occur in it, in such a way as to perceive the difference between the representation that results and the thing represented. Because of this, purely material representations (*Vorstellungen*) can never become thoughts (*Gedanken*).[37]

Wolff's criticisms here do not mean, however, that he was opposed to naturalization as such. As we saw in Chapter 3, it is rarely a question of advocating or rejecting naturalization *tout court*, for what is typically at issue are competing strategies of naturalization. Competing strategies play a role in the present case. Wolff was opposed to the reduction of the mind to the brain, but his promotion of empirical psychology was an advocacy of a form of naturalization. So long as this naturalization respected the sharp distinction between empirical and rational psychology, there was no issue. So, for example, one self-styled Wolffian, the physician Johann Gottlob Krüger, embarked on 'an experimental science of the mind'.[38] Krüger's project was to bring together the methods and results of physics, the knowledge of the brain provided by physiology (which he construed in a largely mechanistic way), and the knowledge of case histories drawn from clinical observation.[39] The experimental advances he had in mind were primarily experiments on animals, but also included those on patients who had suffered brain damage. Questions concerning the nature of the soul and its immortality were not touched upon, and there is no sense in which Krüger's work (which in

36 Wolff, *Anmerckungen*, 396. Olson, 'The *Camera Obscura* and the Nature of the Soul', is, I believe, the first to identify Bucher as Wolff's target, and to appreciate that Wolf's discussion turns on the inadequacy of the *camera obscura* as a model for the mind.

37 Christian Wolff, *Vernünfftige Gedancken von Gott*, §755.

38 Krüger's main work is *Versuch einer Experimental-Seelenlehre* (Halle/Helmstädt, 1756). But see also the earlier *Naturlehre: Nebst Kupfern und vollständigem Register* (3 vols, Halle, 1740–9).

39 On Krüger, see Gary Hatfield, 'Remaking the Science of the Mind', in C. Fox, R. Porter, and R. Wokler, eds, *Inventing Human Science: Eighteenth-Century Domains* (Berkeley, 1995), 184–231: 201–5.

any case did little more than summarize the results of his predecessors and contemporaries) strays from the borders of empirical psychology as defined by Wolff.

From the 1770s onwards, there were a number of works by physicians, in particular, offering contributions to a general anthropology, the most influential being Ernst Platner's *Anthropologie für Ärtze und Weltweise*, which appeared in 1772. But, even though reference is occasionally made to *médecins philosophes* such as Tissot, these works are very different from the kinds of literature being produced in France. Platner, for example, takes his starting point from what is essentially a metaphysical problem, first set out by Leibniz, about the relation between mind and body. At stake here were three views of this interaction, considered to exhaust the possibilities: physical influx theory, by which a change in a substance (such as a body) can genuinely cause a change in another substance (even a different kind of substance, such as a mind); occasionalism, which rules out any causal interaction between things; and pre-established harmony, which rules out causal interaction but where the behaviour of substances runs in parallel as it were.[40] In the 'Preface' to the *Anthropologie*, the starting point is identified as physiological and anatomical observations of human beings, to be combined with what is in effect an empirical psychology, the combination of the two yielding a synthesis of body and soul, which he terms anthropology, although there are a number of metaphysical and theological assumptions built into this. Advocating an influx theory, that is, an account stressing direct causal interaction between mind and brain, he offers a naturalized account of thought in which excessive mental exertion causes brain damage, a view we have encountered in looking at the *médecins philosophes*. But the project here is rather different from theirs, and shows its metaphysical origins. It is more a philosophical theory of mind with an explicitly empirical dimension, rather than what we identified in Chapter 3 as anthropological medicine.[41] Platner's follower Johann Metzger uses the term 'medical psychology', and this sums up the project better. He defines it in these terms:

> Medical psychology is distinguished from philosophical psychology in that the latter isolates the soul; that is, it observes it as an entity separated from the body and investigates its characteristics under that viewpoint. The former [i.e. medical psychology], by contrast,

[40] For details, see John Yolton, *Locke and French Materialism* (Oxford, 1991), ch. 1.

[41] There were German physicians whose work fits into anthropological medicine, as I have characterized it, but their role is marginal compared to the French movement. Zimmerman's *Von der Erfahrung in der Arzneykunst* (1763–4) is probably the leading theoretical statement of anthropological medicine. At a practical level, the medical practice of Marcus Herz was very much in the physiological/moral tradition, and it formed the basis for the journal *Gnosse Seauton: Magazin für Erfahrungsseelenkunde* (1783–1793), edited by Karl Philipp Moritz, who promoted a case-study approach. On these movements see Martin L. Davies, 'Karl Philipp Moritz's *Erfahrungsseelenkund*: Its Social and Intellectual Origins', *Oxford German Studies* 16 (1985), 13–35. Specifically on Herz, see idem, *Identity or History! Marcus Herz and the End of Enlightenment* (Detroit, 1995).

takes as its object the soul bound up with the body . . . as well as the phenomena arising from this unification.[42]

Although there were, from the 1770s to the 1790s, a significant number of publications on anthropology in Germany,[43] most of them are contributions to the 'natural history of man', which will be our topic in Chapter 5, rather than to what I have termed 'philosophical anthropology'. But philosophical anthropology did flourish, above all in the work of Herder.

Herder's naturalism, while it does involve a commitment to the theory of physical influx, is primarily concerned with a different set of questions. What was at issue in influx theory is not simply a metaphysical problem about the relation between mind and body. There was also an issue about where metaphysics fitted into the picture.[44] This question was raised in Flögel's *Geschichte des menschlichen Verstandes*, first published in 1765. Flögel sets out to follow 'the methods of a natural scientist' with a view to providing 'a natural history of human understanding', and avoiding speculation.[45] He rejects metaphysical treatments of the relation between mind and body as fruitless. Mind–body interaction was a fact, he pointed out, and we could make good empirical and pragmatic inroads into knowledge of this fact if we left metaphysical issues to one side.[46] For thinkers such as Herder who were committed to naturalistic projects, the choice of influx theory was not necessarily a commitment to a metaphysical position on the mind–body question, but may, as in Flögel's case, simply be a recognition of a fact, leaving metaphysics to one side.[47]

Herder's primary naturalistic interests lie not in the question of how mind and body can interact, but in uncovering the empirical resources that will enable one to explain what it is that shapes one's cognitive and affective states. As we have seen, in the Wolffian tradition the role of empirical psychology had been above all a necessary preparatory one, organizing the raw materials upon which an a priori structure could then be imposed, and thereby yielding a degree of understanding which empirical psychology in its own right could not aspire to. For Wolff, the

[42] Johann Daniel Metzger, *Medizinisch-philosophische Anthropologie für Aerzte und Nichtarzte. Zum Gebrauch akademischer Vorlesungen* (Weissenfels and Leipzig, 1790), 55. Cited in Steven Lestition, 'Kant's Philosophical Anthropology: Texts and Historical Contexts, Continuity and Change', unpublished PhD dissertation, University of Chicago, 1985, 698.

[43] See Lestition, 'Kant's Philosophical Anthropology', ch. 12.

[44] I am not denying that there were metaphysical issues, and Herder takes his starting point from Leibniz's dynamism, with Spinozan elements coming into the picture. See Beate Dreike, *Herders Naturauffassung in ihrer Beeinflussung durch Leibniz' Philosophie* (Wiesbaden, 1973); and Herbert Lidner, *Das Problem des Spinozismus im Schaffen Goethes und Herders* (Weimar, 1960). The question is what kinds of issue shape the discussion and determine what metaphysical resources are drawn upon.

[45] Carl Friedrich Flögel, *Geschichte des menschlichen Verstandes* (2nd edn, Breslau, 1773), 7.

[46] Ibid., 137–8. See the discussion in Zammito, *Kant, Herder and the Birth of Anthropology*, 238–9.

[47] See Karl Figlio, 'Theories of Perception and the Physiology of Mind in the Late Eighteenth Century', *History of Science* 12 (1975), 177–212.

first stage was as far as one could get with sensibility, so that it was considered that the problem with sensibility was that it could never aspire to anything beyond the kind of limited and necessarily incomplete comprehension offered by empirical psychology. By the second half of the century, however, the second stage—that of providing metaphysical structure—had fallen into disrepute outside what was by this time a very small Wolffian circle, not least because of Hume's onslaught on metaphysics. In its place, what was now effectively being proposed was that the kind of account offered by empirical psychology was really all we had as far as understanding was concerned, and that with an expansion of its resources, it was actually a much more powerful explanatory tool than its treatment at the hands of metaphysicians would suggest. This is the premise underlying the various naturalistically inclined projects, including Herder's. Herder turned not to anatomical or physiological investigation as the primary source for evidence for his naturalizing project, however, still less to the kind of introspection that provided Wolff with his raw materials. He turned instead to language.

LANGUAGE AND ART

The appeal to language transforms sensibility from something inhabiting a private sensory realm into a public manifestation of sensibility which brings with it a rich source of evidence. It amounts to an extension of naturalizing resources to include language, so that language is no longer merely a means of expression of thoughts, but something that goes to the heart of what it is to think. If empirical psychology was to go beyond the role that it played in Wolff's and Baumgarten's metaphysical project, or in the projects of medical psychology—if it was to offer an analysis of thought that had genuine aspirations to clarity and understanding—then thought, in all its distinctiveness, had to be captured somehow. Reason could not simply be rejected in favour of sensibility, along Rousseauean lines, nor could thought simply be reduced to a form of physiological activity, along the reductionist lines that La Mettrie was proposing. If Herder's project was to have any plausibility, thought had to retain some autonomy: there had to remain something distinguishing about it which successful explanation was able to capture.

Language and thought had been associated from antiquity, and Descartes had argued in his *Discours de la méthode* that the fact that human beings have language but animals have none 'shows not only that animals have less reason than men, but that they have none at all'.[48] The close relation between language and rational thought was a given of seventeenth- and eighteenth-century culture. Herder wants and needs something stronger than this however: he needs the analysis of language to be an analysis of thought. A number of resources in this respect were

[48] *Oeuvres de Descartes*, ed. Charles Adam and Paul Tannery (2nd edn, 11 vols, Paris, 1974–86), vi. 58.

available in what was a sophisticated tradition of thinking about language and thought in Germany in the second half of the eighteenth century,[49] which Herder was able to develop to his own ends. This tradition is one in which linguistic, cognitive, hermeneutic, and historical-cultural factors all play a part, and although it flourished in Germany it originated not there but in France, the formative text being Condillac's *Essai sur l'origine des connoissances humaines* of 1746.[50]

For Condillac, linguistic signs were not merely a convenient medium in which to express our thoughts. Although humans, like animals, can have thoughts without language, what makes us human—as opposed to mere animals—is language, for it is linguistic signs that provide us with the means to work on our perceptions of the world in a conscious fashion:

> The principal advantage that derives from the way in which I conceive of the soul is that it can be seen clearly how good sense, the mind, reason, and their contraries each arise equally from the same principle, which is the connection of ideas one with the other; and going up a level, we see that this connection is produced as the result of signs; and that consequently the progress of the human mind depends entirely on the skill with which we make use of language.[51]

In other words, the mind would not be what it is without language, and the increasing sophistication of the mind mirrors the increasing sophistication of language. Here we have the two key issues in Condillac's account: the way in which language provides us with a window on to the activities of the mind; and the importance of the origins of language to the study of its significance.

On the first question, it is a central thesis of Condillac's account that there is a fundamental difference between gestures and the primitive signs designating action, on the one hand, and a fully fledged language on the other. The faculty of reflection was gradually developed as human beings began to free themselves from their environment, to dissociate natural and gestural language from its associative contexts, and to employ arbitrary signs. Natural and gestural signs are immediate and have no internal differentiation, whereas the distinctive feature of fully fledged language is its linearity: language requires the breaking down of thoughts and ordering them in a linear fashion. One can analyse one's thoughts, and those of others, because one can present them in a structured linear

[49] See Avi Lifschitz, *Language and Enlightenment: The Berlin Debates of the Eighteenth Century* (Oxford, 2012).

[50] See Hans Aarsleff, 'The Tradition of Condillac: The Problem of the Origin of Language in the Eighteenth Century and the Debate in the Berlin Academy before Herder', in idem, *From Locke to Saussure: Essays on the Study of Language and Intellectual History* (London, 1982), 146–209.

[51] Etienne Bonnot de Condillac, *Essai sur l'origine des connoissances humaines: ouvrage où l'on réduit à un seul principe tout ce qui concerne l'entendement* (2 vols, Amsterdam, 1746), i. 152–3. Later (posthumous) editions, such as the Paris edition of 1793 and the version presented in the 1798 *Œuvres*, finish after the phrase 'as a result of signs', so that the statement about the progress of the human mind is missing. I am at a loss to understand why.

fashion: linearity is a precondition of the reflection that is distinctive of human thought.[52] In short, language does not simply capture thought, but orders it in a distinctive way that enables one to think reflectively. This in turn provides language users with the means to control the operations of the soul: memory, imagination, and attention. Such an approach was not unprecedented, and there had been an independent stream of thought about language in Germany. In the 'German Metaphysics' of 1724, for example, Wolff developed a theory of signs around the distinction between intuitive and symbolic cognition, arguing that most of our understanding is not intuitive, but mediated by signs, so that signs are the means by which we distinguish ideas from one another.[53] Nevertheless, it was in Condillac's *Essai* that the issues were crystallized, and which stimulated the debates of the second half of the century.

On the second question, that of the importance of the origins of language to the study of language, the issues were directly concerned with the question of what it takes for something to be a language. Rousseau, for example, reflecting on Condillac's account, takes up a worry in Condillac about how the move from primitive sounds to articulate ones could have been possible, and transforms it into a problem that shaped the subsequent discussion of language in the eighteenth century. Rousseau sets out the problem in these terms in Part I of *L'inégalité parmi les hommes*:

> A mother saying to her child the words he was to use in asking her for something, shows how already existing languages are taught, but this cannot be the explanation of how they came into existence. But assume this first difficulty to be resolved. Suppose ourselves, for the moment, to be on this side of the vast gulf that separates a pure state of nature and that in which languages are necessary, and, admitting this necessity, then ask how they could have begun to be established. Here we have a new and far worse problem, for if men need speech in order to learn to think, they must have been in far greater need of knowing how to think if they were to discover the art of speaking. And although we might conceive how vocal sounds had been taken as the conventional means of interpretation of our ideas, it would still remain for us to discover what could have been the means of interpretation of this convention in the case of those ideas that, answering to no sensible objects, could be indicated by neither gesture now voice. Given this, we are barely able to offer any supportable conjectures concerning this art by which we communicate our thoughts and establish a connection between our minds.[54]

[52] Analysis in Condillac goes beyond the ability to break down one's own thoughts, and his theory that science is a well-constructed language, one in which analysis is primary, was to have significant impact in a range of areas of scientific research in the final decades of the eighteenth century: see William Randall Albury, 'The Logic of Condillac and the Structure of French Chemical and Biological Theory, 1780–1801', unpublished PhD dissertation, Johns Hopkins University, 1972.

[53] Wolff had in fact started thinking about the standing of language over twenty years earlier, in his *Disquisitio philosophica de loquela* (Leipzig, 1703).

[54] Rousseau, *Discours sur l'origine et les fondements de l'inégalité parmi les hommes,* 35–6.

This seems straightforward enough, but the dilemma that Rousseau raises here must be considered relative to an explicit assumption that he makes, namely that he is considering a primitive state in which existence is almost exclusively solitary.[55] In Part II, he makes a different assumption, namely that we are now dealing with a rudimentary social organization, involving forms of social interaction, and in this case the problems seem to Rousseau less intractable:

> Inarticulate cries, many gestures and some imitative sounds, must have been for a long time the universal language, to which were added, in every country, some conventional articulate sounds (of which, as I have already intimated, the first institution is not too easy to explain); particular languages were produced, but these were rude and imperfect, much like those now to be found among some savage nations.[56]

Yet to treat conventional articulate sounds as simply an added extra assumes that these are the same kinds of thing as inarticulate cries, gestures, and imitative sounds. It is far from clear that the problems in accounting for the transition from inarticulate cries to articulate language are any less pressing in the case of primitive societies than in the case of an isolated state of nature.

German debates on language were initially quite different, and tended to be pursued within the theological context of the possibility of a literal reading of the Bible. At the centre of the controversy was Hebrew vowel-points, whether these diacritical marks signifying vowels had been added at an early stage (by Ezra in the fifth century BCE) allowing clear meanings from the beginning, as Protestants argued, or whether they were a later interpolation, as Cardinal Bellarmine had suggested, raising the possibility that Rabbinic scholars in the interim could have been motivated by disputes between Christians and Jews to alter the text.[57] But Maupertuis' appointment as President of the Berlin Akademie der Wissenschaften in 1746 meant that French debates on the origins of language were rapidly transferred to Berlin. He drafted his pamphlet *Réflexions philosophiques sur l'origine des langues et la signification des mots* in the 1740s, and in a procedure reminiscent of the use by Condillac and Buffon of the image of a statue to which we add sensory faculties in order to explore our relation to the world, Maupertuis tries to imagine his mind empty of ideas, and to take account of all the sensations with which it is assaulted by assigning a sign to each of these sensations. Because of the vast amount of sensory information, he finds himself having to divide up signs, so that 'I see a tree' becomes two signs—'I see' and 'tree'—instead of one. The first stage is identical for everyone, Maupertuis believes, but the second is the result a

[55] See Aarsleff, 'The Tradition of Condillac', 156.

[56] Rousseau, *Discours sur l'origine et les fondements de l'inégalité parmi les hommes*, 75–6. This is the starting point that he takes up in his posthumously published *Essai sur l'origine des langues* (Geneva, 1781).

[57] See Richard A. Muller, 'The Debate over Vowel Points and the Crisis in Orthodox Hermeneutics', *Journal of Medieval and Renaissance Studies* 10 (1980), 53–72, and Lifschitz, *Language and Enlightenment*, ch. 4.

form of linguistic analysis, which is not fixed by sensation but varies, giving rise to the diversity of languages. But Maupertuis seems to have been aware of the speculative nature of this account, and subsequently made a number of calls for members of the Akademie to follow up the question of the origins of language.[58] By 1756, he had in any case shifted from the earlier sensationalist model to one more in line with Condillac's account.[59] Now it is a case of an early expressive language that evolves into a system of artificial signs. This was Condillac's view: the faculty of imagination is gradually developed as human beings come to dissociate natural and gestural language from its associative contexts and to employ them as arbitrary signs. This is a process in which human beings gradually manage to free themselves from their instinctual response to their environment.

But here we are back with Rousseau's problem: how can human beings with no reflective reason invent conventional signs for abstract notions? Herder bites the bullet: human ability to reflect is constitutive of what it means to be human.[60] As such, it is not, and could not be, the product of an evolutionary development: there is simply no gradual development from natural language to human language. Human beings do not have to free themselves from their instincts because they do not have any instincts. Unlike animals, human beings are not restricted to particular fields of activity or responses.

In his *Fragmente über die neuere deutsche Literatur*, from 1767–8,[61] Herder writes that language is 'a form of cognition not merely in which, but also in accordance with which, thoughts take shape, where in all parts of literature thought is tied to/follows/adheres to [*klebt*] expression, and forms itself in accordance with the latter'.[62] Such linguistic naturalism has an immediate advantage over materialist forms of reductionism, for these latter were widely deemed to commit one to determinism, whereas the linguistic naturalism that Herder is offering here clearly has no such consequences. It simply has no bearing on the question of determinism. Linguistic naturalism also has a positive advantage in that it allows a comparative form of study. By comparing earlier languages with present ones, and by comparing contemporary languages with one another, one is comparing forms of thought. In this respect, we can even perhaps see some of the advantages of considering morality naturalistically, as a particular form of thought. Seventeenth- and eighteenth-century accounts of morality had had great problems in dealing with moral diversity. There was a concerted attempt in Locke, Shaftesbury, and Hutcheson for example, to establish some universality

[58] See Lifschitz, *Language and Enlightenment*, ch. 3.

[59] Pierre-Louis Moreau de Maupertuis, 'Dissertation sur les différens moyens dont les hommes se sont servis pour exprimer leurs idées', *Histoire de l'Academie, 1754* (Berlin, 1756), 349–64.

[60] See Dorothea E. von Mücke, 'Language as the Mark of the Soul: Herder's Narcissistic Subject', in Kurt Mueller-Vollmer, ed., *Herder Today* (Berlin, 1990), 331–44 for a good account of these questions.

[61] Johann Gottfried Herder, *Fragmente über die neuere deutsche Literatur* (Riga, 1766–7).

[62] Herder, *Philosophical Writings*, 48 (translation altered slightly).

in morality. But it is not so much the failure of this project that is interesting, as the inability of prevailing theories of moral sensibility to take into account, in a theoretical way, the fact of widespread moral diversity: at least outside those accounts that simply traced it back to the Fall, where they postulated the emergence of a kind of moral Tower of Babel. I characterized moral sensibility as a form of sensibility earlier, in setting out the understanding of what the term included in the *Encyclopédists* and others, but simply treating it as coming under sensibility as opposed to reason is going to raise fundamental problems about moral responsibility. This is particularly apparent in social reductionist accounts of morality such as that of Helvétius. To the extent to which thought can be incorporated into, but not reduced to, the naturalist schema, Herder's approach has a potentially great advantage. For linguistic variability is ultimately a variability in what would come to be termed mentalities, and moral variability is a form of divergence in mentalities, something that might still be considered problematic and indeed worrying, but no longer treated as bizarre and incomprehensible.

Herder set out a comprehensive account of language in his prize-winning essay, *Ueber den Ursprung der Sprache*,[63] his entry in a competition announced in 1769 by the Berlin Akademie. Responses were invited to the question: 'If human beings had only their natural faculties, would they be able to invent language? And by what means could they invent it'? A recently deceased academician, the demographer J. P. Süßmilch, had set out a comprehensive account of the divine origins of language,[64] and the competitors were invited to discuss Süßmilch's account with a view to presenting the case for the natural origins for language. Süßmilch's target had been Maupertuis' Condillac-inspired account of the transition from a speechless state to the association of arbitrary signs with things. Defenders of this thesis had stressed the length of the period over which it was supposed to have occurred, but Süßmilch argues that this does not help resolve the conceptual problem. The interdependence of signs and human cognition was the only solution to the dilemma posed by Rousseau. Süßmilch assumes that naturalistic theses about the origins of language must follow an evolutionary model,[65] so naturalism is ruled out: human language must have been a product of intelligent design.

Rejecting both the evolutionary model and the hypothesis of intelligent design, Herder's task was to rescue a naturalistic account of language. The idea of a divine origin of language, he writes, 'is useless and extremely harmful. It destroys all efficacy of the human soul, explains nothing, and makes everything, all

[63] Johann Gottfried Herder, *Ueber den Ursprung der Sprache* (Berlin, 1772).

[64] Johann Peter Sußmilch, *Versuch eines Beweises, daß die erste Sprache ihren Ursprung nicht vom Menschen, sondern allein vom Schöpfer erhalten habe* (Berlin, 1766).

[65] As Lifschitz (*Language and Enlightenment*, 84) points out, Süßmilch was aware that some non-naturalistic views also accepted an evolutionary account: Richard Simon, for example, following Gregory of Nyssa, argued that Adam did not receive a fully developed language at creation but merely the capacity to develop language further: Süßmilch, *Versuch eines Beweises*, 5–12.

psychology and all sciences, inexplicable. For have human beings, then, with language received all the seeds of forms of knowledge from God? So *nothing* comes from the human soul? So the *beginning of every art, science*, and form of *knowledge* is always *unintelligible*?'[66] One of his main criticisms of Süßmilch is his lack of any developmental understanding of language, taking a fully formed, rich, contemporary version of language as effectively its only form, ignoring 'a thousand indications in one language, and millions of traces in a variety of languages' that 'peoples have learned gradually to think through using language, and to use language through thinking'.[67] But this does not mean that naturalist alternatives are necessarily on the right track. Herder is quick to criticize those theories which find the origins of language in animal cries. 'I cannot conceal my astonishment', he writes,

> that philosophers, that is, people who seek distinct concepts, were ever able to arrive at the idea of explaining the origin of human language from this cry of the sensations. For is human language not obviously something completely different? All animals, down as far as the dumb fish, sound forth their sensation. But it still is not the case that, just because of that, any animals, even the most perfect, make the slightest real beginning towards a human language. Let one form and refine and organize this cry however much one wants: unless an understanding is added, so as to use this sound with intention, then I do not see how from the preceding natural law human, voluntary language ever comes about. Children utter noises of sensation like the animals, but is not the language that they learn from human beings a quite different language?[68]

Herder identifies Condillac and Rousseau as the prime culprits here. Condillac he accuses of maintaining that children in the desert would immediately communicate through signs expressing their sensations. How, Herder asks, are they supposed to know what kinds of thoughts are to be associated with particular signs? Rousseau, who holds that language came into being with the spontaneous expression of emotion, is criticized on similar grounds. In both cases, the cognitive content of language is ignored, yet this is precisely what we should be attempting to capture in a theory of the natural origins of language.

How then does Herder propose that language originates? He begins with a crucial contrast between animals and human beings. Animals work within a very confined range of activities, and are well served by instinct for this. The human being, by contrast, 'has no such uniform and narrow sphere where only a single sort of work awaits him; a world of occupations and destinies surrounds him. His senses and organization are not sharpened for a single thing; he has senses for everything and hence naturally for each particular thing weaker and duller senses.'[69] What he needs, in these circumstances, is not particulars but general

[66] Herder, *Philosophical Writings*, 163.
[67] Ibid., 57. [68] Ibid., 74–5.
[69] Ibid., 79. Note that adaptability of human beings to all climates was a commonplace of eighteenth-century thinking on the role of environment, and was widely regarded as a crucial

facts. 'Needs' is the key word here, because Herder argues that human beings need language if they are to survive, by contrast with animals, who have no use for it. Language provides a means of organizing our mental life. It orders and clarifies our experience by individuating things in such a way as to reidentify them, by providing a means of storing and conveying general facts. Without this, our distinctively human form of life would not be possible.

One of the distinctive features of Herder's account of language is its rich aesthetic dimension, for as well as a theory of language in which thought is dependent on and bounded by language, he also offers a theory of interpretation, and a theory of translation. The theory of interpretation is based on three principles. First, Herder denies the widespread view that the author of the text to be translated may have thoughts that transcend linguistic expression. Second, the task of the translator, as far as capturing the meaning is concerned, is simply to determine the author's pattern of word usage. But, third, the interpreter must also imaginatively recapture the sensations and affective states that accompany the author's word usage. These considerations also shape Herder's theory of translation. If one is to reproduce the distinctive meanings of an author from a remote historical period or culture for example, the usage of words in the target language must be 'bent' in order to approximate the usage of words in the source language. Moreover, the metre, rhyme, and assonance of the source language have to be reproduced as faithfully as possible, for these express nuances of the author's affective states.[70]

This aesthetic dimension is not an added extra for Herder. Language is not just a means of representation for him, but above all a means of expression, and it is the expressive role of language that is crucial to his account of the shaping of character, which is the end point of his anthropology. Representation alone could never play this role. The importance of language lies in the fact that it provides an embodied medium for expression: self-understanding comes not from the deliberations of a disembodied rational soul, but from concrete forms of expression, and here it is theories of art, rather than theories of language as such, that are able to serve as the model. In this context, Herder develops a conception of aesthetics that Baumgarten did much to promote. Baumgarten had raised the standing of aesthetics, as a general study of sensibility, from sensation to artistic creation, and he had done this largely by drawing attention to faculties associated with sensibility (*Empfindung*) that Wolff had either glossed over or ignored. Because he also insists that sensibility is both a state of the soul and a state of the body, it involves

factor marking out human beings from animals: see for example, William Falconer, *Remarks on the Influence of Climate, Situation, Nature of Country, Population, Nature of Food, and Way of Life, on the Disposition and Temper, Manners and Behaviour, Intellects, Laws and Customs, Form of Government, and Religion, of Mankind* (London, 1781), which is in many respects a compendium of eighteenth-century thinking on environmental questions.

[70] See the discussion in Forster, *After Herder*, ch. 2.

self-consciousness, and he is able to go some way to transforming what in the Wolffian tradition are the lower faculties.[71] Baumgarten's definition of poetry as 'sensually perfect speech' is, Herder argues, particularly important because it enables him to locate poetry in a theory of the human soul.[72]

Although intellectual knowledge—governed by reason—is still the ultimate aim for Baumgarten, and indeed the defining feature of the human character, at the same time he had an essentially epistemological conception of aesthetics: it is a form of knowledge, namely sensible as opposed to intellectual knowledge. As Beiser points out, the arts that fall under his concept of aesthetics are those necessary to acquiring knowledge: the art of attention, the art of abstraction, the art of memory, and the art of imagination, among others.[73] Indeed, aesthetics is not merely a practical discipline but also a moral one, to be judged by the principles of moral philosophy.[74] It is part of character formation, *Bildung*, refining sensibility and giving one's life a direction. Herder took up and developed this conception, deposing reason as the core of philosophy, and transforming aesthetics from a mere guide to taste to the centrepiece of the new anthropologically motivated philosophy:

> The essence of philosophy is to raise to view ideas that lie within us, to bring to the light of distinctness truths that we know only darkly, to develop proofs for which we do not grasp all the intermediate steps clearly . . . Here lies the essence and the creative power of all philosophy, that through it I am empowered to see distinctly truths at least to a level of evidence, of certainty, that I would previously have had or at least never so clearly.[75]

For Herder, aesthetics becomes transformed not merely into an 'abstract science', but the most fruitful of the abstract sciences, a core part of anthropology.[76]

If it was Baumgarten who was the key figure in establishing aesthetics as central to the project of a philosophical anthropology, it was Baumgarten's one-time student Winckelmann who, more than anyone else in Germany, established the importance of understanding something in terms of its place in cultural and historical context. Historical contextualization can be traced back as far as Bacon

[71] Baumgarten, *Metaphysica*, §640. See Stefanie Buchenau, *The Founding of Aesthetics in the German Enlightenment: The Art of Invention and the Invention of Art* (Cambridge, 2013), 170–1.

[72] See Herder, 'Begründung einer Ästhetik in der Auseinandersetzung mit Alexander Gottlieb Baumgartens Schriften', *Werke*, i. 1233–73; especially the final part, 'Bruchstück vom Baumgartens Denkmal', 1269–73.

[73] Beiser, *Diatoma's Children*, 120.

[74] The important works in this respect are *Ethica philosophica* (Halle, 1740), *Initia philosophiae practicae primae* (Halle, 1760), and *Ius Naturae* (Halle, 1763). See the discussion in Buchenau, *The Founding of Aesthetics in the German Enlightenment*, ch. 9.

[75] Herder, 'Kritische Wälder, oder Betrachtung über die Wissenschaft und Kunst des Schönen', *Werke*, ii. 257. Cited in Zammito, *Kant, Herder, and the Birth of Anthropology*, 325. On Herder's aesthetics, see Robert E. Norton, *Herder's Aesthetics and the European Enlightenment* (Ithaca, NY, 1991).

[76] See the sustained discussion of this point in Hans Adler, *Die Prägnanz des Dunklen: Gnoseologie, Ästhetik, Geschichtsphilosophie bei J. G. Herder* (Hamburg, 1990).

at the beginning of the seventeenth century, but the crucial developments were mid-eighteenth century ones. A number of Voltaire's works from mid-century, particularly his *Siècle de Louis XIV* (1751) and the *Essai sur les mœurs* (completed in the late 1750s or early 1760s), provided models of historical contextualization;[77] models that, along with Montesquieu's *De l'esprit des Lois* (1748), Winckelmann would take up.

Wolff and Baumgarten had been concerned with matters of taste and aesthetic judgement, and for this purpose art is something abstract and ahistorical: simply a concrete manifestation of beauty and harmony. Winckelmann by contrast treats works of art as products of particular historical cultures. Unlike Wolff and Baumgarten he was not a university teacher, but someone with a practical understanding of ancient art. In 1755 he was appointed overseer of the antiquities collection of Cardinal Alesandro Albani, in his villa outside Rome, and in 1763 also became prefect of antiquities at the Vatican. He paid two visits to Herculaneum and Pompeii, where excavations had begun in 1738 and 1748 respectively.[78] Integrating accounts of the sculpture newly discovered at these sites,[79] together with his unparalleled knowledge of the sculpture of classical antiquity familiar to him from Rome, as well as the detailed descriptions of art in Pliny's *Natural History*, he builds an account of the art of antiquity as a product of its age and culture, a product that can and should serve as a model for all art since that time. There is no hint of aesthetic relativism here. Quite the contrary, classical antiquity provides what are effectively the only standards of good art, standards that will always remain valid. This is why the study of classical culture itself is so important. As Beiser has pointed out, the next generation of German thinkers influenced by Winckelmann—Schiller, Hölderin, and Friedrich Schlegel—saw Greek antiquity as humanity's childhood, a pagan version of paradise with parallels to Rousseau's state of nature.[80]

A key issue here is the idea of art as a historical-cultural artefact. Winckelmann's survey of art in antiquity, his very influential *Die Geschichte der Kunst des Altertums*, published in 1764, treated the history of art as part of an organic growth of humanity, and it opened up an empirical dimension of art, not as a direct empirical test of aesthetic judgement, but rather an empirical study of the

[77] See Gaukroger, *The Collapse of Mechanism and the Rise of Sensibility*, ch. 12. Had his work been better known, Vico would undoubtedly have joined the list. But even Herder, although he had some passing acquaintance with Vico's work, was unfamiliar with his thinking on questions on which they held strikingly similar views: see Robert T. Clark, Jr, 'Herder, Cesarotti and Vico', *Studies in Philology* 44 (1947), 645–71.

[78] See Christopher Charles Parslow, *Rediscovering Antiquity: Karl Weber and the Excavation of Herculaneum, Pompeii and Stabiae* (Cambridge, 1995).

[79] Winckelmann published two accounts of the newly discovered antiquities: *Sendschreiben von den herculanischen Entdeckungen* (Dresden, 1762), and *Monumenti antichi inediti spiegati ed illustrati* (2 vols, Rome, 1767).

[80] Beiser, *Diatoma's Children*, 192.

artefacts of a historical culture, and the mutual relations between these and the culture in which they were produced.[81] The culture here displaces the artist to some extent (not wholly) in accounting for the production of works of art, and marginalizes the idea of genius promoted by *Stürmer und Dränger* such as Hamann.

More importantly, from the point of view of our present concerns, what Winckelmann sets in train is a way of using art to explore a culture, and culture to explore art. He concentrated on sculpture, but the canonical role of sculpture in the arts was questioned by Lessing in the final sections of his *Laokoon* (1766). Lessing sought to reinstitute literature, and especially drama, as the paradigm form of the arts,[82] but he did not question Winckelmann's cultural-historical approach. In *Laokoon*, the discussion is carried out in terms of what Greek aesthetics did and did not allow Greek sculptors, painters, and writers to do. Again there was no concession to aesthetic relativism, but rather an extension to language of the kind of enquiry that Winckelmann had carried out. It is crucial to understanding what is at issue here to appreciate that both Winckelmann and Lessing are 'aesthetic rationalists': neither of them believes that the idea that different kinds of art emerge from different historico-cultural contexts implies aesthetic relativism.[83] Quite the contrary, they use the art to judge the culture from which it emerges, because they have absolute standards in the former case. Recognizing Herder as the outcome of this aesthetic tradition is important because what it brings to light is that historico-cultural contextualization does not automatically bring cultural relativism in its wake. Various other versions of such contextualization do equate the two, particularly nineteenth- and twentieth-century 'conceptual schemes' varieties, which were in effect introduced by Franz Boas,[84] in which the attention is often focused on the way in which vocabularies in different languages cut across one another, providing different conceptual maps

[81] Note however that the reliability of the details of Winckelmann's account had been called into question by the renowned Göttingen classicist Christian Gottlob Heyne as early as 1778, who judged the *Geschichte* so full of errors as to be practically useless: see Alexander Demandt, 'Winckelmann und die alte Geschichte', in Thomas W. Gaehtgens, ed., *Johann Joachim Winckelmann, 1717–1768* (Hamburg, 1986), 301–13: 302. Winckelman's primary task was to contrast ancient art—he tended to homogenize Greek and Roman art in this respect, treating the latter as a degenerate form of the former—with the art of his day, i.e. the late Baroque styles, of which he was deeply critical. See Otto J. Brendel, *Prologomena to the Study of Roman Art* (New Haven, 1979), 19–24.

[82] The displacement of visual canons by verbal ones is a general phenomenon in the eighteenth century: see Barbara Maria Stafford, *Artful Science: Enlightenment Entertainment and the Eclipse of Visual Education* (Cambridge, Mass., 1994).

[83] This is Beiser's core argument in *Diotoma's Children*, although he does not mention Herder in this context.

[84] See George W. Stocking, *Race, Culture and Evolution: Essays in the History of Anthropology* (New York, 1968), 201–14. Stocking and others have traced Boas' influences back to Herder, but this seems to me a misreading of Herder, for the kind of reasons set out un David Denby, 'Herder: Culture, Anthropology and the Enlightenment', *History of the Human Sciences* 18 (2005), 55–76.

of the world as it were.[85] But a project of historico-cultural contextualization does not in itself commit one to following the path of cultural relativism.[86]

This kind of contextualization is absolutely crucial in Herder, and he insists that the greatest commentator is one 'who does not *transform* his subject into someone who is in accord with the taste of the present century, but rather explains him in terms of all the nuances of his own time',[87] for 'it is absolutely essential that one draw from an author what belonged to *his time* or *the world preceding him*, and what he leaves behind to posterity . . . The more he makes himself of service to his world, the more he must accommodate himself to it, and he must penetrate its ways of thinking if he is to shape them.'[88] Herder's criticisms of Winckelmann focus on his use of a criterion derived from one culture to assess the art of other cultures. In his early *Auch eine Philosophie der Geschichte zur Bildung der Menschheit*, he writes that:

> the best historian of art, *Winckelmann*, obviously only passed judgement on the artworks of the Egyptians according to a Greek criterion, hence depicted them very well *negatively*, but so little according to *their own nature and manner* that with almost every one of his sentences in this most important matter the obviously one-sided and sidewards-glancing aspects glares forth.[89]

Even ethics, very generally construed, is a matter of shaping oneself in the context of a particular culture:

> If human nature is no *independent divinity* in goodness—it has to *learn* everything, to be *formed* through *progressions*, *step* ever further in *gradual struggle*—then naturally it is formed *most* or *only* on those *sides* where it has such *occasions* for virtue, for struggle, for progression. Therefore in a certain respect each human perfection, is *national*, *generational*, and, considered most exactly, *individual*. People form to greater fullness only what *time*, *environment* [*Klima*], *need*, *world*, *fate* gives occasion for.[90]

[85] There is no problem with the idea of languages providing different conceptual maps of the world but profound difficulties arise once these are taken to be constitutive of our experience of the world and incommensurable with one another: see my *Objectivity: A Very Short Introduction* (Oxford, 2012), ch. 5.

[86] On Herder's rejection of relativism, cf. Michael L. Frazer, *The Enlightenment of Sympathy: Justice and the Moral Sentiments in the Eighteenth Century and Today* (Oxford, 2010), 162–3.

[87] 'Über Thomas Abbs Schriften', *Werke*, ii. 577.

[88] Ibid., 579.

[89] Herder, *Werke*, iv. 23; *Philosophical Writings*, 283. Cf. *Ideen*: 'The difference between the so-called enlightened and the unenlightened, or between the cultured and the uncultured, is not a difference of kind but one of degree. Pictures of nations come in an infinite variety of shades, dependent on place and time. But as in all pictures, everything depends on our point of view or perspective. If we take the idea of European culture for our standard, it will only be applicable to Europe. But if we establish arbitrary distinctions between cultures and modes of enlightenment, we will likely lose ourselves in the clouds.' *Werke*, vi. 340. On the political consequences of this pluralism for Herder's politics, particularly on the iniquities of imperalism, see Sankar Muthu, *Enlightenment against Empire* (Princeton, 2003), ch. 6.

[90] Herder, *Werke*, iv. 35; *Philosophical Writings*, 294.

It is important to note, however, that when Herder thinks of different cultures and physical circumstances shaping humans in different ways, he gives historical examples, not synchronic ones, and his descriptions of these imply a sequence with a definite direction:

One can also see why in reality no nation after *another*, even with *all the accoutrements* of the latter, ever *became* what the other was. Even if all *the resources of their cultures* were the same, *their culture was never the same* because always all the influences of the old, now-*changed* nature needed for that were already missing. . . . Also, into *no land* has civilization [*Bildung*] been able to take a *step back* so as to *become a second time* what it *was*: the path of fate is iron and strict. The scene of *that* time, *that* world, is already over; the *purposes* for which they were destined to exist, are past; can today become yesterday? Since God's *course among the nations* proceeds forth with giant steps, will *childish backward paths* be capable of being effected by *human forces*?[91]

This developmental strategy is crucial to Herder's understanding of the relations between cultures. It is very much part of the Enlightenment notion of development towards maturity, and the diachronic modelling of the relation between cultures shapes how one thinks of coexisting cultures. But the last stage in the sequence does not provide the kind of vantage point that it does for Diderot and d'Alembert. As Zammito points out, the aim of Herder's 'history of mankind' was not to trace the trajectory of 'progress', but 'to discriminate the varieties of human excellence'.[92] Moreover, he does not assume—as did Diderot and d'Alembert—that there is a single sequence for all cultures. For Herder and his German contemporaries, what was important was that *their* culture, perhaps uniquely, could trace a direct lineage to that of classical antiquity, and that is why the values of classical antiquity were so important to it.[93]

Here it is worth noting an important feature of Herder's anthropological history of mankind, which marks it out from contemporary schemes in France for example. Although Herder's approach is one of naturalization, translating questions that had been treated as purely conceptual into an empirical form, this does not mean that he assimilated cognitive values to scientific ones. Quite the contrary, the value of a civilization lay in its broadly aesthetic culture, not in its scientific culture, because in its most crucial respects a civilization was a manifestation of its aesthetic culture. To the extent to which Herder exalts some historical or contemporary societies over others, what matters are cultural achievements and not—as with d'Alembert, Diderot, or Condorcet for example—scientific ones.

[91] Herder, *Werke*, iv. 87–8; *Philosophical Writings*, 339–40.

[92] Zammito, *Kant, Herder and the Birth of Anthropology*, 333. See his discussion of Herder's essay on Shakespeare as a paradigm instance of Herder's conception of his project, 342–4.

[93] See Walter Leifer, *Hellas in deutschen Geistesleben* (Bad Herrenalb, 1963); Suzanne L. Marchand, *Down from Olympus: Archaeology and Philhellenism in Germany, 1750–1970* (Princeton, 1996); Jacques Taminaux, *La nostaglie de la Grèce à l'aube de l'idéalisme allemand* (The Hague, 1967); George S. Williamson, *The Longing for Myth in Germany: Religion and Aesthetic Culture from Romanticism to Nietzsche* (Chicago, 2004).

Here our earlier stress on the importance of distinguishing between naturalization as the opening up of questions to empirical treatment, by contrast with their reduction or assimilation to the procedures or aims of the natural sciences, is crucial. Nothing could be further from Herder's aims than the latter.

THE SHAPING OF EXPERIENCE

In his *Scienza Nuova*, which first appeared in 1725, Vico, whose prescient work parallels Herder's in a number of respects, distinguished between philosophy and philology: 'philosophy contemplates reason, whence comes the knowledge of the true; philology observes that of which human choice is author, whence comes consciousness of the certain'.[94] Philosophy explores the universal forms of understanding common to all experience in an abstract way. Philology, by contrast, explores the empirical world that is a product of choice: languages, customs, and the actions of peoples making up civil society and its history. The one grasps the universal, the other the contingent and, for Vico, neither can be pursued fruitfully in isolation from the other. We can think of Herder's project in terms of an incorporation of 'philological' enquiry, in Vico's sense,[95] into philosophical anthropology, where the skill comes in identifying where the resources of philosophical enquiry run dry, and in identifying just what philological resources can be drawn upon and what exactly they must do.

This skill in putting together very diverse sets of resources in a novel and productive way is evident in the three elements that we have distinguished in Herder's anthropology. The first is the claim that the route to the understanding of the mind is that of empirical psychology. The idea of an empirical psychology comes from Wolff, as we have seen, and its scope is widened considerably and its standing enhanced in Baumgarten. Herder takes Baumgarten's innovations further, making empirical psychology not only autonomous with respect to rational psychology, but able to accomplish all the tasks that rational psychology had claimed as its own, not least those of organizing and making sense of experience. The second is the idea that language is integral to thought—shaping it, not merely being an expression of it. This conception owes a great deal to Condillac, above all in the idea that thinking in language orders one's thoughts in a linear fashion, and this is a prerequisite for the kind of intellectual analysis that is distinctive of human beings. But this approach required a workable account of the origin

[94] Giambattista Vico, *The New Science of Giambattista Vico*, trans. T. G. Bergin and M. H. Fisch (Ithaca, NY, 1968), §138. The work originally appeared as *Principj di una Scienza Nuova intorno alla natura delle nazioni, per la quale si ritruovano i principj di altro sistema del Diritto Naturale delle genti* (2 vols, Naples, 1725). The Bergin and Fisch translation uses the third (1744) edition.

[95] Vico's use of the term 'philology' here is somewhat idiosyncratic, though it does capture in a general way a sense of what underpins philological enquiry in its more specific forms, on which see James Turner, *Philology: The Forgotten Origins of the Modern Humanities* (Princeton, 2014).

of language, and this Condillac failed to provide. Herder identifies the problem and argues for a non-evolutionary but nevertheless naturalistic account of language, tightening the connection between thought and language by stressing the idea that the cognitive dimension of language is integral to it, not something that it could have existed without at some stage in some postulated evolution. Language becomes the vehicle for cognition.[96] The third ingredient is the idea that language is a historico-cultural phenomenon. It is from the aesthetic tradition of Winckelmann and Lessing that Herder takes this idea. Winckelmann establishes the compatibility of aesthetic rationalism—the belief in absolute aesthetic values—with viewing the plastic arts as made possible by, and as distinctive products of, definite historical cultures. Lessing set out to topple the plastic arts from the role of aesthetic standard, reinstating literature in this role, but he followed Winckelmann in the idea that it was the aesthetic products of the culture of classical antiquity that set the standard. Aesthetics is crucial to Herder's project, and he operates with a highly aestheticized conception of language in which expression rather than just representation is crucial. This conception is complemented by his understanding of different languages as expressing different conceptions in his treatment of interpretation and translation, where the concern is with retaining the aesthetic qualities of the original. It is the aesthetic tradition, effectively starting with Baumgarten, that provides Herder with what becomes an appreciation of different mentalities.[97]

These three elements together help to provide the basis on which to explore what different mentalities tell us about the human condition: in many respects, they are constitutive of Herder's anthropology.

A core issue here—one that will separate Herder from Kant, and one that bears directly on the way in which our beliefs, aspirations, and claims to knowledge are shaped—is that of the origins of, and sources of, our knowledge. 'Philosophers have exalted human reason to a supreme standing, something independent of the senses and organs', Herder writes in *Ideen*. Yet, he continues, 'just as there is no such thing as an isolated faculty of reason, so there is no one who has become everything he is through his own efforts alone, as he is inclined to imagine in his dream of life'.[98] The 'senses and organs' are crucial here, and it is important to distinguish sensation, as envisaged by Locke for example, from sensation as envisaged by Herder. Like Locke, Herder believes that 'all our concepts are sensuous',[99] but sensation (*Empfindung*) for him is a highly active process: 'When one considers the number of hidden connections and separations,

[96] As Herder puts it in his *Metakritik zur Kritik der reinen Vernunft*, 'the human soul thinks with words': *Werke*, viii. 324–5.

[97] It also leads him to a very intellectualized notion of culture: economic, commercial, legal aspects of material culture are entirely missing. See Gerald Broce, 'Herder and Ethnography', *Journal of the History of the Behavioural Sciences* 22 (1986), 150–70.

[98] Herder, *Werke*, vi. 336.

[99] Ibid., i. 10.

judgements and conclusions that a developing human must make, in order to establish within himself the first idea of an external body, of *figure*, *shape*, *size*, and *dimension*, it is astonishing'.[100] Moreover, sensation is intimately tied up with a more general notion of sensibility. This means that the idea that sensation is free from any input from the sensing mind, that it is a neutral source of information, is rejected.

Here Herder and Kant are at one. The difference comes on the nature of the input. For Herder this input takes the form of *Besonnenheit*, 'reflectiveness'. By contrast with animals, who rely on instincts, conceived as a forcing of attention on an environment over which one has no control, human beings lack these instincts but instead are able to stand back from and be aware of their own mental operations.[101] Through the natural application of *Besonnenheit* to our sensory ideas we are able to form abstract ideas. But to the extent to which the formation of abstract ideas is a part of the generation of reason, note that for Herder reason is not something separate from sensibility. It could never be purely cognitive, for example, but always involves an emotional response: 'The extent to which we participate in what surrounds us, how deeply love and hate, disgust and revulsion, vexation and pleasure, plant their roots in us: this tunes the string-play of our thoughts, this makes us the human beings we are'.[102] Rather than being 'an infallible oracle, self-established, eternal, and independent of everything', reason in fact emerges from our experience of the world,[103] and because ideas are just the interpretations of the experiences of particular communities, they are variable: 'the representations that we have been accustomed to seeing as the most universal principles of human reason disappear depending on physical environment'.[104] There must be a universal human understanding, Herder writes, just as there must be a universal understanding for each species of animal, for example, but no individual is well placed to describe it. However much we talk about universal reason, we have hardly yet 'explained what this actually is, and where it resides, where our reason has released itself from, where people diverge and where they converge. Universal human reason, as we are pleased to understand the term, is a cover for our favourite little whims, superstitions, blindness, and inertia.'[105]

Such statements about reason clearly raise the stakes in debates about reason and sensibility, and were rightly perceived as part of the naturalization package. To thinkers like Mendelssohn and Kant, they amounted to a disparagement of

[100] Ibid., ii. 275.

[101] See the discussion in Sonia Sikka, 'Herder's Critique of Pure Reason', *Review of Metaphysics* 61 (2007), 31–50; and Anik Waldow, 'Back to the Facts: Herder on the Normative Role of Sensibility and Imagination', in Martin Lenz and Anik Waldow, eds, *Contemporary Perspectives on Early Modern Philosophy* (Dordrecht, 2013), 115–34.

[102] Herder, *Werke*, iv. 339–40. Translation from Herder, *Philosophical Writings*, 196.

[103] Herder, *Ideen zur Philosophie der Geschichte der Menscheit*, *Werke*, vi. 294.

[104] Ibid., 304.

[105] Herder, *Von Erkennen und Empfinded der menschlichen Seele*, in *Werke*, iv. 372.

reason, and were part of the unacceptable price one had to pay for naturalization. The effectiveness of the kind of empirical probing that Herder advocates works on the assumption that the understanding of the mind can exclusively take the form of an empirical enquiry. This is what Kant denies, and indeed it is the standing of sensibility that is a key issue that separates the naturalizing anthropological tradition from Kant, who, in the *Kritik der reinen Vernunft*, remarks:

The Germans are the only one who now employ the word 'aesthetics' to designate that which others call the critique of taste. The ground for this is a failed hope, held by the excellent analyst Baumgarten, of bringing the critical estimation of the beautiful under principles of reason, and elevating its rules to a science. But this effort is futile. For the putative rules or criteria are merely empirical as far as their sources are concerned, and can therefore never serve as a priori rules according to which our judgement of taste must be directed: rather the latter constitutes the genuine touchstone of the correctness of the former.[106]

The point is repeated on the opening page of the *Kritik der Urtheilskraft*:

In order to decide whether or not something is beautiful, we do not relate the representation by means of understanding to the object for cognition, but rather relate it by means of the imagination (perhaps combined with the understanding) to the subject and its feeling of pleasure and displeasure. The judgement of taste is therefore not a cognitive judgement, hence not a logical one, but is rather aesthetic, by which is understood one whose determining ground *cannot* be *other than subjective*.[107]

We have seen that developments in aesthetics play an important role in Herder's anthropological thinking. Kant, in making it clear that, on his account, aesthetics could not possibly play such a role, opens up an unbridgeable gulf between Herder's conception of anthropology and his own. For Kant, aesthetics cannot be brought under reason because it is empirical and, in a view reminiscent of Wolff's approach to empirical psychology, cannot therefore make claims to knowledge. But the reasons that Kant has for refusing to give a merely empirical discipline the standing of knowledge cannot be the same as those of Wolff, because it is not a question of having failed to apply a priori rules to them, for example, and as a consequence having failed to transform them into something more secure, since for Kant they are not objects of cognition and can never be transformed into objects of cognition.

[106] Kant, *Kritik der reinen Vernunft*, A21/B35–6, footnote; *Critique of Pure Reason*, 156.

[107] Kant, *Gesammelte Schriften*, v. 203–4; *Critique of the Power of Judgement* (Cambridge, 2000), 89–90. Note that to say that the judgement is subjective is not to say that it is relative. For Kant, aesthetic judgements are subjective in that they depend on feelings of pleasure or displeasure in the subject: were there no such feelings that motivated our judgements, these would simply not be aesthetic judgements. This has nothing to do with relativism, and Kant makes it clear that 'When we call something beautiful, the pleasure that we feel is expected of everyone else in the judgement of taste as necessary, without relation to the feeling of the subject.' *Gesammelte Schriften*, v. 218; *Critique of the Power of Judgement*, 103.

What is at stake here is nothing less than the sensibilization of reason that underlies not only Herder's philosophical anthropology, but also to a large extent the anthropological medicine that we examined in Chapter 3. If Kant is right, anthropological understanding either must take a very different form, or cannot have the kind of status that Herder claims for it. This raises the questions of the standing of anthropology of the kind that Herder is seeking, and what a Kantian alternative would look like.

THE LEGITIMACY OF ANTHROPOLOGICAL EXPLANATION

Kant claimed that anthropology is the highest form of understanding. All the fundamental questions of metaphysics, ethics, and religion, he points out in the *Jäsche Logik*, provide answers to the question 'What is a human being? and as such can all be counted as anthropology.'[108] On the face of it, such an assessment is very much one Herder would have shared. But their conceptions of what anthropology is and what it can do, and above all their conceptions of its standing, differ radically. Herder construes anthropology as having real explanatory power. Kant, by contrast, associated explanatory power with scientific demonstration, and anthropology could never achieve that on his account: it has delusions of grandeur, like the pre-critical metaphysics of Wolff, and sets out an unrealizable programme.

Whether or not it had delusions, Herder's scheme certainly had grandeur, and it is one of the points in our story at which the naturalization of the human and the humanization of nature come together most closely.[109] What he was proposing was a conception of nature in which it is both organic and in a continuous state of development, something that, for Herder, has precedents in Leibniz, Spinoza, and Shaftesbury.[110] Shaftesbury played a particularly important role in offering an explicit account of nature as a living whole unified by a single spiritual principle, and it is instructive how important religious, aesthetic, and moral considerations were in shaping this conception.[111] Shaftesbury conceives of the universe as purposive, and in order to understand any phenomenon we must

[108] *Gesammelte Schriften*, ix. 25; *Lectures on Logic*, 538.

[109] As Marion Heinz notes, Herder's philosophy has a double structure: that of the naturalization of the spiritual and the spiritualization of the natural. 'Grundzüge von Herders Psychologie. "Uebers Erkennen und Empfinden der Menschlichen Seele" (1774)', in W. Koepke, ed., *J. G. Herder: Academic Disciplines of Knowledge and the Pursuit of Knowledge* (New York, 1996), 137–51: 145.

[110] See Herder to Jacobi, 6 February 1784, in *Herders Briefe*, ed. Wilhelm Dobbek (Weimar, 1959), 227.

[111] See, for example, Ernst Cassirer, *The Philosophy of the Enlightenment* (Princeton, 1951), 312–23.

discern its purpose. He uses the watchmaker analogy, for example, to argue the case for a functional understanding of nature:

IF A Passenger shou'd turn by chance into a Watchmaker's Shop, and thinking to inform himself concerning *Watches,* shou'd inquire, of what Metal, or what Matter, each Part was compos'd; what gave the Colours, or what made the Sounds; without examining what the real Use was of such an Instrument; or by what Movements its *End* was best attain'd, and its Perfection acquir'd: 'tis plain that such an Examiner as this, wou'd come short of any Understanding in the real Nature of the Instrument. Shou'd a Philosopher, after the same manner, employing himself in the Study of human Nature, discover only, what Effects each Passion wrought upon the Body; what change of Aspect or Feature they produc'd; and in what different manner they affected the Limbs and Muscles; this might possibly qualify him to give Advice to an Anatomist or a Limner, but not to *Mankind* or to *Himself:* Since according to this Survey he consider'd not the real Operation or Energy of his Subject, nor contemplated the *Man,* as *real* Man, and as a human Agent; but as a *Watch* or common *Machine.*[112]

The thrust of Shaftesbury's account is not merely that any explanation of a natural phenomenon that does not include an account of purposes or ends is necessarily incomplete. It is also that such purpose in nature explains virtue and beauty, for human beings exist for the sake of the good of the whole, and study of the whole—enquiry into what is truly natural for each creature[113]—leads one in the direction of virtue and beauty.[114] This holistic conception is transformed in Herder into something that is intended to direct scientific understanding of the natural world. In a fragment from 1770, he writes:

Everywhere two forces oppose one another but they must, nevertheless, work together, for it is only by the combined and measured influence of both that there emerges the higher reality of a wise excellence, ordering, development, organization, and life. All life arose in such a manner from death, and from death all lesser life: all organization involving destruction and metamorphosis arose from these lesser forces; everything that is ordered and designed from light and shadow, arose from diverging, mutually opposed forces, where the higher positive law, which limits and transcends both, alone inaugurates and harmonizes *kosmos,* world, plan, whole, highest good, and communal happiness. Mathematics, physics, chemistry and the physiology of living things all seem to me to provide evidence for this everywhere.[115]

Herder reworks the conception of nature in Shaftesbury into something with a much more organic and biological focus, something that can be traced back, in the modern era, to Leibniz.[116] But the idea of two forces is something that Herder

[112] Earl of Shaftesbury, *Characteristicks of Men, Manners, Opinions, Times* (3 vols, London, 1737), i. 294.

[113] Ibid., iii. 312.

[114] Shaftesbury's account was taken up and developed in an influential French work: Louis-Jean Lévesque de Pouilly, *Théorie des sentimens agréables* (Geneva, 1747).

[115] Herder, 'Über die dem Menschen angeborene Lüge', *Werke,* iv. 398–9.

[116] See Justin E. H. Smith, *Divine Machines: Leibniz and the Sciences of Life* (Princeton, 2011).

takes from Kant's pre-critical *Allgemeine Naturgeschichte und Theorie des Himmels*, the second part of which not only develops a two-force theory, but also insists on the importance of analogy in moving from the known to the unknown.[117] Herder develops these ideas in a radical way. On the question of explaining phenomena in terms of two forces, in his defence of Spinozism, *Gott. Einige Gersprache*, he argues that all matter must be active on the basis of a two-force theory. Dead matter, he writes:

is not dead, but lives. For in it and conforming to its outer and inner organs, a thousand living, manifold forces are at work. The more we learn about matter the more forces we discover in it, and as a result the empty conception of a dead extension completely disappears. Just in recent times, what numerous and different forces have been discovered in the atmosphere! How many different forces of attraction, union, dissolution and repulsion, has not modern chemistry already found in all bodies? Before the magnetic and electrical forces were discovered, who would have suspected their existence in bodies, and what countless others may still lie dormant and undetected in them?[118]

On analogy, he takes a similarly radical position, arguing that all knowledge is ultimately on the basis of analogy with ourselves:

In everything we call dead nature we know no inner condition. We daily express the words *mass*, *impact*, *fall*, *motion*, *rest*, even *force* of *inertia*, and who knows what they mean within the thing itself? But the more we thoughtfully observe the great drama of effective forces in nature, the less we can avoid everywhere experiencing similarity with ourselves, enlivening everything with our sensation. We speak of action and rest, of our own or received, of a force remaining with us or reproducing, dead or living, and entirely from our souls. *Mass* seems to us a yearning for a mid-point, for the goal and place of rest; *inertia* the small partial rest on a thing's mid-point through its connection with itself; *motion* a foreign drive, a communicated and onwardly effective striving which overcomes rest, disturbs the rest of foreign things, until it finds its own rest again.[119]

In a passage in his review of Herder's *Ideen*, Kant opens up a set of questions that bear directly on the legitimacy of the kind of anthropological explanations envisaged by Herder:

But now as far as that invisible realm of effective and self-sufficient forces is concerned, it is difficult to see why the author, since he believes he could reliably deduce the existence of such a realm from organic generation/creation, did not transfer the thinking principle in the human being, which is of a purely spiritual nature, to this realm directly, without first raising it up out of chaos through the structure of organization. Unless, that is, he were to

[117] For example: *Gesammelte Schriften*, i. 250 and 253; *Natural Science* (Cambridge, 2012), 217, 219.

[118] Herder, *Werke*, iv. 711.

[119] Herder, 'Vom Erkennen und Empfinden der menschlichen Seele', *Werke*, iv. 329. On Herder's use of analogy, see Hans Dieter Irmscher, 'Beobachtungen zur Funktion der Analogie im Denken Herders'. *Deutsche Vierteljahrshrift für Literaturwissenschaft und Geschichte* 55 (1981), 64–98.

take these spiritual forces to be something wholly other than the human soul, and regarded the latter not as a particular substance, but merely as the effect of an invisible universal nature that works on matter and animates it. But we have reservations about ascribing this opinion to him. Yet what is one to think in general about the hypothesis of invisible forces, effecting organization, and therefore about wanting to explain *what one does not comprehend* from *what one comprehends even less*? At least with respect to the former we can become acquainted with its laws through experience, even though their causes will remain unknown; but with respect to the latter we lack any experience, and now what can the philosopher adduce here in justification of his assertion, except just the despair about ever finding the disclosure in any cognition of nature and the decision he is forced into of seeking it in the fruitful field of his poetic power? Also this is metaphysics, indeed even a very dogmatic one, however much our writer denies it because that is what the fashion wills.[120]

There are two principal issues here: the viability of Kant's understanding of scientific explanation, and the contrast between Kant's anthropology and Herder's.

Questions of teleology in the late eighteenth century hinged on the issue of whether there could be a fundamental force in nature that accounted not just for the behaviour of matter, but also for the living realm, that is, on the problem of whether the material world could harbour ends or aims. It should be said from the outset that part of the problem with the way in which Kant and his contemporaries argued about the nature of life lay in the assumption that it was matter theory to which one must turn to account for the behaviour of living things. Cell theory, as set out by Schleiden and Schwann in 1839, would change this: not by solving the question of the nature of life, but by specifying where in the biological realm life is located. This would remove an inappropriate and indeed misleading responsibility from matter theory but, before cell theory, the debate had to centre around matter theory.

This question lay at the core of the second part of Kant's third *Kritik*, and it became a much contested issue in the 1790s. As we have seen, the understanding of matter that he was advocating was in crucial respects a development of that devised by seventeenth-century mechanists, and reworked in the rational mechanics tradition, most notably Euler, and then repackaged as a theory of material properties. Its defining features, for Euler, were its occupation of space, its impenetrability, and its inertia. These were established by him, as we have seen, on purely conceptual grounds, not empirical ones, and he worked on the assumption—which was never more than a promissory note—that mechanics lay at the basis of all physics, from gravitation onwards, and perhaps even all of natural sciences. This is the standpoint that Kant takes up, and the crucial claim is

[120] Kant, *Gesammelte Schriften*, viii. 53–4; *Anthropology, History, and Education* (Cambridge, 2007), 131–2.

that the physical world has a fundamental structure which is to be explored conceptually, not empirically.

In comparing his own position with that of Herder, he sets out a contrast between mechanics and the life sciences. This contrast is supposed to pit matter theory against something that invokes processes that work towards ends, but since mechanics had been displaced by chemistry as the canonical form of matter theory by the middle of the century, what he should be comparing with the life sciences is not mechanics but chemistry. But of course he cannot do that because chemistry, unlike mechanics, is not an a priori discipline. For his argument to work, he needs to align the kind of certainty that he believes derives from a priori demonstration with the kind of fundamental standing due to a discipline that explains the behaviour of a body in terms of its material constituents. But no such alignment is possible. Consider his account of chemistry in the *Metaphysische Anfangsgründe der Naturwissenschaften*. Here Kant treats chemistry as an 'improper science':

> What can be called proper science is only that whose certainty is apodictic; cognition that can contain mere empirical certainty is only improperly called *knowledge*. Any whole of cognition that is systematic can, for this reason, already be called *science*, and, if the connection of cognition in this system is an interconnection of grounds or consequences, even *rational* science. If however the grounds or principles themselves are still in the end merely empirical, as in chemistry, for example, and the laws from which the given facts are explained through reason are mere laws of experience, then they carry with them no consciousness of their *necessity* (they are not apodictically certain), and thus the whole of cognition does not deserve the name of a science in the strict sense; chemistry should therefore be called a systematic art rather than a science.[121]

What would be needed to convert chemistry from an art into a science for Kant? Chemical experiments yielded systematic knowledge of chemical affinities, that is, the degrees to which chemical substances will interact and displace one another. Kant's claim is that, for chemistry to be properly scientific, we should be able to deduce these affinities from 'the approach or withdrawal of the parts of matter (as, say, in proportion to their densities and suchlike)', that is, the interaction between the attractive and repulsive forces that he believes exhaustively determine the properties of matter.[122] Far from this demand for reduction to mechanics following a Newtonian path, as Kant believed, it is exactly the kind of thing that Newton rejected, for example, in the call by his opponents to account for the

[121] Kant, *Gesammelte Schriften*, iv. 468; 'The Metaphysical Foundations of Natural Science', in *Theoretical Philosophy after 1781*, 184. This exclusion of chemistry is at odds with anything I have encountered in any other writer from the period: compare, for example, Karsten's 1783 standard textbook—*Entleitung zur gemeinnützlichen Kenntniß der Natur* (Halle, 1783)—which divides science, as was customary, into natural history, natural philosophy (i.e. basically mechanics), and chemistry, and not only rejects as implausible the outdated idea that natural philosophy is at the basis of chemistry, but also the idea that it relies wholly on mathematics.

[122] Kant, *Gesammelte Schriften*, iv. 471; *Theoretical Philosophy after 1781*, 186.

colour spectrum in micro-corpuscularian terms. Reduction to a mechanically describable state is not a general desideratum for Newton, any more than it was for his successors working in areas such as chemistry and electricity. Kant is in effect asking for the translation of the empirical results of chemical experiment into the language of mechanics, with no attempt to specify how we would be able to move from 'densities and suchlike' to the complexities of chemical behaviour. From the point of view of explanation and understanding, it simply bore no relation to what was happening in either chemistry or mechanics.

From the beginnings of the critical philosophy up to 1790, one gets the impression that any part of the natural realm that fails Kant's test of systematic derivation, propositional understanding, and reason, simply falls outside legitimate scientific enquiry. But his preoccupation with the mechanics that he believed embodied these ideals was marginal to the point of eccentricity in physical enquiry (i.e. outside mathematical circles) in the second half of the eighteenth century. Kant's attitude changes in 1790, however, and he finally offers a response to at least some of the developments that he has ignored up to this point.[123] The second part of the *Kritik der Urtheilskraft*, published in that year, deals with 'the critique of the teleological power of judgement', and here Kant offers what he takes to be an account of natural phenomena that fall outside the purview of mechanical explanation. One especially problematic feature of his account is the way in which he treats the notion of active matter in the life sciences as if this exhausted the areas that fall outside a mechanical conception. The behaviour of living things by no means exhausts the range of problematic phenomena for which he needs to account. As we have seen, virtually every material process of interest to those working in chemistry and electricity for example is inexplicable in mechanical terms.

Kant treats the advocacy of active matter in terms of allowing goal-directedness into the behaviour of matter, into its very essence. His alternative to this is to remove any element of teleology from matter itself, and to account for it instead in terms of how we perceive such activity.[124] The idea is that the behaviour of everything in the natural realm is mechanical, but that living things also need to be accounted for in terms of final causes, that is, teleological explanations. Kant sees in this an antinomy, which is set out as follows:

The *first maxim* of the power of judgement is the *thesis*: All generation of material things and their forms must be judged as possible in accordance with merely mechanical laws.

[123] That is, ignored in the context of incorporation into an account of the critical philosophy. He had taken some interest in natural history, especially on the question of preformationism versus epigenesis. See for example Jennifer Mensch, *Kant's Organicism: Epigensis and the Development of Critical Philosophy* (Chicago, 2013).

[124] See the 'Analytics of the Teleological Power of Judgement', *Gesammelte Schriften*, v. 362–83; *Critique of the Power of Judgement*, 235–55. On these questions see the extended discussion in Rachel Zuckert, *Kant on Beauty and Biology: An Interpretation of the* Critique of Judgment (Cambridge, 2007), chs 3 and 4.

The *second maxim* is the *antithesis*: Some products of material nature cannot be judged as possible according to merely mechanical laws (judging them requires an entirely different law of causality, namely that of final causes).[125]

The antinomy is resolved only when we realize that these principles are not about the world, but about our judgements about the world. For Kant, the need for teleological explanations tells us something about us, not something about the world. In terms of the distinction he makes in the 'Introduction' to the *Kritik der Urtheilskraft*, they involve 'reflecting' (*reflectirend*) as opposed to 'determining' (*bestimmend*) uses of judgement.[126] This is a distinction that Kant introduces at a late stage in his writings.[127] Up to this point, he had treated judgement in terms of what he is now distinguishing as determining judgement, that is, subsuming a particular under a universal that has already been given. Reflective judgements are the opposite of this, in that here we start with a given particular and must find a universal—a concept or a rule—under which to subsume it. Determinative judgement deals with the order of things, reflective judgement with the order of concepts.

Kant characterizes what he opposes as hylozoism, and hylozoism figures as one of the four basic categories of matter that he identifies.[128] These categories come in pairs, and the first pair works on the basis that 'all purposiveness in nature is unintentional'. The first option postulates 'lifeless matter', where a mechanical causality is all that is involved, and Kant associates this position with ancient atomism. The second invokes 'a lifeless God', a position he associates with Spinoza. He argues that Spinoza's God lacks intelligence, purposiveness, freedom, and causality, and as a consequence Spinoza is committed to fatalism.[129] The second pair of alternatives works on the basis that 'some purposiveness in nature (in organized beings) is intentional'. One option asserts the existence of a living (as opposed to a merely Spinozean) God, but this is 'incapable of dogmatically establishing the possibility of natural ends as the key to teleology'. The other option is hylozoism. This is rejected on the grounds that:

even the most daring hypothesis can be authorized only if at least the *possibility* of that which is assumed to be its ground is *certain*, and one must be able to ensure the objective reality of its concept. However, the possibility of a living matter (the concept of which contains a contradiction, because lifelessness, *inertia*, constitutes its essential characteristic), cannot even be conceived; the possibility of an animated matter and of the whole of

[125] *Gesammelte Schriften*, v. 388; *Critique of the Power of Judgement*, 259.

[126] *Gesammelte Schriften*, v. 179–81; *Critique of the Power of Judgement*, 66–8.

[127] See the 'Editor's Introduction' to *Critique of the Power of Judgement*, xxi–xxiv.

[128] *Gesammelte Schriften*, v. 391–5; *Critique of the Power of Judgement*, 262–6.

[129] On the correctness of this assessment see, for example, Henry Allison, 'Kant's Critique of Spinoza', in Richard Kennington, ed., *The Philosophy of Baruch Spinoza* (Washington, DC, 1980), 199–227; Zammito, *The Genesis of Kant's Critique of Judgment* (Chicago, 1992), ch. 12. More generally, see Beth Lord, *Kant and Spinozism: Transcendental Idealism and Immanence from Jacobi to Deleuze* (London, 2011).

nature as an animal can be used at all only insofar as it is revealed to us (for the sake of an hypothesis of purposiveness in nature at large), in experience, in the organization of nature in the small, but its possibility can by no means be understood a priori.[130]

Kant seems to be doing little more here than stating that anything at variance with a mechanical conception of matter, one in which 'lifelessness, *inertia*, constitutes its essential characteristic', is simply not viable. The claim can only be on metaphysical grounds. On observational grounds, almost everyone working in a physical discipline had come to the opposite conclusion by the second half of the seventeenth century. The reproductive behaviour of the polyp, for example, looked like direct evidence of living matter. How else could one account for regeneration of a whole animal from thin slices? On this view, as we have seen, apparently inert matter is just living matter that has died, and in the process lost all its interesting properties. This dead matter is then mistakenly taken as the only form of matter, life being an added extra which has to be accounted for outside matter theory.

Just what metaphysical account one would give of this is an open question, but it cannot be ruled out on a priori grounds as Kant proposes. Practically speaking, one might treat the polyp case and the phenomenon of muscular irritability, for example, as compelling evidence that at least some matter is living, however one accounts for it, and reject metaphysical speculation as to the reasons for this as futile, in much the same way that mind–body interaction was treated, that is, it was taken as a fact, and any metaphysical account that denied this fact was obviously flawed. One of the most pressing problems for the Kantian account is that it provided a metaphysical defence of a conception of matter that was, for scientific purposes, useless. As chemists from the early decades of the eighteenth century onwards had realized, material properties are not reducible to mechanics. The attempt to subsume matter theory under mechanics—so that any form of matter not fully characterizable in mechanical terms is excluded—was simply misconceived, and something that metaphysics could not possibly establish.

Up to 1790, Kant had concentrated on a mechanical model of the natural realm, only to come to realize that such a model was apparently at odds with the life sciences. Unable to abandon his mechanical model, he offers an account of teleology in which all the problematic features are removed from matter itself and instead made into questions about how we conceive of the world. But there are many areas that have nothing to do with the question of life, areas where the mechanical conception is equally manifestly inadequate. To make it a matter of mechanics versus teleology is a highly misleading strategy. No one was claiming that substances involved in chemical reactions were acting teleologically, for example, but their mechanical properties take us nowhere in understanding their chemical properties, or their material properties more generally. Even had

[130] *Gesammelte Schriften*, v. 394; *Critique of the Power of Judgement*, 265.

Kant's account of teleology worked for biological entities, he would still manifestly lack a general theory of matter of the kind he claims to be presenting. It is instructive, in this context, to note that for Kant's successors, that is, those thinkers who saw themselves as being in the Kantian tradition, it was not just the existence of things in themselves that caused them to worry but, even more decisively, Kant's mechanical conception of nature. The principal representatives of Kantianism in the mid-nineteenth century (before the rise of the neo-Kantian schools), Adolf Trendelenburg and Hermann Lotze, both insisted that although the world is governed by mechanical laws, as Kant had argued, it is also governed by teleology or purpose, something they considered necessary on a number of grounds, not least the need to sustain ethical values in the face of science.[131]

It is difficult to avoid the conclusion that, for Kant, the attack on what he considers to be Herder's hylozoism goes to the core of the difference between him and Herder on questions of naturalization and anthropology. But the difference does not lie so much in matter theory as strategies of explanation. Herder can include the developments in the life sciences, and the moves towards unifying the life sciences, in his conception of scientific understanding, and on such an understanding, one can see naturalization as a fruitful strategy: that is, one can see how it might be part of a more general programme, how it might be able to draw on a wide range of new resources, and how it might be connected with other developments. On Kant's conception, by contrast, naturalization as a programme of explaining human behaviour is simply misconceived. This does not rule out piecemeal and very limited forms of naturalistic explanation in the case of race and physical geography, as we shall see. But whatever his commitment to these claims, they are for him speculative, not scientific. In short, the argument about the possibility of a philosophical anthropology is one about the limits of scientific explanation, something that hinges crucially on the model of science that one deploys.

Kant's model of science is the source of significant problems in his anthropological project. The issues arise from the fact that his philosophy forms a highly integrated and comprehensive system. This brings with it problems when something that he wants to pursue falls outside his system. And because it is his anthropology—the form of understanding that he values most highly—that has difficulty finding a place in the critical philosophy, the problems become fundamental ones, for it is now difficult to see what role if any it can play in our thinking about the world and our place in it. What is at issue here is Kant's conception of anthropology, but also his conception of the critical philosophy and in particular why this excludes the kind of anthropological project that Herder takes up: a project of the kind to which Kant himself had been sympathetic in his pre-critical days.

[131] See Frederick Beiser, *Late German Idealism: Trendelenburg and Lotze* (Oxford, 2013).

Kant's mature critical philosophy offers what he conceives as a philosophical, systematic solution to Hume's dilemma about the role and standing of metaphysics. Systematic derivation, propositional understanding, and reason are aligned here, just as earlier thinkers had tried to align them in metaphysics, but Kant, as we have seen, develops complex and detailed arguments to show that metaphysics must necessarily fail in this task. In doing this, he is going beyond Hume in significant respects. Hume's criticism of the limits of metaphysics did not identify the point at which it ceases to work, if for no other reason that he would not have thought that there was any such single point. For Hume the balance between thinking in metaphysical terms and bypassing metaphysics in favour of 'the common life' was a matter of weighing up what one wanted to achieve in particular cases, and it was ultimately a question of personal judgement. Kant takes the question of the limits of metaphysics in a staunchly metaphysical way. He uses metaphysics to reveal it own limits, in a more comprehensive fashion than Hume ever did, and sets out to show exactly where those limits are.

Hume believed that metaphysics, in spite of the absurdities to which it led if pursued to the end, was not something that we could do without, because it was our means of probing fundamental questions, and without it we would be back in a state of pre-civilization. Kant sometimes associates the term 'metaphysics' with the discredited Wolffian metaphysics, which had presented itself as a comprehensive account. But there is a more traditional sense of the term in which it is simply that form of thought in which system, propositional understanding, and reason were legitimately aligned, and Kant set out to establish the content of such a discipline by means of the derivation of principles from transcendental truths: this is the prime meaning of 'metaphysics' for Kant after 1781. Once he has systematically established the limits of this form of thought, he argues that we are not in fact left without a means of probing fundamental questions that fall outside it. In considering questions that lie outside the legitimate limits of pure reason, there are various strategies we can explore, but there is no single form of understanding to which all of them can aspire. The kinds of understanding that we have in considering the nature of morality, religion, the critique of taste, and anthropology differ significantly.

As far as anthropology is concerned, it is instructive that Rousseau had offered an anthropology in *Emile*, one that entirely eschewed metaphysical considerations and considerations of deductive rationality. It is the kind of thing that would have fallen under Kant's conception of a pragmatic anthropology. Many of Kant's contemporaries who were concerned to develop an anthropology, not least Herder, thought of anthropology as something that was above all practical, and they used the term 'pragmatic' in this context.[132] But their usage was different

[132] The term 'pragmatisch' has a complex set of meanings in its eighteenth-century usage, among which is something 'that unearths causes and effects of important events': see the discussion of 'pragmatic history' in Reill, *The German Enlightenment and the Rise of Historicism*, 42–5.

from Kant's. For Kant, 'practical' was opposed to 'scientific' or 'theoretical', whereas for his contemporaries it was not. For Herder, for example, his anthropology was practical in the sense that it was crucial to the shaping of character, *Bildung*, but it was something theoretical or scientific that had been developed with a view to practical consequences, not something *merely* practical.

Kant is concerned with something exclusively practical. In his characterization of his own anthropology, in *Anthropologie in pragmatischer Hinsicht*—the textbook he compiled in 1798 from his lectures on anthropology delivered every year over twenty-four years—the contrast he draws is between an anthropology based on empirical psychology (he has that of Platner in mind) and one based on a 'freely acting human being'. One might think the former scientific and the latter metaphysical, but it is the latter that Kant defends here:

> A doctrine of the knowledge of the human being, systematically formulated (anthropology), can exist either in a physiological or in a pragmatic point of view. Physiological knowledge of the human being concerns the investigation of what *nature* makes of the human being; pragmatic, the investigation of what *he* as a free-acting being makes of himself, or can or should make of himself. He who ponders natural phenomena, such as what the causes of the faculty of memory may rest on, can speculate back and forth (like Descartes) over the traces of impressions remaining in the brain, but in doing so he must admit that in this play of his representations he is a mere observer and must let nature run its course, for he does not know the cranial nerves and fibres, nor does he understand how to put them to use for his purposes. Therefore all theoretical speculation about this is a pure waste of time. But if he uses observations concerning what has been found to hinder or stimulate memory, in order to augment it or make it more responsive, and if he requires knowledge of the human being for this, then this would be part of an anthropology with a *pragmatic* purpose, and this is precisely what concerns us here.[133]

On the face of it, this is incompatible with his characterization of anthropology in the *Kritik der reinen Vernunft*, where he explains why anthropology must be physiological and must exclude freedom:

> Thus the ability to choose of every human being has an empirical character, which is just a certain causality of his reason, insofar as, in its effects in appearance, this reason exhibits a rule, in accordance with which one could derive the rational grounds and the actions themselves according to their kind and degree, and estimate the subjective principles of a person's ability to choose. Because this empirical character itself must be drawn from appearances as effect, and from the rule which experience provides, all the actions of the human being in appearance are determined in accord with the order of nature by his empirical character and the other cooperating causes; and if we could investigate all the appearances of his ability to choose down to their basis, then there would be no human action that we could not predict with certainty, and recognize as necessary given its preceding conditions. Thus in regard to this empirical character there is no freedom, and according to this character we can consider the human being solely by *observing*, and,

[133] Kant, *Gesammelte Schriften*, vii. 119–20; *Anthropology, History, and Education*, 231.

as happens in anthropology, by trying to investigate the moving causes of his actions physiologically.[134]

The conflict is more apparent than real, however, for anthropology is not a unitary category, still less an integrated one in Kant, and we must distinguish three different types of anthropological enquiry. Physical geography, which deals with human beings in terms of what they are physically—as one 'type of thing' that inhabits the earth—is an area that falls under 'the science of man'. Physiological anthropology, which deals with how human beings function, Kant regards as too underdeveloped to be of use and to be of no philosophical significance, although he allows that physicians may have occasional recourse to it. Finally, there is pragmatic anthropology, which deals with human beings as freely acting entities.[135]

Pragmatic anthropology is marked out from theoretical investigation in that the latter delivers understanding, whereas pragmatic observation is a form of interaction with other human beings, a form of 'participation'.[136] A distinctive feature of pragmatic anthropology is that, in contrast to scientific investigation, it employs a teleological form of enquiry. Kant's treatment of teleology is elaborate and complex. His account of purpose in Part II of the *Kritik der Urtheilskraft*, on the critique of teleological judgement, mirrors his account of beauty in Part I, on the critique of aesthetic judgement.[137] In the first Part, Kant argued in detail against what he called the 'subreption of beauty', that is, the tendency to ascribe beauty, which on his account is a response of the subject, to the object itself. In the second, he mounts a case against a parallel 'error of judgement', namely the ascription of purpose to nature, whereas, he argues, it is actually a feature of our judgement. The ubiquity of purpose is not in question. 'Everything in the world is good for something', he writes, 'nothing in it is in vain; and by means of the example that nature gives in its organic products, one is justified, indeed called upon, to expect nothing in nature and its laws but what is purposive in the whole'.[138] The crucial question for Kant is whether nature itself harbours these purposes, or whether they are simply a necessary means by which to investigate nature, and he comes down firmly in favour of the latter:

Objective purposiveness [*Zweckmäßigkeit*], as a principle of the possibility of the things in nature, is—far from being *necessarily* connected with the concept of the latter—precisely that to which one refers above all in order to prove the contingency of nature and of its

[134] *Kritik der reinen Vernunft*, A550/B578. Essentially the same point is repeated in the *Kritik der praktischen Vernunft*, where we are told that 'at the time I act I am never free', and that if it were possible to have deep insight into someone's character then 'his future conduct would be predicted with as great a certainty as the occurrence of a solar or lunar eclipse.' *Gesammelte Schriften*, v. 94, and 99; *Practical Philosophy*, 216 and 219 respectively.

[135] See the comprehensive treatment in Sturm, *Kant und die Wissenschaften vom Menschen*, chs 6–8.

[136] *Gesammelte Schriften*, vii. 120; *Anthropology, History, and Education*, 232.

[137] See John H. Zammito, *The Genesis of Kant's Critique of Judgment*, 151–8.

[138] *Gesammelte Schriften*, v. 379; *Critique of the Power of Judgement*, 250.

form. For if one adduces, e.g. the structure of a bird, the hollowness of its bones, the placement of its wings for movement and of its tail for steering, etc., one says that, given the mere nexus of efficient causes in nature, without the help of a special kind of causality, namely that of final causes, this is all in the highest degree contingent: i.e. that nature, considered as a mere mechanism, could have formed itself in a thousand different ways without hitting precisely upon the unity in accordance with such a rule, and that it is therefore only outside the concept of nature, not within it, that one could have even the least ground a priori for hoping to find such a principle. Nevertheless, teleological judging is rightly drawn into our research into nature, at least problematically, but only in order to bring it under principles of observation and research in *analogy* with final causes, without presuming thereby to explain it.[139]

The point here is that although there are no purposes in nature, we cannot make any judgements about nature unless we posit an end in nature, for otherwise we would not be able to make sense of the parts. Perhaps the best way to think of this is to compare it with what he could have said in his pre-critical phase. There, the point would be that nature has no intrinsic ends, but it does have extrinsic ones: God has created nature with a purpose. This was a commonplace view. But once Kant has developed the critical philosophy, with its severe restrictions on what we can and cannot know, he could no longer talk about God playing this role. Since the study of nature cannot proceed without purposes or ends, one thing he could have done was to make these extrinsic purposes intrinsic, but that would have been incompatible with his mechanical conception of nature. The only other option was to remove real independent ends and simply grant them a regulative standing, so that the question of whether they are extrinsic or intrinsic cannot arise: we act as if there were purposes, because of something about us, not something about nature.[140]

The question remains in what sense we are investigating free agents in Kant's pragmatic anthropology, since it is nature here that is purposive (that is, we must treat it as being purposive, even if it harbours no purposes as such), not the human subjects under investigation. One suggestion is that his pragmatic anthropology is an investigation into nature's purposes for the human species, from which we would hope to learn how we might use what nature has endowed us with in our capacity as biological organisms requiring all kinds of things if we are to survive. Such natural constraints and needs restrict our actions, but they do not deny to us

[139] *Gesammelte Schriften*, v. 360; *Critique of the Power of Judgement*, 233–4. Cf. *Gesammelte Schriften*, v. 246, and *Critique of the Power of Judgement*, 129–30, for a parallel argument in the case of beauty, which is treated as human artifice projected on to nature.

[140] But see Dalia Nassar, 'Sensibility and Organic Unity: Kant, Goethe, and the Plasticity of Cognition', *Intellectual History Review* 25 (2015), 311–26, who points out that this is Kant's explicit line of argumentation, but that at least implicitly he sometimes seems to maintain that nature gives us clues as to how it should be judged—either mechanically or teleologically—and in this case there is more to our making these judgments than the fact that it has to do with our cognitive structuring of the world.

the kind of freedom that is characteristic of us as free agents.[141] Nevertheless, at best, this just gives us an anthropology that is not incompatible with free agency: it is not an anthropology that tells us anything about that free agency.

For this, Kant needs to combine his pragmatic anthropology with his critical philosophy, and the point of contact between the two would seem to be his account of morality. He distinguishes two forms of enquiry into morals: the metaphysics of morals and moral anthropology. As regards the former, he explains in his lectures on morality that:

> The imperative of morality abstracts from all inclinations. The motivating ground is not drawn from sense, or from happiness, but given solely by pure reason. The motivating grounds and the law itself must be a priori. . . . The metaphysics of morals, or *metaphysica pura*, is only the first part of morality; the second part is *philosophia moralis applicata*, moral anthropology, to which the empirical principles belong. . . . Moral anthropology is morality applied to men. *Moralia pura* is based upon necessary laws, and hence it cannot be founded upon the particular constitution of a rational being, such as man.[142]

The point is made in very similar terms in the *Metaphysik der Sitten*:

> A metaphysics of morals cannot be based upon anthropology but can still be applied to it. The counterpart of a metaphysics of morals, the other member of the division of practical philosophy as a whole, would be moral anthropology, which, however, would deal only with the subjective conditions in human nature that hinder people or help them in *fulfilling* the laws of a metaphysics of morals.[143]

The most serious problem with this proposal is that the inability of anthropology to fall under critical philosophy doesn't just sideline it, it effectively rids it of any serious content. It is quite unlike Rousseau's rich anthropology for example, and matches Herder's characterization of something comprising 'our favourite little whims, superstitions, blindness, and inertia'. It is not surprising that commentators, and proponents of a philosophical anthropology, have not given Kant's anthropology a favourable reception. Schleiermacher, for example, in an essay on Kant's anthropology from 1799, argues that while it has the appearance of a collection of trivialities, it should in fact be read 'not as anthropology, but rather as negation of all anthropology, as at once claim and proof that such a thing according to the idea established by Kant, through him and in his way of thinking, is absolutely impossible, intentionally put in place just as he expressly sets up and peculiarly construes the empty faculties in the division of the sciences or its objects'.[144] And

[141] This is the kind of account offered in Alix Cohen, *Kant and the Human Sciences: Biology, Anthropology and History* (London, 2009), ch. 4.

[142] *Gesammelte Schriften*, xxix. 599; *Practical Philosophy*, 226–7. On this distinction see Robert B. Louden, 'The Second Part of Morals', in Brian Jacobs and Patrick Kain, eds, *Essays on Kant's Anthropology* (Cambridge, 2003), 60–84.

[143] *Gesammelte Schriften*, vi. 217; *Practical Philosophy*, 372.

[144] Quoted in Brian Jacobs, 'Kantian Character and the Science of Humanity', in Brian Jacobs and Patrick Kain, eds, *Essays on Kant's Anthropology* (Cambridge, 2003), 105–34: 118.

indeed it is remarkable not only how poorly integrated his anthropology is into his critical philosophy, but how little he is able to do with it, when compared to Locke's use of the travel literature in the first Book of the *Essay*, for example, where an analogous purely practical and descriptive literature is used to open up deep problems for moral theory and, ultimately, for epistemology.[145]

Recently, there have been attempts to revive Kantian anthropology, by establishing connections with his metaphysics of morals.[146] But the metaphysics of morals plays such a dominant role in these attempts, and so much of the content of Kant's own anthropology is jettisoned or ignored, that it is more a case of getting his moral philosophy to yield an anthropology, than it is of reconciling it with his anthropology as it stands. Indeed, it has even been questioned whether there is any moral focus at all in his pragmatic anthropology,[147] which would make the exercise futile as a piece of Kant interpretation. Underlying this is an even more intractable question. There were two kinds of anthropological understanding tied up in both Herder's and Kant's projects, philosophical anthropology and the 'natural history of man'. These are in many respects the forebears of what came to be known as social anthropology and physical anthropology respectively. In the first volume of his *Ideen*, Herder had set out to integrate into his philosophical anthropology a mass of empirical material from recorded history, physical geography, and work on the human species. The project was one in which organic matter arises out of inorganic matter, species evolve towards higher species, and where human beings emerge from purely natural forces. In his review of *Ideen*, Kant was adamant that philosophical anthropology should be kept out of physical anthropology. Recognizing a conceptual distinction between the two is indeed crucial, but this is not what Kant has in mind. He wants in effect to rid philosophical anthropology of the kind of guiding role that it has in Herder. He was concerned to foster physical anthropology, but as a collection of empirical facts about the world, not in the integrated way that we find in Herder.[148]

This leaves the role and standing of philosophical anthropology in Kant uncertain. It also raises the question whether Kant does in fact import various metaphysical and theological assumptions into his physical geography or science of man. In 1785 there appeared a two-part piece criticizing Kant's view on race by the distinguished naturalist and ethnologist Georg Forster.[149] Forster had

[145] See Gaukroger, *The Collapse of Mechanism and the Rise of Sensibility*, ch. 4.

[146] See, for example, Louden, 'The Second Part of Morals', and Allen W. Wood, 'Kant and the Problem of Human Nature', in Brian Jacobs and Patrick Kain, *Essays on Kant's Anthropology*, 38–59.

[147] See Reinhard Brandt, *Kritischer Kommentar zu Kants Anthropologie in pragmatischer Hinsicht (1789)* (Hamburg, 1999).

[148] The anthropological projects of Kant's followers are distinctive in this respect. Perhaps the best known, Christoph Meiners' *Grundriss der Geschichte der Menschheit* (Lemgo, 1785), for example, is really just an annotated bibliography of current travel writing and ethnography.

[149] Georg Forster, 'Noch etwas über die Menschenraßen,' *Teutsche Merkur* (Fourth Quarter, 1786), 57–86, 150–66. On Forster's response to Kant see Zammito, *The Genesis of Kant's Critique of Judgment*, 207–9.

impeccable empirical credentials in the area. Having spent his youth accompanying his father on a number of voyages of discovery, in 1777 he published an account of Captain's Cook's second voyage to the Pacific,[150] offering a path-breaking ethnological account of Polynesia, on the strength of which he was elected, at the age of twenty-two, to the Royal Society.

Kant, in his writings on race, had maintained that although human beings constitute a single species, the racial differentiation that had originally emerged through climate and other factors was irreversible, and had been the outcome of providence.[151] Forster takes the existence of distinct racial groups, Polynesians and Melanesians, living under the same conditions as a problem for single origin theories. More importantly, he opposes any role for providence in accounting for the development of the human species, and generally rejects the teleology that such an approach brings with it as wholly inappropriate to what should be a purely scientific and empirical venture. A crucial point at issue is just what guides Kant's anthropological account. He gives the impression that it is a purely empirical matter, but Forster's charge that he is being guided by metaphysical considerations elicits from Kant the reply that what guides him is not metaphysics as such but hypotheses, which are inevitable and designed simply to clarify the problem.[152] But then of course the question arises what guides these hypotheses. Forster points out that empirical enquiry seeks efficient causes, not final causes. Kant agrees, but insists that where efficient causes cannot be found, we must seek final causes. In other words, Kant in effect concedes that teleological hypotheses—where metaphysical and theological assumptions are difficult to avoid—are what guides his account.

[150] Georg Forster, *A Voyage Round the World in His Majesty's Sloop, Resolution, commanded by Capt. James Cook, during the Years 1772, 3, 4, 5* (London, 1777). The *Voyage* was actually designed to present the results of his father's findings. Georg was not yet eighteen, but his father was forbidden by the admiralty from publishing a separate account of the voyage from that of Cook. He did in fact publish an account the year after his son's work appeared: Johann Reinhold Forster, *Observations Made During a Voyage Round the World,* (London, 1778).

[151] There are two main writings. The first, dating from 1775, with a second edition appearing in 1777, is 'Von den verschiedenen Racen der Amenschen zur Anküdigung der Vorlesungen der physichen Geographie im Sommerhalbenjahre 1775' (Of the Different Races of Human Beings): *Gesammelte Schriften*, ii. 427–43; *Anthropology, History, and Education*, 84–97. The second, 'Determination of the Concept of a Human Race' dates from 1785: 'Bestimmung des Begriffs einer Menschenrace': *Gesammelte Schriften*, viii. 89–106; *Anthropology, History, and Education*, 145–59. In his *Anthropologie in pragmatischer Hinsicht—Gesammelte Schriften*, vii. 320; *Anthropology, History, and Education*, 415—Kant refers the reader to the extended treatment of questions of race in the work of his disciple Christoph Girtanner, *Über das Kantische Prinzip für die Naturgeschichte. Ein Versuch diese Wissenschaft philosophisch zu behandeln* (Göttingen, 1796). Another of Kant's disciples ranked races on moral grounds in a way that did not have Kant's support: Meiners, *Grundriss der Geschichte der Menschheit*, 43.

[152] Kant's response, 'Über den Gebrauch teleologisscher Principien in der Philosophie' ('On the Use of Teleogical Principles in Philosophy'), appeared in *Teutsche Merkur* in early 1788: *Gesammelte Schriften*, viii. 158–84; *Anthropology, History, and Education*, 195–218.

Kant claimed that anthropology was a central part of philosophical enquiry, yet it is difficult to understand in what way he thought it could play any significant role, given his account of what anthropology could achieve. In terms of his overall understanding of the critical philosophy, and by contrast not only with those areas that fell within 'theoretical knowledge' but also those that fell outside it, such as God, morality, and aesthetics, it appears very much as an afterthought—'a paddling pool for non-swimmers', as one commentator described it[153]—which at a theoretical level was largely cobbled together in response to Herder. Unlike his moral theory, with its very influential account of freedom, it had little traction in Kant's own time, and attempts to repair it, first in the 1860s in the 'South-Western' school of neo-Kantians,[154] and then more recently in attempts to develop an anthropology from Kant's moral philosophy, have strayed far not only from Kant's own approach but also from his results.[155] Herder's philosophical anthropology, by contrast, fed directly into late eighteenth- and early nineteenth-century Romanticism. As with the *philosophes*, it was not just a question of a theoretical programme but something that had a practical cutting edge, if one very different from that of the *philosophes*. Herder's anthropology was a central ingredient in Schiller's account of the shaping of character in the mid-1790s, which was to offer the basis for one of the most promising alternatives to a religious conception of the human being, one that would continue to exercise an influence from that time onwards, and to which we shall return in Chapter 7.

[153] Michael Landmann, *Einführung in die philosophische Anthropologie. Menschliche Selbstdeutung in Geschichte und Gegenwart* (Berlin, 1964), 42.

[154] See Thomas E. Willey, *Back to Kant: The Revival of Kantianism in German Social and Historical Thought, 1860–1914* (Detroit, 1978).

[155] This is because Kant's critical philosophy and his anthropology are not complementary but stand in a 'mutually supplementary relation': see Günther Zöller, 'Kant's Political Anthropology', *Kant Jahrbuch* 3 (2011), 131–61. Cf. Paul Menzer, *Kants Lehre von der Entwicklung in Natur und Geschichte* (Berlin, 1911).

5

The Natural History of Man

Go, proud reasoner, and call the worm thy sister![1]

The third form of naturalization of the human, to which we now turn, is 'the natural history of man', a form of comparative natural history. Although it was treated in varying degrees by later eighteenth-century thinkers as part of the anthropological project that we considered in Chapter 4, and indeed was largely assimilated to it in Kant, it is directed towards different questions, and deploys very different resources. The two kinds of project point in different directions, and it is important that we keep them distinct if we are to assess their respective contributions to understanding the human condition.

In the Aristotelian categorization of forms of enquiry, natural history could not aspire to a scientific standing and it was considered a decidedly inferior form of enquiry. It was concerned only with classification, not with grasping things in terms of their essential principles. While the standing of natural history underwent a positive reassessment in the sixteenth and seventeenth centuries, the question of whether classification was simply a matter of a conventional grouping or whether it should and could capture natural differentiations remained a contested question.[2] By the eighteenth century, the question had become a fundamental one, on which hinged the nature of our understanding of the world and our place in it, a concern renewed in the nineteenth century. As the geologist Charles Lyell was to put it in 1857:

> The ordinary naturalist is not sufficiently aware that when dogmatizing on what species are, he is grappling with the whole question of the organic world and its connection with the time past & with Man; that it involves the question of Man & his relation to the brutes, of instinct, intelligence & reason, of Creation, transmutation & progressive improvement or development.[3]

[1] Erasmus Darwin, *Zoonomia; or, The Laws of Organic Life* (2 vols, Dublin, 1794–1806), i. 198. As it happens, nearly a century later, his grandson Charles Darwin was to defend the intelligence of earthworms in his *The Formation of Vegetable Mould, Through the Action of Worms* (London, 1881).

[2] See Gaukroger, *The Emergence of a Scientific Culture*, 188–96.

[3] Charles Lyell, *Scientific Journals on the Species Question*, ed. L. G. Wilson (New Haven, 1970), 164.

One of the issues that prompted a rethinking of the aims of classification, was that of whether the distinction between animals and plants, the most basic division in natural history, was as secure as it had traditionally been taken to be. The purpose of classification was to gather together similar things and separate dissimilar ones. The behaviour of the freshwater polyp undermined the traditional distinction between animals and plants, for it was an animal by three of the traditional criteria of nutrition, sensation, and physiology, as set out in Aristotle's *Historium Animalium*. Its tentacles moved independently of the water in which it was immersed, for example, and movement could be stimulated in response to touching the tentacles or shaking the container; moreover, anatomically and physiologically, it was typically animal, having a mouth and a stomach. Yet as far as the fourth criterion, reproduction by emission of 'semen', was concerned it was wholly anomalous. Zoophytes—organisms that behaved like both animals and plants—had in fact been known since antiquity,[4] but had always been of marginal interest. This changed in the eighteenth century,[5] as the investigation of freshwater polyps, sponges, coral, starfish, carnivorous plants and other anomalous organisms challenged the distinction between vegetable and animal.[6] Coral was a particularly difficult case, classified by some writers as lying between a plant and a stone,[7] whereas others wondered whether coral flowers should (along with sensitive plants) be classified as animal.[8] No less problematic from the point of view of classification were the 'animalcules' seen

[4] See, for example, G. E. R. Lloyd, 'Fuzzy natures?', in idem, *Aristotelian Explorations* (Cambridge, 1996), 67–82.

[5] At least in the early stages, i.e. mid-century, it was not natural historians, but natural philosophers more generally who took an interest in these phenomena. As Susannah Gibson points out, 'In an age when much natural history centred on aesthetics, on cabinets and collections, on public displays of the wonders of nature, the zoophyte could not compete for popular attention. Compared to the beauty of a butterfly, the majesty of a lion, or the wonder of a kangaroo, the zoophyte seemed small, dull and ugly to many. It was primarily scholarly naturalists, natural philosophers and theorists who took an interest in the zoophyte.' Susannah C. Gibson, 'The Pursuit of Nature: Defining Natural Histories in Eighteenth-Century Britain', unpublished PhD thesis, University of Cambridge, 2011, 49.

[6] See the discussion in Ritterbush, *Overtures to Biology*, 122–41. Added complications arise from the fact that sponges and coral, for example, are 'stationary animals', that is, they lack the power of self-motion typical of animals, even though they are deemed capable of sensation: see Aristotle, *De anima*, 432b19–26. Recent discoveries have blurred the distinction between plant and animal further. Learning, for example, has always been considered specific to animals, but there is now clear evidence that the plant *Mimosa pudica* exhibits an elementary form of learning. See Monica Gagliano et al., 'Experience Teaches Plants to Learn Faster and Forget Slower in Environments Where It Matters', *Oecologia* 175/1 (2014), 63–72.

[7] See, for example, Étienne François Geoffroy, 'Observations sur les analyses du corail & de quelques autres plantes pierreuses, faites par M. le Comte Marsigli', *Mémoires de l'Académie royale des sciences* (1709), 102–5.

[8] René Antoine Ferchault de Réaumur, 'Observations sur la formation du corail, & des autres productions appellées plantes pierreuses', *Mémoires de l'Académie royale des sciences* (1728), 269–81. On sensation in plants, see Janet Browne, 'Botany for Gentlemen: Erasmus Darwin and *The Loves of Plants*', *Isis* 80 (1989), 593–621.

through microscopes,[9] for it was unclear how something so small could be incorporated into standard systems of classification, or even what it meant to ascribe life to something on such an unimagined scale of existence. More generally, it was a mystery how things lacking organs could even be considered to be alive.

These kinds of questions provoked a rethinking of just what the aim of classification should be, and Buffon was a crucial figure in this rethinking, shifting classification from a concern with nature's divisions to a theory of the interconnections between living things. It was Buffon who highlighted the question whether the similarities and dissimilarities employed by systems of zoological and botanical classification could be deemed natural, or whether they were merely conventional. He raised the issue in stark terms, and gave an uncompromising answer. Were there systems of classification in natural history that captured distinctions there in nature, as Linnaeus was claiming in the case of botanical families, and his more ambitious followers were claiming in the case of botanical genera? Capturing such distinctions 'in nature' meant identifying what were deemed to be unique characteristics, and Buffon regarded this as arbitrary, insisting that all features must be taken into account. The idea that there are natural groupings presupposes discrete entities, whereas in fact in nature there is a continuous shading of features:

It is clear that it is impossible to provide a general system, a perfect method, not only for the whole of natural history, but even for a single one of its branches: for in order to have a system, an arrangement, in a word a general method, it is necessary that it include everything; and this whole must be divided into different classes, which must be separated into genera, and the genera sub-divided into species, and all this according to an ordering which is necessarily arbitrary. But nature proceeds by indistinguishable gradations, and consequently it cannot lend itself wholly to these divisions, since it passes from one species to another, and often from one genus to another, by means of imperceptible nuances, in such a way that there are a great number of intermediate species and things that would fit in to different species, so that one does not know where to place them, and this of necessity undermines the project of a general system.[10]

[9] On some the problems raised by microscopy see Ratcliffe, *The Quest for the Invisible*, ch. 2; cf. Catherine Wilson, *The Invisible World: Early Modern Philosophy and the Invention of the Microscope* (Princeton, 1995).

[10] Buffon, *Œuvres complètes*, i. 63–4. The practical upshot of this for botanical classifications is clear from Lamarck's work on botany, the publication of which was sponsored by Buffon. Lamarck argued that there is nothing 'natural' about genera, which are subject to constant revision, and genera could just as usefully be named after great botanists since they disclose nothing about the nature of the plant. The most natural arrangement would be one confined to species: Jean-Baptiste-Pierre-Antoine de Monet de Lamarck, *Flore Françoise, ou description succincte de toutes les plantes qui croissent naturellement en France, disposée selon une nouvelle méthode d'analyse à laquelle on a joint la citation de leurs vertus les moins équivoques en médecine, et de leur utilité dans les arts* (3 vols, Paris, 1779), i. lxxxiv. Cf. Daubenton's view that only individuals exist in nature, and these can be compared with one another for the purposes of determining species: 'Introduction à l'histoire naturelle', *Encyclopédie méthodique. Histoire naturelle des animaux* (4 vols, Paris, 1782–9), i. i–xv.

On this view, if there is to be classification, as there must be for natural history to proceed, one has to give up the idea of natural groupings and instead think of classification as relative to human interests. It is these human interests that will then determine what we pick out as salient to distinguishing and grouping things. Buffon's use of morphological criteria for the identification of species is purely pragmatic, for example, as is his insistence on the ability to interbreed as the basic criterion.[11] Ability to interbreed had been taken by earlier natural historians as a sign of a natural affinity, but for Buffon it is a purely functional criterion, guided by the kinds of reasons that we have for distinguishing species in the first place, not something that mirrors any essential similarities or differences.

General statements of the importance of human interests can be found in a number of writers. For d'Holbach, for example:

> It is therefore only in his imagination that man finds the model for what he calls order or confusion, which, like all his abstract metaphysical ideas, supposes nothing beyond his reach. But order is nothing else than the faculty of putting himself in conformity with the things by which he is surrounded, or with the whole of which he is a part.[12]

For Diderot, it is not simply a matter of classification but one of what significance we give to natural phenomena. The vantage point that we must adopt, he argues, is not one fixed in nature, for there is no such thing, but one relative to our interests as human beings:

> One consideration above all must not be lost sight of, and that is that if man—or the thinking observing being—is banished from the surface of the earth, this moving and sublime spectacle of nature is nothing but a sad and silent scene. The universe is dumb; silence and night overtake it. Everything becomes a vast solitude where unobserved phenomena occur in a dark and mute manner. It is the presence of man that gives interest to the existence of beings; and we cannot do better in undertaking the history of those beings than to yield to this consideration. Why not introduce man into our work, as he is placed in the universe? Why not make a common centre of him? Is there any better point in infinite space from which the immense lines that we propose to extend to all other points could originate?[13]

Even if Buffon's insistence on the conventionality of species classifications were rejected, or at least not acted upon (and most naturalists continued to follow the Linnaean system because it was wholly adequate to their practical

[11] Buffon, *Œuvres complètes*, x. 285. The criterion was not pragmatic enough for everyone however: Blumenbach did not believe that all domestic dogs, from Great Danes to Pekinese, could belong to the same species, for example, and he revised Buffon's criterion.

[12] [Paul Henri Thiery, Baron d'Holbach], *Système de la Nature ou Des Loix du Monde Physique* (2 vols, [Paris], 1770), i. ch. 5.

[13] *Encyclopédie*, xii. 382–3. Aristotle had argued that we must begin the treatment of animals with an account of man, but for entirely different reasons: 'To begin with, we must take into consideration the parts of man. For, just as each country reckons by that monetary standard with which it is most familiar, so must we do in other matters. And, of course, man is the animal with which we are all most familiar.' *Historia animalium*, 491a19–23.

purposes and provided clear usable guidance on which there could be general agreement[14]), there was an even more profound set of classificatory problems for which there was no pragmatic solution. This was the question of those things that seemed to fall between the broadest categories of classification. Trembley's discovery of the reproductive behaviour of the polyp provoked intense discussion of the division between plant and animal, raising fundamental questions about the nature of life. Such classificatory uncertainty was not confined to the divisions between plants and animals however: it also encompassed the mineral realm.

THE PROBLEM OF CLASSIFICATION

In 1726, the historian Johann David Köhler wrote that:

> the best chronologists date the beginning of the world to 26 October 1657 years before the Flood, and 3947 years before Christ's birth. To be sure, the ancient Egyptians and Chaldeans, as well as the modern Chinese, make the world many thousand of years older, but Holy Scripture is more believable that all the other books of heathen fables founded on the ancients' vain search for fame.[15]

It was not a reassessment of the chronologies of the Egyptians or the Chinese that undermined confidence in this dating, but investigation of the mineral realm. The study of rocks, minerals, and fossils[16] gave rise to profound problems in the late eighteenth and early nineteenth centuries, as questions began to be asked about the extent to which geological features were the outcome of changes so long term that they contradicted not only traditional Christian views about the age of the earth, but also traditional Christian views about how it could have been created for the benefit of mankind. To appreciate the enormity of the challenge, consider the parallel with the change in notions of space. In the sixteenth century, Copernicanism had upset received notions of space. If the earth was indeed to revolve around the sun, rather than vice versa, then the lack of any observable parallax in the fixed stars indicated that they would have to be at a distance so

[14] Above all, its system of citations and technical language joined naturalists into a working community. See Mary Windsor, 'The Development of Linnaean Insect Classification', *Taxon* 25 (1976), 57–67. In fact Linnaeus' binomials for flora and fauna were originally introduced by him as a stopgap measure for pedagogic purposes: see Koerner, *Linnaeus*, ch. 2.

[15] Johann David Köhler, *Sculptura Historiarum et Temporum Memoratrix: Das ist, Gedächtnüß-Hülfiche Bilder-Lust der Merwürdigsten Welt-Geschichten aller Zeiten von Erschaffung der Welt biß auf gegenwärtige* (Nüremberg, 1726), 2.

[16] Linnaeus had provided the standard classification of the mineral realm, subdividing it into *Petrae* (rocks), *Minerae* (minerals), and *Fossilia* (petrifactions). This roughly corresponds to the modern differentiation between rocks (aggregates of minerals or non-minerals), minerals (naturally occurring abiogenic solids having an ordered atomic structure), and fossils (preserved remains of organic material).

immense as to be almost inconceivable. The eighteenth-century investigation of the mineral realm harboured similarly radical consequences, for it upset received Judeo-Christian notions of time,[17] postulating periods of time that were almost inconceivable in their length. It was not just that the earth had existed for a time that was many degrees of magnitude longer than the period described in the Bible. While fixity of species remained a basic assumption, this meant either that the earth pre-existed, by a huge margin, the creation of Adam and Eve, or that the biblical account was misleading and that human prehistory occupied a far longer period than human history. Whether it was a question of prehuman history or human prehistory, the role of what was in effect an indefinitely long earlier period was difficult to fathom on the providential models of history that were current.

In the second half of the eighteenth century, there were a number of geological observations suggesting that the biblical timescale was simply not of the right order of magnitude.[18] Volcanoes were a striking case of the inadequacy of a geological history of four or five thousand years. In the late 1760s, William Hamilton, British Ambassador to the Kingdom of Naples, observed eruptions of Vesuvius and Etna. He realized that if volcanoes had, as he believed, been built up by a succession of eruptions at the same rate and on the same scale as those recorded from Pliny onwards, the total age of the volcano stretched well beyond recorded history. Another indication of the longevity of the earth was river valleys, at least some of which could be attributed to erosion, and the time required for such erosion was again immense. Finally, there was the existence of secondary rock strata, which earlier investigators had assumed could have been laid down all at the same time, but by the late eighteenth century it had become evident that they must have been deposited layer by layer, in which case the timescale involved was again immense.

The realization of the great age of the earth threw doubt on the idea that it had been created for the benefit of mankind. But some managed to incorporate this into a providential story. One of the formative figures in late eighteenth-century geology, James Hutton, had no doubts that the timescales involved were incalculable, but he questioned the conclusions some had drawn from this:

> For having, in the natural history of this earth, seen a succession of worlds, we may from this conclude that there is a system in nature; in like manner as, from seeing revolutions of the planets, it is concluded, that there is a system by which they are intended to continue these revolutions. But if the succession of worlds is established in the system of nature, it is

[17] 'Judeo-Christian' notions of time, but not those of classical antiquity. Aristotle had assumed, along with his contemporaries, that the world had existed from eternity. On the difference between classical and Judeo-Christian notions of time, see Garry W. Trompf, *The Idea of Historical Recurrence in Western Thought: From Antiquity to the Reformation* (Berkeley, 1979).

[18] See Martin J. S. Rudwick, *Bursting the Limits of Time: The Reconstruction of Geohistory in the Age of Revolution* (Chicago, 2005), 119–24.

in vain to look for any thing higher in the origin of the earth. The result, therefore, of this physical enquiry is, that we find no vestige of a beginning,—no prospect of an end.[19]

To make sense of the geological evidence, he argued, we need a providential understanding, in which the world gradually becomes reformed. Although this is guided by divine providence, it acts wholly through natural processes, involving the action of water and the internal heat of the earth, into a human habitation:

When we trace the parts of which this terrestrial system is composed, and when we view the general connection of those several parts, the whole presents a machine of a peculiar construction by which it is adapted to a certain end. We perceive a fabric, erected in wisdom, to obtain a purpose worthy of the power that is apparent in the production of it.[20]

Unlike Hutton, not everyone treated the age of the earth as incalculable, and the internal heat of the earth was seen as the key. Buffon, in particular, had gone to considerable effort to offer a quantitative dating on this basis. His starting point was an argument in the fourth (1749) edition of the *Dissertation sur la glace* of Jean-Jacques Dortous de Mairan, subsequently repeated in a paper presented to the Académie des Sciences in 1765, showing that temperature differences in summer and winter did not correspond to differences in sunshine. If sunshine were the only factor, winters would be much colder, which suggests that the earth itself has a fund of heat. Buffon, adding his own observations of the increase in temperature as one descends into mines, and the constant temperature of the depth of oceans where sunlight does not penetrate, concludes that if the earth's core was hot, then perhaps this heat was that remaining from the original heat of the solar matter. This opens up the question of the age of the earth, for if it were still cooling, and we had some idea of at least the order of magnitude of its initial temperature, then, if we could determine the rate of cooling, we could calculate its age. Buffon sets out the details in a number of places, most notably in his *Époches de la Nature* of 1778. Using an industrial forge set up on his estate, he had been experimenting on the rates of cooling of iron spheres for some time, and, reviewing Newton's experiments and calculations, he had concluded that an iron ball the size of the earth would take just over 96,670 years to cool to its present temperature. However, the earth had not cooled in air but in a vacuum, and it was not an iron ball; rather 'in reality, the principal materials of which the earth is made, such as clays, sandstone, stones etc., must cool in less time than iron'. After 150 pages of calculations the published figure that he comes up with is 74,832 years, although he continued to recalculate and in manuscript drafts he concludes that the timescale involved is probably of the order of three million years.[21] Nor does he stop at the age of the earth. Knowing the sizes of planets and

[19] James Hutton, *Theory of the Earth, with Proofs and Illustrations* (2 vols, Edinburgh, 1795), i. 200.
[20] Ibid., 3.
[21] See the discussion in Jacques Roger, *Buffon: A Life in Natural History* (Ithaca, NY, 1997), 409–13.

CONSOLIDÉES jusqu'au CENTRE.	REFROIDIES à pouvoir LES TOUCHER.	REFROIDIES à la température ACTUELLE.	REFROIDIES à $\frac{1}{25}$ de la température ACTUELLE.
ans.	ans.	ans.	ans.
	LA TERRE.		
En...... 2936	En... $34270\frac{1}{2}$	En.... 74832	En.... 168123
	LA LUNE.		
En...... 644	En... 7515	En.... 16409	En.... 72514
	MERCURE.		
En...... 2127	En... 24813	En.... 54192	En.... 187765
	VÉNUS.		
En...... 3596	En... 41969	En.... 91643	En.... 128540
	MARS.		
En...... 1130	En... 13034	En.... 28538	En.... 60326
	JUPITER.		
En...... 9433	En... 110118	En.... 240451	En.... 483121
	Ier SATELLITE.		
En...... 8886	En... 101376	En.... 222203	En.... 444406
	II[d] SATELLITE.		
En...... 7496	En... 87500	En.... 193090	En.... 386180
	III[e] SATELLITE.		
En...... 6821	En... 80700	En.... 176212	En.... 352424
	IV[e] SATELLITE.		
En...... 2758	En... 32194	En.... 70296	En.... 140542
	SATURNE.		
En...... 5140	En... 59911	En.... 130821	En.... 262020
	ANNEAU DE SATURNE.		
En...... 6558	En... 76512	En.... 126473	En.... 252946
	I[er] SATELLITE.		
En...... 4891	En... 57011	En.... 124490	En.... 248980
	II[d] SATELLITE.		
En...... 4688	En... 54774	En.... 119607	En.... 239214
	III[e] SATELLITE.		
En...... 4533	En... 51108	En.... 111580	En.... 223160
	IV[e] SATELLITE.		
En...... 2138	En... 24962	En.... 54505	En.... 109010
	V[e] SATELLITE.		
En...... 600	En... 7003	En.... 15298	En.... 67747

Fig. 5.1 Table for the cooling of the earth. Buffon, *Époches de la Nature* (1778).

their satellites, he goes on to provide a table (Fig. 5.1) of the time taken for each of these to cool enough to be solid (column 1), to reach a temperature cool enough to be touched by the hand (column 2), and to reach the temperature of the earth (column 3).[22]

The second column is particularly important, for it represents the temperature at which life would have been able to emerge on a cooling earth. The discovery of the tusks and bones of warm-climate animals such as elephants, hippopotamuses, and rhinoceroses in Northern Europe and Siberia indicated that these regions had once been as warm as present-day tropical climates. Buffon argues that life emerged only once the temperature dropped sufficiently to support it, through the effects of heat on matter.[23] Not only did animals and plants come into existence through purely natural means, but they emerged at a comparatively late stage in the development of the earth.

Buffon's account is not an evolutionary one. He believed that fully formed animals and plants emerged spontaneously at the point at which the temperature dropped sufficiently to support life. But it was difficult to understand what possible justification could have been offered for this assumption. The sudden

[22] Buffon, 'Histoire de Minéraux, Partie Hypothetique', *Œuvres complètes de Buffon,* iv. 266–481. Fig. from p. 413.

[23] Buffon's account is sketchy and has to be reconstructed. See Roger, *Buffon*, 414–17.

emergence of fully formed elephants and other large quadrupeds, for example, was a great price to pay for his terrestrial cooling theory. An alternative was to embrace the idea of transformation of species, and by the last decade of the century, transformationism was being proposed by a number of writers. This approach was developed into a theory of the emergence of species, for example, by Erasmus Darwin in 1794: 'When we revolve in our minds the great similarity of structure, which obtains in all the warm-blooded animals, as well as quadrupeds, birds, and amphibious animals, as in mankind; from the mouse to the bat to the elephant and whale; one is led to conclude, that they have alike been produced from a similar living filament.'[24] After detailing a number of other examples, he continues:

> From thus meditating on the great similarity of structure of the warm-blooded animals, and at the same time of the great changes they undergo both before and after their nativity; and by considering in how minute a portion of time many of the changes in animals above described have been produced; would it be too bold to imagine, that in the great length of time, since the earth began to exist, perhaps millions of ages before the commencement of the history of mankind, would it be too bold to imagine, that all warm-blooded animals have arisen from one living filament, which THE GREAT FIRST CAUSE endued with animality, with the power of acquiring new parts, attended with new propensities, directed by irritations, sensations, volitions, and associations; and thus possessing the faculty of continuing to improve by its own improvements by generation to its posterity, worlds without end![25]

Darwin's approach was unusual in that it included human beings in the postulated evolutionary development, but not in other respects. From the second half of the last decade of the eighteenth century in France, there was a burgeoning of theories of the emergence of species.[26] In 1795, Jean-Claude Delamétherie had set out an elaborate theory of the formation of minerals and soils, the spontaneous generation of prototype plant and animal species, and the transformation of these as a result of slow environmental changes.[27] Two years later, Philippe Bertrand offered an alternative account, criticizing Delamétherie's theory as depending exclusively on chemical processes. In its place, he offered a cosmogonic account in which life first appeared in a primitive ocean, noting that what we can now detect are not differences in the principle of life itself, but only the varied 'modes of life', which have developed as animals adapt to new environments, so that their structure changed as a result.[28] In 1802, Cabanis, in his *Rapports du physique et du moral de l'homme*, offered a transformationist account of how species can undergo modification:

[24] Darwin, *Zoonomia*, i. 549–50. [25] Ibid., 551–2.
[26] See Corsi, *The Age of Lamarck*, 76–84, to which I am indebted here.
[27] Jean-Claude Delamétherie, *Théorie de la terre* (3 vols, Paris, 1795).
[28] Philippe Bertrand, *Nouveaux principes de géologie, comparés et opposés à ceux des philosophes anciens et modernes, notamment de J.-C. Delamétherie, qui les a tous analysés dans sa théorie de la terre* (Paris, 1797).

But the empire of habits is not confined to the profound and eradicable imprints made on each individual, for these are at least partially transmissable by way of generation. The great ability of certain organs to act, to produce certain movements, to carry out certain functions, in a word particular faculties developed to the highest degree can be propagated from generation to generation.[29]

In the same year, Lamarck published his first comprehensive statement of transformationism, a combination of spontaneous generation (typified, he believed, in the reproduction of the polyp) and a theory of adaption to environment through the use and disuse of certain characteristics, for example in the shape and size of certain external organs, which can be transmitted in reproduction:

It is not the organs—the nature and form of the parts of an animal's body—that are responsible for the animal's habits and particular faculties. Quite the reverse, it is the animal's habits, its mode of life, and the circumstances in which its individual ancestors found themselves that, over time, have determined the form of its body, the number and state of its organs, and finally the faculties with which it is endowed'.[30]

There was no shortage of speculative accounts of how species might develop and be transformed, but how was one to identify the stages in this transformation? More importantly, how did one decide in what way the fossils and bones of animals that were no longer extant were related to modern species? Here, a powerful tool was at hand: the newly emerging discipline of comparative anatomy. For our purposes, the realization that there was a need for comparative anatomy can be dated to 1795, when the geologist Horace-Benédict de Saussure set out in detail a number of desiderata for investigating the age of species. He argued that investigators needed to find out whether there were fossil shells found in older mountains but not in newer ones, and to estimate the relative ages and dates of the species; they needed to compare fossil shells, plants, etc., with living analogues, to determine whether or not there was an exact resemblance, and if not whether it was a case of different varieties or different species.[31] Cuvier quickly took up Saussure's challenge to verify the claim that the fossil bones of quadrupeds such as elephants, rhinoceroses, oxen, and deer were different from those of their extant counterparts. Comparative anatomy, he argued, showed beyond any

[29] *Rapports du physique et du moral de l'homme: neuvième mémoire*, § III, *Oeuvres complètes de Cabanis* (5 vols, Paris, 1823–5), iv. 146–7.

[30] Lamarck, *Recherches sur l'organisation des corps vivans et particulièrement sur leur origine, sur la cause de ses développemens et des progrès de sa composition, et sur celle qui, tendant continuellement à la détruire dans chaque individu, amène nécessairement sa mort; précédé du discours d'ouverture du cours de zoologie, donné dans le Muséum national d'Histoire Naturelle* (Paris, 1802), 50. See Corsi, *The Age of Lamarck*, chs 2–4. As Corsi points out (122), the *Recherches* contained all the basic theses subsequently to be developed in his more famous *Philosophie zoologique, ou exposition des considérations relatives à l'histoire naturelle des animaux* (2 vols, Paris, 1809).

[31] Horace-Benédict de Saussure, 'Agenda, ou tableau général des observations et des recherches dont les résultats doivent servir de base à la théorie de la terre', *Journal des mines* 4/20 (1795), 1–70: 32–3.

doubt that these were different species from the extant ones.[32] One of the first to build on Cuvier's work was Eugène-Louis-Melchior Patrin, who, in a comprehensive treatise on mineralogy published in 1801,[33] argued that the oldest organized bodies were generated spontaneously in the sea, and with the gradual fall in sea level more complex bodies arose. He then went on to develop Cuvier's thesis that the older the stratum, the greater the difference between the fossils and bones found there and modern species.

ANTHROPOLOGY AND NATURAL HISTORY

Comparative anatomy quickly came to play a crucial role in comparisons between human beings and other species, but before we examine it, it will be helpful to look briefly at an anthropological tradition that predated comparative anatomy yet which treated anatomical and physiological questions as being part of a more general understanding of the human being.

The first treatise describing itself in its title as being concerned with anthropology was Magnus Hundt's *Anthropologia, de hominis dignitate, nature et proprietatibus* (1501).[34] Hundt was an anatomist, but as well as an account of human anatomy and physiology, in the *Anthropologia* he offered a theory of human psychology, and a speculative history of human origins. The aim was to present elements of a picture of the human being that integrated the Greco-Roman medical traditions with the very different kinds of concerns fostered in Christian teaching on human nature. At the end of the century, the humanist Otto Casmann published his *Psychologia anthropologica sive animae humanae doctrina* (1594), which, defining anthropology as 'die Lehre von der menschlichen Natur' (the study of human nature), aspired to establish a systematic unity in the various different traditions of enquiry into human beings, exploring the idea of a spiritual and a corporeal nature. Yet the two are accounted for in different terms, the first relying on classical philosophical texts and Christian teaching, the other on the medical tradition. The division is further emphasized in his *Nucleus mysteriorum*

[32] Georges Cuvier, 'Mémoire sur les espèces d'élephants vivantes et fossiles. Lu le premier pluvoîse an IV', *Mémoires de l'Institut national des sciences et des arts. Sciences mathématiques et physiques* 2 (1799), 1–22. On Cuvier's palaeontology see William Coleman, *Georges Cuvier Zoologist: A Study in the History of Evolution Theory* (Cambridge, Mass., 1962), ch. 5.

[33] Eugène-Louis-Melchior Patrin, *Histoire naturelle des minéraux, contenant leur description, celle de leur gîte, la théorie de leur formation, leurs rapports avec la géologie ou l'histoire de la terre, le détail de leurs propriétés et de leurs usages, leur analyse chimique* (5 vols, Paris, 1801).

[34] On the early development of anthropology see Mareta Linden, *Untersuchungen zum Anthropologiebegriff des 18. Jahrhunderts* (Bern, 1976); Werner Sombart, 'Beiträge zur Geschichte der wissenschaftliche Anthropologie,' *Sitzunesberichte der Preußische Akademie der Wissenschaften philosophisch-historische Klasse* 13 (1938), 96–130; and Claude Blanckaert, 'L'*Anthropologie* en France: Le mot et l'histoire (XVI-XIX siècle)', *Bulletin et mémoires de la Société d'Anthropologie de Paris* 1 (1989), 13–44.

Naturae enucleatus (1605), where anthropology is included under natural philosophy, by contrast with psychology (of the *de Anima* variety) or moral philosophy. In broad terms, this is the approach followed by Descartes, in his *L'Homme*, composed in the early 1630s, which treats human beings in physiological terms, and indeed this was to provide the model for subsequent treatments of the human being in biomechanical terms. Yet in *Les Passions de l'âme* (1648) he suggests a completely different approach, attempting to integrate physiology with more traditional humanist treatments of the soul.[35] The ambiguity continues into the next century. The exclusively anatomical and physiological use of the term 'anthropology' was revived in James Drake's *Antropologia nova* (1707), for example, which offered an anatomico-medical account of the human body. But by the middle of the century, the term is being used again—for example, in Wilhelm Friedrich Struve's *Anthropologia naturalis sublimior* (1754)—in the context of a project that attempts to unify the various anatomical and moral disciplines.

The anthropological medicine tradition that we looked at in Chapter 3 can be seen as furthering this programme, through its medicalization of the human being, so that moral, social, and political issues come ultimately to be joined to the highly socialized form of medicine that it promotes. The 'natural history of man' is a different development of this programme. Under the general rubric of natural history, it naturalizes the study of human nature not by locating it in a medical context, but rather by exploring continuities between and differences from other animals, or by exploring continuities between and differences from various physical 'types' of human being. It is an essentially comparative exercise, and this marks it out from earlier anthropological projects. In both anthropological medicine and the natural history of man, the 'joining' of two concerns that were traditionally quite distinct is a delicate matter, fraught with concerns over the appropriateness of various reductive and comparative strategies.

The specific strategies at play in the natural history of man turned primarily on three kinds of comparisons: of physical features of human and animals; of the role of environmental factors on the development and behaviour of animals and humans; and of what were regarded as distinctive human types with one another. It is not just the comparative nature of the investigation that makes these cases distinctive, but the kind of information sought from the comparisons. Comparisons of human anatomy and that of other animals was a feature of anatomical practice from the origins of the discipline, and taboos on the anatomical investigation of human corpses meant that information about human anatomy was extrapolated from that of other mammals. But the aim of comparative anatomy as it developed in the late eighteenth century went well beyond instructive analogies, and pointed instead to the understanding of the human being as an

[35] See Gaukroger, *Descartes, An Intellectual Biography*, chs 7 and 10.

animal for purposes that transcended shared anatomy.[36] Although the natural histories of classical antiquity, notably that of Aristotle, had treated the human being as an animal, for example, this was on the basis of anatomical analogies, not systematic anatomical comparisons, still less on the basis of a zoological taxonomy. Moreover there was considered to be an unbridgeable gulf separating human beings from other animals. Human beings alone were considered to possess reason, and despite numerous anatomical similarities, this in effect put them in a class of their own, as a very different kind of creature from animals.

This is just what now became problematic, and the problems were evident as early as the work of seventeenth-century anatomists such as Thomas Willis. On the basis of anatomical investigation of the nervous system in humans and in animals, Willis demonstrated that the human nervous system was much more complicated than that of any other animal, and he took this as proof that God had created human beings as rational creatures and animals as creatures lacking reason.[37] Yet he also discovered that the nervous systems of humans and quadrupeds were analogously constructed, and were so similar in form and function that an immaterial principle needed to be introduced to account for the differences in mental behaviour.[38] But the problems were not just those of comparative anatomy. More generally, we witness the extension of methods of studying animals, such as differences arising from climate and habitat, which would traditionally have been considered inappropriate for the understanding of human behaviour.[39] What in the eighteenth century was termed 'physical geography', for example, attempted to identify the environmental factors that shape human behaviour, in such a way that there is continuous transition from studies of the role of animal environment to that of human environment. Nor is it just comparison of animals and humans that is at issue here. The comparative approach is applied to human variation, particularly racial variation, with a view to generating fundamental insights into human nature. But it is also at the basis of a number of distinctive nineteenth-century disciplines—physiognomy, craniology, and phrenology—which set out to identify the external marks of character. Physiognomy in particular was considered more revealing of human character than language by its proponents, since the latter may involve lying,

[36] The first comprehensive treatise on comparative anatomy was Georges Cuvier, *Leçons d'anatomie comparée* (5 vols, Paris, 1800–5), and the first textbook was Johann Friedrich Blumenbach, *Handbuch der vergleichenden Anatomie* (Göttingen, 1805).

[37] Thomas Willis, *Cerebri anatome: cui accessit nervorum descriptio et usus* (London, 1664).

[38] See William F. Bynum, 'The Anatomical Method, Natural Theology, and the Functions of the Brain', *Isis* 64 (1973), 444–68: 447. As Bynum notes (451), Willis' comparative anatomy is different from that of Harvey, who confined his attention primarily to single species.

[39] For details of why it would have been inappropriate, see Dennis Des Chene, '"Animal" as a Category in Bayle's *Rorarius*', in Justin E. H. Smith, ed., *The Problem of Animal Generation in Early Modern Philosophy* (Cambridge, 2006), 215–32.

whereas physiognomy is something over which one has no control.[40] Indeed in Switzerland, Germany, France, and England, there arose the practice of household servants being selected on the basis of a comparison of their features with Lavater's descriptions and engravings of facial and cranial features.[41] Charles Darwin was almost refused his position on the *Beagle* because Captain Fitzroy 'was an ardent disciple of Lavater, and was convinced that he could judge of a man's character by the outline of his features; and he doubted whether any one with my nose could possess sufficient energy and determination for the voyage'.[42] It is also worth noting that physiognomy had, towards the end of the eighteenth century, initiated a new form of commemoration: Lessing's death mask, made in 1781, marked the beginning, in the modern period, of a practice of revering the dead through their physiognomy.[43]

I have mentioned that philosophical anthropology and the natural history of man were not clearly distinguished in the eighteenth and early nineteenth centuries. Nevertheless, the two proceed in entirely different ways from one another. Note, for example, the conceptual gulf between the assumption behind various forms of physical geography, namely that there is a continuity between the role of animal environment and that of human environment, and Herder's conception of the difference between animals and humans that we looked at in Chapter 4, which Herder locates in the contrast between the way in which animals respond to their environment and that in which human beings respond.

Buffon was the first to draw out into the open the philosophical consequences of the zoological treatment of human beings as animals. 'Natural history', he writes in his *Premier Discours de la manière d'étudier et de traiter l'histoire naturelle* (1749), 'when taken in its entirety, is an immense history, embracing all the objects that the universe presents to us'.[44] Yet the study of natural history teaches us very specific lessons, and:

The first truth yielded by this sober examination of nature is one that is perhaps humbling to man. This is that he ought to range himself in the class of animals, which he resembles in

[40] See Christoph Siegrist, '"Letters of the Divine Alphabet": Lavater's Concept of Physiognomy', in Ellis Schookman, ed., *The Faces of Physiognomy: Interdisciplinary Approaches to Johann Caspar Lavater* (Columbia, South Carolina, 1993), 25–39: 29.

[41] See John Graham, *Lavater's Essays on Physiognomy: A Study in the History of Ideas* (Bern, 1979), 62. Lavater himself secured subscriptions to his own books by promising subscribers that, if they sent him their silhouette, he would analyse their physiogonomy in a future volume: see Gudrun Swoboda, 'Die Sammlung Johann Caspar Lavater in Wien', in G. Mraz and U. Schölg, eds, *Das Kunstkabinett des Johannes Caspar Lavater* (Vienna, 1999), 75–95: 93.

[42] Francis Darwin, ed., *The Life and Letters of Charles Darwin* (3 vols, London, 1887), i. 59–60.

[43] See Gode-von Aesch, *Natural Science in German Romanticism*, 222.

[44] Buffon, *Œuvres complètes*, i. 53. His 36 volume *Histoire Naturelle* appeared between 1749 and 1789: George Louis Leclerc de Buffon, *Histoire naturelle, générale et particulière* (15 vols, Paris, 1749–67); *Histoire naturelle des oiseaux* (9 vols, Paris, 1770–83); *Histoire naturelle, générale et particulière. Supplément* (7 vols, Paris, 1774–89); *Histoire naturelle des minéraux* (5 vols, Paris, 1783–8); all are collected in the *Œuvres complètes*.

everything material; and even their instincts will perhaps appear more certain than his own reason, and their industry more admirable than his arts. Then, as he examines successively and in order the various objects that make up the universe, while placing himself at the head of all created beings, he will see to his astonishment that it is possible to descend by almost imperceptible degrees from the most perfect creature to the most formless matter, from the most perfectly organized animal to the most brute matter. He will recognize that these imperceptible gradations are the great work of nature, and will find them not only in the size and shape of things, but in the motion, generation, and successions of each species.[45]

The 'natural history of man' was, then, very much part of a general concern with natural history. Among other things, it covered the nature of mind and the history of language and culture, as well as the bodily form and structure of humans and non-humans. Its most significant feature from the point of view of our present interests is that it brought the concerns of traditional natural history into the domain of civil history.[46] 'History' here should be understood in the eighteenth-century sense: it is as much comparative geography as it is comparative history in the modern sense, and that distinctive eighteenth-century exercise, conjectural history, which sets out to trace the origins of human morality and society,[47] appeals as much to contemporary primitive states as to those of earlier times. Moreover, the 'comparative' is important here, for this is effectively constitutive of the discipline. That is to say, there is no prior circumscribing of what it might encompass: any comparison that can throw light on human beings is relevant, from anatomy to religion, from the age of rocks to statistics on births, deaths, and marriages. It is a way of building up understanding from discerning similarities and dissimilarities.

Similarities and dissimilarities lie at the core of the natural history approach, and this is nowhere more striking than in the question of the standing of various feral humans, who had developed in isolation from human society, and who seemed to observers to have, in part, the character of animals. In later editions of his *Systema naturae*, Linnaeus, in his division of *Homo sapiens* into varieties, had included, along with his principal racial varieties and a category of 'monstrosities', a sixth variety: *Homo sapiens ferus*. Feral humans, unlike the other varieties, comprise individual cases.[48] The earliest reported case to be included in the list was one recorded by the sixteenth-century naturalist Conrad Gesner of a feral human

[45] *Œuvres complètes*, i. 62–3.

[46] This has forerunners in principle, if not significantly in substance. One might think that 'man' would traditionally have been included under civil history, but Hobbes' formative discussion in 'Of Man' in *Leviathan* (London, 1651) divides 'Science, that is Knowledge of Consequences' into 'Natural Philosophy' and 'Civil Philosophy', the latter drawing consequences from the institution of commonwealths of rights and duties of subjects and sovereigns. Various subdivisions of natural philosophy lead from 'Consequences from qualities of bodies permanent' to 'Consequences from the Qualities of Animals' and to 'Qualities of men in Special'. See the fold-out chart facing page 40 of *Leviathan*.

[47] The precedents were set by Hobbes, Pufendorf, and Montesquieu.

[48] See Franck Tinland, *L'Homme sauvage: Homo ferus et Homo sylvestris, de l'animal à l'homme sauvage* (Paris, 2003), ch. 2.

quadruped inhabiting the forest of Hanesbergium in the parish of Salzburg.[49] But the seventeenth- and eighteenth-century cases tended to be children who had grown up in extreme isolation: *Johannes Leodicensis*, reported in the forests outside Liège in the early seventeenth century;[50] *Juvenis Ovinus hibernus*, captured in Ireland, reputedly raised among sheep, exhibited as an adolescent in Amsterdam, and included in the Linnaean taxonomy by Tulp, who had examined him,[51] as a feral human in 1672; Peter the wild boy of Hamelin, described in 1724 (aged around twelve), who had been living wild in the woods for an unknown period, was unable to learn to speak, and adopted a quadruped posture;[52] *Puella Campanica*, found in the forest of Songi in the region of Champagne in 1731, and the subject of an extensive account by La Condamine in 1755;[53] and *Juvenis Averionensis*, 'Victor' the wild boy of Aveyron (see Fig. 5.2), the most extensively studied of all the feral children.

Victor was first spotted in 1797: a naked boy, aged around eleven, searching for acorns and roots in the woods and hills in the region of Lacaune in south-central France. He was captured a couple of times only to escape, but in January 1800 came to a local village looking for food—he ate acorns and roots but refused bread and meat—and was taken in by a local lawyer.[54] He very quickly became the subject of interest of a number of naturalists and educational reformers. When word of Victor reached Abbé Pierre Joseph Bonnaterre, a professor of natural history at the Central School for Aveyron, he immediately applied for permission from the government commissioner to examine him. This was the first of a number of examinations to which Victor was subject. One thing of particular interest was the fact that he was mute at first, but subsequently began to utter various cries. Given the close association between language and being human, the question of whether he could be taught to speak was a pressing one. The Abbé Roch-Ambroise Cucurron Sicard, a member of the recently formed *Société des observateurs de l'homme*, a society for the study of anthropology, set out to try to teach Victor to speak. He was the principal of a school for the deaf, and managed

[49] Conrad Gesner, *Historia animalium* (5 vols, Zurich, 1551–87), i. 979. There are in fact cases of human quadrupeds, sufferers of a rare form of cerebellar ataraxia which causes balance problems making bipedalism almost impossible. If the sufferers have learned to crawl in the appropriate way, they come to adopt a quadrupedal posture and move using feet and wrists (not knuckles, as in apes). See Nicholas Humphrey, John. R. Skoyles, and Roger Keynes, 'Human Hand-Walkers: Five Siblings Who Never Stood Up', *LSE Research Online*, 2005.

[50] He is described in Kenelm Digby, *Two Treatises: In the One of Which, The Nature of Bodies; In the Other, The Nature of Mans Soule, is Looked into: In Way of Discovery of the Immortality of Reasonable Soules* (Paris, 1644), [i.] 310–12.

[51] Tulp, *Observationum medicarum. Editio Nova* (Amsterdam, 1672), 296–8.

[52] Contemporary reports on Peter are given in Lord Monboddo, *Antient Metaphysics* (6 vols, Edinburgh, 1779–99), iii. 367–77.

[53] Charles-Marie de La Condamine, *Histoire d'une jeune fille sauvage trouvée dans les bois à l'âge de dix ans* (Paris, 1755).

[54] I take details from the comprehensive account in Harlan Lane, *The Wild Boy of Aveyron* (London, 1977).

Fig. 5.2 'Victor' the wild boy of Aveyron. Itard, *De l'éducation. d'un homme sauvage* (1801).

to get Victor transferred there, not because he believed him to be deaf, but because teaching children how to speak was one of the aims of the institution. Victor developed a very elementary capacity for understanding language, but failed to progress beyond that level. He subsequently went to live with a physician, Jean Itard, who had interests not only in the treatment of deaf-mutes but also in the care and education of the severely impaired. Itard observed Victor closely over several years and compiled two comprehensive reports on his development, a detailed initial report and a no less detailed follow-up study five years later.[55] Itard saw the key to human behaviour in appropriately Rousseauean terms as lying in language and empathy, and he was to have some limited success in developing sociable behaviour in Victor, but less success with language.

For many naturalists, feral children threw into doubt traditional criteria, such as linguistic ability, for inclusion in the human species. At the same time, tailless apes, such as the chimpanzee and the orangutan, threw into doubt traditional criteria for exclusion from the human species. The brother of Georges Cuvier, Frédéric, who supervised the menagerie at the Muséum national d'histoire naturelle, was able to perform a number of intelligence tests on a young female orangutan who had been able to manipulate the lock to her cage and escape, and

[55] Jean Itard, *De l'éducation d'un homme sauvage, ou, Des premiers développemens physiques et moraux du jeune sauvage de l'Aveyron* (Paris, 1801); idem, *Rapport fait à son excellence le ministre de l'interieur sur les nouveaux développements de l'état actuel du sauvage de l'Aveyron* (Paris, 1807).

he credited her with an 'extensive use of reason'.[56] Monboddo went further, explicitly comparing the behaviour and intelligence of orangutans and feral children, and concluding that there was little if any difference. Appendix III to the third volume of his *Antient Metaphysics* is entitled 'The Difference between *Man* and *Brute* further illustrated and explained, with additional Facts and Observations concerning the Oran Outang and Peter the Wild Boy'. He begins: 'That it is by the Intellectual part of our nature we are so distinguished, every body will admit. If, therefore, it be true that the Brutes have Intellect as well as we, there is an end of the distinction betwixt us and them'.[57] His aim is to show that if we count Peter the Wild Boy as human we must also treat the orangutan as having human capacities. Monboddo had already given an extensive account of the behaviour of the orangutan in the first volume of his *Of the Origin and Progress of Language*,[58] and in *Antient Metaphysics* he provides a summary account. In their native island of Sumatra, he tells us, orangutans are regarded by the indigenous population as human. They live in societies, they are quick to learn commands, and they display 'melancholy and discontent' when captive, and the males are able to copulate with women and produce fertile offspring.[59] We are not concerned here with the accuracy of these reports,[60] but with the way the comparison of feral children and orangutans can be used to prize open the question of the characteristics by which we decide whether or not an animal is human.

Feral children were not just of interest to naturalists, but also to metaphysicians, educators, and physicians. The way in which they approached the questions often differed. But when a physician like Itard subjects Victor to careful examination, the anthropological conclusions he draws are not those of a *médecin philosophe*. Although they may look analogous in certain respects, the way in which the 'natural history of man' naturalizes the human is very different from the way in which this occurs in anthropological medicine. The key to the former project lies in a rejection of the idea of order in nature and its replacement by natural relations as guided by, and indeed constituted by, human interests. The human being is embedded in a wholly natural realm and the relationship is reciprocal: the human being gives meaning and significance to nature, but nature in its turn gives meaning and significance to human existence.

[56] Frédéric Cuvier, 'Description d'un orangoutang, et observations sur ses facultés intellectuelles', *Annales du Muséum d'histoire naturelle* 16 (1810), 46–65: 62. It should be noted, however, that he subsequently revised this claim.

[57] Monboddo, *Antient Metaphysics*, iii. 337.

[58] Monboddo, *Of the Origin and Progress of Language* (2nd edn, 6 vols, Edinburgh, 1774–1809), i. 270–361.

[59] Monboddo, *Antient Metaphysics*, iii. 359–61.

[60] The source of the last is difficult to discern, but James L. Larson, *Interpreting Nature: The Science of Living Form from Linnaeus to Kant* (Baltimore, 1994), 85, suggests that Buffon hinted at fertile crosses between Africans and apes. On accounts of male apes ravishing women, see Londa Schiebinger, *Nature's Body: Gender in the Making of Modern Science* (Boston, Mass., 1993), 94–8.

Buffon is of particular interest in this respect. The natural history of man is the culmination of Buffon's 'Histoire des animaux', even though, in line with the majority of his contemporaries, he saw a gulf between humans and other animals.[61] Buffon extended natural history in the way parallel to that in which the *médecins philosophes* extended medicine: and indeed the two share some concerns, such as an interest in the 'four stages of life'. Buffon treats the development through birth, childhood, puberty, adulthood, old age, and death as a core part of the understanding of human behaviour, supplemented by the analysis of over one hundred pages of statistical tables on births, deaths, and marriages.[62] By incorporating the study of human behaviour into natural history, Buffon is explicitly naturalizing it; and in introducing statistical tables into the programme, he is refining this naturalization at a methodological level. In these respects, his natural history of man marks a crucial early move in a transition from religious and humanistic understandings of human behaviour to 'anthropology'.

During his lifetime, Buffon was the dominant figure in natural history in France above all because his carefully fostered connections with the French court,[63] as well as those of other countries, meant that he effectively controlled funding and research in the discipline. His death in 1788 elicited a number of hostile responses to his 'literary' approach to natural history from figures such as Condorcet and Cuvier.[64] In compiling his *Histoire Naturelle*, Buffon had relied generally on morphology, although he had employed his assistant Daubenton to provide anatomical detail, and in the case of quadrupeds, for example, Daubenton dissected and provided anatomical details of 182 species. But anatomical detail was the exception rather than the rule in the *Histoire*, and it is completely lacking in Buffon's extensive treatment of birds, where the distinguishing features

[61] Rousseau is the most significant exception in this respect: see his *Discours sur l'origine et les fondemens de l'inegalité parmi les hommes*.

[62] Cf. Diderot's treatment in the entry 'Homme (hist. nat.)', *Encyclopédie*, xvii. 652–60.

[63] For example, in his 'Épochs de la nature', Buffon interpreted each of the seven days of creation in terms of vast spans of time. Although others had done this, the popularity of the work meant that the claim was perceived as a threat by the theologians of the Sorbonne, but powerful as they were, the King's protection meant that he could respond to their complaint in a perfunctory and scornful way without fear of censorship. See Jean Stengers, 'Buffon et la Sorbonne', *Études sur le XVIIIe Siècle* 1 (1974), 97–127.

[64] See Corsi, *The Age of Lamarck*, ch. 1. As Corsi notes, Buffon's project was not simply abandoned at his death. Charles-Nicholas-Sigisbert Sonnini de Manoncourt brought out a vastly expanded and reordered edition of the *Histoire Naturelle* over a ten-year period, in which he not only added large volumes on plants, and on animals that Buffon had not been able to cover, but also extensive notes by himself, the pharmacist and physician Julien-Joseph Virey, the entomologist Pierre-André Latreille, and the zoologist Bernard Germain de Lacepède: Buffon, *Histoire naturelle, générale et particulière, par Leclerc de Buffon. Nouvelle édition, accompagné de notes, dans lesquelles les supplémens sont insérés dans la première texte, à la place qui leur convient. L'on y a ajouté l'histoire naturelle des quadrupèdes et des oiseaux découverts depuis la mort de Buffon, celle des reptiles, des poissons, des insects et des vers; enfin, l'histoire des plantes dont ce grand naturaliste n'a pas eu de tempts de s'occuper. Ouvrage formant un cours complet d'histoire naturelle. Rédigé par C. S. Sonnini* (127 vols, Paris, 1789–1808).

were so visually striking that morphology was all that was needed to capture the distinctions.

For followers of Buffon, taxonomy was in a very fluid and uncertain state after his death. He had made no impact on botanical taxonomy, which remained Linnaean. Much of his own zoological taxonomy was morphological, the *Histoire des oiseaux* exclusively so. Moreover there was a reversion to Linnaeism in the ranks of his closest followers in many cases: it was Daubenton himself who advised Lacepède, an avid follower of Buffon, to adopt Linnaean taxonomy for reptiles and fish, for example. Finally, there was Daubenton's use of comparative anatomy in establishing the distinguishing features of species of quadrupeds. While the Linnaean system prevailed for many practical purposes, the search for a deeper understanding of the differences between species hinged on the choice prompted by Buffon's *Histoire*, namely that between morphology and comparative anatomy. The choice was crucial, for it was far easier for morphology to mirror traditional distinctions between humans and other animals than it was for comparative anatomy to fall in line with traditional distinctions. Moreover, morphology required one to choose those features on which to base classification, and this was often something in dispute,[65] whereas the criteria for anatomical differentiation, even though decisions had to be made about what characters to subordinate to others, were far more secure. More importantly, while it was possible, on morphological grounds, to think of humans as part of a grouping that included other animals, there was no particular reason to do this, and there was no reason why it should be a pressing issue. Comparative anatomy, by contrast, forced open the question.

COMPARATIVE ANATOMY

The natural history of man was an essentially comparative exercise, and we can identify four contrast classes for comparative purposes: feral children, apes, non-European races, and women. The first two prompt questions about the defining characteristics of human beings and, as we have seen, they raise problems about what to include and what to exclude from the human species. The second two point to differences within the human species, and raise problems about the significance of these.[66] Except in the case of feral children, the comparisons are primarily anatomical, and it is comparative anatomy that generates the most significant issues.

[65] The most notable disputes in the eighteenth century were in botany, where there were constant calls for an increase in the number of plant features used in classifications. See, for example, Michel Adanson, *Familles des plantes* (2 vols, Paris, 1763) and Antoine-Laurent de Jussieu, *Genera plantorum, secundum ordines naturales disposita, juxta methodum in Horto regio parisiensi exaratum* (Paris, 1789).

[66] It seems that not quite everyone had included women within the human species. In the entry on 'Femme (anthropologie)' in the *Encyclopédie* we are informed: 'Everyone had heard speak of an

The importance of comparative anatomy can be gauged from the article on natural history in the *Encyclopédie*, which describes the subject in these terms:

The naturalist only considers a thing in order to compare it with others; he gives priority to those features in each thing that distinguish it from others, and he strives to discern the course of nature in each of its products. The anatomist by contrast contemplates each thing in itself; he develops each of its parts to discern the least evident, and he employs his whole art so as to recognize the first material agents, and all the means that nature employs for animate bodies to move. Up to the present, anatomy had hardly had any other object but man; this is certainly the principal object, but the human body does not encompass all models of the mechanisms of the animal economy. There are, in animals, structures that are very different from those in man, who has more developed parts. In comparing them to one another and to man, one may come to a better understanding of man in particular and the mechanism of nature in general. This great aim is that of comparative anatomy, which has a more immediate relation to *natural history* than simple anatomy, because one can only draw observations of details from the latter, whereas the former furnishes general results and facts which form the core of the *natural history* of animals.[67]

Parallels between humans and animals had not been that uncommon in the earlier history of anatomy. Pierre Belon, in his 1555 *L'Histoire de la nature des oyseaux*, for example, offered striking illustrations of the comparative anatomy of bird and human skeletons (see Fig. 5.3). But no direct links were suggested in this case, any more than they were in other anatomical accounts of the era. It was only with the appearance of Edward Tyson's *Orang-Outang, sive Homo Sylvestris* of 1699 that such links were suggested. Tyson's was the first work to include human beings in the ape family on anatomical grounds. What was distinctive about apes was their bipedalism. It had been a commonplace of natural history that quadrupeds were paradigm brutes: there was no more certain correlate of physical characteristics and mental capability than the ability to walk upright, which was considered unique to humans. This certainty was shattered when, in 1698, a young chimpanzee walked off a ship newly arrived in London from the Angolan coastline.[68] The chimpanzee died soon afterwards, and Tyson, a surgeon and fellow of the Royal Society, although he had not been able to observe it alive, obtained the cadaver and took upon himself the task of examining its anatomy in detail. The illustrations that accompanied *Orang-Outang* (see Fig. 5.4),

anonymous *dissertation*, in which it is claimed that women are not part of humankind [*le genre humain*], *mulieres homines non esse*.' Diderot and d'Alembert, *Encyclopédie*, xiii. 954.

[67] Ibid., xvii. 566–7 (Histoire naturelle).

[68] The youth of the chimpanzee is important as far as parallels with humans are concerned. Apes transported to Europe were captured only after their mother had been shot, and immature apes have a human appearance that they subsequently loose: see Stephen Jay Gould, 'Chimp on a Chain', *Natural History* 92/12 (1983), 18–26.

served to help establish its bipedalism.[69] But at the same time he set out to show that one could not ascribe language to chimpanzees, thereby severing the traditional connection between bipedalism and language. In doing so, he opened up the question as to just what the relation was between apes and human beings.[70]

This set in train a number of such comparisons, one of the most famous being that of the botanist and natural historian Richard Bradley, who illustrated the similarities between monkey and human skeletons (see Fig. 5.5).[71] But it was the

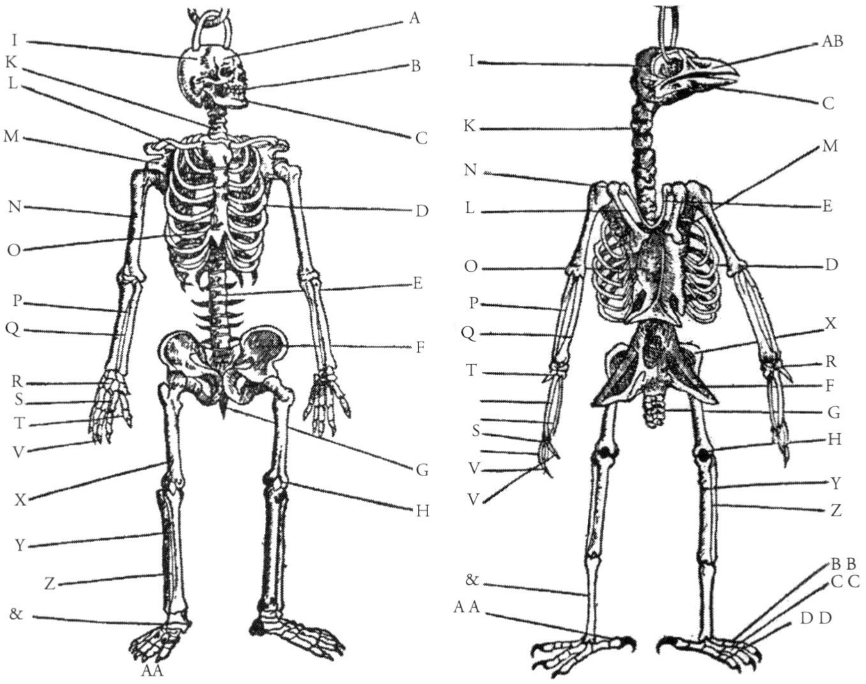

Fig. 5.3 Comparison of bird and human skeletons. Belon, *L'Histoire de la nature des oyseaux* (1555).

[69] Tyson believed only bipedalism and quadrupedalism were natural, so dismissed the reports of the knuckle-walking of the chimpanzee as a temporary aberration or caused by deformity of the bones.

[70] Edward Tyson, *Orang-Outang, sive Homo Sylvestris: or, The Anatomy of a Pymgie Compared with that of a Monkey, an Ape, and a Man* (London, 1699). On Tyson, see Justin E. H. Smith, 'Language, Bipedalism, and the Mind–Body Problem in Edward Tyson's Orang-Outang (1699),' *Intellectual History Review* 17 (2007): 291–304.

[71] Plate from Richard Bradley, *A Philosophical Account of the Works of Nature. Endeavouring to set forth the several Gradations Remarkable in the Mineral, Vegetable, and Animal Parts of the Creation. Tending to the Composition of a Scale of Life* (London, 1721).

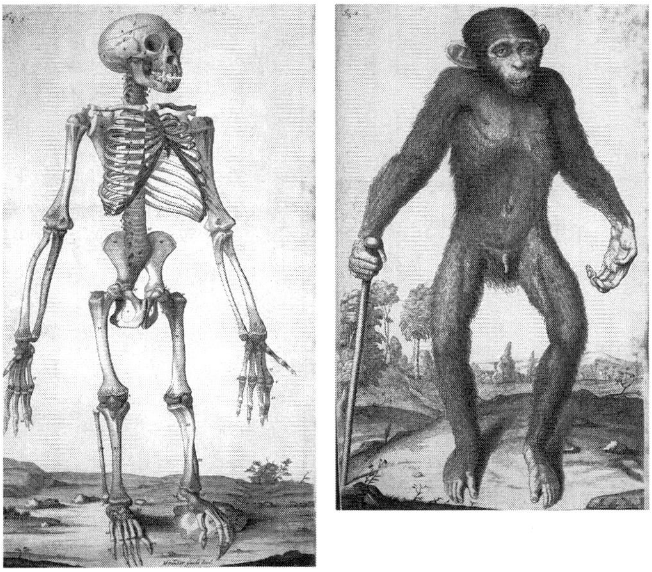

Fig. 5.4 Chimpanzee bipedalism. *Tyson, Orang-Outang, sive Homo Sylvestris* (1699).

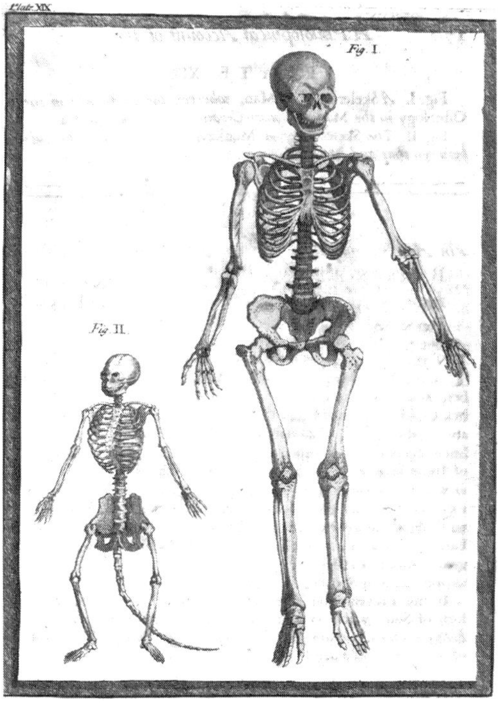

Fig. 5.5 Comparative anatomy of monkeys and humans. Bradley, *A Philosophical Account of the Works of Nature* (1721).

Fig. 5.6 An ape. Tulp, *Observationum medicarum* (1641).

first edition of Linnaeus' *Systema Naturæ* (1735) that established and applied a general descriptive method whereby human beings could be described in the same way as other animals, and established it as an autonomous area of enquiry, not just an adjunct to medicine.[72] He placed man (*Homo*) in the class *Quadrupedia*, in the same order as the ape (*Simia*) and the sloth (*Bradypus*). In the tenth edition (1758), he introduced the class *Primates*, and *Homo* became a genus that included two species *H. sapiens* and *H. troglodytes*, the latter of which included the 'orang-outang'

72 On the long-term importance of this see Bernard Balan, *L'ordre et le temps: l'anatomie comparée et l'histoire des vivants au XIXe siècle* (Paris, 1979). The *Systema* comprised thirteen folio pages, dividing the three kingdoms into subordinate Kingdoms and (with newly coined names) Orders, Classes, Genera, and Species.

(actually a chimpanzee).[73] It is worth noting in this connection the very humanized illustration of a genial ape (Fig. 5.6) in the widely consulted *Observationum medicarum* (1641) by the famous Dutch physician and surgeon, Nicolaes Tulp.

Linnaeus had first-hand familiarity with apes, keeping some himself in his zoological collection at Uppsala, and he used the teeth of apes as evidence for establishing that human beings should be classed as carnivores.[74] Indeed, he considered apes and humans to be almost indistinguishable in terms of basic anatomical features:

I scarcely know one feature by which man can be distinguished from apes, if it be not that all the apes have a gap between their fangs and their other teeth. . . . Indeed, in the uvula, the womb, and the muscles they are similar to us—even though they do not talk—and different from *bruta*. They often walk upright, standing only on their hind legs. They also take their food in their hands and carry it to their mouths, they drink from their cupped palm, and where there is no water they dig wells.[75]

This claim was by no means unchallenged, and the inclusion of humans with apes attracted significant critical attention. Some editors of Linnaeus' *Systema*, for example, either simply omitted the offending material,[76] or quietly changed *Homo troglodytes* into *Simia troglodytes*[77] so that, as one commentator has put it, 'with a stroke of the classifier's pen, the chimpanzee became an ape'.[78] At the same time, other critics, such as the chemist Johan Gottschalk Wallerius and the zoologist Jacob Theodor Klein, denied on anatomical and physiological grounds that humans were related to quadrupeds.[79] Those with no interest in natural history simply denied that it could have any bearing on human understanding. Adam Ferguson, for example, writes that:

in opposition to what has dropped from the pens of eminent writers, we are obliged to observe, that men have always appeared among animals a distinct and a superior race; that

[73] See Tinland, *L'Homme sauvage*, ch. 3. The true (Asiatic) orangutan was only distinguished from the (African) chimpanzee in the 1780s, by the anatomist Robert Camper.

[74] See Gunnar Broberg, '*Homo sapiens*: Linnaeus's Classification of Man', in Tore Frängsyr, ed., *Linnaeus: The Man and his Work* (Berkeley, 1983), 156–94: 169.

[75] Quoted in ibid., 166. Cf. Linnaeus' remark in the 'Foreword' to *Fauna Svecica* (Stockholm, 1746): 'As a natural historian I have yet to find any characteristics which enable man to be distinguished on scientific principles from the ape.' Quoted in Broberg, '*Homo sapiens*', 170. Koerner reports that Linnaeus believed that there was a continuum between some primitive people and apes, and that some apes were able to play backgammon: *Linnaeus*, 87.

[76] An example is John Martyn, *Elements of Natural History* (Cambridge, 1775).

[77] See, for example, Robert Kerr's English translation of Gmelin's edition: Linnaeus, *The Animal Kingdom or Zoological System . . . being a translation of that part of the Systema Naturae, as lately published . . . by Professor Gmelin* (Edinburgh 1792), 55.

[78] William F. Bynum, 'Time's Noblest Offspring: The Problem of Man in the British Natural Historical Sciences, 1800–1863', unpublished PhD dissertation, University of Cambridge, 1974: 25.

[79] Johan Gottschalk Wallerius, *Decades Binae Thesium Medicarum* (Uppsala, 1741); Jacob Theodor Klein, *Summa Dubiorum Circa Classes Quadrupedum et Amphibium in Linnaei Systemate Naturae* (Danzig, 1743). See Broberg, '*Homo sapiens*', 170–2; and more generally, Tinland, *L'Homme sauvage*, ch. 3.

neither the possession of similar organs, nor the approximation of shape, nor the use of the hand nor the continued intercourse with this sovereign artist, has enabled any other species to blend their nature or their inventions with his; that in his rudest state, he is found to be above them; and in his greatest degeneracy, never descends to their level. He is, in short, a man in every condition; and we can learn nothing of his nature from the analogy of other animals.[80]

Despite Ferguson's protestations, however, one thing was clear by the last decade of the eighteenth century: comparative anatomy had begun to play a crucial role in the debates over whether humans were unique and what this uniqueness consisted in.

Comparative anatomy differed from traditional anatomy in a number of important respects. Anatomy had traditionally comprised the description of the internal parts of living things, and many anatomists had considered this an autonomous exercise. Although there had been some precedents, notably Vicq d'Azyr's classification of organs according to function,[81] it was Cuvier who, more than anyone else, was instrumental in the reform of anatomy, integrating it into physiology and subordinating it to functional considerations. Cuvier had a teleological conception of life, in which it was integrated into creation, and this formed the basic premise of his zoology, which he spells out in his late systematic account of zoology, *Le règne animal*:

Natural history has, moreover, a principle on which to reason, which is peculiar to it, and which it employs advantageously on many occasions; it is that of the *conditions of existence*, commonly termed *final causes*. As nothing can exist without the concurrence of those conditions which render its existence possible, the component parts of each must be so arranged as to render possible the whole living being, not only with regard to itself, but to its surrounding relations; and the analysis of these conditions frequently conducts to general laws, as demonstrable as those which are derived from calculation or experiment.[82]

The vital processes in animals were directed, and this directedness was expressed in their physiology. Physiology thus held the key to the discovery of the structural bases and relative importance of animal functions.[83] Despite morphological similarities between crustacea and insects for example, the presence of a heart and brachia in crustacea indicates a mode of life different from that of insects. This is what is crucial for classification, and it is these animal functions that

[80] Adam Ferguson, *An Essay on the History of Civil Society* (4th edn, London, 1773), 8–9.

[81] Félix Vicq d'Azyr, *Traité d'anatomie et de physiologie, avec des planches coloriées* (Paris, 1786).

[82] Baron Cuvier, *The Animal Kingdom arranged after its Organization; forming a Natural History of Animals, and an Introduction to Comparative Anatomy*, new edn (London, 1863), 2–3. Originally published as *Le règne animal distribué d'après son organisation pout servir de base à l'histoire naturelle des animaux et l'introduction à l'anatomie comparée* (5 vols, Paris, 1829–30).

[83] See Coleman, *Georges Cuvier Zoologist*, ch. 2.

provide the basis for comparative anatomy, which Cuvier conceives as the ultimate scientific aim of the exercise:

The most effectual mode of observing is by comparison. This consists in successively studying the same bodies in different positions in which Nature places them, or in a comparison of different bodies together, until constant relations are recognized between their structures and the phenomena which they manifest. These various bodies are kinds of experiments ready prepared by Nature, who adds to or subtracts from each of them different parts, just as we might wish to do in our laboratories, and shows us herself the results of such additions or retrenchments.[84]

Despite Cuvier's confidence in the autonomy of comparative anatomy, however, it did not usually work in isolation from other questions, and its results were open to interpretation depending on a number of other assumptions. In particular, the claims of links between apes and human beings were reinforced by theories of a continuous transition from the one to the other. As William Smellie, editor of the *Encyclopedia Brittanica* and translator of Buffon's *Histoire naturelle*, writes in his *Philosophy of Natural History* of 1790:

Were we to consider the characters, the manners, and the genius of different nations, of different provinces and towns, and even the members of the same family, we should imagine that the species of men were as various as the number of individuals. How many gradations may be traced between a stupid Huron, or a Hottentot, and a profound philosopher? Here the distance is immense; but Nature has occupied the whole by almost infinite shades of discrimination. In descending the scale of animation, the next step, it is humiliating to remark, is very short. Man, in his lowest condition, is evidently linked, both in the form of his body and the capacity of his mind, to the large and small orang-outang. These again, by another slight gradation, are connected to the apes, who, like the former, have no tails. It is wonderful that Linnaeus, and many other naturalists, should have overlooked this gradation in the scale of animals.[85]

Smellie's treatise was not atypical. There was no shortage of works arguing that there was a continuity between human beings and apes, with Negro races forming the link between the two: in Germany, for example, Samuel Soemmerring's *Ueber die körperliche Verschiedenheit des Negers vom Europäer* (1785), in Britain Charles White's *Account of the Regular Gradation in Man* (1799), and in France Julien-Joseph Virey's *Histoire naturelle du genre humain* (1800).[86] White's work was

[84] Cuvier, *The Animal Kingdom*, 3.

[85] William Smellie, *The Philosophy of Natural History* (2 vols, Edinburgh, 1790), 522–3. Note however that while critics of Smellie's book complained that it was derivative, no one suggested it challenged conventional views of human beings. See Bynum, 'Time's Noblest Offspring', 2.

[86] Samuel Thomas Soemmerring, *Ueber die körperliche Verschiedenheit des Negers vom Europäer* (Frankfurt am Main, 1785); Charles White, *An Account of the Regular Gradation in Man, and in Different Animals and Vegetables, and from the Former to the Latter* (London, 1799); Julien-Joseph Virey, *Histoire naturelle du genre humain* (Paris, 1800).

particularly influential, and Virey's *Histoire* is especially indebted to it.[87] White combined three resources: Lavater's physiognomy, which attempted to read character, in animals and in humans, from the shape of the head;[88] Camper's anatomical accounts of jutting jaws in primates; and the reports of naturalists who had visited Africa.[89] He concluded that anatomical and physiological differences exist between races, which can be used to arrange them in a graduated series, which in turn can be extrapolated to apes.[90] Gradation does not mean continuity for White, however, and, contrary to the direction in which his gradation argument might seem to lead, he denies continuity between humans and apes, and instead postulates four separate and discontinuous major human types: European, Asiatic, American, and African.

The greatest challenge to gradationist and continuist theories came from Blumenbach, especially his *De generis humani varietate nativa*, first published in 1775, with an extensively revised and expanded third edition appearing in 1795.[91] Blumenbach's account of human species and varieties is anchored firmly within comparative anatomy.[92] He had compared a large numbers of skulls of apes with one another and with those of human beings of different races, and it was primarily on this basis that he approached the question of the relation between apes and humans. Linnaeus' division of humans had, Blumenbach believed, been superseded by the results of scientific voyages, and Linnaeus had been too restrictive in classifying mammals by their teeth.[93] Rather, he considered, what was needed was a conception of nature as an active whole producing an infinite number of varieties within each species. Comparative anatomy lies at the core of the 'natural' system that replaced Linnaeus' artificial classification, a classification that includes not just morphology, but also considerations of physiological functions of the body in the process of adapting to its environment. It is in this context that we must consider Blumenbach's account of racial

[87] See the discussion in Corsi, *The Age of Lamarck*, 146–7.

[88] Physiognomy had been around since the pseudo-Aristotelian *Physiognomica*, and received a new lease of life with the publication of Della Porta's *De humana physiognomonia* in 1586, but Lavater refined it to a considerable extent in his *Physiognomische Fragmente, zur Beförderung der Menschenkenntniß und Menschenliebe* (4 vols, Leipzig and Winterthur, 1775–8): see Rotraut Fischer and Gabriele Stumpp, 'Die Allegorisierung des Individuums in der Physiognomik Johann Caspar Lavaters und Carl Gustav Carus', in R. Fischer, G. Schrader, and G. Stumpp, eds, *Natur nach Mass: Physiognomik zwischen Wissenschaft une Ästhetik* (Marburg, 1989), 11–58; Karl Pestalozzi, 'Physiognomische Methodik', in A. Finck and G. Gréciano, eds, *Germanistik aus interkultureller Perspektive* (Strasbourg, 1988), 137–53; and more generally Richard T. Gray, *About Face: German Physiognomic Thought from Lavater to Auschwitz* (Detroit, 2004), chs 1–3.

[89] See Bynum, 'Time's Noblest Offspring', 32–3.

[90] White, *An Account of the Regular Gradation in Man*, 81–4.

[91] Johann Friedrich Blumenbach, *De generis humani varietate nativa* (Göttingen, 1775). References are to the third edition (Göttingen, 1795).

[92] See Stefano Bertoletti, 'The Anthropological Theory of Johann Friedrich Blumenbach', in S. Poggi and M. Bossi, eds, *Romanticism in Science* (Dordrecht, 1994), 103–25, to which I am indebted here.

[93] Blumenbach, *De generis*, viii–xiii.

varieties. Buffon had held that there was a single human species, and that races were formed through degradation from a primordial source. Blumenbach rejects such genealogical criteria, if only as useless in practice. They cannot tell us whether the African elephant and the Indian elephant are different species for example, whereas comparative anatomy can.[94] It is comparative anatomy, not genealogy, that establishes the unity of the human species. Racial differences are to be explained by degeneration (*Ausartung*),[95] as in Buffon. But in that case, the problem arises how one holds on to monogenism—a single origin for the human species—without thereby having to account for the emergence of subsequent racial variation in transformationist terms. The idea of the creation of new species by hybridization will not work because it would contravene the principle of the immutability of species. But as Larson notes, in a more general context:

> the essentialist species concept, which established the reality, constancy, and separateness of true species, simultaneously established the varietal concept as its necessary complement. In theory, the varieties absorbed the nonessential changes produced in the unchanging species by accidental causes—that is, climatic and edaphic conditions.[96]

For Blumenbach, the degeneration that lies at the heart of the origin of races from a single human species is due to environmental factors such as climate and lifestyle, and is transmitted hereditarily down the generations. In other words, there must be transmissibility of acquired features,[97] although this cannot be a straightforward matter if it is to account for specific racial features such as shape of skull and eye colour.[98]

As he was developing his ideas, Blumenbach was building up a large collection of human skulls, labelling each one in terms of geographical origin, environmental conditions, anatomy, and morphology. He treated racial differences as determined by such factors as climate, food, and mode of life, and as varying on a continuous scale, so that if the factors changed the variation from the original

[94] Ibid., 71.

[95] In eighteenth- and early nineteenth-century usage, 'degeneration' is used to mean variation within a species due to climate and food. Theories of generation were developed to explain the diversity of flora and fauna across continents, and only subsequently applied to racial variation.

[96] Larson, *Interpreting Nature*, 97.

[97] Blumenbach, *De generis*, 108. Degeneration from a single source had been the traditional reading of biblical sources: Noah's son Ham, and Ham's son Cush, were identified with dark skin colour and slavery. As Aaron Garrett points out, by the middle of the eighteenth century this explanation was dismissed in a number of works: 'Human Nature', in Knud Haakonssen, ed., *The Cambridge History of Eighteenth-Century Philosophy* (Cambridge, 2006), 160–233: 184–5. See, for example, the works cited by Garrett: Claude Nicolas Le Cat, *Traité de la couleur de la peau humanine en général et de celle des nègres en particulier, et de la métamorphose d'une de ces couleurs en l'autre, soit de naissance, soit actuallement* (Amsterdam, 1765), 5–23 ('Origine des diverses especes d'Hommes'); Georg Forster, 'Leitfaden zu einer künftigen Geschichte der Menscheit', in *Werke: Sämtliche Schriften, Tagesbücher, Briefe* (Berlin, 1958–), viii. 193; Benito Gerónimo Feijóo y Montenegro, 'Color etiópico', in *Teatro crítico universal* (8 vols, Madrid, 1726–36) vii., Discourse 3, §§3–5.

[98] Blumenbach, *De generis*, 113.

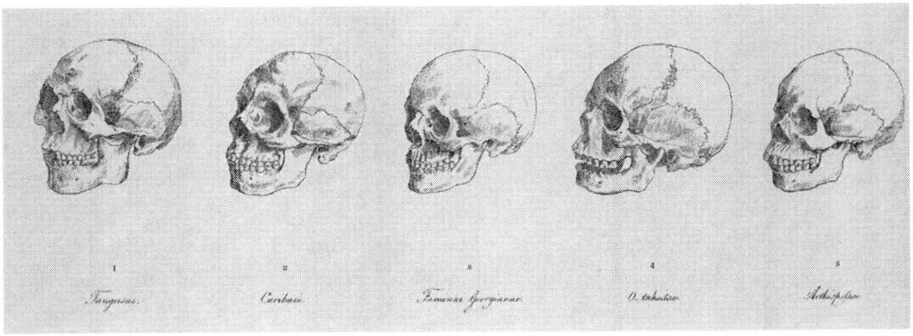

Fig. 5.7 Skulls of the five races. *Blumenbach, De Generis Humani Varietate Nativa* (1795).

stock was reversible. Nevertheless, he distinguished five races by geographical region and skull shape: Oriental, American Indian, Caucasian, Malay, and African (1 to 5 respectively in Fig. 5.7). Oriental (Mongolian) and African (Ethiopian) exhibited extreme features, in Blumenbach's view, while Caucasian skulls exhibited a perfect mid-point. The Leiden anatomist Peter Camper had compared skulls of different races and ages from his own large collection in terms of facial angle: the interior angle between the line running from forehead to chin and that running from the base of the ear to the chin. He found the facial angle to vary with race. Blumenbach built on this, adding measurements of cranial capacity. Important as skull shape was to his classification, however, he also included two more traditional factors: skin colour, with its connection to hair type and eye colour; and facial structure. Treating Caucasians as the original stock, evidently on aesthetic grounds,[99] he distinguished two lines of degeneration, one through American Indian to Oriental, and the other through Malay to African. Note that such degeneration is not a form of transmutation. Rather, it is a series of environmentally induced modifications of the *Bildungstrieb*: it is nature that is variable, not the human species as such. His conception of what marks humans out from other animals parallels that of Herder: human beings were not originally in a savage condition but have always been domestic, and unlike other animals they were created to suit all types of lifestyle, climate, and food.[100] This, Blumenbach considers, has allowed humans to spread over the whole world, in the greatest range of climates and food sources, for example, requiring the greatest degree of adaptability, with the result that they have undergone a greater

[99] See the discussion in Stephen Jay Gould, *The Mismeasure of Man* (rev. edn, New York, 1996), 391–424. Buffon also invoked aesthetic criteria, which he associates with moral qualities, e.g. in the cases of Bengals and Lapps: see his discussion of racial features 'Variétés dans l'Espèce Humaine', *Œuvres complètes*, xii. 192–480.

[100] Blumenbach, *Beyträge zur Naturgeschichte* (2 vols, 2nd edn, Göttingen, 1806), i. 40.

degree of degeneration than other animals. At the same time, in contrast to other animals, they are without instinct, but they compensate for this through the use of reason and language.[101]

It is worth noting in this context that, although craniology plays a significant part in distinguishing different human races, Blumenbach invokes a far greater range of factors in accounting for what distinguishes humans from animals. Outward morphology, internal structure, the functions of the animal economy, spiritual life, and illnesses peculiar to the human being, are all listed as criteria.[102] What lies behind his differentiation, however, is the belief that human beings are, by contrast with every other animal, intrinsically domestic, and so cannot be continuous with other animals.

As well as the contrast between humans and animals, there is also a contrast between different races, driven or at least reinforced by considerations from comparative anatomy, as we have just seen. But craniology and other forms of comparative anatomy offered a yet finer form of differentiation, that between males and females not only of the same species but of the same race. Women now form a contrast class, taking the idea of a natural history of *man* in a literal sense.

In the course of the eighteenth century, anatomical comparisons of men and women shifted focus in quite a radical way. Already by the end of the sixteenth century, the Galenic comparison of male and female in terms of degrees of perfection, where women were imperfect men and the uterus was an inadequate penis, was being called into question.[103] Comparison of male and female organs came to be rejected as misguided by the end of the eighteenth century. The point was that the uterus had no analogue with the male organ, so comparison could serve no purpose.[104] A more general differentiating feature of women was needed, and this was based on the female skeleton. Before 1730, skeletons in anatomy had all been of the male,[105] but female skeletons were illustrated by William Cheseldon in 1733, by Pierre Tarin in 1753, Marie Thiroux d'Arconville in 1753, and Soemmerring in 1796.[106] As Londa Schiebinger points out, anatomists began to look to the skeleton

[101] Blumenbach, *De generis*, 52–5.

[102] Ibid., 1–3. Cf. Soemmerring, who advocates categorizing skeletons by age, sex, 'nationality' (what will subsequently be called race), nourishment, susceptibility to illnesses, lifestyle, and clothing: *Vom Baue des menschlichen Körpers* (5 vols, Frankfurt, 1791–6).

[103] See Thomas Laqueur, 'Organism, Generation, and the Politics of Reproductive Biology', in C. Gallagher and T. Laqueur, eds, *The Making of the Modern Body: Sexuality and Society in the Nineteenth Century* (Berkeley, 1987), 1–41.

[104] See Gabriel Jouard, *Nouvel Essai sur la femme considerée comparativement à l'homme* (Paris, 1804), 51–5; Carl Ludwig Klose, *Über den Einfluss des Geschlechts-Unterschiedes auf Ausbildung und Heilung von Krankenheiten* (Stendal, 1829), 28–30.

[105] As Schiebinger notes (*The Mind Has No Sex?*, 194–5), Drake's *Anthropologia Nova* of 1707 depicts a female skeleton but it is not compared to a male one: rather, he uses a male skeleton to show a view from the front and a female one to show a view from the back.

[106] William Cheseldon, *Osteographia or the Anatomy of the Bones* (London, 1733); Pierre Tarin, *Ostéographie, ou Description des os de l'adult* (Paris, 1753); Marie Thiroux d'Arconville, illustrations

to provide a 'ground plan' for the body and give a 'certain and natural' direction to the muscles and other parts of the body attached to it. If sex differences could be found in the skeleton, then sexual identity would no longer depend on differing degrees of heat (as the ancients had taught), nor would it be a matter of sex organs appended to a neutral human body (as Versalius had thought). Instead, sexuality would be seen as penetrating every muscle, vein, and organ attached to and molded by the skeleton.[107]

The sexual differences believed to be revealed by the skeleton focused on the pelvis, the cranium, and to a lesser extent to width of the rib cage.[108] The pelvis was proportionately larger in the female skeleton than in the male one, and there was a general, if not universal, view that this difference enabled one to distinguish the male from the female.[109] By contrast, the cranium was considered to be proportionately smaller in women, and this was in line with the view that cranium size provided an objective measure of mental capacity. Indeed, the larger size of the pelvis in women was considered by some to be directly correlated with the need to accommodate the large cranium of the European male in the birth canal.[110] The problem was that women's skulls actually turned out to be proportionately heavier than men's, relative to body weight, raising the possibility that women were actually 'more adept than men in intellectual pursuits'.[111] As Schiebinger points out, however, the gradual acknowledgement of the greater cranial capacity of women relative to size did not mean that this conclusion was conceded. Quite the contrary, a different form of anatomical comparison was brought to bear to show that the size of women's brains was a sign of their arrested development. Women's brains resemble those of children, whose crania are large relative to body size. This line of argument was pursued most vigorously in the 1830s by the Edinburgh anatomist John Barclay, whose treatise on anatomy[112] included an illustration of a 'skeleton family' (Fig. 5.8). What earlier anatomists such as Soemmerring had described as peculiar to the female skull were in fact, Barclay argued, distinctive of the foetal skeleton. In general,

for her French translation of Alexander Monro's *The Anatomy of Human Bones* (Edinburgh, 1726), which appeared as *Traité de ostéologie* (Paris, 1759); Samuel Thomas Soemmerring, *Tabula sceleti feminini juncta descriptione* (Frankfurt am Main and Mainz, 1796).

107 Schiebinger, *The Mind Has No Sex?*, 191.

108 Note however the view that individual variation meant that one could find 'feminine' skulls and breastbones in men: see for example Joseph Wenzel's introduction to his translation of Jacob Ackermann, *Über die körperliche Verschiendenhiet des Mannes vom Weibe ausser Geschlechtstheilen* (Koblenz, 1788), 5.

109 A dissenting view was presented in Exupère-Joseph Bertin, *Traité d'ostéologie* (4 vols, Paris, 1754), iii. 211–12. Note also the classification of pelvis size and shape along racial lines: oval in Europeans, round in Americans, square in Mongolians, and oblong in Africans. See Moritz Weber, *Die Lehre von den Ur- und Racen-Formen der Schädel und Becken des Menschen* (Düsseldorf, 1830).

110 See Schiebinger, *The Mind Has No Sex?*, 209.

111 Ackermann, *Über die körperliche Verschiendenhiet*, 146.

112 John Barclay, *The Anatomy of the Bones of the Human Body* (Edinburgh, 1829).

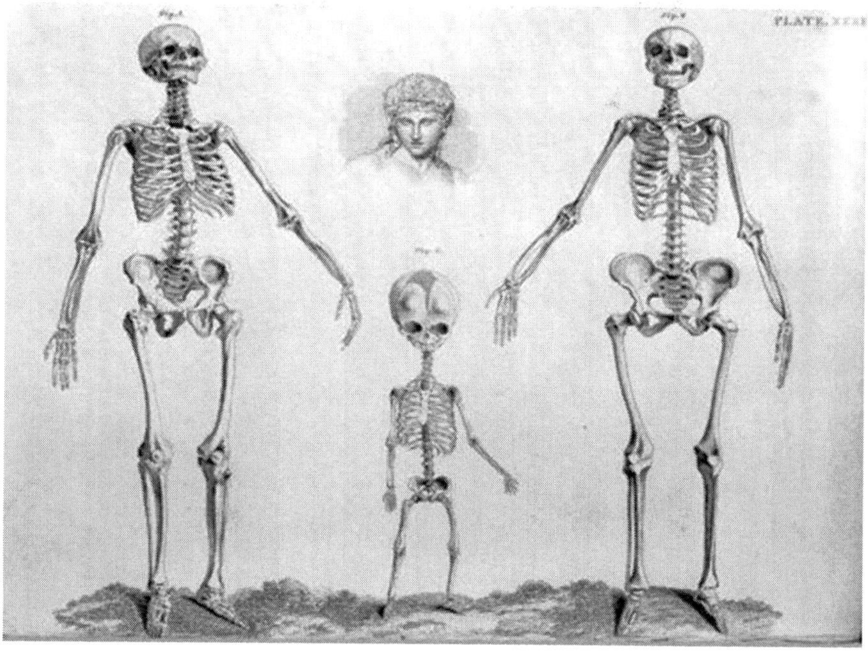

Fig. 5.8 Skeleton family. Barclay, *The Anatomy of the Bones of the Human Body* (1829).

women were closer to children not just in bone size, but also in size of rib cage, jaw shape, and feet size.[113]

Part of the problem with such approaches is whether comparative anatomy is a sufficiently fine-tuned resource to allow contentious differentiations within species to be made. The proposed anatomical rationalizations of racial and gender differences—based as they were on the unquestioning assumption of the European male not only as the standard against which everything else must be judged, but as the pinnacle of development in the animal world—show how easily the autonomy of comparative anatomy could be undermined, and how easily it could function to mask unexamined assumptions and prejudices. But as a tool for examining relations between species, comparative anatomy became one of the most powerful—if never wholly uncontentious—resources of the nineteenth-century life sciences.

[113] This corresponded of course to contemporary caricatures of women as childlike. Note, in this context, Diderot's argument in his *Sur les femmes* (*Œuvres complètes*, ii. 251–62) that much of women's behaviour derived from the fact that men treated them as idiot children.

COMPARATIVE GEOGRAPHY

Comparative anatomy was not the only form of comparative naturalization deployed to account for the distinctiveness of human capacities and behaviour. Comparative history and geography offered alternative, and in some respects complementary, forms of investigation, and they come close to the kind of thing that Hume had in mind in writing about the natural history of man. The comparative nature of the exercise was crucial, for this is where the power of ethnography lies: in highlighting what is common to all human societies and what is a local phenomenon. The connection between comparative geography and comparative history was often an intimate one, and accounts of contemporary 'primitive' societies were typically couched in terms of comparisons with ancient societies. This is evident in the way in which, in the course of the seventeenth century for example, there was a constant use of classical sources in understanding American Indians, and vice versa,[114] a tradition continued in the eighteenth century, evident just from the titles of popular works such as La Créquinière's *Conformité des coutumes des Indiens orientaux, avec celles des Juifs & des autres peuples de l'antiquité* (1704) and Lafitau's *Mœurs des sauvages ameriquains, comparé aux mœurs des premiers temps* (1724). Indeed, there were some who thought that comparative geography could easily substitute for comparative history. In 1777, for example, Edmund Burke wrote to William Robertson that we 'no longer need to go to history' to study human nature, because 'now the great map of mankind is unrolld at once; and there is no state or tradition of barbarism, and no mode of refinement which we have not at the same instant under our View'.[115]

Here a form of synchronic enquiry stands in for comparative history, accomplishing its tasks by different means. But it was possible to consider geographical questions as autonomous, purely from a natural history point of view, without regard to questions of comparative history, as Buffon did in the opening sections of the *Histoire naturelle*, and as Kant, following Buffon quite closely, did in his lectures on geography. The path there takes one from geological considerations, through various geographical features of the earth (typically articulated through the differences between frigid, temperate, and tropical zones), to human races. Kant was particularly concerned to exclude comparative human history from his lectures on geography, possibly for pedagogic reasons, possibly because he was following the early sections of Buffon's *Histoire naturelle*, but doubtless

[114] See Alfonzo Maurizio Iacono, 'The American Indians and the Ancients of Europe: The Idea of Comparison and the Construction of Historical TIme in the Eighteenth Century', in W. Haase and M. Reinhold, eds, *The Classical Tradition and the Americas, Volume One: European Images of the Americas and the Classical Tradition* (Berlin, 1994), 658–81; and Frank E. Manuel, *The Eighteenth Century Confronts the Gods* (Cambridge, Mass., 1959), 15–19.

[115] *The Correspondence of Edmund Burke*, ed. T. H. Copeland (10 vols, Cambridge, 1958–78), iii. 350–1.

also because the combination of synchronic and diachronic forms of ethnographical investigation in writers like Herder offered a potentially far more powerful resource than Kant was prepared to accept, given his eagerness to deny philosophical anthropology the status of scientific knowledge. Above all, as his review of Herder's *Ideen* makes clear, he is incensed by Herder's dismissal of physical factors such as race playing any part in shaping human groups, and his insistence on culture being the central factor in shaping them.[116] Keeping his geography lectures free of considerations from human history helps Kant to portray geography as a self-contained and purely descriptive endeavour.[117]

But there is nothing purely descriptive about the formative eighteenth-century work of comparative geography, Montesquieu's *De l'esprit des lois* (1748), which offered a provocative account of the influence of climate on human behaviour and beliefs. Indeed, in some respects comparative geography was a discipline that encouraged reductive claims, and it had had a radical cutting edge before Montesquieu. As Voltaire pointed out in his entry on 'climat' in his *Dictionnaire philosophique*, Montesquieu failed to mention three predecessors—Fontenelle, Jean-Baptiste Chardin, and Jean-Baptiste Dubos—although he acknowledges that he goes beyond them in several respects.[118] In his *Digressions sur les anciens et les modernes* (1688), Fontenelle takes a traditional view, deriving from Polybius, that climate (in the broad sense of environment) determines mental qualities in humans' primitive state, seemingly on the grounds that they are immersed in a natural realm and there is nothing else that could influence what and how they think. But once they reach a social state, people influence how and what one another thinks, and the influence of climate drops out: in particular, differences between ancient Greeks and Romans, for example, and ourselves cannot be due to differences in climate, because the climates are just too similar.[119] Chardin's very popular account of his travels to the East, especially Persia,[120] offers, by contrast, a defence of the influence of climate. The backwardness of the Persians, especially

[116] Kant, 'Recensionen von J. G. Herders Ideen zu Philosophie der Geschichte der Menschheit': *Gesammelte Schriften*, viii. 43–66; *Anthropology, History, and Education*, 121–42.

[117] Kant does pursue history elsewhere, notably in the *Idee zu einer Allgemeinen Geschichte* (*Idea for a Universal History*), but it has no connection with what he sees as the subject matter of physical geography.

[118] François-Marie Arouet de Voltaire, *Œuvres complètes* (72 vols, Gotha, 1784–90) xxxix. 31–40. In fact a fuller list of Montesquieu's predecessors would also include Jean Bodin, *Methodus ad facilem historiarum cognitionem* (Paris, 1566), and Giovanni Botero, *Della Ragion di Stato* (Venice, 1589). See Clarence J. Glacken, *Traces on the Rhodian Shore: Nature and Culture in Western Thought from Ancient Times to the End of the Eighteenth Century* (Berkeley, 1967). Glacken's book, while questionable on questions of interpretation, is nevertheless useful in identifying and summarizing sources.

[119] Bernard le Bovier de Fontenelle, *Digressions sur les anciens et les modernes* (Paris, 1688).

[120] The first volume of his best-selling book appeared as *Journal du Voiage du Chevalier Chardin en Perse & aux Indes Orientales, par la Mer Noire & par la Colchide*, published simultaneously (with slight variations in the title page) in Amsterdam, Paris, and London in 1686. Twelve volumes eventually appeared.

their conservatism and the consequent lack of any progress in their society, are put down to the hot climate:

The third Observation is, That the hot Climates enervate the Mind as well as the Body, lay the quickness of the Fancy, necessary for the invention and improvement of Arts. In those Climates the Men are not capable of Night Watchings, and of close Application, which brings forth the valuable Works of the Liberal, and of the Mechanick Arts. 'Tis by the same Reason likewise, that the Knowledge of the *Asiatics* is so restrained, that it consists only in learning and repeating what is contain'd in the books of the Ancients; and that their Industry lies Fallow and Untill'd, if I may so express myself.[121]

Jean-Baptiste Dubos, despite being employed on several senior diplomatic missions, did not travel outside Europe, but he was an avid reader of the travel literature, especially Chardin's volumes. His *Réflexions critiques sur la poésie*, first published in 1719, examines the determinants of artistic creativity, in particular what determines the geographical distribution of gifted artists.[122] Moral and social causes, such as stability of government and an interest in the arts on the part of the ruler and the citizens, certainly provide a nurturing environment for the cultivation of the arts but, he points out, they cannot produce creativity. For Dubos, the geographical localization of great art in Europe has to do with climate. This he believes is demonstrated by the rise of sculpture and painting in Egypt, where climate favoured the development of certain art forms, and indeed will generally produce them given time, but they cannot be transported to regions with a different climate. The crucial feature of climate in this account is the composition of the air, which transfers qualities to the blood via the lungs in respiration. The air links the surface of the earth, rivers and streams, rain and snow, human beings, and all forms of organic life. It contains, in varying degrees, emanations absorbed from the earth, as well as animalcules and their seeds, and it is these, acting through the four bodily humours, that determine what he effectively treats as the active ingredients in the air. This account of the importance of earth or dust has, of course, a biblical resonance, but it can be found in authors hostile to biblical accounts. Buffon, for example, in his entry on the ox in the *Histoire naturelle*, writes:

The surface of the Earth, shorn of its greenery, is the inexhaustible and common source on which men and animals subsist. Everything that lives in nature lives off what grows in the ground, and plants in their turn live off the remains of everything that has lived and grown. In order to live it is necessary to destroy, and it is only in destroying other things that animals can feed themselves and multiply. God, when he created the first individuals of

[121] Chardin, *The Travels of Sir John Chardin in Persia* (2 vols, London, 1720), ii. 257.

[122] Jean-Baptiste Dubos, *Réflexions critiques sur la poésie et sur la peinture: Ut Pictura Poesis* (2 vols, Paris, 1719).

each species of animal and plant, not only bestowed form on the dust of the earth, but gave it animation, by infusing these individuals with a greater or lesser quantity of active principles, of living organic particles, which are indestructible and common to every organized being. These particles pass from body to body, and are equally causes of life, of the continuation of species, of growth, and of nutrition.[123]

The physical basis of Dubos' argument on the physical effects of climate is certainly not unusual at the time. But his move from physical to mental effects was. If climate can affect such gross physical features as skin colour and physique, he asks, why then should it not affect the far more subtle intellectual and creative capacities? Cornelius de Pauw, in his popular 1768 *Recherches philosophiques sur les Américains*,[124] had no hesitation in substituting intelligence for skin colour in classifying racial differences. Earlier, John Arbuthnot in his 1733 *Essay concerning the Effects of Air on Human Bodies*, not only correlated bracing climates and industrious and courageous activities,[125] but went on to offer what is undoubtedly the most radical claim for climatic theories, when he argued that language differences were due to climate ones, maintaining that the

serrated close way of Speaking of Northern Nations, may be owing to their Reluctance to open their Mouth wide in cold Air, which must make their Language abound in Consonants; whereas from a contrary Cause, the Inhabitants of warmer Climates opening their mouths, must form a softer Language, abounding in Vowels.[126]

We are not particularly concerned here with the idea of the effects of air on the physical and mental constitutions, since we have already looked at more nuanced and elaborate versions in the context of eighteenth-century medicine in Chapter 3. Rather, consideration of Dubos, along with Fontenelle and Chardin, enables us to set the scene for Montesquieu, whose general account of the effects of climate in Books 14 to 17 of *L'Esprit des Lois* was to be widely cited throughout the rest of the eighteenth century, almost always critically. He opens Book 14 with a statement of the programme: 'If it is true that the character of the mind and the passions of the heart vary widely in different climates, the *Laws* should be in relation both to the variety of these passions and the variety of these characters.'[127] That character and passions varied from region to region was commonplace, but that the laws should accommodate such differences was dangerously relativistic. That these differences were due to variations in climate was even more galling to his critics, who particularly objected to what they took to be his physical determinism, with climatic factors determining the development of philosophy, law, and religion. To make matters worse, Montesquieu construes climate in a

123 Buffon, *Œuvres complètes*, xiv. 136–7.

124 Cornelius de Pauw, *Recherches philosophiques sur les Américains ou Mémoires intéressant pour servir à l'histoire de l'espèce humaine* (2 vols, Berlin, 1768–9).

125 Arbuthnot, *Essay Concerning the Effects of Air*, 148–9.

126 Ibid., 153–4.

127 Montesquieu, *L'Esprit des Lois*, book 14, ch. 1.

very restrictive sense, basically as temperature, with the effect of heat and cold on nerve fibres playing a crucial role.[128] In the northern regions, we find people with few vices and many virtues, as well as frankness and sincerity, whereas the further south we travel (presumably in the northern hemisphere), the less these qualities are in evidence.[129] In a chapter on Buddhism, he explicitly puts this religion down to the warm climate of India,[130] raising the delicate question of whether all religious differences are merely a question of differences of climate.[131]

Montesquieu's climatic theory had a number of followers, particularly in Germany. In his study of Mosaic law,[132] for example, Johann David Michaelis took climate seriously as an explanation for the variety of laws in different regions. Johann Lorenz von Mosheim, in his ecclesiastical history,[133] traced liberty to colder climates. Isaak Iselin, contrasted the strong, unimaginative, stubborn men of the North with the weak, complacent, sensuous men of the South.[134] But the overwhelming reaction to Montesquieu was critical. Climatic theory was crude and counterexamples to particular claims were easily found: Hume, for example, although not unsympathetic to the idea of the influence of physical environment on customs, beliefs, and abilities, gives nine counterexamples to the claim that climate could account for variations in national characteristics in his essay 'Of National Characters'. Moreover, while there was agreement that natural causes shaped character in primitive societies, there was a widespread view that reason was autonomous, and once reason began to predominate, natural causes ceased in this role.[135] Nevertheless, against this, there was a view, for example in the Scottish medical community, that the effects of such causes could not be discounted.[136] The issues were far from settled.

[128] From Book 18 onwards, other factors such as fertility, surface features, and relative isolation are discussed, but these play a diminished role in comparison to climate. In Book 19, chapter 4, the strongly deterministic tone of chapters 14–17 is less in evidence, and, among the various causes that influence mankind, climate, religion, laws, precedents, morals, and customs are mentioned. But since he has argued in preceding chapters that these are largely if not exclusively a result of climate, it is a little hard to tell whether he is in fact watering down his climate determinism. The nerve fibre theory is set out in the *Essai sur les causes qui peuvent affecter les esprits et les caractères*, composed in the late 1630s or early 1640s, but unpublished until the nineteenth century.

[129] Ibid., book 14, ch. 2.

[130] Ibid., book 14, ch. 5.

[131] Montesquieu replied to his critics that Christianity was a 'revealed religion' and because of this was not subject to physical causes: 'Défense de l'Esprit des Lois', Part II, 'climat', *Œuvres de Montesquieu* (7 vols, Lyon, 1805), iv. 238–41. His critics had been concerned that he occupied himself wholly with natural religion to the detriment of revealed religion.

[132] Johann David Michaelis, *Mosaisches Recht* (6 vols, Frankfurt, 1770–5).

[133] Johann Lorenz von Mosheim, *Institutionum historiæ ecclesiasticæ antiqui et recentioris quattuor* (Helmstedt, 1755).

[134] Isaak Iselin, *Über die Geschichte der Menscheit* (Zürich, 1764).

[135] For the most developed defence of this view, see James Dunbar, *Essays on the History of Mankind in Rude and Cultivated Ages* (London, 1780).

[136] Alexander Wilson, *Some Observations relative to the Influence of Climate on Vegetable and Animal Bodies* (London, 1780); Falconer, *Remarks on the Influence of Climate*.

One of the most pressing criticisms of Montesquieu came from Voltaire, whose *Essai sur les Moeurs et l'esprit des nations* (1756–62) offered an account of European and non-European nations, but in a historical context that was in direct contrast to Montesquieu.[137] An important part of Voltaire's project was the replacement of a Judeo-Christian framework for understanding history by one that explored social, political, and economic causes. The early sections, for example, deal with the history and cultural achievements of China and India, and very effectively distance the reader from the idea that Western developments lie at the core of our understanding of society and culture, while above all reminding us that there are civilizations older than those of the West and the Middle East. Not just a Christocentric model of development, but Judeo-Christian chronology is abandoned. Voltaire develops Dubos' four-age model for Western development: the age of Philip and Alexander, where Aristotle and Praxiteles flourished; the age of Caesar and Augustus, where the key figures are Cicero, Lucretius, Livy, Virgil, Horace, and Ovid; the age of the Medici; and that of Louis XIV. Although Voltaire sees little evolution in the human society, beyond the move from a primitive state, the idea of developmental stages is important to his account. It is therefore not surprising that one of his principal objections to Montesquieu's treatment is that it cannot account for social and cultural change.[138] Social and cultural change are at the core of Voltaire's understanding of what the human condition consists in. Montesquieu had fragments of a historical theory, but they never amounted to more than that,[139] whereas what was needed to develop any plausible anthropological understanding was more powerful historical resources.

The understanding of physical geography and what we could learn from it was transformed in the early decades of the nineteenth century by Alexander von Humboldt, whose five-volume *Kosmos* (1845–62)[140] synthesized work he had been pursuing since the end of the eighteenth century. Humboldt did not so much treat regions as parts of the globe, but rather took the globe to be composed of regions that had unique characters which were the result of complex networks of interlocking forces. The unity of nature was not some a priori premise that one brought to the study of natural history, but was rather something that one constructed from detailed comparisons of natural processes at work in

[137] See the extensive treatment of Voltaire's historical writings in Pocock, *Barbarism and Religion*, ii. 72–159.

[138] See, for example, the 'climat' entry in the *Dictionnaire philosophique*, *Œuvres complètes*, xxxix. 31–40.

[139] See Reill, *The German Enlightenment and the Rise of Historicism*, 56.

[140] Alexander von Humboldt, *Kosmos: Entwurf einer physischen Weltbeschreibung* (5 vols, Stuttgart, 1845–62); translated as *Cosmos: A Sketch of a Physical Description of the Universe* (5 vols, London, 1849–58). See also his *Ansichten der Natur: mit wissenschaftlichen Erläuterungen* (Stuttgart, 1849).

the different regions that went to make up the world. As Daston points out, Humboldt's idea of natural law

> implied regularity, universality, but *not* uniformity. In contrast to Newton's law of gravitation and its Enlightenment elaborations . . . Humboldtian natural laws did not assume that nature was, in its essentials, everywhere and always the same. Humboldtian nature was *bunt* [variegated/multicoloured], both literally and figuratively. Varying patterns of vegetation and geological formations coloured each region differently and distinctively; Humboldt's endless series of world maps were ingeniously coloured to show how an order emerged from this patchwork.[141]

One of the crucial things that emerged from this patchwork were isotherms.[142] Climbing the mountains outside Caracas in 1800, he had been struck by the similarity between local flora—such as rhododendrons and the willows—and those with which he was familiar from his European journeys, something confirmed by other travellers, who had noted similarities between the flora of the Alps and Pyrenees and those of Australia and Tenerife. Soil and atmospheric pressure were insufficient to explain the striking similarities, and distance from the equator varied significantly. The crucial factor was temperature, and this varied as a function of distance from the equator and altitude. Humboldt connected the different regions with similar flora in terms of an isothermal map, which connected the regions by lines (Fig 5.9), showing how vegetation changed with increasing altitude in exactly the same way that it changed at sea level with distance from the equator. The twisting patterns of isothermal lines—created by obtaining the values of physical variables (in this case temperature), representing them as points on a map, and connecting those having the same values with a line—are the result of global causes acting locally. As Humboldt describes it:

> Temperature and magnetism are not like those phenomena which, derived from a single cause or a central force, can be freed of the influence of disturbing causes by restricting attention to the mean results of a great number of observations, in which these foreign effects reciprocally counteract and destroy one another. The distribution of heat, like the declination and inclination of the compass needle or the intensity of magnetic force, is

[141] Lorraine Daston, 'The Humboldtian Gaze', in M. Epple and C Zittel, eds, *Science as Cultural Practice. Volume 1: Cultures and Politics of Research from the Early Modern Period to the Age of Extremes* (Berlin, 2010), 45–60: 46. See also idem, 'Unruly Weather: Natural Law Confronts Natural Variability', in L. Daston and M. Stolleis, eds, *Natural Laws and Laws of Nature in Early Modern Europe* (Aldershot, 2008), 233–48. Compare Humboldt's approach with that of William Robertson in his 1777 history of America: 'In a general history of America, it would be highly improper to describe the condition of every petty community, or to investigate every minute circumstance which contributes to form the character of its members. Such an inquiry would lead to details of immeasurable and tiresome extent. The qualities belonging to the people of all the different tribes have such a near resemblance, that they may be painted with the same features.' *The History of America, volume the first* (Dublin, 1777), 283.

[142] Humboldt, 'Des lignes isothermes et de la distribution de la chaleur sur le globe', *Mémoires de physique et de chimie de la Société d'Arcueil* 3 (1817), 462–602.

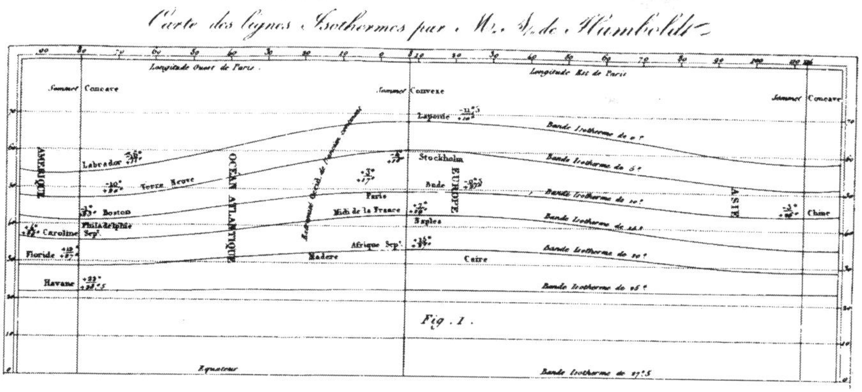

Fig. 5.9 Isothermal map. Humboldt, 'Des lignes isothermes et de la distribution de la chaleur sur le globe', *Mémoires de physique et de chimie de la Société d'Arcueil* (1817).

essentially conditioned by location, composition of the soil, by the proper ability of the earth's surface to radiate heat.[143]

The representation of natural processes by lines might not seem so unusual: after all, Kepler's elliptical orbits and Galileo's parabolic trajectories for projectiles were well known. But Humboldt's maps of the distribution of climatic zones were novel and of central importance to the study of climate. As Norton Wise reminds us, at the time Humboldt was writing, 'there were no magnetic lines of force, no supply and demand curves, no bell-shaped curves for the propensity to commit suicide with a knife, no Carnot diagrams for steam engines, no curves of rotating storm systems, and no blood pressure recorders, to say nothing of ekg's, seismographs, and plots of the Dow-Jones Industrial Average over the past week'.[144] Humboldt was quick to extend the method of mapping isothermal lines, setting out to discover whether 'air pressure, the amounts of rain falling from the air, the relative frequency of prevailing winds, and the direction of isothermal lines, like the distribution of magnetism across the earth, are subject to secular variations'.[145] One particularly important application of the idea was

[143] Ibid., quoted in Michael Dettlebach, 'Humboldtian Science', in N. Jardine, J. A. Secord, and E. C. Spary, eds, *Cultures of Natural History* (Cambridge, 1996), 287–304: 295.

[144] M. Norton Wise, 'What's in a Line?', in M. Epple and C Zittel, eds, *Science as Cultural Practice. Volume 1: Cultures and Politics of Research from the Early Modern Period to the Age of Extremes* (Berlin, 2010), 61–102: 61.

[145] Humboldt, 'Ueber der Mittel, die Ergründung einiger Phänomene des tellurischen Magnetismus zu erleichtern', *Annalen der Physik und Chemie* 91 (1829) 319–36: 319. Humboldt's method of isothermal lines was used as the basis for the first modern meteorology in Heinrich Wilhelm Dove, initially in his 'Einige meteorologische Untersuchungen über der Wind', *Annalen der Physik und Chemie* 11 (1827), 545–90; and then in a more comprehensive form in his *Meteorologische Untersuchungen* (Berlin, 1837).

to independent geological formations, where a formation is independent if could be recognized in different circumstances and in different regions, but always had the same structural position relative to other formations with the same status.[146] Again the procedure is to start from particular regional cases and to work out, by establishing connections along paths that become evident only from the investigation of these particular cases, not by enquiring how a pre-given notion of the natural realm applies in particular regions. Dettelbach sums up the situation nicely when he writes that for Humboldt, 'nature's lawfulness emerged gradually and progressively from laborious observing, averaging, and mapping over increasingly extended areas. Indeed, for Humboldt, averaging *itself* was the paradigm of lawful behaviour, and mean values were "the final object, the expression of physical laws; they show us the constant amidst the flux and flight of phenomena".'[147]

COMPARATIVE HISTORY

If regional differences, whether caused by climate solely or in combination with other factors, lie at the core of comparative geography, then it is stadial differences—differences in human development that are explained in terms of the stage of development of the society—that lie at the core of comparative history. A good deal of stadial history is what Dugald Stewart, looking back in 1793 on his predecessors, characterized as conjectural history, and he provides the rationale for it in these terms:

When, in such as period of society as that in which we live, we compare our intellectual acquirements, our opinions, manners, and institutions, with those which prevail among rude tribes, it cannot fail to occur to us as an interesting question, by what gradual steps the transition has been made from the first simple efforts of uncultivated nature, to a state of things so wonderfully artificial and complicated. Whence has arisen that systematical beauty which we admire in the structure of a cultivated language . . . ? Whence the origin of the different sciences and the different arts . . . ? Whence the astonishing fabric of the political union . . . ? On most of these subjects very little information is to be expected from history, for long before that stage of society when men began to think of recording their transactions, many of the most important steps of their progress have been made. A few insulated facts may perhaps be collected from the causal observations of travellers, who have viewed the

[146] Humboldt, *Essai géognostique sur le gisement des roches dans les deux hémisphères* (Paris, 1823). See the discussion in Martin J. S. Rudwick, *Worlds before Adam: The Reconstruction of Geohistory in the Age of Reform* (Chicago, 2008), 36–9.

[147] Dettlebach, 'Humboldtian Science', 298. See also Janet Browne, 'Biogeography and Empire', in N. Jardine, J. A. Secord, and E. C. Spary, eds, *Cultures of Natural History* (Cambridge, 1996), 305–21.

arrangements of rude nations; but nothing, it is evident, can be obtained in this way, which approaches to a regular and connected detail of human improvement. In this want of direct evidence, we are under a necessity of supplying the place of fact by conjecture; and when we are unable to ascertain how men have actually conducted themselves upon particular occasions, of considering in what manner they are likely to have proceeded, from the principles of their nature, and the circumstances of their external situation.[148]

Comparative history offers a richer vein than comparative geography for understanding human behaviour, and the attempt to discern stages in the development of human history is a distinctive feature of historical enquiry from the middle decades of the eighteenth century onwards. It is not that earlier writers had not thought in terms of historical stages, but rather that, generally speaking, they had not thought of people being products of their times, in the way that those pursuing comparative geography, as we have just seen, often thought of people as being products of the regions from which they came, as in racial theories for example.

Before the sixteenth century, the closest earlier thinkers came to this kind of historical thinking was in the idea of a 'Golden Age', in which particular figures are taken as representative of a particular time and culture in which things were possible which are no longer possible. For philosophers, for example, classical Athens and Rome were treated as the Golden Age, with Plato, Aristotle, Cicero, and Seneca accomplishing things that were no longer achievable in the present age. In the second half of the sixteenth century, the idea of a Golden Age was inverted, with writers like Bodin, Le Roy, and Popelinière arguing for the superiority of the modern era.[149] In place of Golden Age conceptions, they offered a model of growth from infancy to maturity and wisdom. But it was Bacon who, in his *Redargutio Philosophiarum* of 1608, offered the first comprehensive account of how the distinctive characteristics of an age enable us to evaluate the cultural products of that age. For Bacon, different ages offer different potentialities for human development, and the potential increases as ages succeed one another.[150] Nevertheless, it is only from the mid-eighteenth century that stadial accounts come into their own, and from then onwards they quickly take over much historical thinking.

We have already come across a number of such accounts, in d'Alembert, Winckelmann, and Herder, for example. It is important to appreciate from the

[148] Dugald Stewart, *The Collected Works of Dugald Stewart*, ed. William Hamilton (11 vols, Edinburgh, 1854–60), x. 33–4.

[149] Jean Bodin, *Methodus ad facilem historiarum cognitionem*; Loys Le Roy, *De la vicissitude ou variété des choses en l'univers, et concurrence des armes et des lettres par les premieres et plus illustres nations du monde, depuis le temps où a commencé la ciuilité, & mémoire humaine iusques à présent* (Paris, 1575); Henri Lancelot du Voisin de la Popelinière, *L'Histoire des Histoires, avec l'idée de l'Histoire accomplie* (Paris, 1599).

[150] See Gaukroger, *Francis Bacon*, 110–18.

outset, however, that stadial accounts depend on what the stages are designed to reveal, and those that we have already encountered do not exhaust the variety of interests that regulate how stages are marked out. Although there were commonalities—such as opposition to providential readings of history, with a particularly strong anti-clerical agenda in France—stadial accounts took a number of different forms, and the stages were distinguished in line with various socio-economic, political, intellectual, cultural, and other interests, which sometimes cut across one another.

A distinctive four-stage model was developed in political economy, from mid-century onwards. Turgot, in a short essay 'Reflexions sur les langues' of 1750,[151] which is largely a review of Maupertuis' *Réflexions philosophiques sur l'origine des langues*, offered a periodization based on linguistic criteria: hunters have very few unconnected words, shepherds progress to a more refined language, and husbandmen to one that is grammatically coherent. By the beginning of 1752, he had developed a four-stage theory in his discourses on universal history.[152] The first stage was that of hunters who, in the absence of cultivation, are forced to lead a nomadic lifestyle, which Turgot equates to that of contemporary American Indians. In regions where domesticable animals are to be found, a transition to the next stage, that of a pastoral way of life (unavailable to American Indians because of the lack of domesticable animals), is an easy one, as it brings with it greater abundance of food as well as a greater reliability of food sources. The move to the third stage, an agricultural society, comes about because of the increased benefits that agriculture provides over pastoralism. Finally, with the surpluses produced in agricultural society, we witness the development of trade, useful arts, and towns: in short a 'commercial society' emerges.[153]

It was in Scotland, rather than in France, that socio-economic models flourished, although they did not have the field to themselves. Hume's *History of England* (1788) and William Robertson's *History of Scotland* (1759) had taken the writing of history to a new level of historiographical sophistication, but neither of them offered stadial histories.[154] They both had a strong sense of a three-stage transition from savagery to barbarism to modern civilization, and they had ideas about what retarded this transition in some societies and what eased it in others. But although they have been rightly thought of as writers of 'philosophical' histories, in that they dealt with political issues to which philosophers had devoted their energies, the kind of single overriding causes that stadial accounts offered

[151] *Œuvres de Turgot*, ii. 752–6. Later in the same year, in his Sorbonne lecture 'Sur les progrès successifs de l'esprit humain' (ibid., 597–611), he expands somewhat on the historical progress in ideas.

[152] Ibid., 626–75.

[153] It should be noted, however, that Turgot questions whether, despite the progress in knowledge and science that this development brings, it can be associated with an improvement of the human condition, asking whether 'men rise only in order to fall'. Ibid., 606.

[154] See Ronald L. Meek, *Social Science and the Ignoble Savage* (Cambridge, 1976), 30–1, 136–46.

would have fitted poorly with their commitment to exploring the empirical contingencies underlying historical events. Moreover there were those, such as Lord Kames, who rejected stadial histories on the grounds that they assumed that everyone had the same sort of character in the first historical stage. Kames believed there were intrinsic racial differences (fitting different peoples to different climates), and this fundamental racial variation would have existed right from the beginning: he even went so far as to posit a multiplicity of Adam and Eves to account for different racial origins.[155]

In those Scottish writers who offer stadial histories, notably Adam Ferguson and John Millar, there is a shared socio-economic motivation which shapes how stages of development are discerned, and both work with a four-stage series set out by Adam Smith in his 1762–3 Lectures on Jurisprudence in these terms:

> Before we consider exactly this or any other methods by which property is acquired it will be proper to observe that the regulations concerning them must vary considerably according to the state or age society is in that time. There are four distinct states which mankind pass thro:-1st, the Age of Hunters; 2dly, the Age of Shepherds; 3dly, the Age of Agriculture; and 4^{thly}, the Age of Commerce.[156]

Although he has a long-term interest in socio-economic periodization, dating back to his 1752 Glasgow moral philosophy lectures, Smith nevertheless never probes the four-stage theory.[157] This was left to Ferguson, in his *Essay on the History of Civil Society*, which first appeared in 1767, and which sets out the four stages—hunting and gathering, pastoralism and nomadism, agricultural society, and finally commercial society—in more detail. The elaboration of the stages was not a straightforward matter, and Ronald Meek has pointed out that we can trace at least three elements in this account: Montesquieu's identification of savages with hunters and barbarians with herdsmen; the dynamization of these categories; and the concern with changes in property which derives from the Grotius–Pufendorf–Locke tradition.[158]

Given that such accounts generally renounced providence, or at least renounced it as an explanation of the transition between stages, two basic problems they faced were how the transition from one stage to another was effected, and why the transition was from one particular stage to another particular stage—from pastoralism to an agricultural society for example, and not from pastoralism to a commercial society. Turgot had touched on these questions, but Ferguson offers

[155] Lord Kames, *Sketches of a History of Man* (4 vols, Edinburgh, 1788), i. 22.

[156] Smith, *Works and Correspondence*, v. 14.

[157] There is a widespread view that Smith's predecessors, among whom Pufendorf is a key figure, were familiar only with the first three stages, and that the introduction of a commercial society as a fourth stage was due to Smith, but see Istvan Hont, *Jealousy of Trade: International Competition and the Nation-State in Historical Perspective* (Cambridge, Mass., 2005), 159–84.

[158] Meek, *Social Science and the Ignoble Savage*, 153.

something more elaborate. His solution to the first is to invoke the idea of unintended consequences:

Mankind, in following the present sense of their minds, in striving to remove inconveniencies, or to gain apparent and contiguous advantages, arrive at ends which even their imagination could not anticipate, and pass on, like other animals, in the track of their nature, without perceiving its end. He who first said, 'I will appropriate this field; I will leave it to my heirs'; did not perceive, that he was laying the foundation of civil laws and political establishments. . . . Every step and every movement of the multitude, even in what are termed enlightened ages, are made with equal blindness to the future; and nations stumble upon establishments, which are indeed the result of human action, but not the execution of any human design.[159]

But if unintended consequences provide the motor for change, they certainly do not provide it with a direction. Although he does not put it explicitly in the form of a solution to this problem, for Ferguson the succession of stages in a particular direction is driven by economic interests, which are to be found even in the most primitive social states:

When the individual no longer finds among his associates the same inclination to commit every subject to public use, he is seized with concern for his personal fortune; and he is alarmed by the cares which every person entertains for himself. . . . He suffers considerations of interest to rest on his mind, and when every present appetite is sufficiently satisfied, he can act with a view to futurity . . . Thus mankind acquires industry by many and by slow degrees. They are taught to regard their interest; they are refrained from rapine; and they are secured in the possession of what they fairly obtain; by these methods the habits of the labourer, the mechanic, and the trader, are gradually formed.[160]

Although there is progress in the transition from one stage to the next on Ferguson's account, from a moral point of view there is not wholly unqualified progress: primitive societies had virtues that may have been lost, and commercial society is subject to decay and corruption.[161] But of course the transition is driven not by moral factors but by economic ones: this does cause great problems for eighteenth-century thinkers, as we shall see in Chapter 6, but it does not affect the validity of the four-stage model.

The most ambitious version of the four-stage model was that of John Millar. He uses a far wider range of sources than his contemporaries, delving into travel books, and classical and modern histories. At the same time, he deploys his four-stage theory to account not just for the development of law and politics, but also uses it to explain family and gender relations, manners and morals, literature, art, and science. The general direction of the development of society he puts down to the fact that

[159] Ferguson, *An Essay on the History of Civil Society*, 204–5.

[160] Ibid., 161. Cf. his statement later in the book that 'the prospect of being able to exchange one commodity for another, turns, by degrees, the hunter and the warrior into a tradesman and a merchant.' Ibid., 302.

[161] Ibid., part v (341–93).

there is 'in man a disposition and capacity for improving his condition, by the exertion of which, he is carried on from one degree of advancement to another'. This lifts him out of a state of savagery, and by 'gradual advances in rendering their situation more comfortable, the most important alterations are produced in the state and condition of a people'.[162] Millar does not restrict his causal account to socio-economic considerations, acknowledging a role for factors such as climate, but whereas other accounts of the rise of literature and the arts, for example, had been restricted to the conditions under which they could flourish, Millar sees them more as a direct expression of socio-economic conditions.

Socio-economic factors play no such fundamental role in the two particularly influential mid-century French accounts, those of d'Alembert and Voltaire. This is in large part because their aim is not to understand how commercial society arose, so much as to counter the standard providential histories of Catholic thinkers, typified in Bossuet's *Histoire universelle* (1681),[163] where the idea of an unbroken chain of tradition is claimed as a basis for belief. In his genealogy of knowledge in the preliminary discourse of the *Encyclopédie*, d'Alembert adopts a quasi-religious strategy, a retelling of a story of the triumph of good over evil, except that roles are reversed, so that the Church now becomes the evil to be overcome. What he offers is a history of the triumph of knowledge over ignorance, on the model of a transition from infancy to adulthood. Such a model, Baconian in its orientation, is very much one shared by eighteenth-century writers generally, but no one else quite makes it do the work that d'Alembert does. One of the most interesting features of d'Alembert's approach is his account of the obstacles to knowledge. Bacon had pioneered this kind of study in his account of the 'idols', but d'Alembert goes beyond Bacon in significant respects. Unlike the study of knowledge as such, which follows an epistemological path from the seventeenth century onwards, accounts of obstacles to knowledge are more concerned with psychological factors. This is particularly true of Bacon,[164] and d'Alembert, in uncovering possible sources of error, does not ignore the psychological effects of a host of institutional and cultural factors, such as the shift in the way in which the natural realm was studied from direct observation of nature to exclusive concern with other people's reports about nature with the invention of printing, and the way in which religion took upon itself the mantle of philosophy.[165] The thrust of d'Alembert's historiography is not so much that

[162] John Millar, *Observations concerning the Distinction of Ranks in Society* (2nd enlarged edn. London, 1773), 2–4. See the account in Meek, *Social Science and the Ignoble Savage*, 160–77.

[163] Jacques-Bénigne Bossuet, *Discours sur l'histoire universelle a monseigneur le dauphin pour expliquer la fuite de la religion et les changemens des empires* (Paris, 1681). Providential history was also defended by a number of German historians of the time. See, in particular, Johann Hübner, *Kurtze Fragen aus der politischen Historie biß auf gegenwärtige Zeit continuiret* (10 vols, Leipzig, 1715).

[164] See Gaukroger, *Francis Bacon*, ch. 4

[165] See Gaukroger, *The Collapse of Mechanism and the Rise of Sensibility*, 433–7.

religion is to be identified with early stages of development, but rather that it acts in particular, specifiable ways at particular stages in this development to hinder the progress of learning.

Like d'Alembert, Voltaire sees Christianity as having acted contrary to reason, having a deleterious effect on the progress of the sciences, and on emancipation, and he is motivated by many of the same concerns as d'Alembert, particularly with countering the Christian historiography that was still shaping thinking about history. His approach is not to attack religion as holding up the progress of knowledge, however, but rather to relativize Christianity: it is one religion among many, and the culture in which it is fostered is one culture among many, certainly not the oldest and noblest, and not something that could ever provide a standard against which other cultures, or other eras, can be judged. For Voltaire, a form of enquiry that integrates historical and geographical perspectives in a coherent way enables us to gain insight into the make up of societies and cultures by disabusing us of the idea that our parochial values are universal and absolute ones. As I have already indicated, the periodization that he employs for the West is based on that of Dubos: the four periods are classical Athens, classical Rome, the age of Medici, and Voltaire's own age, that of Louis XIV.

While Scottish and French stadial theories are motivated socially and politically, this is not a *sine qua non* of such theories. One of the most original contributions to historical periodization was that of Vico. In Book II of *Scienza Nuova* he writes that his new science, 'comes to be at once a history of the ideas, the customs, the deeds of mankind. From these three we shall derive the principles of the history of human nature, which we shall show to be the principles of universal history, which principles it seems hitherto to have lacked.'[166] Vico's account is unique among stadial accounts, in that although he identifies three successive historical periods—divine, heroic, and human—it is far from clear that there is anything in the earlier stages that brings about the later ones, although the later ones must have proceeded through the earlier ones. If anything is responsible for the transition it is providence, which continues to play a role in his account.[167] Each period plays a causal role in shaping the culture of that period, however, with natural law, jurisprudence, language, and system of government all being distinctive of one of the three periods. Vico refers to his history as 'ideal' and, unlike the Scottish four-stage accounts, there is no causal role for socio-economic factors. But comparison with such accounts is in any case difficult because Vico's project is motivated by concerns quite alien to

[166] *The New Science of Giambattista Vico*, §368.

[167] In 1722, Vico had written to Cardinal Lorenzo Corsini (the future Pope Clement XII) that he intended *Scienza Nuova* to provide a decisive refutation of those, such as Hobbes and Bayle, who had argued that nations could be ruled without the benefit of a providential deity. At the same time, he added, it refuted theories of natural law—particularly those of Grotius, Pufendorf, and Selden—which ignored the operation of divine providence. See Robertson, *The Case for Enlightenment*, 201–2.

them. He is in effect reviving Renaissance humanism, placing the humanities at the centre of knowledge, and marginalizing science. His advocacy of Renaissance humanism, with its concern with philology and history rather than natural philosophy, proceeds in an unlikely way, via a response to Cartesian scepticism.[168] Accepting Descartes' geometrical model for certainty, he argues that assimilation of physics to geometry could not be successful because what delivers certainty in geometry is the fact that we construct its objects ourselves, whereas we do not do this with the objects of physics, which we must investigate through observation. The general principle linking the fact that we construct geometry ourselves and the degree of certainty that we attach to the results of our investigation of geometry is the *verum ipsum factum* or 'maker's knowledge' principle. Geometry is not unique in this respect, however. There is a sense in which the political and social worlds fall under the maker's knowledge principle. Locke had argued that the special kind of knowledge that the maker's knowledge principle delivers derives not from the making or doing as such but from knowing the archetype or model in accordance with which something is made or done.[169] Vico takes up this idea of the superiority of moral and civil knowledge over knowledge of the natural world, and his view is that while geometry deals straightforwardly with abstractions, the humanities deal with humanly constructed realities. The certain knowledge that we can have of civil society takes the form of the three-stage periodization, which is designed to display the commonalities of human nature in its three possible forms.

Vico's central concern with humanistic education[170] parallels the preoccupation of German philosophers later in the century with *Bildung*. It mirrors the anthropological projects of Herder and Schiller in that the ultimate aim is the shaping of human beings. The kinds of philological and linguistic concerns that Vico and Herder bring to periodization, as well as the focus on literary matters, not least on the literary power of myths, could not be further from the project of socio-economically based periodizations. This was very much in line with earlier germanophone writings, and we can confine our attention to two examples.

In a comprehensive if rather eclectic 'history of mankind' first published 1764,[171] the Swiss philosopher Isaak Iselin identified the *Seelenkräfte*—a human drive based on the faculties of sensation (*Empfindung*), imagination (*Einbildung*), and cognition (*Verstand*)—as the organizing basis of historical

[168] See Stephen Gaukroger, 'Vico and the Maker's Knowledge Principle', *History of Philosophy Quarterly* 3 (1986), 29–44.

[169] See James Tully, *A Discourse on Property* (Cambridge, 1980), 23.

[170] Vico, *Le Orazioni Inaugurali I–IV*, ed. G. G. Visconti (Bologna, 1982); idem, *De Nostri temporis studiorum ratione* (Naples, 1709).

[171] See the account in Reill, *The German Enlightenment and the Rise of Historicism*, 65–9, to which I am indebted here.

development. There are two variables in this account. First, the drive is retarded or restricted by environmental forces, whether physical or social. Second, the balance between the three elements varies since they do not exist in fixed proportions. These variables form the basis for Iselin's three-stage account of historical development. In particular, the stages correspond to the dominance of one faculty over the others: sensation in the primitive savage state; imagination in the second, the age of cultivation, in which language, poetry, and myth predominate; and finally the third age, that of reason, which produces a harmony between the three faculties. A similar three-stage periodization is to be found in Lessing's influential 1780 work on the education of the human race,[172] but this has an entirely different focus. It is the development of spiritual or religious ideas that drives historical development for Lessing. The sequence is one of increasing moral awareness, an awareness achieved through rational insight.[173] In the first stage, there is a belief that there is a single god who rewards and punishes one while one is alive. In the second stage, the belief develops that human beings are immortal, and that divinely instituted punishment and rewards occur in an afterlife. Finally, in the third stage, there is a realization that true morality is its own reward, and that notions of punishment and rewards in an afterlife are inappropriate.

We shall return to the reworkings of religious ideas in Chapter 7, when we come to explore the naturalization and humanization of religion. My aim here has been to draw attention the diversity of motivations behind stadial histories. This diversity did not dilute the idea of history proceeding in stages, but rather served to establish it as a form of common ground for discussing those antecedents that shaped present behaviour. The combination of comparative history and comparative geography offered an increasingly important resource for probing the nature of modern society. This is nowhere more evident than in the analysis of the socio-economic basis of modern society, a concern that takes us directly to the fourth kind of naturalization of the human: social arithmetic.

[172] Lessing, *Die Erziehung des Menschengeschlechts* (Berlin, 1780). See the account in Henry Allison, *Lessing and the Enlightenment.*

[173] Lessing, *Die Erziehung des Menschengeschlechts*, §7, and §§11–12.

6

Social Arithmetic

Traditional accounts of what came to be termed the human condition—those accounts that had been pursued within metaphysics, theology, and civic humanism—considered human behaviour in terms of what their proponents took to be universal human attributes. Whether these universal human attributes were in part derived from observed human behaviour, or whether they were arrived at on a priori or scriptural grounds, as regulative precepts, what was at issue was an interplay between individual and universal characteristics. This was an understanding initially developed in the hierarchically ordered societies of the medieval West, where distinctions of rank and office governed questions of social roles and responsibilities. Morality was in large part a function of whether one satisfied the responsibilities of one's office or rank. When this conception was subsequently revamped along democratic lines, moral responsibility was relocated in the individual, irrespective of rank or office, so that moral demands were the same for everyone. It was above all Kant who was responsible for this reworking of moral agency. He thought of human development in terms of 'inherently rational creatures', and he construed history in terms of the unfolding movement towards a fully rational form of human life lived in peace with others.[1] Although this has often been taken as a canonical form of Enlightenment thinking,[2] Kant's individual self-understanding is in reality a somewhat limited way of thinking about social and political questions. It does not work at all well for the kinds of societies that were emerging in the eighteenth century, or before that. In England, as early as the decades following the Civil War, questions of social order and political stability arose that were in many respects quite new, but what was needed was not an account centred on the moral responsibilities of an abstract rational agent, but rather one that focused on how various different kinds of moral and political responsibility were generated, and whether some account could be provided of how what was due in one capacity might conflict with what was due in another.[3]

[1] See, for example, *Idee zu einer allgemeinen Geschichte in weltbürgerlicher Absicht*: *Gesammelte Schriften*, viii. 17–31; *Anthropology, History and Education*, 107–20.

[2] See, for example, the very influential work of Jürgen Habermas, especially his *Erkenntnis und Interesse* (Berlin, 1968); translated as *Knowledge and Human Interests* (London, 1972).

[3] See Conal Condren, *Argument and Authority in Early Modern England: The Presupposition of Oaths and Offices* (Cambridge, 2006).

We can find one of the first signs of such a new approach in the seventeenth century in the work of William Petty. Petty, rejecting the utopian schemes of more radical elements in the English Civil War, concluded that individual behaviour is not something that can be collectively reformed: 'That some are poorer than others, ever was and ever will be; And that many are naturally querulous and envious, is an Evil as old as the World'.[4] Consequently it was not a matter of discovering some inherent order in the world, as earlier political thinkers had tried to do, but rather one of creating order: something was needed to enable the ruler to remove the obstacles that 'hinder the operations of Authority'.[5] How, then, should one understand political and social stability, and what form should this understanding take? Petty writes:

> The Method I take to do this, is not yet very usual; for instead of using only comparative and superlative Words, and intellectual Arguments, I have taken the course (as a Specimen of the Political Arithmetick I have long aimed at) to express myself in Terms of *Number*, *Weight*, or *Measure*; to use only Arguments of Sense, and to consider only such Causes, as have visible Foundations in Nature; leaving those that depend upon the mutable Minds, Opinions, Appetites, and Passions of particular Men, to the Consideration of others.[6]

On this account, what one needs to understand in order to govern properly is collective behaviour, not individual behaviour. There are a number of significant elements in this approach. Petty rejects the idea that the study of human behaviour is simply the study of the behaviour of individual human beings, and correlatively there is an implicit rejection of the idea that individual reason provides the key to understanding human behaviour. At the same time, there emerges the idea that collective behaviour is something that is not only different from individual behaviour, but also that the methods by which it should be explored are different from those that have been used to understand individual behaviour. The new methods are naturalistic and quantitative, something inappropriate in the case of individuals, whose behaviour can only be accounted for in terms of reason, will, and affective faculties.[7]

These are major developments, and lie at the core of our concern in this chapter. Collective behaviour is the subject of social arithmetic, the generalized

[4] William Petty, *Political Arithmetick* (London, 1690), preface.

[5] See Peter Buck, 'Seventeenth-Century Political Arithmetic: Civil Strife and Vital Statistics', *Isis* 68 (1977), 67–84; and idem, 'People Who Counted: Political Arithmetic in the Eighteenth Century', *Isis* 73 (1982), 28–45. Conal Condren has pointed out to me that the theme of order being created goes back to Hobbes' *Leviathan*, and Harrington and Petty were in part commenting on this, criticizing his a priori reasoning.

[6] Petty, *Political Arithmetick*, preface.

[7] Petty had a general philosophical agenda, and from an early stage in his career he had advocated the reform of natural history, conceived along roughly Baconian lines, in terms of 'Number, Weight and Measure'. See for example William Petty, *A Treatise of Taxes and Contributions* (London, 1662), and *A Discourse Made Before the Royal Society . . . Concerning the Use of Duplicate Proportion* (London, 1674). Generally on Petty, see Ted McCormick, *William Petty: And the Ambitions of Political Arithmetic* (Oxford, 2009).

successor to Petty's 'political arithmetick'. It is the fourth kind of naturalization of the human that we shall be examining, and what is distinctive about it is that it deals with collective properties. Unlike distributive properties—those properties of the whole that are the properties of each of the parts—collective properties are not mirrored at the individual level, and they cannot be arrived at by abstraction from individual behaviour. An increasing number of writers recognized this in the eighteenth century, particularly those composing 'philosophical' histories such as Voltaire, Montesquieu, and Hume. Hume, for example, in developing the idea that the relation between cause and effect is a matter of inference rather than direct perception, argues that such connections are more easily established when changes in human conditions produce changes in large-scale human behaviour, as in the case of the rise and progress of the arts and sciences, or that of the rise and progress of commerce, rather than on an individual level, and he distinguishes two forms of history corresponding respectively to cultural change and to individual actions.[8] Montesquieu, likewise, writes that causes become less arbitrary to the extent that they have a more general effect: we know better what shapes the achievements of societies that have adopted a given way of life than we do what shapes the lives of individuals.[9]

Note the stress on causal explanation here. If one wants to account for social, political, or economic questions in causal terms, one has in effect to abandon individual behaviour and turn instead to the behaviour of collective or aggregate entities. And to the extent to which rendering phenomena amenable to causal explanation is part of a process of naturalization, then, in the shift from understanding behaviour at the personal level to the aggregate level, one is moving to a different kind of understanding of human behaviour. The significance of the move is evident when one considers that causal accounts in the case of individuals were associated with determinism. Individual behaviour—and insofar as we are concerned with typically human behaviour here we are dealing with behaviour for which the person is responsible—is characterizable primarily in terms of intentions rather than causes. It is far from clear that the language of intentionality is applicable to collective or aggregate behaviour. The question that then arises is whether such behaviour lacks a moral dimension. Indeed, collective behaviour is distinctive and problematic in a number of fundamental respects. As we saw in Chapter 1, when contrasting Condorcet's and Wolff's defences of reason, both were committed to the idea that reason was the sole key to understanding human behaviour. But whereas Wolffian metaphysics was concerned to establish the grounding of individual behaviour in reason, Condorcet's concern was with what account we can give of the rationality of collective or aggregate behaviour. If a notion like knowledge is to be applicable to collective reason it is certainly not

[8] Hume, 'Of the Rise and Progress of the Arts and Sciences', *Essays and Treatises*, i. 111–37.

[9] Montesquieu, 'Essay sur les causes qui peuvent affecter les esprits et les caractères' (1734), *Œuvres complètes*, ed. R. Caillois (2 vols, Paris, 1949–51), ii. 39.

in terms of the kind of self-understanding that had traditionally been invoked,[10] and the questions of propositional and non-propositional forms of understanding that we explored earlier cannot be raised in this context.

Many of the issues raised by approaching human behaviour in terms of collective activities had been brought to a head in the early decades of the eighteenth century by Mandeville. 'We have no Regard for the Persons themselves, but only the Benefit they may be of to the Publick,' he wrote in 1720, 'and they are only look'd upon as Parts and Members of the whole Society.'[11] The identification of collective behaviour as a discrete area of enquiry generated a host of deep problems for the understanding of human behaviour, problems that were to play a crucial role in moral, political, and economic thinking from the middle of the eighteenth century onwards. In particular, the moral standing of collective behaviour was deeply problematic. In the case of commercial behaviour, it was associated with values that were in open conflict with those of individual morality.

THE FABLE OF THE BEES

We saw in Chapter 3 how Bordeu compared the human body to a swarm of bees: the organs of the body are each alive, and do not have to be brought under some centralized principle of life. All activity is under localized control, and the organs cohere into a whole in something like the way bees cohere into a swarm. Analogies with bees had in fact been used from the sixteenth century onwards to account for various forms of collective behaviour, especially in political contexts, and the same images that were used in the early seventeenth century to defend the monarchy had been transformed into defences of republicanism by the eighteenth century.[12] When Mandeville first published his *The Grumbling Hive, or Knaves Turn'd Honest* in 1705, he used the bee analogy in a much more radical way. The first part of the poem emphasizes the economic benefits that follow from a society having morally corrupt members, whereas the second part explores the deleterious consequences of a society that purges its immoral and immoderate members. As one commentator has put it, 'Mandeville successfully fashioned an intuitively compelling picture of a world dramatically transformed by new mechanisms of exchange, a world that frighteningly demanded from its members the relegation of

[10] Note also that just how effective such self-understanding is can be questioned. Consider for example J. B. S. Haldane's remark: 'I have come to the conclusion that my subjective account of my own motivation is largely mythical on almost all occasions. . . . I don't know why I do things.' 'Discussion: Future of the Mind', in Gordon E. W. Wolstenholme ed., *Man and his Future*, Ciba Foundation Seminar (London, 1963), 322–7: 326.

[11] Mandeville, *Free Thoughts on Religion*, 253.

[12] Compare, for example, Charles Butler, *The Feminine Monarchy, or, A Treatise concerning bees, and the due ordering of them* (Oxford, 1609), and Rousseau's 1762 *Du Contrat Social*, Book I, chs 6 and 7.

their civic ideals to the realm of nostalgia, and the adoption of an intransigently egoistic morality.'[13] Working with the French notion of *politesse*, Mandeville explored the gap between genuine virtue and techniques for self-preservation, between personal and social identity.[14]

Mandeville was born into an anti-clerical republican family in Rotterdam and had studied medicine at Leiden. Caught up in the political upheavals of the early 1690s—his father was banished from the city in 1693—he emigrated to London in 1694, where he set up a medical practice specializing in nervous and stomach disorders. In his *Treatise of the Hypochondriack and Hysterick Disorders* of 1711, he defended a strongly Hippocratic view of medicine, of the kind associated with Sydenham, that was in line with the phenomenalist position defended in the experimental natural philosophy tradition, subsequently rationalized by Locke. The main protagonist in the *Treatise* (which despite its title uses the dialogue form) is Philopirio, who refuses to indulge in hypotheses, or 'to reason about the Causes and Seat of the Distemper', because medicine is about observation, not offering theories about causes.[15] Although this was a contentious view, it was a widely held one. A more distinctive and far more radical feature of the *Treatise* is Mandeville's characterization of the qualities needed by the physician. What should drive the physician is not reason, Mandeville argues, but pride, reputation, and gain, because it is these that in reality motivate the physician to generate the results from which we benefit in medicine.[16]

These aspects of Mandeville's account of medicine derive from a more general theory about a lack of fit between individual behaviour and collective results. The theory is set out in progressive editions of his *Fable of the Bees*, which, beginning as a four-page broadsheet poem in 1705, evolved over a decade and was republished in a fuller form in 1714, where a prose commentary was added for the first time, and where the whole first took the title of *The Fable of the Bees*.[17] Neither version attracted much notice, but in 1723 another edition appeared, which included *An Essay on Charity and Charity Schools*, which did provoke a reaction. The essay rejected the newly founded charity schools, which were instituted as a means of educating and instilling virtue in the poor. Mandeville rejected both aims. Educating the poor, he claimed, would only encourage them to want more material goods, which would be counterproductive, given the original aims of the charity schools. The idea that children of the poor might be instilled with

[13] Edward G. Hundert, *The Enlightenment's* Fable: *Bernard Mandeville and the Discovery of Society* (Cambridge, 1994), 13.

[14] Ibid., 118–19.

[15] Mandeville, *Treatise of the Hypochondriack and Hysterick Disorders* (London, 1711), 55.

[16] But at the same time he condemns the egoism of his fellow physicians. See the discussion in Phillip Hilton, *Bitter Honey: Recuperating the Medical and Scientific Context of Bernard Mandeville* (Bern, 2011).

[17] I shall be using the 1723 edition as given in Mandeville, *The Fable of the Bees: Or, Private Vices, Publick Benefits. By Bernard Mandeville. With a Commentary Critical, Historical, and Explanatory by F. B. Kaye* (2 vols, Oxford, 1924). On the various editions, see ibid., i. pp. xxxiii–xxxviii.

virtue in such schools presupposed that those educating them, the rich and their representatives, were virtuous, but Mandeville had no doubt that vice was ubiquitous, not a prerogative of the poor, so charity schools could achieve nothing in this respect. During the course of the first three decades of the century, Mandeville published a number of provocative books and pamphlets, from his sexually charged prose poem *The Virgin Unmask'd* of 1709,[18] to his 1724 proposal for public houses of prostitution,[19] which did not endear him to the authorities, and in 1729 *The Fable of the Bees* was the subject of a prosecution for immorality, although there is no evidence that Mandeville was at all concerned by this development.[20]

The core argument of *The Fable of the Bees* is indicated by its subtitle: private vices, public virtues. Mandeville's view is that avarice, in particular, has a positive social function for any society seeking to increase its level of prosperity. This is the moral price of commercial prosperity. Private virtues are of no use in the public realm:

It is certain that the fewer Desires a Man has and the less he covets, the more easy he is to himself; the more active he is to supply his own Wants, and the less he requires to be waited upon, the more he will be beloved and the less trouble he is in a Family; the more he loves Peace and Concord, the more Charity he has for his neighbour, and the more he shines in real Virtue, there is no doubt but that in proportion he is acceptable to God and Man. But let us be Just, what Benefit can these things be of, or what earthly Good can they do, to promote the Wealth, the Glory and worldly Greatness of Nations? It is the sensual Courtier that sets no limits to his Luxury; . . . the profuse Rake and lavish Heir, that scatter about their Money without Wit or Judgment, buy every thing they see, and either destroy or give it away the next Day, the Covetous and perjur'd Villain that squeez'd an immense Treasure from the tears of Widows and Orphans, and left the Prodigals the Money to spend: It is these that are the Prey and proper Food of a full grown Leviathan; or in other words, such is the calamitous Condition of Human Affairs that we stand in need of the Plagues and Monsters I named to have all the Variety of Labour perform'd, which the Skill of men is capable of inventing in order to procure an honest Livelihood to the vast Multitudes of working poor, that are required to make a large Society: And it is folly to imagine that Great and Wealthy Nations can subsist, and be at once Powerful and Polite without.[21]

[18] Bernard Mandeville, *The Virgin Unmask'd: or, Female Dialogues betwixt an Elderly Maiden Lady, and her Niece* (London, 1709). The book was later reissued under the title *Mysteries of Virginity*. The work has a political element, being strongly critical of the policies of Louis XIV.

[19] Mandeville, *A Modest Defence of Public Stews* (London, 1724).

[20] In his *Autobiography*, Benjamin Franklin, in one of the few contemporary descriptions of Mandeville, called him 'a most facetious, entertaining companion': Benjamin Franklin, *Works*, ed. John Bigelow (12 vols New York, 1904), i. 92.

[21] Mandeville, *The Fable of the Bees*, i. 355–6. Cf. his *A Letter to Dion*, where we are told that in civil society, 'the Avarice of Some and the Profaneness of Others, together with the Pride and Envy of most Individuals are absolutely necessary to raise them to a great and powerful, and, in the Language of the World, polite Nation. It seems still to be a greater Paradox, that natural as well as moral Evil and the very Calamities we pray against, do not only contribute to this worldly Greatness, but a

Despite appearances, the point here is not a negative or sceptical one. Mandeville is suggesting that the kind of behaviour needed to maintain a large-scale commercial society is significantly different from that of traditional societies, and that norms that are generally appropriate in traditional societies have, at best, a limited appropriateness in large-scale commercial societies.

The basic dilemma raised by Mandeville's account is that the demands placed on individual behaviour, by those collective goals that bring with them economic and political well-being, are in conflict with the moral demands that shape the individual behaviour that we value most. This is a problem that goes back at least to Machiavelli, who had warned that virtue, humility, and modesty would lead to disastrous results in collective civic action. To believe otherwise, Mandeville argues, is to follow those 'silly People who imagine that the Good of the whole is consistent with the Good of every Individual'.[22] A crucial point here is that responses to Mandeville's account were generated within attempts to come to terms with commercial societies. As Phillipson notes in the case of Smith's *Theory of Moral Sentiments*, for example, the aim was to 'instruct young men of middling rank in their duties as men and citizens of a modern commercial polity'.[23]

The core question at stake is the self-interest that Mandeville believes motivates individual behaviour in the first place. The discrepancy between collective and individual goals, and its consequences for our understanding of morality, turned out to be a fundamental problem for political economy. Mandeville is the ultimate source for the attempts of nineteenth-century political economists such as Bentham and Mill, for example, to rethink the nature of morality, and develop a philosophical account of morality—utilitarianism—that was consequentialist, that is, it held that the morality of an act depends on its consequences, not on the intrinsic nature of the act, or on the character of the agent. More generally, whatever the solution proposed to Mandeville's dilemma, it was clear to all those who took it seriously that a fundamental rethinking of the nature of morality was required. The problem was that such a rethinking did not result in any single conception, but rather in a divergence in conceptions of what the role and nature of morality was, and this was something whose impact was far more profound than that of the abandonment of exclusively religious notions of morality.

Clearly a great deal hinges on whether collective activity has a moral dimension. If it does not, then there can be no moral conflict between individual behaviour

certain Proportion of them is so necessary to all Nations that it is not to be conceiv'd how Society should subsist upon Earth exempt from all Evil, both natural and Moral.' *A Letter to Dion* (London, 1732), 21.

[22] *The Fable of the Bees*, 49. Cf. Arbuthnot's argument that political systems work only when they reject private morality: John Arbuthnot, *Proposals for Printing A Very Curious Discourse . . . intitled ΨΕΥΔΟΛΟΓΓΑ ΠΟΛΙΤΙΚΗ; or, a Treatise on the Art of Political Lying* (London, 1712).

[23] Nicholas Phillipson, 'Adam Smith as Civic Moralist', in I. Hont and M. Ignatieff, eds, *Wealth and Virtue: The Shaping of Political Economy in the Scottish Enlightenment* (Cambridge, 1983), 179–202: 179.

and collective behaviour. The traditional view, if only implicit, had been that only individual behaviour has a moral dimension. Individuals are able to form intentions and act upon these, and this would seem a prerequisite of moral behaviour, yet the role of intentionality in collective behaviour is difficult to fathom. The natural home of morality, so to speak, lies in the realm of the individual. On such a view, while it is true that empires, nations, and states can act in various ways that attract moral sanction or approval, and can be subject to moral maxims such as that of the doctrine of just war, moral responsibility is properly ascribed not to a state as such but to the ruler. For the kind of approach taken by Mandeville, by contrast, there is a need to establish that collective activity has a moral dimension, reducible neither to the behaviour of the individuals making it up, nor that of their representatives. Given that the collective values that Mandeville argues are associated with commercial activity are manifestly in conflict with those traditionally understood to characterize personal morality, the question arises how they are generated as *moral* values.

One way to deal with this issue is to raise the question of just what it is that makes any value a moral one in the first place. It is here that the project of naturalization comes into play. Mandeville's strategy is to pursue a form of naturalization through historicization, seeking an understanding of morality through an account of its social origins. Morality thereby becomes an empirically describable historical phenomenon. This is very much in line with the approaches of Bayle and Locke, who drew on travel books and other sources to establish the existence of moral diversity.[24] Moral diversity is simply an empirically established fact of life on Mandeville's account, and he emphatically rejects Shaftesbury's definition of virtues and vices as 'permanent realities'. Shaftesbury's conceptions of virtue and vice are 'capable by the help of a little Enthusiasm of Inspiring us with the most Noble Sentiments concerning the Dignity of our exalted Nature; What Pity it is that they are not true'.[25] He then sets out examples showing that there is no certainty in morals:

> Plurality of Wives is odious among Christians, and all the Wit and Learning of a Great Genius in defence of it has been rejected with contempt: But Polygamy is not shocking to a Mohametan. What men have learned from their Infancy enslaves them, and the Force of Custom warps Nature, and at the same time imitates her in such a manner, that it is often difficult to know which of the two we are influenced by. In the *East* formerly Sisters married Brothers, and it was meritorious for a Man to marry his Mother. Such Alliances are abominable; but it is certain that, whatever Horror we conceive at the Thoughts of them, there is nothing in Nature repugnant against them, but what is built upon Mode

[24] Mandeville's contemporaries placed him very much in a Lockean tradition. See, for example, George Blewitt, *An Enquiry whether a General Practice of Virtue tends to the Wealth or Poverty, Benefit or Disadvantage of a People?* (London, 1725), 87–8; John Brown, *Essays on the Characteristics of Lord Shaftesbury* (London, 1751), 139; Philip Skelton, *Ophiomaches; or, Deism Revealed* (2 vols, London, 1749), i. 125. I take the references from Daniel Carey, *Locke, Shaftesbury, and Hutcheson* (Cambridge, 2006), 172.

[25] Mandeville, *The Fable of the Bees*, i. 324.

and Custom . . . Which is the best Religion? is a Question that has caused more mischief than all other Questions together. Ask it at *Peking*, at *Constantinople*, and at *Rome*, and you'll receive three distinct Answers extremely different from one another, yet all of them equally positive and peremptory.[26]

What this line of argument suggests is that custom has been mistaken for morality, with religion, traditionally taken as an absolute source of moral instruction, being assimilated to custom.[27] There had been a seventeenth-century tradition of such arguments,[28] but Mandeville pursues the point in detail. Such a source of moral instruction, he argues, simply imposes morality from outside, without probing the sources of the behaviour that fall under moral characterization. Because of this, moral discourse, at least as traditionally conceived, is wholly self-deceiving. By denying impulse and passions as forms of motivation, and accounting for human action in terms of a rationalizing ethics, it disqualifies itself from offering a realistic theory of motivation, and ignores the actual conditions under which practical decisions, moral and otherwise, are made:

All the Passions and Instincts in general were given to all Animals for some wise End, tending to the Preservation and Happiness either of themselves or their Species: It is our Duty to hinder them from being detrimental or offensive to any Part of the Society; but why should we be ashamed of having them? The Instinct of high Value, which every Individual has for himself, is a very useful Passion: but a Passion it is, and though I could demonstrate, that we should be miserable Creatures without it, yet, when it is excessive, it often is the Cause of endless Mischiefs.[29]

Reason is not autonomous on this account, and 'we are ever pushing our Reason which way soever we feel Passion to draw it'.[30]

But if moral behaviour is generated in this way, how has morality come to have the standing that it has, namely as something that can regulate behaviour in absolute terms? Personal rectitude as a source of public good, Mandeville maintains, has always been merely a ruse whereby ruling elites lay claim to private virtues, in order to disguise their own self-seeking and to obtain the respect of subordinates. The crucial point is that for Mandeville this is not a criticism of such societies but a fact of life, since without such rulers there would be no peace, rule of law, or political enlightenment, but rather a state of strife and anarchy. In their behaviour, people are driven by their passions and the force with which they feel these passions. Mandeville is clear that there is nothing one can do about this, at

[26] Ibid., 330–1.

[27] In his *An Enquiry into the Origin of Honour* (London, 1732), Mandeville goes furthest in construing religion as a form of politics or custom.

[28] See Condren, *Argument and Authority in Early Modern England*, ch. 2.

[29] *The Fable of the Bees*, ii. 91. On the question of motivation, see Mandeville's discussion of 'self-liking' in his *An Enquiry into the Origin of Honour*. See the discussion in Hundert, *The Enlightenment's* Fable, ch. 2.

[30] Mandeville, *The Fable of the Bees*, i. 333.

least at the level of moral and political instruction. But what one can do is shape the communal expression of these passions, and it was this, rather than reform at the level of the individual, that Mandeville believed marked out successful, socially stable societies from barbarous anarchy. Note that rulers are needed to do the channelling: it is not a case of collective effects in their own right naturally producing something good. Rulers are successful to the extent that they are able to neutralize inherently antagonistic passions and, through strict rules of justice as well as other means, channel them into a productive outcome.[31]

For Mandeville, there is not the stark contrast between naked self-interest and morally regulated behaviour that we find in other moralists. The 'natural' morality that we find in savages is simply a matter of acting so as to relieve one's compassion, which Mandeville treats as a form of self-interest.[32] The apparently altruistic morality of civilized society is likewise treated in terms of self-interest: it is simply love of praise or fear of blame.[33] But there are differences between the two stages of society. What appears in the latter case to be a contrast between external political constraints and an internal, moral form of constraint turns out to be more complex. The internal form of constraint is actually an *internalized* form of constraint, not qualitatively distinct from external political constraint as it might at first seem. In broad terms, the argument is that there is a natural state of humanity, characterized by self-interest, which is motivated throughout the development of society, from its most primitive forms onwards, by competition for scarce resources.[34] Self-interested behaviour must be regulated in particular ways depending on what ends the behaviour is directed towards. The regulation is of necessity always initially extrinsic, but it can become internalized. This extrinsic nature of the regulation of behaviour is important because it means that one can no longer identify greed and avarice, for example, as virtues and vices in an absolute sense: there are no absolute values. Virtues and vices arise only in the context of forms of regulation of behaviour appropriate to competition for scarce resources, a regulation that does not aspire to capture 'natural' forms of morality, *à la* Shaftesbury and Hutcheson.

Over and above Mandeville's arguments that motivation is always self-regarding and that the rules of morality are conventional, there is a crucial claim. This is that the values of civilized society, particularly moral values, are a secondary effect—albeit a necessary one—of the civilizing process. As he puts it, 'the nearer we search into human Nature, the more we shall be convinced, that the Moral Virtues are the Political Offspring which Flattery begot upon Pride'.[35] More specifically, these values emerge only at an advanced stage in its evolution, which he associates with the advent of political society, telling us that:

the first Rudiments of Morality, broach'd by skilful Politicians, to render Men useful to each other as well as tractable, were chiefly contrived that the Ambitious might reap the

[31] Ibid., ii. 78–9. [32] Ibid., i. 256–9. [33] Ibid., i. 68.
[34] Ibid., ii. 238–40. [35] Ibid., i. 51.

more Benefit from, and govern vast Numbers of them with greater Ease and Security. This Foundation of Politicks being once laid, it is impossible that Man should long remain uncivilized: For even those who only strove to gratify their Appetites, being continually cross'd by others of the same Stamp, could not but observe, that whenever they check'd their Inclinations or but followed them with more Circumspection, they avoided a world of Troubles, and often escap'd many of the Calamities that generally attended the too eager Pursuit after Pleasure.[36]

A problem that many of Mandeville's critics—Hutcheson, Law, Ferguson, Rousseau, Smith—perceived here was the emergence of a dystopia. Economic self-interest, envy, and the drive for profit may well promote commercial activity, but the problem was that there was no reason to believe that they would be confined to commercial activity. The risk here is that commercial values would, if promoted, come to corrupt and dominate all other aspects of communal life.[37] Rousseau is typically forthright on the question. Accepting that modern commercial society relied on the kind of behaviour that Mandeville describes, he concluded in his *Discours sur l'origine et les fondements de l'inégalité* that a society of self-regarding individuals was exactly the result to which modernity led, and against which he had warned.

THE NATURALIZATION OF MORALITY

The basic dilemma raised by Mandeville's account is that the demands placed on individual behaviour by collective goals are in conflict with individual morality. More generally, the existence of parallel, conflicting systems of morality within a single form of society raises a number of deep problems. There are two main problem areas: how to deal with competing systems of morality, and whether the solution to the kind of competition to which Mandeville draws attention lies in a revision and broadening of the scope of individual morality, or whether it requires accounting for the collective values in their own right.

A number of critics of Mandeville had questioned the existence of moral diversity, but this was an unpromising strategy. To the extent to which it was based upon the assumption of secure moral guidance provided by religion, the problem was that religion had similar problems, since religious diversity—not just in the cases that Mandeville raises, but even, as Locke pointed out, within a single creed such as Anglicanism—was a pressing problem to which there was no satisfactory non-dogmatic answer.[38] The project of establishing a 'natural' morality, independent of religion, was similarly fraught, and gave every appearance of

[36] Ibid., i. 47.

[37] See Hundert, *The Enlightenment's* Fable, 178–9. Mandeville replaces discussion of avarice with envy in later editions of the *Fable*, probably as a means of deflecting criticisms: see ibid., 176.

[38] See Gaukroger, *The Collapse of Mechanism and the Rise of Sensibility*, 46–52.

begging the questions in favour of Christianity.[39] Moral diversity was in fact widely accepted by the eighteenth century. For more naturalistically inclined writers—that is, those who did not accept a wholly religious determination of moral values—morality answered to human needs, aims, and goals. As Ferguson put it, 'before we can ascertain rules of morality for mankind, the history of man's nature, his dispositions, his specific enjoyments, and sufferings, his condition and future prospects, should be known'.[40] The question was whether these human needs, aims, and goals were so fundamental as to be universal. If one allowed a degree of contingency in these—as many eighteenth-century thinkers routinely did, for example between the demands of primitive societies and civilized ones—then one would expect the moral values that accompanied them to vary.

From the way in which Mandeville describes the transition to a commercial society, it might seem that the question is simply that of choosing a new set of values for society: replacing one set of values, which were appropriate to one kind of society, by others which are appropriate to a different, commercial kind of society. But there is an ambiguity in his account in this respect. On the one hand, he does at times seem to assume that self-interest and commercial values have come to permeate all professions in a commercial society, as some of his critics feared. For example, he argues that it is absurd to think that a barrister toils to secure the property of others, that a doctor works day and night to secure the health of his patients, or that a clergyman who holds several livings does so in order to minister to many souls.[41] But the idea that commercial societies *of necessity* foster avarice and luxury in every aspect of those societies was a problematic notion. As one of Mandeville's fiercest critics, George Blewitt, pointed out, one of the world's most successful commercial nations, the Netherlands, rejected luxury and avarice and exhibited a significant level of frugality.[42] The luxuriance of the Spanish, combined with their sliding into decadence and material decline, provided the obverse counterexample, as did parallel developments in absolutist France. Mandeville was well aware of these cases, and strove to establish that their commercial success or failure was due to different kinds of reasons,[43] but they at least loosen the connection between certain sets of values and commercial success.

There is, however, also a different kind of approach which is suggested by other passages in Mandeville. This proceeds not in terms of one set of values gradually eclipsing another, but rather associates different values with different offices. This is a question that has a significant history before Mandeville. In the sixteenth and seventeenth centuries, moral agency was tied to realms of responsibility: mothers, judges, and princes, for example, had different responsibilities, and the moral

[39] See ibid., 402–9.

[40] Adam Ferguson, *Institutes of Moral Philosophy. For the Use of Students in the College of Edinburgh* (2nd edn, Edinburgh, 1773), 11.

[41] Mandeville, *The Fable of the Bees,* ii. 52.

[42] Blewitt, *An Enquiry*, 48–9.

[43] See Hundert, *The Enlightenment's* Fable, 198–205.

demands placed on them derived from their office. There was an occasional awareness that what was due in one capacity might conflict with what was due in another,[44] and one form of response to this were the various forms of moral pluralism, in Hume and Smith for example. Mandeville's argument is that we should evaluate the behaviour of people by different criteria depending on 'the Usefulness and dignity of their callings, their capacities, with all qualifications required for the exercise or performance of their functions'.[45] Extreme examples of such callings would include military and ecclesiastical professions. Military values had occasionally become dominant throughout societies in times of war,[46] and there was a persistent view in the eighteenth and nineteenth centuries that war was necessary to revive the life force of societies that had been made effete by urban luxuries.[47] But this was a minority view, and while Smith offers a defence of 'the martial spirit' in the *Wealth of Nations* for example, he is not calling for the adoption of military values throughout society.[48] At the other extreme, there was a rejection of what were conceived of as monastic values for society in general, because even though, for many in Catholic societies, these values embodied the highest ideals of Christianity, they were deemed inappropriate in the general case. In both cases the relevant values have immediate moral implications, with which everyday individual morality may conflict, and with which commercial values may conflict.

Smith offers a form of moral pluralism, and his discussion takes place in the context of two sets of issues. The first is consideration of the traditional question of the unity of virtue. The second is consideration of the merits of 'sympathy' on the one hand, and 'self-interest' on the other. Following Hume, he rejects the unity of virtue telling us that: 'By running up all the different virtues too to this one species of propriety, Epicurus indulged a propensity, which is natural to all men, but which philosophers in particular are apt to cultivate with a peculiar fondness, as the great means of displaying their ingenuity, the propensity to account for all appearances from as few principles as possible.'[49] But if the various virtues do not form a unity, nor are they wholly independent of one another. The

44 See Condren, *Argument and Authority in Early Modern England.*

45 Mandeville, *Free Thoughts on Religion*, 282.

46 See, for example, the discussion of the rise and decline of military values in early modern England in Keith Thomas, *The Ends of Life: Roads to Fulfilment in Early Modern England* (Oxford, 2009), ch. 2.

47 See, for example, Daniel Pick, *War Machine: The Rationalization of Slaughter in the Modern Age* (New Haven, Conn., 1993).

48 Adam Smith, *An Inquiry into Nature and Causes of The Wealth of Nations*, V.i.f.60.

49 Adam Smith, *The Theory of Moral Sentiments*, VII.ii.2.14. See the discussion in Michael B. Gill, *Humean Moral Pluralism* (Oxford, 2014), ch. 4. A thoroughgoing, and compelling, defence of moral pluralism more generally is to be found in Charles Larmore, *Patterns of Moral Complexity* (Cambridge, 1987). Cf. the attempt to combine an intention-based ethics with utilitarianism and the doctrine of association of ideas, on the basis that the former delves into the source of morality whereas the latter are concerned with teaching mortality, in James Mackintosh, *Dissertation on the Progress of Ethical Philosophy* (Edinburgh, 1836).

question arises, for the moral pluralist, of what the balance between the various virtues consists in, and here Smith parts company with Hume:

That system which places virtue in utility, coincides too with that which makes it consist in propriety. According to this system, all those qualities of the mind which are agreeable or advantageous, either to the person himself or to others, are approved of as virtuous, and the contrary disapproved of as vicious. But the agreeableness or utility of any affection depends upon the degree to which it is allowed to subsist in. Every affection is useful when it is confined to a certain degree of moderation; and every affection is disadvantageous when it exceeds the proper bounds. According to this system therefore, virtue consists not in any one affection, but in the proper degree of all the affections. The only difference between it and that which I have been endeavouring to establish, is, that it makes utility, and not sympathy, or the correspondent affection of the spectator, the natural and original measure of this proper degree.[50]

It is not surprising that Smith's only extended explicit discussion of Mandeville immediately follows this passage, in the chapter entitled 'Of Licentious Systems'. In countering Mandeville's promotion of self-interest and avarice by advocating moral pluralism, the crucial thing for Smith to provide is some way of balancing the competing claims of various virtues. For Smith, it is sympathy—a capacity through which we understand, and identify with, what others are feeling—that lies at the basis of our behaviour as moral agents, and it is sympathy that acts as the common measure of the virtues:

Our sensibility to the feelings of others, so far from being inconsistent with the manhood of self-command, is the very principle upon which that manhood is founded. The very same principle or instinct which, in the misfortune of our neighbour, prompts us to compassionate his sorrow; in our own misfortune, prompts us to restrain the abject and miserable lamentations of our own sorrow. The same principle or instinct which, in his prosperity and success, prompts us to congratulate his joy; in our own prosperity and success, prompts us to restrain the levity and intemperance of our own joy. In both cases, the propriety of our own sentiments and feelings seems to be exactly in proportion to the vivacity and force with which we enter into and conceive his sentiments and feelings.[51]

Sympathy rests above all on the ability to place oneself in the position of another. In this respect, it would seem to be the exact opposite of self-interest. Yet at times Smith makes not sympathy, but prudence, to be the foundation of values. In the *Wealth of Nations*, he writes that 'though the principles of common prudence do not always govern the conduct of every individual, they always influence that of

[50] *The Theory of Moral Sentiments*, VII.ii.3.21.

[51] Ibid., III.iii.34. On sympathy see in particular discussions in Christel Fricke, 'Adam Smith: The Sympathetic Process and the Origin and Function of Conscience', in C. J. Berry, M. P. Paganelli, and C. Smith eds, *The Oxford Handbook of Adam Smith* (Oxford, 2013), 177–200; and Charles L. Griswold, Jr, *Adam Smith and the Virtues of Enlightenment* (Cambridge, 1999), ch. 2.

the majority of every class or order'.[52] If prudence is equated with self-interest,[53] then it would seem that Smith's statement later in the *Wealth of Nations* should be taken as effectively a restatement of Mandeville:

> The natural effort of every individual to better his own condition, when suffered to exert itself with freedom and security, is so powerful a principle, that it is alone, and without any assistance, not only capable of carrying on the society to wealth and prosperity, but of surmounting a hundred impertinent obstructions with which the folly of human laws too often incumbers its operations; though the effect of these obstructions is always more or less either to encroach upon its freedom, or to diminish its security.[54]

But in fact the association is questionable: self-interest is certainly a species of prudence, but not constitutive of it, and in any case, it is only one among a number of interests that Smith identifies as influencing people's behaviour.[55] What then is the relationship between prudence and sympathy? On Smith's account they have a different standing. Prudence, along with justice, humanity, generosity, and public-spiritedness for example, is a virtue. But sympathy is not a virtue as such. Rather, it provides a criterion by which one judges the appropriateness of behaviour. Whether self-interested behaviour is appropriate in particular circumstances is something to be determined on the basis of sympathy, in the form of the judgement at which an impartial spectator would arrive.[56] In particular, what is essential to any society, including a commercial one, is not a virtue such as benevolence, but justice: this is what the impartial spectator provides.[57]

This approach enables one to 'tame' prudence, in the form of self-interest, rather than simply denying it a role, by controlling the role it can play. Self-interest is subject to a moral constraint which balances various interests and virtues. The problem is that a fixed range of moral values, with a single form of control—sympathy as manifest in the impartial spectator—regulating the appropriate balance, does not appear to leave much room for variation. Moreover, his form of moral pluralism is too limited to take account of the moral diversity uncovered in the travel and historical literature. The difficulty becomes more evident when we consider that the dilemma of conflicting values that Mandeville poses is a problem generated by a transition from an agricultural society to a commercial one, and the shift in values that this transition brings with it, rather

52 Smith, *An Inquiry into Nature and Causes of The Wealth of Nations*, II.2.36.

53 As in George J. Stigler, 'Smith's Travels on the Ship of State', *History of Political Economy* 3 (1971), 265–77.

54 Smith, *An Inquiry into Nature and Causes of The Wealth of Nations*, IV.v.b.43.

55 See Amartya Sen, 'Adam Smith's Prudence', in S. Lall and F. Stewart, eds, *Theory and Reality in Development* (London, 1986), 28–37.

56 See Pratap Bhanu Mehta, 'Self-Interest and Other Interests', in Knud Haakonssen, ed., *The Cambridge Companion to Adam Smith* (Cambridge, 2006), 246–69: 246.

57 Smith, *The Theory of Moral Sentiments*, II.ii.3.3. See the discussion in Eugene Heath, 'Adam Smith and Self-Interest', in C. J. Berry, M. P. Paganelli, and C. Smith eds, *The Oxford Handbook of Adam Smith* (Oxford, 2013), 241–64.

than a problem that has a purely synchronic dimension, as if it were confined to a conflict of values within a particular society. Yet Smith's solution is very much within the latter genre, paralleled by the fact that he treats morality in terms of individual morality.

I want to compare Smith's response with what is in many respects the diametrically opposed response of Bentham, who devises an account of collective morality and then applies it to individual morality, avoiding many of the difficulties that Smith's account faces, but offering a solution that comes at a great cost. But to give some idea of the range of possible responses, it will help to look first, albeit briefly, at an account that takes synchronic questions seriously, and takes seriously the idea of collective entities generating their own values, but conceives of these antagonistic values ultimately being resolved in the form of a set of universal individual values which are compatible with those that Smith envisages.

Like Rousseau, Marx believed that communities had evolved towards the kind of commercial society, capitalism, that Mandeville describes. Moreover, he takes seriously the idea of collective entities having distinct sets of values and interests. A crucial difference from Mandeville, however, lies in his insistence that capitalism is characterized not by a single collective entity, but by a number of economically defined classes, the antagonism between two of which define the nature of the society. One of these is the economically and politically dominant class. The other is the class identified with wage-labourers, that is, those who earn their living by selling their ability to labour. For Marx, capitalism is not the end point of the evolution of society, and it is the dynamics of the relation between these two classes that determines not only the fundamental shape of capitalist society, but also the possibility of its future transformation. This future transformation is largely a function of two factors. First, what lies at the heart of the antagonism between the two main classes is the kind of economic relation in which they stand to one another, namely the purchase by members of one class, not of the products of labour, but of the ability to labour for a specified length of time. This results in a specific kind of dependence of one class on the other which, Marx considers, inevitably generates fundamental antagonisms. Second, Marx has a conception of labour as a form of self-fulfilling activity, which we will be looking at in Chapter 7. This acts as an absolute model, on which a satisfactory form of society can be envisaged, one in which one's labour can be genuinely fulfilling and one's life is not prevented from being satisfying by the kinds of factor that militate against this in capitalist society. Marx conceives this form of society not only as a genuine goal, but also as a direction in which the antagonisms between capital and labour will inevitably lead.[58] Note that it is a society in which morality effectively returns to the individual level, one in which non-antagonistic interests replace

[58] In this sketch I have run together some things from different stages of the development in Marx's thought. For a more nuanced account, see Stephen Gaukroger, 'Romanticism and Decommodification: Marx's Conception of Socialism', *Economy and Society* 15 (1986), 287–333.

those of class. In short, on this view moral diversity arises only while society is evolving, but it diminishes as moral values now themselves come to reflect genuine human needs and goals. The problem is that, until such a state of affairs is brought about (assuming it ever can be), political considerations effectively replace moral ones, with potentially dire consequences.[59]

One important feature of Mandeville's account is the way in which he refuses to subordinate the collective values of a commercial society to individual morality. This is also a features of Bentham's account, which provides a solution to Mandeville's dilemma by granting collective values the autonomy that individual values traditionally had. Rather than trying to make sense of collective goals in a vocabulary borrowed from the study of individual morality, Bentham sets out to formulate an independent account of collective values. The account has a number of striking features. First, unlike Mandeville and Smith, for example, it rejects the idea that sensibility, or sentiment, or passions are what motivate and regulate morality, making it depend entirely upon reason: a move in the opposite direction to that in which British thinkers had gone, but in a direction that has echoes in Turgot and especially Condorcet. Second, it severs the connection between morality and intentionality, which had been one of the fundamental premises of moral philosophy, incorporated into law as the principle of *mens rea*. Third, in severing this connection it opens up morality to a quantitative assessment. It is Bentham who, at the end of the eighteenth century, consolidates in a novel and unexpected way a programme that Petty had initiated at the end of the seventeenth century, where the catalyst is not general philosophical reflection on the nature of morality, but the specific problem of how one thinks about the values of commercial society. In particular, it is not something that derives from within an autonomous philosophical tradition, as later defenders of utilitarianism in general and consequentialism in particular have urged, in an attempt to decontextualize utilitarianism and make it a purely conceptual ahistorical form of ethics.[60] The move to commercial society has a general political dimension: it is a manifestation of the move from the idea of governments conforming to absolute standards of virtue and justice to *raison d'état* or 'national interest' conceptions whereby the concern of government is with the preservation and enlargement of the state.[61] It is the central role of commerce in the latter that makes the shift to commercial society part of a fundamental transformation.

Bentham's primary concerns were with the improvement of the decisions of politicians and legislators—with punishment and reward, legal reform and

[59] See ibid., 287–92 on the problems of 'socialist morality'.

[60] It was Henry Sidgwick who began the attempt to construct an autonomous philosophical genealogy for consequentialism in his *The Methods of Ethics* (London, 1874).

[61] See Maurizio Viroli, *From Politics to Reason of State: The Acquisition and Transformation of the Language of Politics, 1250–1600* (Cambridge, 1992). Cf. Richard Tuck, *Philosophy and Government, 1572–1651* (Cambridge, 1993); and John G. A. Pocock, *Virtue, Commerce, and History* (Cambridge, 1985), ch. 6.

codification—within a broadly moral context, as the title of his best-known work, *An Introduction to the Principles of Morals and Legislation*,[62] indicates. Although first published in 1789, it received little attention in England until the following century. It was in France that Bentham's works were more immediately known and celebrated: a number of these, including some not previously published in English, appeared in translation, and he was made an honorary citizen of the French Republic in 1792. In the *Principles of Morals and Legislation*, Bentham set out a number of distinctive positions. A key argument, one on which a good deal else depends, is his rejection of natural law theories, which he labels 'nonsense on stilts'.[63] Natural law theories typically held that moral values derived ultimately from the nature of human beings and their interaction with the world: the task of all accounts of human behaviour, from moral theory to legislation, is to identify and promote these values.[64] Moreover, legislation is intimately tied to moral philosophy, in that its principles must be consonant with those that moral theory derives from natural law. This where the link between personal morality and legislation traditionally came, for the principles enunciated in theories of natural law were very much those of individual morality, and this was what secured the authority of laws. Bentham severs the connection between an independently given morality and legislation, arguing that the authority of law cannot be due to its satisfying moral constraints deriving from some assumed natural law. The law needs no external source of legitimacy for Bentham, whether natural or divine. Rights, in particular, do not exist prior to the legislation of the sovereign or government: they are created by the command of the sovereign or the government.[65] Theories that posit an initial social contract from which basic rights derive, Bentham argues, are not only completely fictional but get matters the wrong way round, ignoring the fact that such rights could not be enforced unless there were already a government.

If the rules that govern the political and legal realms can only be conventional (by contrast with divine and natural law theories), then these cannot rely on any anterior principles of personal morality. Moreover, since they are features of state behaviour, which is collective rather than individual (and clearly not

[62] Jeremy Bentham, *An Introduction to the Principles of Morals and Legislation* (London, 1789). The work appeared in the collected works as *An Introduction to the Principles of Morals and Legislation*, ed. J. H. Burns and H. L. A. Hart (Oxford, 1970). This new collected works is still in progress. The only full collected edition to date is *The Works of Jeremy Bentham*, ed. J. Brownring (11 vols, Edinburgh, 1838–43). The *Principles of Morals* is in vol. i.

[63] Bentham, 'Anarchical Fallacies' (1843) in Bhikho Parekh, ed., *Bentham's Political Thought* (New York, 1973), 269. See the discussion in Ross Harrison, *Bentham* (London, 1983), chs 2 and 4.

[64] Having said this, note that there is a significant variation in just what 'natural law' was taken to mean in the seventeenth and eighteenth centuries: see Lorraine Daston and Michael Stolleis, 'Introduction: Nature, Law and Natural Law in Early Modern Europe', in L. Daston and M. Stolleis, eds, *Natural Laws and Laws of Nature in Early Modern Europe* (Aldershot, 2008), 1–12.

[65] I am simplifying someone here: there turn out to be a small number of rights that cannot be commanded. See David Lyons, *In the Interest of the Governed: A Study of Bentham's Philosophy of Utility and Law* (Oxford, 1973).

representative of any universal norms on Bentham's account), they are autonomous with respect to individual morality. Individual morality not only loses its power to regulate the kind of collective behaviour at stake in political and legal decision making, however. It also loses its conceptual centrality in our understanding of what morality is. To the extent to which collective decisions or agreements on political and legislative matters have a moral dimension, a model taken from individual morality is clearly inappropriate. Yet collective activity does have a moral standing for Bentham. The key lies not in intentionality, but in consequences. The assessment of whether an action is right or not is simply a function of the consequences of that action. It is a matter of using the consequences of various forms of behaviour to appraise their moral standing.

Such an approach depends on two factors. We need to identify what it is that we assess the consequences of behaviour with respect to. And, something distinctive of consequentialism, we need a means of determining how successfully we have achieved this. On the first question, Bentham adopts a criterion—pleasure and pain—that had been adumbrated at various times in the history of moral philosophy, most notably by Epicurus in antiquity, and by Hobbes and Hume in the modern era. But the way in which he deploys it is quite original. He combines this with a second criterion, 'the greatest happiness of the greatest number', again not original,[66] but again deployed in a novel way. On the account he proposes, the moral value of an act not only depends on its consequences, but the value of these consequences is a function of the pleasure or pain that they provide. This connects directly with his answer to the second question, for the importance of pleasure and pain in Bentham lies in the fact that he considers them to be, if not straightforwardly quantifiable, then at least amenable to quantitative comparison, and this allows him a comparative measure of value. Accordingly, he can then argue that those actions are best that maximize benefits. However, in determining the value of some action, we do not restrict ourselves to benefits for us: it is a fundamental principle of his utilitarianism that benefits for us must count equally, must be of equal value, as similar benefits for any other person.

Seen as an answer to Mandeville's dilemma, this approach has a number of strengths. Not least, it attempts to allocate moral value to collective behaviour in a coherent way. It is not a question of a triumph of commercial values at the collective level, but rather a matter of how one accounts for how values are generated outside the individual realm. Moreover, in accounting for the generation of collective values, Bentham can avail himself of the fact that collective behaviour is quantifiable in a way that individual behaviour is not. This gives him an absolute standard, sweeping to one side worries about moral relativity, because there now is a single procedure for the moral worth of actions that is universal. Because ethical behaviour, as in Kant, is a function of a set of abstract universal

[66] It had been proposed earlier in Cesare Beccaria, *Dei delitti e delle pene* ([Milan], 1784).

rules or maxims, rather being concerned with concrete prohibitions or injunctions, problems of moral diversity do not arise, because there is no varying content that they have to come to terms with.

Collective behaviour is naturalizable and quantifiable in a way that individual behaviour is not. As such, it is able to offer a new route to the unification of ethics that was not open to Mandeville, one that avoids Mandeville's dilemma. Instead of trying, *per impossibile*, to extrapolate from the individual level to collective values, consequentialism works in the opposite direction, extrapolating from collective to individual values. Just how far Bentham himself was prepared to follow the path that he sets out is a matter of dispute. He saw the need for a deontology in the case of personal morality, in order to do justice to pressing questions of motivation that arise there, and although his account of it was notoriously unsatisfactory,[67] David Lyons has made the case that Bentham had a dual standard, 'with community interest the criterion of right and wrong in public or political affairs and personal interest the proper standard for "private ethics"'.[68] But whatever Bentham's own hesitation, a utilitarian consequentialism of the kind that Bentham sets out was subsequently developed into a general account of ethics by later utilitarians such as James Mill, John Stuart Mill, and Henry Sidgwick. This account was general in the sense that it covered collective and individual morality without distinction. In this unified understanding of morality, it is an account that is developed to resolve problems in the collective case that becomes the general model. Questions of intentions and character are jettisoned in favour of something far more straightforwardly observable, namely the consequences of acts.

Modelling human behaviour in terms of collective properties, as opposed to individual or universal ones, generated a new kind of conceptual space which brings with it a rethinking of traditional conceptions. When Petty talked of examining human behaviour in terms of 'number, weight, or measure', it is unlikely that he foresaw that moral discourse would be completely reconfigured along these lines. Consequentialism offered a form of naturalization of moral discourse, one that brought with it the promise of quantitative techniques for the assessment of moral worth that bypassed religious and humanistic instruction on the cultivation of a moral *persona*. At the same time, the transformation of moral discourse, by Bentham's successors in the course of the nineteenth century, was mirrored in the rise of concerns about what the quantitative study of collective properties can reveal to us about human nature, and what shapes or constrains human behaviour. Everyday life slowly becomes naturalized, and the vehicle for this naturalization is quantification.

[67] See Harrison, *Bentham*, ch. 10.

[68] Lyons, *In the Interest of the Governed*, vii.

THE QUANTIFICATION OF EVERYDAY LIFE

In the last two decades of the eighteenth century, we witness the beginnings of the extensive use of social arithmetic in dealing with political and social questions. We have just seen how Bentham was proposing a form of calculation of the morality of acts. At the same time, Condorcet was defending the inevitability of progress and showing how the route to this lay in the transformation of societal choice into a rational decision-making calculative process. There was no unanimity here on where calculation was appropriate. Commenting on Condorcet's 'Jury Theorem' (which states that the likelihood of a collective decision being correct converges towards 100 per cent as the number of participants increases indefinitely), Bentham comments that a 'prejudice' becomes the 'axiom' that: 'if any instance of the ends of judicature failed of being fulfilled, it was for want of a sufficiently great multitude of judges. We have a book, my Lord, on this subject, by *Condorcet*: a quarto volume with 460 well-filled pages in it; all algebra, all demonstration, and this axiom (Preface, p. 24) a basis of it.'[69] But if they could not agree on its application, there is no doubt that both Condorcet and Bentham were committed to social arithmetic. Similarly with those who questioned the conclusions of this type of reasoning: Malthus, in response to the social-arithmetical arguments of Condorcet and William Godwin for the inevitability of progress, for example, used the same kind of reasoning to yield the opposite conclusion, offering a mathematical argument against the idea of inevitable progress.

This was not the first time that social arithmetic had been proposed as a way to decide political and social programmes. In the last decades of the seventeenth century, Petty had calculated the number of clergy requisite for England based solely on land area, and he had proposed, on the basis of a calculation of the productivity of land in Ireland, that the whole population, with the exception of a few cowherds, should be shipped to England.[70] In 1702, Daniel Defoe proposed a national pension office in which the population—that is, 'all sorts of people, who are labouring people, and of honest repute'—would be obliged to register by name, occupation, and place of residence and make regular contributions to a central fund.[71] Nor was it the first time notions of calculation had been referred to in connection with morality: in 1728, Hutcheson writes about 'computing the *Quantities* of Good or Evil'.[72] What was distinctive about the late

[69] 'Letters to Lord Greville on the Proposed Reform in the Administration of Civil Justice in Scotland', Letter II. Bentham, *Works*, v. 19.

[70] Porter, *The Rise of Statistical Thinking*, 19–20. See also Julian Hoppit, 'Political Arithmetic In Eighteenth-Century England', *Economic History Review* 49 (1996), 516–40.

[71] Daniel Defoe, *Essays upon Several Projects, or Effectual Ways, for Advancing the Interests of the Nation* (London, 1702), 146–69.

[72] Hutcheson, *An Essay on the Nature and Conduct of the Passions and Affections*, 40.

eighteenth-century arguments was rather just how many of them turned on questions of social arithmetic in one form or another. More generally, there was a quantification of everyday life, in which statistics played a key role, and I want to focus here on two aspects of everyday life that came to be investigated in statistical terms: population and health.

Malthus was famously concerned with the problem of overpopulation, but throughout the eighteenth century there was no agreement on the direction in which population numbers were tending. In fact, what worried those concerned with questions of polity and economy most was underpopulation. The question of population was intimately connected with that of wealth. From the sixteenth to the mid-eighteenth century the dominant theory of wealth was mercantilism, which, in its simplest form, was the view that the wealth of a nation lay in the amount of specie or bullion that it had. Countries with gold and silver mines either in their own territories or in their colonies or conquered territories clearly had reserves of specie, but so too did countries like England, which had no significant mines but did have a large trade surplus. More developed versions of mercantilism focused on strengthening the nation's economic well-being by establishing a positive balance of trade, primarily through high tariffs. Mercantilism was really a form of absolutism, with the economy under strict control of the monarch, and one distinctive form of regulation, complementing high tariffs and export subsidies, was the lowering of prices in times of shortage. In mid-eighteenth-century France, there were a growing number of criticisms of these policies as economically counterproductive: lowering of prices for example discouraged production, which was exactly the opposite of what was needed in times of shortage. The theoretical basis for an alternative account of the generation of wealth was proposed by Quesnay, in his *Tableau Économique* of 1758,[73] and Quesnay and Turgot became the main proponents of a movement they called physiocracy. The term literally meant 'government of nature'. Quesnay advocated what he considered to be a natural division of the population to a class of landowners, a 'productive' class of agricultural labourers, and a 'sterile' class of artisans and merchants. Income flow in the economy depended on the balance between rent, wages, and purchases, and the expenditure of the landlords on agricultural surpluses and the products of artisans provided the motor of the economy. The key to everything here was the ability of agricultural labour to produce surpluses.

It was agricultural labour, as opposed to artisanal labour for example, that mattered because eighteenth-century Europe was an 'advanced organic' economy, as opposed to the mineral-based economy that developed there in the nineteenth century, when energy needs began to be met by the far more efficient burning of

[73] François Quesnay, *Le Tableau économique* (Paris, 40.1758). On Quesnay, see Elizabeth Fox-Genovese, *The Origins of Physiocracy* (Ithaca, NY, 1976).

fossil fuels.[74] It depended on agrarian production for its food and energy needs. Population was clearly important: a decline in the number of agricultural labourers, in particular, meant a decline in the economy. This much was common ground. As Rousseau remarks:

> What is the end of political association? The preservation and prosperity of its members. And what is the surest mark of their preservation and prosperity? Their numbers and population. Seek then nowhere else this mark that is in dispute. Other things being equal, the government under which, without external aids, without naturalization or colonies, the citizens increase and multiply most, is beyond question the best. The government under which a people wanes and diminishes is the worst. Calculators, it is left for you to count, to measure, to compare.[75]

This prompts the question of how population had changed over the course of time. Montesquieu, for one, believed that it had diminished drastically:

> There are some who claim that the city of Rome alone used to contain more people than the greatest kingdom in Europe does now. There were some Roman citizens who had ten, and even twenty thousand slaves, without counting those who worked at their country houses: and as they estimate that there were four or five hundred thousand citizens, we cannot fix the number of its inhabitants, without shocking imagination itself. In Sicily there were formerly powerful kingdoms, and numerous nations, who have since disappeared. This island is now noted for nothing but its volcanoes. Greece is so deserted, that it doesn't have the hundredth of its ancient inhabitants. Spain, formerly so well-populated, now has almost only uninhabited areas, and the population of France is nothing in comparison with that ancient Gaul as described by Caesar. The northern countries are very depleted. They no longer need to divide and to send out, like swarms, colonies and whole nations to look for new places to live. Poland, and the European part of Turkey have hardly any population. In America we find not even a fiftieth of the men who once formed such great empires there. Asia is scarcely in a better state.[76]

Montesquieu's estimates of modern populations are little more than uninformed guesses and he alters them quite radically between editions of the *Lettres persanes*, but the claim of a drastic decrease in population remains. The problem was how one determined in what direction population numbers were tending. Before the emergence of regular national censuses, this was difficult to determine, although there was some progress as notice began to be paid to the conditions under which populations vary. Mid-century authors such as John Brown and Robert Wallace

[74] See Edward Anthony Wrigley, *Continuity, Chance and Change: The Character of the Industrial Revolution* (Cambridge, 1988); Craig Muldrew, *Food, Energy, and the Creation of Industriousness: Work and Material Culture in Agrarian England, 1550–1780* (Cambridge, 2011). On the vagaries to which agricultural economies were subject see Geoffrey Parker, *Global Crisis: War, Climate Change and Catastrophe in the Seventeenth Century* (New Haven, Conn., 2013). On the effects of the failure to move to a mineral-based economy in the East, see Kenneth Pomeranz, *The Great Divergence: China, Europe, and the Making of the Modern World Economy* (Princeton, 2000).

[75] Rousseau, *Du Contrat Social*, Book III.9.

[76] Montesquieu, *Lettres persanes* ('Cologne' [i.e. Amsterdam], 1721), Lettre XCII.

used London Bills of Mortality to argue that the luxuries of big cities led to a decline and enfeeblement of their populations,[77] while Mirabeau and Benjamin Franklin, reflecting on the wide open spaces of North America, noted the almost unlimited opportunities for population growth there.[78] Such considerations posit a relation between two variables, population and environment, but the latter incorporates, as well as space, a third variable that was also crucial to discussions from mid-century onwards, namely availability of means of sustenance. A more specific question is now able to be raised: what is the relation between rate of growth of population and rate of growth of resources?

This is the question that Süßmilch—whom we have already encountered in Chapter 4 on the question of the origins of language—set himself in a 1741 work attempting to establish a divine order in changes in population.[79] In his *Die göttliche Ordnung,* Süßmilch calculates that the population at mid-century was around one billion, and that the fully habitable surface, used as efficiently as possible, would sustain a population of between four and five billion; making various assumptions about the rate of population increase over each generation, he concluded that environmental limits would not be reached until some time in the distant future, and so should not be a source of concern.

Building on Süßmilch's work, Malthus, in his *On the Principle of Population*,[80] drew some novel and unsettling contrary conclusions from his study of population. The book's principal target was the doctrine of the perfectibility of man, as defended in Paris by Condorcet and in London by Richard Price and William Godwin.[81] In his *Esquisse*, published posthumously in 1795, Condorcet had foreseen a world in which reason had finally overcome the obstacles to human improvement, and he had seen the path to this as lying in the application of mathematical reasoning to social problems. Demography and vital statistics played a crucial role in this, as they had for his patron Turgot, who had striven to reform public administration, one of his most notable policies being the advocacy of free markets to alleviate food shortages. In an effort to introduce the rational reform of social and political policies, Condorcet sought statistical information on a wide range of phenomena from population and health to weather and productivity. One of his earliest supporters in England was the radical Christian polymath Richard Price. Among Price's many interests was

[77] John Brown, *Estimate of the Manners and Principles of the Times* (2 vols, London, 1757–8); Robert Wallace, *Dissertation on the Numbers of Mankind in Ancient and Modern Times, in which the superior populousness of Antiquity is maintained* (Edinburgh, 1753).

[78] See Robert J. Mayhew, *Malthus: The Life and Legacies of an Untimely Prophet* (Cambridge, Mass., 2014), ch. 1, to which I am indebted here. See also Charles E. Strangeland, *Pre-Malthusian Doctrines of Population: A Study in the History of Economic Though*t (New York, 1904).

[79] Johann Peter Süßmilch, *Die göttliche Ordnung in den Veränderungen des menschlichen Geschlechts, aus der Geburt, Tod, und Fortpflanzung desselben erwiesen* (Berlin, 1741).

[80] Thomas Malthus, *On the Principle of Population, as it affects the Future Improveent of Society. With Remarks on the Speculations of Mr. Godwin, M. Condorcet, and other writers* (London, 1798).

[81] See Mayhew, *Malthus*, ch. 2.

mathematics, and from around 1766 he was employed by the first mutual insurer in Britain, the Equitable Life Assurance Society, to quantify mortality rates so that life insurance schemes could be based on mathematically sound tables of life expectancy.[82] In his *Observations on Reversionary Payments* (1771),[83] he argues—on the basis of an extensive examination of the only records available in the absence of government bodies collecting statistics,[84] namely the parish registers of rural clergy and the London mortality bills of metropolitan physicians—that the population in England, and in London in particular, was in serious decline. He identified three reasons for this: in the case of the countryside, the taking over of agricultural land from yeomen who worked smallholdings for sustenance by squires whose motivation was self-interest; in the case of cities the unhygienic conditions, which led to high mortality rates; and finally, the luxuries of the city, which left people selfish and debased. Finally, there was William Godwin, whose *Enquiry Concerning Political Justice* aimed to re-establish connections between morality and politics. In the third edition,[85] while optimistic about a permanent balance between population, environment, and economy, he presented arguments to show that, were there to be any misalignment between the three, these could be circumvented by improvements in agricultural technology and in medicine.

But if Malthus' targets were such radical advocates of the perfectibility of man, his conclusions ended up alienating radicals and conservatives alike. The most famous argument of *On the Principle of Population* was that while food production tends to increase arithmetically (increments are generated by adding), population tends to increase naturally at a geometric rate (increments are generated by multiplying), meaning that, *ceteris paribus*, food production will not keep up with population growth.[86] A consequence of this is a severing of the long-standing association between an economically successful nation and a large population. Unlike Süßmilch, Malthus does not treat overpopulation as a hypothetical possibility whose realization is too far in the future to require measures to be

[82] See ibid., 305 on Price.

[83] Richard Price, *Observations on reversionary payments on schemes for providing annuities for widows, and for persons in old age; on the method of calculating the values of assurances on lives; and on the national debt* (London, 1771).

[84] Record keeping by the British government in the eighteenth century was very lax. This even extends to financial records. Although parishes were required to keep records of expenditure on poor relief from 1601 onwards, for example, parish officials were not required to submit such records until parliamentary committees of 1775 and 1786 insisted on this. See Dorothy Marshall, *The English Poor in the Eighteenth Century: A Study in Social and Administrative History* (London, 1926), 77.

[85] William Godwin, *Enquiry Concerning Political Justice and its Influence on Morals and Happiness* (3rd edn, London, 1797).

[86] The principle of geometrical population growth compared with arithmetic agricultural growth was only stated in the first edition, whereas it was argued for in detail on a statistical basis in the second edition. The first edition was somewhat sketchy in a number of respects, but this was remedied in later editions, where the size of the book grew significantly with the addition of a mass of statistical information and references.

taken now. Quite the contrary, he poses the problem in immediate terms. The idea of increasing food production to meet the problems, he argued, is simply not viable: reclaiming unused land and more intensive farming are very limited in what they can achieve, with rapidly diminishing returns. Bearing in mind that he (and his contemporary opponents) were working on the model of an advanced organic economy, even though this was something that was just about to change, his reasoning was sound. Agricultural produce is a very inefficient source of energy. The answer must lie therefore, he reasoned, in checking population growth: through late marriage, contraception, emigration, or by having numbers reduced through events over which the general population had no control, such as wars and epidemics. Questions of population had never been treated on the basis of such calculations before, ones devoid of any moral or religious considerations.[87] Malthus completely rejected Süßmilch's view that God—whose command was 'be fruitful and multiply'—balances questions of population, environment, and economy in a harmonious way, for example, and his reasoning does not engage questions of morality. Just as there was a fierce reaction to Mandeville's argument that we have to embrace self-interest, morally repugnant as this may be in our daily lives, if we are to reap the benefits of a commercial society, so too there was an equally fierce response to Malthus' argument that we have to embrace population control, by means that violate traditional family norms and values, if we are to avoid starving to death.

On the Principle of Population had one immediate effect: the urgent passing, after decades of procrastination, of the first Census Act in Britain, in 1800. National population now became a pressing issue, mirroring a parallel development in medicine, namely concern with the health of the nation. As we saw in Chapter 3, the terms of reference of the commission founded in 1776 by the French government to investigate the recent cattle plague included human and animal epidemics, as well as sanitary conditions and endemic and occupational diseases, and in 1778 it was formed into the Société Royale de Médecine with the aim of compiling a medical topography. The Société also dealt with broader questions such as environmental factors affecting health, including water quality and manner of life, and this was facilitated through the establishment of a network of national correspondents who reported on climatic and meteorological factors as well as social and domestic ones. In the hospital reforms following the Revolution, the hospitals were transformed into the locus of medical activity and research, and medical statistics began to be used extensively in the analysis of case histories and the evaluation of therapy.

[87] Nevertheless, in his dispute with Ricardo over the latter's construal of political economy as a mathematical discipline, Malthus insisted it was closer to the science of morals and politics. On the dispute, see Stefan Collini, Donald Winch, and John Burrow, *That Noble Science of Politics: A Study in Nineteenth-Century Intellectual History* (Cambridge, 1983), ch. 2.

The first use of medical statistics in deciding matters of health policy predates these developments, however. Smallpox epidemics peaked in England around 1720 and in France around 1750, and the standard procedure was the same as it had been earlier for the plague: quarantine and fleeing to less risky environments. Smallpox inoculation—which was already in use in China, Turkey, and parts of Africa—introduced a preventative measure, but there were intense debates over its effectiveness. It consisted in making a scratch in two limbs and placing a thread covered in pus from the virulent pock on the incisions.[88] A fever would subsequently develop along with some pock marks, and the patient would recover after about fifteen or sixteen days, with a significant measure of immunity against future smallpox epidemics, although in rare cases a severe and deadly case of smallpox was contracted.

Smallpox inoculation was first introduced in England in 1721, when six prisoners were inoculated (in exchange for regaining their freedom if they survived, which they did) as an experiment, before members of the royal family received inoculations.[89] Despite the absence of ill effects, there followed a long heated controversy.[90] The procedure involved giving a healthy patient an illness, violating the Hippocratic maxim to do no harm. James Jurin, physician and mathematician, Secretary to the Royal Society from 1721 to 1727, and publisher of their *Philosophical Transactions* over four years, published a number of pamphlets between 1723 and 1727 using statistical arguments to press the case for inoculation.[91] Taking numbers of natural smallpox deaths from the London morality bills, and making a number of judicious adjustments to correct for factors that skewed the figures, he calculated that the rate of death from natural smallpox in the first two decades of the century was one in seven or eight persons. He then gathered information—from the Royal Society records and from reports from Boston, Massachusetts where there had been an inoculation programme in 1721—on the number of deaths from inoculated smallpox, for the first years of inoculation, 1721 and 1722, and calculated the mortality rate as one in ninety-one.[92] The calculations were not simple. As Andrea Rusnock has noted, the creation of standard narratives 'from the numerous case histories involved the

[88] See Emanuele Timoni, 'An Account, or History, of the Procuring of the Small Pox by Incision, or Inoculation; As It Has For Some Time Been Practised at Constantinople. Being the Extract of a Letter from Emanuel Timonius, Oxon. & Patav. M.D. F.R.S. Dated at Constantinople, December 1713. Communicated to the Royal Society by John Woodward, M.D. Profes. Med. Gresh. and F.R.S.', *Philosophical Transactions* 29 (1714), 72–82.

[89] In this paragraph I draw on the account in Andrea A. Rusnock, *Vital Accounts: Quantifying Health and Population in Eighteenth-Century England and France* (Cambridge, 2002), ch. 2.

[90] See Geneviève Miller, *The Adoption of Inoculation for Smallpox in England and France* (Philadelphia, 1957).

[91] See Rusnock, *Vital Accounts*, 51 n.24 for a list of the pamphlets.

[92] James Jurin, 'A Letter to the Learned Caleb Cotesworth, MD, FRS, of the College of Physicians, and Physician to St. Thomas' Hospital, Containing a Comparison between the Mortality of the Natural Small Pox, and That Given by Inoculation', *Philosophical Transactions* 32 (1722–3), 213–24.

extraction of consistent, quantifiable information, which then had to be tallied and tabulated in established categories. Jurin's ratios of the number of inoculated and natural smallpox cases were thus the result of an arduous and, at times, controversial process of soliciting, selecting, and sorting varied case histories.'[93]

In France, it was the *philosophes*, notably Voltaire and La Condamine, rather than physicians, who were the advocates of smallpox inoculation. Moreover the arguments were rarely pursued in statistical terms, and on the one famous occasion on which they were, in the debate between d'Alembert and Daniel Bernoulli at the Académie des Sciences in 1760,[94] no agreement could be reached because, despite both d'Alembert and Bernoulli being advocates of inoculation, they could not agree on how to calculate the risks involved. On the death of Louis XIV from smallpox in 1774, the tide turned in favour of inoculation, but statistics played a little role in this.[95] By the early decades of the nineteenth century, however, the use of statistics was becoming accepted by many in the medical community, and Quetelet's statistical work in particular was being drawn upon by the 1820s. Quetelet insisted that in medicine the basic distinction was between what was normal and what was pathological, and that normality was something that must be established on statistical grounds.[96] The resistance to this approach, which was significant, is particularly revealing in what it brings to light not just about the assumptions of the anthropological approach to medicine that still existed in the 1820s, but more generally about just what medical practice is based on.[97]

The chief opponents of statistical approaches were vitalists, who envisaged, at best, a very limited role for statistics. As Elizabeth Williams points out, for the vitalist, 'medical phenomena were characterized by particularity, spontaneity, and variability. Since each case of disease was individual, medicine had to recognize and interpret minute nuances'.[98] Here the question of medical judgement arises. Is it just a question of keeping statistics so one can ascertain which treatment has

93 Rusnock, *Vital Accounts*, 55.

94 Jean d'Alembert: 'Onzième mémoire, Sur l'application du calcul des probabilités à l'inoculation de la petite vérole', *Opuscules mathématiques* (8 vols, Paris, 1761–80), ii. 26–95. Daniel Bernoulli, 'Essai d'une nouvelle analyse de la mortalité causée par la petite vérole et des avantages de l'inoculation pour le prévenir', *Mémoires de l'Académie Royale des Sciences, 1760* (Paris, 1765), Part 2, 1–79.

95 See Rusnock, *Vital Accounts*, ch. 3.

96 On the relation between biometric norms and statistical averages see Georges Canguilhem, *The Normal and the Pathological* (New York, 1991), 115–229.

97 The most comprehensive account of the debates is J. Rosser Matthews, *Quantification and the Quest for Medical Certainty* (Princeton, 1995). See also Bernard-Pierre Lécuyer, 'Probability in Vital and Social Statistics: Quetelet, Farr, and the Bertillons', in L. Krüger, L. Daston, and M. Heiderlberger, eds, *The Probabilistic Revolution, Volume 1: Ideas in History* (Cambridge, Mass., 1987), 317–35.

98 See Williams, *The Physical and the Moral*, 155. One of the chief spokesmen for this view was Frédéric Dubois d'Amiens, 'Discussion sur la statistique médicale', *Bulletin de l'Académie royale de médecine* 13 (1837), 684–714.

been the most effective in the past, or is it rather a question of the physician using his training and skills to exercise his professional judgement on individual cases? As Cabanis noted, medical practice could only have been built up from its beginnings in the ancient world through trial and error, building on successes and transforming what had been merely probable into virtual certainties.[99] But illness was not an unvarying phenomenon; even in modern medicine diagnosis was constantly faced with exceptions to the rule, and this was all the more reason to value medical training reinforced by long experience. Medicine was like art, where 'certainties are found in the very sensations of the artist, rather than in the principles of the art'.[100] The practice of medicine, he argues, requires all kinds of procedures wholly different from those that can be regulated by a simple formula: unlike physics, where calculated probabilities may have a place.[101] But there was also an especially delicate problem for anthropological medicine. On the kind of holistic model to which it aspired, with its explicit connections between the treatment of mental ailments and physical ones, it would be impossible to quarantine a statistical approach so that it was confined to bodily ailments. Anthropological medicine would then be open to the charge of materialism, and the reduction of mental processes to numerical operations. This ran completely contrary to the thrust of anthropological medicine, which was not a form of materialism.

The difficulties that arise here have parallels with Mandeville's dilemma. Statistical methods are applicable to collective properties, but collective properties, by contrast with distributive ones, are quite unlike individual ones. Yet there is an expectation that the study of collective properties throws light on individual ones. We have seen how Bentham offered a collective model for individual ethics. But in 1835, in his *Sur l'homme*,[102] Quetelet offered an even more ambitious programme for the modelling of the individual on aggregate properties, proposing that social and medical enquiry be centred around the idea of the 'average man'. It is in Quetelet, even more than in Bentham, that Petty's exhortation to count, weigh, and measure finds a comprehensive vindication and development.

THE AVERAGE MAN

Quetelet studied pure mathematics at the University of Ghent, but in 1820 was active in the move to set up a new observatory in Brussels, and while in Paris in the early 1820s consulting Arago and others on astronomy and meteorology in preparation for this, he learned about probability theory and its applications from

[99] Cabanis, *Du degré de certitude de la médecine* (Paris, 1798), 40.
[100] Ibid., 117. [101] Ibid., 87.
[102] Lambert Adolphe Jacques Quetelet, *Sur l'homme et le développement de ses facultés, ou essai de physique sociale* (2 vols, Paris, 1835).

Fourier and Laplace. In 1826 he became a regional correspondent for the Statistical Bureau of the Low Countries (i.e. modern day Belgium and Holland), working on population data for a planned 1829 census, and applying techniques from probability theory that he had learned from Laplace, using procedures that were later to be refined as random sampling to save time and effort in collecting statistics.[103] From 1827 onwards he set out to compile tables and draw graphs correlating all kinds of pairs of variables: birth and death rates by month and city, by temperature and by time of day; month of conception compared to time of marriage; age of mortality by profession, season, and profession as well as among prisoners and hospital patients; and so on. The *Annales d'hygiène publique et de médecine légale* was publishing Quetelet's work, and in 1831 an extract from his study of crime appeared. Quetelet had compiled and analysed statistics for various forms of antisocial behaviour, notably crime and drunkenness, in particular statistically defined groups, and he had produced figures for the propensity to crime in various groups, noting for example that, with respect to age groups, propensity for crime peaked in the twenty to twenty-five age range. In 1837 a new journal, *Mémoires de la Société médicale d'observation* appeared, founded by Pierre Charles Alexandre Louis, who had compiled statistics on large groups of patients with pneumonia at La Charité hospital, and on the basis of these had rejected the use of bloodletting as a form of treatment.[104] Louis promoted the use of statistical methods in medicine, and Quetelet, who was regarded in the medical community as the leading authority on the use of statistics by this time, was a regular contributor to the *Mémoires*.

Quetelet brought his various studies together in his book *Sur l'homme et le développement de ses facultés, ou essai de physique sociale*.[105] It was quickly translated into English, although the reference to 'social physics', something perhaps invoking Condorcet and the programmes of the French Revolution although in fact it is borrowed from Comte, was not included in the English title.[106] The work begins with an account of human physical characteristics and gradually moves up to human moral and intellectual qualities. This conforms to a standard model of the time, especially in the tradition of the 'natural history of man'. But there are two distinctive features of Quetelet's treatment. First, it is true that elementary forms of statistics (such as population estimates) had been used as aids to enquiry, but these were incidental, whereas for Quetelet they are constitutive of the enterprise in the first two sections of the work: 'Development of the Physical Qualities of

[103] He subsequently changed his mind, being convinced (wrongly) by arguments against basing population estimates on incomplete samples: see Stigler, *The History of Statistics*, 163–9.

[104] Pierre Charles Alexandre Louis, *Recherches Sur Les Effets De La Saignée Dans Quelques Maladies Inflammatoires Et Sur L'action De L'émétique Et Des Vésicatoires Dans La Pneumonie* (Paris, 1835).

[105] See Maurice Halbwachs, *La théorie de l'homme moyen: Essai sur Quetelet et la statistique morale* (Paris, 1912).

[106] The translation appeared as *A Treatise on Man and the Development of His Faculties* (Edinburgh, 1842).

Man', 'Development of Stature, Weight, Strength etc'. Second, the 'natural history of man' treatments restricted statistics to physical features: there was never any sense that the statistics alone could tell us anything about moral and intellectual qualities. But in the third and fourth sections, on 'Development of the Moral and Intellectual Qualities of Man', and 'Of the Properties of the Average Man, of the Social System, and of the Final Advancement of this Study', this is precisely what Quetelet does. His endpoint marks a radical departure from traditional conceptions of the goals of enquiry into the human condition. Moreover, it is not just that the kind of understanding sought in *Sur l'homme* is different from that which earlier investigators had sought, but that it aims to displace other forms of understanding.

In describing his procedure in the first section, Quetelet writes:

> In order to keep faithfully by the plan which I had chalked out, I have viewed the individual from the hour of his birth; I have sought to determine, for that epoch, the different relations of bulk, subsisting between the various parts of his frame; and to ascertain how far these relations become modified during his development, what they are in the flower of his age, and in what position they remain up to the instant of decay. It is only by long and laborious study, and by the comparison of a vast number of individuals, that it will be possible to succeed in establishing correct average proportions for each age, and in settling the limits betwixt which they can be made to vary, without ceasing to be accurate and faithful to nature—our first and great guide in this difficult study.[107]

For Quetelet, in proceeding in this way, he is adhering to a scientific method. 'In giving to my work the title of Social Physics', he writes, 'I have had no other aim than to collect, in a uniform order, the phenomena affecting man, nearly as physical science brings together the phenomena appertaining to the material world. If certain deplorable facts present themselves with an alarming regularity, to whom is blame to be ascribed?'[108] Nevertheless, there is some circularity in his procedure. He attempts to establish norms statistically, but his criteria of selection harbour all kinds of assumptions about what is normal and what deviates from the normal. The range from which material is selected for his study is limited by what seem to be largely aesthetic conceptions: 'Although *artistical* limits will always be less extended than the *natural* limits, yet it is to be observed that, by the term natural limits, I understand those within which the human proportions may vary, not only without constituting deformities and monstrous aberrations from nature, but also without wounding the eye by want of harmony.'[109]

[107] Ibid., vi. [108] Ibid., vii.

[109] Ibid. 10. Later he tells us that artists and others must 'contrast the most different physiognomies and characters; but the truth must always lie between the extremes which they present to our view, and these extremes themselves lie within *limits* defined by nature. Going beyond, we only create fantastic beings and monstrosities' (98).

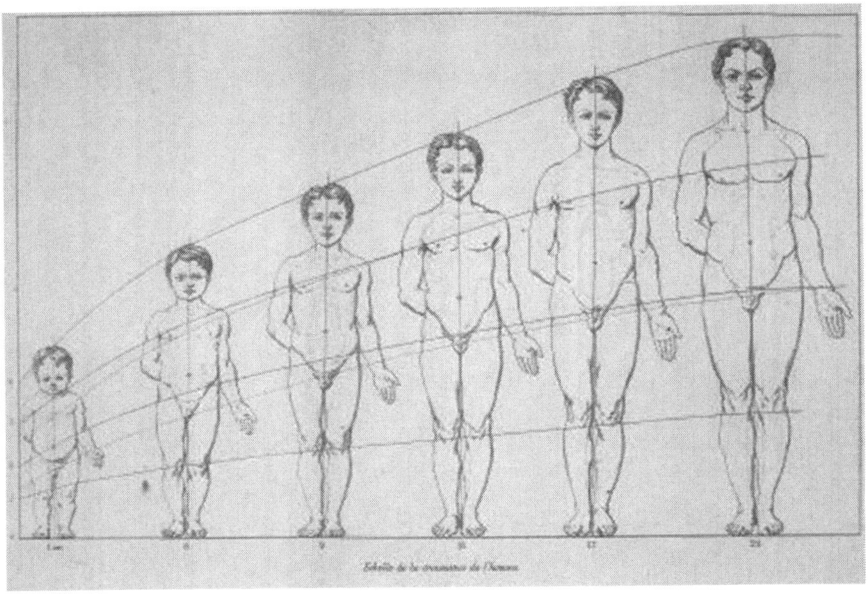

Fig. 6.1 Anthropometric curve. Quetelet, *Anthropométrie* (1871).

The kind of thing he has in mind is well illustrated later in *Anthropométrie* (1871), where he talks of a 'general curve' which represents a pattern of distribution for any quality—height (see Fig. 6.1), weight, strength, and speed for example—across any population, and for 'moral and intellectual qualities', as well as 'the law of growth: to man, to animals, to plants'.[110] The aesthetic conception suggests, however, that what Quetelet is seeking from his study is something that is more like an ideal type than an average. This is in fact in keeping with much scientific work of the time, in the representation of anatomical and botanical specimens for example. Anatomical illustrations of skeletons never showed the broken limbs and bone deformities that characterized the skeleton of a significant number of the population but only perfectly healthy specimens: and botanical illustrations in flora never showed the dead or wilting leaves, broken stems, or various forms of idiosyncratic asymmetry which are features of many plants that one observes in the field.[111] There are of course reasons for this, which turn on what one wants the representations for: instruction in identifying features, for example, would certainly be hampered by 'warts and all' illustrations. In this respect, Quetelet shares many of the simplifying assumptions of earlier accounts

[110] *Anthropométrie ou mesure des différentes facultés de l'homme* (Brussels, 1871), 8.

[111] See the discussion of these issues in Lorraine Daston and Peter Galison, *Objectivity* (New York, 2007).

that worked in terms of the 'rational man': the fact that information is presented in terms of graphs and tables rather than prose alters nothing. Above all, note the implication behind the aesthetic vocabulary: being closer to the average is to display a form of superiority. Whereas superiority was at one time almost by definition the possession of a few, it is now associated with the norm, as the average becomes the norm.

Quetelet's approach defines normality as the average over a predetermined range. The fact that the range is predetermined is less significant than the new association between normality and averages. His approach differs from the 'rational man' account in a number of other respects. For one thing, he confines any claims about understanding human beings to what can be directly observed. 'This study is a difficult one', he writes: 'Speculative philosophy has long been occupied with it; but there are questions not able to be resolved by such means; speculation has its limits, as observation also has.'[112] Quetelet's study is not directly of the human being, but 'the analysis of the moral man through his actions, and of the intellectual man through his productions'.[113] His account proceeds in terms of inferences from aggregate facts. This is crucial, for these allow a causal account in the way that consideration of individual facts does not. He strives, he tells us,

> never to lose sight of those scientific principles which ought to guide the observer in all his investigations. My aim is not to defend systems, or bolster up theories; I confine myself to the citation of facts, such as society presents to our view. If these facts be legitimately established, it follows that we must accept of and accommodate our reason to them. Now, what do these facts teach us? I repeat, that in a given state of society, resting under the influence of certain causes, regular effects are produced, which oscillate, as it were, around a fixed mean point, without undergoing any sensible alterations.[114]

Second, it extends well beyond physical characteristics. Quetelet sought uniformities in human behaviour, showing in what respects human beings were subject to the same kinds of unvarying causes that determined animal behaviour. In showing that rates of marriage, crimes, and suicides—what, in the vocabulary of the time, were termed moral phenomena—remained constant from year to year, he uses aggregate facts to uncover basic and unexpected features of human behaviour. Anthropological medicine, in its various forms, had always insisted on the integration of physical and moral phenomena, and as we have seen, devised the treatment of psychological and physical symptoms accordingly, but Quetelet is not moving between psychological and physical phenomena in the way that the *médecins philosophes* were concerned with these, for example. He is subordinating them both to methods of averaging over aggregates: his methods abstract from content in such a way that they do not vary whether he is dealing with the effectiveness of bloodletting in pneumonia patients or proclivity to suicide in the

[112] *A Treatise on Man*, viii. [113] Ibid., viii. [114] Ibid., vii.

twenty to twenty-five age range. And in the case of diagnosis he is firm: 'the physician is obliged to have recourse to the common standard, and compare his patient with the average man'.[115]

A long anonymous review of *Sur l'Homme* in the English journal *Athenaeum* in 1835 concluded: 'We consider the appearance of these volumes as forming an epoch in the literary history of civilization.'[116] Indeed they did, and we can single out two particularly significant nineteenth-century developments.[117] The first lies in the fact that the various averages that Quetelet works out are not averages for the human species, but averages for particular nations or races. As he puts it in his *Recherches statistiques*: 'One may ask if there exists, in a people, *un homme type*, a man who represents this people by height, and in relation to which all other men of the same nation must be considered as offering deviations that are more or less large.'[118] Quetelet introduced a new quantitative criterion of a race. Skull measurement and racial identification on the basis of typical skull shapes and sizes had been a staple of human comparative anatomy since the last decades of the eighteenth century, as we have seen, but Quetelet is offering something of a more systematically quantitative nature: a race is now characterized by the measurement of its physical and moral qualities, as represented in the 'average man' of that race. As I have indicated, his principles of selection of data can tend to generate ideal types, and in the case of race, these ideal types are reinforced and rendered 'objective' through quantitative techniques of averaging. This will prove a crucial spur to the development of eugenics later in the nineteenth century.

The second development concerns the transformation of an average, such as average height, which we can think of as an abstract result of an arithmetical operation, into a real feature of a population. This happens in the move from the average height of men to the height of the average man.[119] In 1817, for example, a report was published documenting differences in height and chest size in eleven Scottish regiments.[120] Concentrating on chest measurements, Quetelet obtained a distribution for 5,738 chests with maxima at 1,073 soldiers at thirty-nine inches and 1,079 soldiers at forty inches. He then makes the crucial move, maintaining that this is the same as taking 5,738 measurements of a single Scottish soldier. Not every single measurement will be exactly the same—just as in measuring the

[115] Ibid., 99. Quetelet qualifies this, but in a way that establishes it as the standard procedure.

[116] Anon, Review of Quetelet, *Sur l'Homme*, *Athenaeum* 8 August 1835, 593–5; 15 August 1835, 611–13; 29 August 1835, 658–61: 661.

[117] See Ian Hacking, *The Taming of Chance* (Cambridge, 1990), ch. 13, to which I am indebted here.

[118] Quetelet, *Recherches statistiques* (Brussels, 1844), 54. Compare *Anthropométrie*: 'The average man characterizes the nation to which he belongs' (16).

[119] As Hacking puts it, what happens is that Quetelet 'transformed the theory of measuring unknown physical quantities, with a definite probable error, into the theory of measuring ideal or abstract properties of a population. Because these could be subjected to the same formal techniques they became real quantities.' *The Taming of Chance*, 108.

[120] For details, see Stigler, *The History of Statistics*, 206–9.

positions of astronomical bodies, as we saw in Chapter 1, there will be a range of error determined by a tool for calculating standard deviation. The idea is that, in both the case of the 5,738 soldiers and that of the single soldier measured 5,738 times, there should be an identical bell-shaped curve. In the case of the single soldier, the value generated by the statistical methods is his real height. So too in the case of the regiments: what is generated by the statistical methods is something real. The significance of this lies in the way in which mathematically generated outcomes can become basic social laws. Statistical laws that were merely descriptive of large-scale generalities become transformed into physical and social laws that deal with real properties and events and causal processes.

Bentham and Quetelet both illustrate the way in which qualities that had been thought to be specific to the individual—moral responsibility, time of marriage, proclivity to crime or suicide—can be reworked in such a way that the study of collective or aggregate properties allows the elucidation of individual qualities. What makes these collective properties valuable as an object of study varies from case to case, but the way in which it allows causal accounts of behaviour is crucial. In the process, as I have tried to show, what happens is that this form of study can go beyond simply elucidating the study of individual human behaviour and actually replaces it. In doing so, it will provide resources for economic and political theory in the nineteenth century. In particular, with the replacement of political economy by neoclassical economics, beginning with the work of Jevons in the 1860s, the scope of economic theory narrowed. As Winch has put it, economics 'became a quasi-mathematical discipline in which important questions were posed as scarcity or choice-problems involving the maximisation or minimisation of strategic economic quantities under specified conditions'.[121] Here notions of rational agents modelled on aggregate behaviour came to the fore. But the study of collective or aggregate properties would also provide resources for the natural history of man, enabling physical anthropology to claim a 'scientific' standing for itself and to eclipse philosophical anthropology for much of the nineteenth century.

[121] Donald Winch, *The Emergence of Economics as a Science, 1750–1870* (London, 1971), 63.

PART III

7
The Naturalization of Religion

In reformulating questions in such a way that they become amenable to empirical investigation, the various naturalizing programmes that we have been considering of necessity replaced traditional concerns with new ones. It could never have been simply a matter of translating questions from a speculative form into an empirical one. Naturalization depends on what one identifies as the naturalizing resources and what one identifies as the appropriate sources of evidence, and these shape the questions one asks. This will be evident from the four forms of naturalization of the human that we have examined, for these not only pose different kinds of questions but are typically at odds with one another on fundamental issues. While there is no such thing as naturalization per se, however, there were commonalities: the four naturalizing strategies covered, in varying degrees, physical, intellectual, and moral life, and, as we have seen, in each case there were concerted attempts to integrate these in one way or another. But there was also a crucial difference between earlier eighteenth-century notions of the human being and the new naturalized ones. The earlier treatments had uniformly included an explicitly religious dimension in their accounts, not just in respect of morality, but in terms of understanding one's place in the realm of things, specifically in accounting for our non-propositional relation to the world, that is, how we encounter and come to terms with the world by means of such things as aspirations, desires, fears, raw beliefs, anxieties, responsibilities, and goals. Religious treatment of these questions was conspicuously absent from the naturalizing accounts.

There is one general category of non-propositional understanding that includes religion, but is broader than it: myth. Bringing non-propositional understanding under the rubric of myth will help us contextualize religious understanding and explore the alternatives that emerged in the early decades of the nineteenth century. Moreover, by thinking of the emergence of myth in terms of a recognition of the need for non-propositional forms of understanding the world, forms that were distinct from those of traditional Christianity, we can begin to understand how the 'age of reason' (a description that we have already had reason to question) could, at least in its Kantian form, have generated such a need.

It was in an explicit dismissal of myth that disciplines such as history (in Thucydides) and philosophy (in Plato) emerged in antiquity. Myth was rejected in favour of a form of discourse that had two features: it used argument and evidence to establish a case, and refused to accept ambiguity, trying instead to resolve conflicting accounts into contradictions between facts.[1] What, in more positive terms, it was that was being rejected here is not straightforward, as there are different characterizations of the nature and role of myth,[2] but for our purposes a characterization that has been proposed in the context of an examination of the role of mythology in classical culture also serves our needs well:

> (1) Myths rework, pare down, clarify and exaggerate experience; to say that they 'reflect' experience is quite inadequate. (2) Clarification is not only not incompatible with ambiguity, but can actually bring it into sharper relief . . . (3) Perceptions reworked in mythology feed back into ordinary life, even if the way in which this happens can be hard to specify. (4) In ritual, behaviour is articulated through symbols with a comparable selectivity to that found in myths. The two symbolic languages contrast with and complement each other.[3]

This conception of myth—which, in a condensed form, mirrors those devised from the late eighteenth century onwards, as we shall see—allows us to highlight two features that are central to the argument of this chapter. The first is that it gives us some idea of what a non-propositional form of engagement with the world might look like, and how it might achieve a measure of coherence, however limited, in contrast to the conception of non-propositional understanding as an intrinsically fragmented set of differently motivated attitudes to the world, something encouraged by a view of propositional knowledge as the ultimate form of understanding. In particular, it is not that non-propositional understanding lacks something that only propositional understanding can provide. The second is that it allows us to include those devotional forms of myth that can be termed religions, as well as non-religious kinds that might take a literary or other aesthetic form.

We can appreciate some of the complexities here by considering the fact that Christianity had always laid a claim to both non-propositional and propositional forms of understanding. Simplifying somewhat—and ignoring for the moment the potential inner complexity of religions that makes a simple separation of components problematic—the former traditionally took the form of devotional practices and the latter the form of theological dogma. But in the second half of the seventeenth century, in the work of Boyle for example, this division was abandoned. In seeking to avoid the sectarianism that he believed inevitably arose in disputes over doctrinal questions, Boyle had advocated that natural philosophy

[1] See Marcel Detienne, *Maîtres de vérité dans la Grèce archaïque* (Paris, 1990); and the summary account in Gaukroger, *The Emergence of a Scientific Culture*, 229–30.

[2] There have been particularly comprehensive and probing discussions of myth in a German context: Hans Blumenberg, *Arbeit am Mythos* (Frankfurt am Main, 1979); Kurt Hübner, *Die Wahrheit des Mythos* (Munich, 1985); Manfred Franck, *Vorlesungen über die neue Mythologie* (2 vols, Frankfurt am Main, 1982–8).

[3] Richard Buxton, *Imaginary Greece: The Contexts of Mythology* (Cambridge, 1994), 96. Buxton adds a fifth criterion which is specific to ancient mythology. Cf. Marcel Detienne, *L'invention de mythologie* (Paris, 1981).

be treated as a non-partisan means of exploring God's purpose in creation, an approach that peaked in the late seventeenth and early eighteenth centuries. But it is not just that natural theology comes to replace dogma in certain respects. For Boyle—as for many of his Protestant contemporaries and successors who, like Boyle, were at the forefront of natural philosophy, not mere onlookers—natural philosophy that was devoted to uncovering God's purposes in his creation was a form of devotion: nature becomes a 'temple' and its investigators 'priests'.[4] Clearly in this case, what was at issue was not the abandonment of a mythological understanding of the world: quite the contrary. More generally, the idea that the abandonment of a mythological understanding of the world is characteristic of the seventeenth and eighteenth centuries is questionable on a number of grounds. In the first place, such a view assumes that scientific and mythological understanding can simply be separated from one another, whereas once one rises above the level of the generation of particular results it becomes evident that—with very rare exceptions such as the project of the *Encyclopédie*—they are part of integrated programmes, which may differ from one another in fundamentals, but which share a common understanding that, if natural philosophy is to go beyond simply understanding natural processes and to aspire to be part of a project that throws light on our place in nature, then it must ally itself with a Christian understanding. Second, some of the most radical criticisms of Christianity in the eighteenth century, from Hume to Fichte, attempted to separate out early types of religion which are non-propositional from later types that have become propositionalized in the form of philosophy and theology, accepting the former while rejecting the latter. Here there is a reinforcement of the value of the mythological content of religion rather than a rejection of a mythological understanding of the world. Third, there is an assumption that traffic between the religious and the secular is one way. This is not the case. There are indeed projects that attempted to secularize Christian culture,[5] but Boyle's is an example of a project that attempts to Christianize a secular culture. Indeed, sometimes there is an ambiguity, and one can see the same programme as both secularization of the Christian and Christianization of the secular. Matters become especially complicated in the nineteenth century, where concepts of the sacred are imported from the philological and aesthetic study of myth into theology. In Germany there was an added complication as a dispute developed between those who saw theology as a university subject, and those who saw academic theology as a thorn in the side

[4] See Gaukroger, *The Emergence of a Scientific Culture*, 151–3. Cf. Peter Harrison, 'Experimental Religion and Experimental Science in Early Modern England', *Intellectual History Review* 21 (2011), 413–33.

[5] 'Secularization' in the modern sense—by contrast with traditional distinctions between secular and regular clergy for example—is a nineteenth-century notion, and what we are discussing here would not have been described in these terms in the eighteenth century. See Ian Hunter, 'Secularization: The Birth of a Modern Combat Concept', *Modern Intellectual History*, 12 (2015), 1–32.

of the Church. Proponents of the former view, such as Harnack, argued that theology conducted under ecclesiastical auspices was positively injurious to true learning, and that the state had an abiding interest in protecting theology against excessive ecclesiastical meddling.[6] In this case there is an ownership dispute as it were over theology, between secular and ecclesiastical authorities, but what was at issue was not secularization in any usual sense of the word.

In this chapter, we shall be concerned with changes in the mythological understanding of the world that accompanied the naturalization of the human. We shall see that, while naturalization could lead to an undermining of a mythological understanding of the world, what in fact tended to happen in cases where there was a significant shift in fundamental thinking about our place in the world was rather a move from one form of mythological understanding to another: simplifying considerably, from a Christian to an aesthetic conception. We need to take account of mythological factors because human self-understanding, in the period with which we are concerned, meant a self-understanding in relation to the natural realm and in relation to the supernatural realm. The four forms of naturalization of the human that we have examined were concerned with the former, and only exceptionally and tangentially dealt with the latter. To understand what naturalization amounts to in the latter case, we need to examine the principal tool of naturalization: historicization.

Our first concern will be to identify how the need for historicization arises. Kant, in an influential argument about the limits of theoretical reason, argued that religion came not under theoretical reason but under practical reason. This seemed to a whole generation to resolve the problems of reason versus revelation, but what it yielded was at best a rationale for morality, and while it was considered to offer some minimal understanding of the divine, this did not extend to any of the basic tenets of Christian dogma. The question of how these were to be explored fell by default to historical forms of enquiry, and Gibbon's account of the circumstances in which Christian doctrines had been formulated, which made no concessions to 'sacred history', provoked a massive backlash. At the same time historical investigation into the life of Jesus came to the fore, and Strauss' work in particular was considered to have de-sacralized Jesus, and met a reaction on a par with that which Gibbon's work elicited. In a separate set of developments, comparative studies of religion began to appear, effectively starting with de Brosses' book on African fetish religion in 1760, and the editions and commentaries on ancient texts by the Göttingen philologist Christian Gottlob Heyne from the 1760s onwards, who treated myth as the first stage of any human culture. Here there emerged an account of religions in historical quasi-evolutionary terms, ranking them from primitive to the most developed. At the apex of this development came Christianity, although the question subsequently began to be raised

[6] See Thomas Albert Howard, *Protestant Theology and the Making of the Modern German University* (Oxford, 2006), 15–16.

whether Christianity might in fact only be the penultimate stage, to be transformed into, or superseded by, humanism.

The upshot of these developments was that the understanding of the divine became a form of historical understanding of how our aspirations, hopes, fears, and raw beliefs are formed. One might think of this in terms of a naturalization of non-propositional understanding of the world. The means that it deploys are historical, and their use results in a historicization that has a powerfully reductive cutting edge. We shall be focusing on something that can be considered both a humanization and a naturalization of the divine: in Strauss and then in the work of Feuerbach, the reduction of the divine to reified human aspirations. In Feuerbach, such reduction is doubtless designed to remove one kind of mythological relation to the world, a religious one, but it would be misleading to regard it as a wholesale demythologization, for human aspirations do not thereby necessarily become part of a rational—that is, propositional—form of understanding of the world. Such ways of engaging with the world as hopes and aspirations resist being transformed into propositional forms of understanding. Their naturalization did not mean that they became effectively transmuted into knowledge of the world, but rather required that they be given some new coherence, the leading contender for which was an aestheticized humanist conception of the aims and meaning of life.

THE LIMITS OF REASON

In 1792, Fichte, who was to become Reinhold's successor as Professor of Philosophy at Jena, and was seen as the successor to Kant, published his *Versuch einer Kritik aller Offenbarung*, 'A Critique of All Revelation'.[7] Fichte had begun his studies at the Jena theological seminary in 1780, before coming under the influence of Kant, to whom he showed his 'Critique' in late 1791, and Kant responded by helping him find a publisher for it. This should be seen as part of a new strategy on Kant's part, in line with his announcement in the 1787 preface to the second edition of his *Kritik der reinen Vernunft*, that he would no longer engage in disputes with critics. This he would leave to others, and from that time onwards he strongly encouraged his students and friends to defend his system.[8] He not only gave his blessing to Fichte's work, but the fact that the book appeared anonymously,[9] and

[7] Johann Gottlieb Fichte, *Versuch einer Kritik aller Offenbarung* (Königsberg, 1792). Translated as Fichte, *Attempt at a Critique of All Revelation*, ed. Allen Wood (Cambridge, 2010).

[8] See Manfred Kuehn, *Kant: A Biography* (Cambridge, 2001), 329.

[9] In 1788, the Rosicrucian Johann Christoph Wöllner was appointed as minister of ecclesiastical affairs by Frederick Wilhelm II (whom he had induced into the Rosicrucian order), and he issued edicts on religion and censorship which suppressed everything that was not strictly orthodox. In this repressive climate, anonymity became a *sine qua non* for publication of serious philosophical works on religion. See ibid., 339–40.

that Kant himself was known to be writing a work on religion—his *Die Religion innerhalb der Grenzen der bloßen Vernuft* ('religion within the limits of reason alone') appeared later, in 1793—led to a widespread belief that Kant himself had written it. On the first appearance of the *Versuch* at the 1792 Easter Book Fair in Leipzig, a note was published in the *Allegemeine Literatur Zeitung* stating that anyone who had read 'even the minor writings by means of which the Königsberg philosopher [i.e. Kant] has obtained the eternal gratitude of mankind will immediately recognize the distinguished author of this work'.[10] Kant responded by disclaiming authorship but praising the work and its author. But although Fichte subsequently set out to develop the critical philosophy, he developed it in a new way, just as Reinhold had, and took it further in an idealist direction than Kant himself envisaged. Coming increasingly under attack from more orthodox Kantians, and from Kant himself, who dissociated himself from Fichte's subjective idealism, as well as from the charge of atheism or pantheism, Fichte tried to clarify his position in a short essay on God in 1798.[11] But the intellectual climate had changed, and the reception of this essay, although it barely went beyond the *Versuch*, was uniformly critical.[12]

Above all, Kant now dissociated himself from Fichte's treatment of religion. This should be seen in the context of the critical reception that had been accorded to his own *Die Religion*. It had initially been subject to censorship. Book I had passed the censor in 1791, being regarded as mainly philosophical in character, but Book II was refused permission. However, the philosophy and theology faculties at Königsberg, along with those in several other German universities, had the right to authorize publication of books dealing with religious subjects, and in 1793 Kant circumvented the Berlin censor by submitting the book as a whole to the philosophy faculty, which gave it its *imprimatur*. On 1 October 1794, the King, Friedrich Wilhelm II, wrote to Kant disciplining him:

> Our most high person has long observed with great displeasure how you misuse your philosophy to distort and disparage many of the cardinal and foundational teachings of the Holy Scriptures and of Christianity; how you have done this specifically in your book *Die Religion innerhalb der Grenzen der bloßen Vernuft*, and similarly in other shorter treatises. We expected better of you, since you yourself must see how irresponsibly you have acted against your duty as a teacher of youth and against our sovereign purposes, of which you are well aware. We demand that you immediately give a conscientious vindication of your actions, and we expect that in the future, to avoid our highest disfavour, you will be guilty of no such fault, but rather, in keeping with your duty, apply your authority and your talents to the progressive realization of our sovereign purpose. Failing this, you must expect unpleasant measures for your continuing obstinacy.[13]

[10] Quoted in ibid., 363.

[11] Fichte, *Ueber den Grund unseres Glaubens an eine göttliche Weltregierung* (Jena, 1798).

[12] See Anthony La Vopa, *Fichte: The Self and the Calling of Philosophy, 1762–1799* (Cambridge, 2001), 348–424.

[13] Kant, *Gesammelte Schriften*, vii. 7: *Religion and Rational Theology* (Cambridge, 1996), 240.

Kant did indeed offer a vindication and promised not to publish anything on religion again. Although he changed his mind after the death of Friedrich Wilhelm and subsequently published his response in 1798,[14] his wariness may explain his wanting to distance himself from the Fichte case.

Whatever the explanation for his response, however, the crucial point is that the reaction to Fichte was to a Kantian critique of religion, if not Kant's own version then one very close to it, and the charge was heterodoxy. Yet, for many, Kant had been (and continued to be) the saviour of Christian orthodoxy from the excesses of unthinking fideism on the one hand and sceptical rationalism on the other. To understand how the Kantian account could turn out to have the kind of radical consequences condemned by critics of Fichte, we need first to understand what Kant was responding to.

An especially pressing question in German thought from the 1750s onwards was that of the limits of reason in defending religion.[15] One group, the 'rationalists', predominantly (but not always) Wolffians, effectively based religion on natural theology, holding that fundamental theological questions, such as the existence of God and the doctrine of personal immortality, were open to rational demonstration, and that scripture was at best allegorical, and at worst mythical.[16] The 'pietist' or 'fideist' response to this approach was to argue that such questions were a matter of faith, and quite independent of reason, and consequently could only be illuminated through the study of scripture and revelation. Everything changes with the addition of history to reason and faith as a means of exploring religious sensibilities, but to appreciate just what was at issue we need first to understand the problems that the reason/faith dichotomy generated.

The most prominent defender of reason in the Leibniz/Wolff metaphysical tradition in the 1880s was Moses Mendelssohn. His aim was to offer a middle way between materialism on the one hand, and zealotry, irrationality, and counter-Enlightenment on the other.[17] For Mendelssohn, all judgements—whether theological, aesthetic, exegetical, or philosophical—have a single form, for their conclusion/predicate only ever makes explicit what is contained in the premises/

[14] It forms the 'Preface' to his *Streit der Facultäten*: *Gesammelte Schriften*, vii. 5–11: *Religion and Rational Theology*, 239–43.

[15] See Lewis White Beck, *Early German Philosophy: Kant and his Predecessors* (Cambridge, Mass., 1969), 288–305.

[16] Wolf's impact on natural theology was significant. His own major contribution to the question was his *Theologia naturalis pertractata* (2 vols, Frankfurt and Leipzig, 1736–7). But the issues had been eagerly taken up before this by two of Wolff's pupils: Bernhard Bilfinger, *Dilucidationes philosophiae de Deo, anima humana, mundo et generalibus rerum affectionibus* (Tübingen, 1725); and Israel Theophil Canz, *Philosophiae Leibnitianae et Wolffianae usus in theologia* (Frankfurt and Leipzig, 1733).

[17] Moses Mendelssohn, *Morgenstunden: oder, Vorlesungen über das daseyn Gottes* (Berlin, 1785), 5. Mendelssohn was aware that reviving Wolffian philosophy would be seen as an outdated way of doing this: ibid., 3–4. This was indeed one of the criticisms that was made of his work, for example by the early Kant supporter Ludwig Heinrich Jakob in his *Prüfung der Mendelssohnischen* Morgenstunden *oder aller spekulativen beweise für des Daseyn Gottes* (Leipzig, 1786), ii–iii.

subject. Of course, judgements invariably seem to reveal to us something new, but on Mendelssohn's account this is only because we have not analysed them fully: such an analysis would show that, in true judgements, the conclusion is always contained in the premises, whereas in false ones it is not. Because of this, reason in the form of metaphysics is able, simply through analysis, to establish the truth or falsity of any judgement. The problem was that this seemed to put in jeopardy principles that we know to be necessary to moral conduct or to our happiness, but whose truth we cannot establish. This would in turn undermine the duty that we have to follow those principles. Mendelssohn responds by distinguishing sharply between moral and intellectual standards. The reasons that we have to find something morally compelling are simply incommensurable with those reasons we have for believing something true or false. The former are dependent on our beliefs and on acts of will, whereas the latter are not, for whether something is true or not is completely independent of us. In contrast to his opponents, Mendelssohn dissociates moral questions from theological ones. Theological questions—those concerning God's existence, personal immortality, and providence for example—*are* matters of truth and falsity, and they are a matter for metaphysics, even if metaphysics is unable to provide definitive answers to such intractable questions.

The most radical consequences of rationalism came not on questions concerning morality or metaphysics, however, but on the issue of the standing of revelation. The 'neologist' school of Biblical interpretation rejected the idea that revelation formed a coherent whole, and instead set out to examine dogmas one by one. Original sin, predestination, supernaturally efficacious grace, the damnation of heathens, Christ's atonement for human sin, and the eternality of punishment were all thrown into question in Johann August Eberhard's *Neue Apologie des Sokrates* of 1772.[18] But the most notorious rationalist account of scripture was that of the orientalist Hermann Samuel Reimarus, in his manuscript *Apologie oder Schutzschrift für die vernünftigen Verehrer Gottes* (apology or defence of the rational worshippers of God), which was never intended for publication, but which his daughter discovered and let Lessing, who was a family friend, have after the death of her father in 1768. In the manuscript, Reimarus questioned the concept of a historical revelation, denying that the Bible is a doctrinal text, but simply a collection of ambiguous and contradictory documents. He criticized the Old Testament on the basis of its teachings, and on the grounds of the immorality of its main figures. The New Testament he attacked as comprising manifest lies and deceptions, motivated by self-importance and unchecked enthusiasm on the part of Jesus' disciples, resulting in the invention of ridiculous stories about his resurrection.[19] After testing the

[18] Johann August Eberhard, *Neue Apologie des Sokrates oder Untersuchung der Lehre von der Seligkeit der Heiden* (Berlin, 1772). See Karl Aner, *Die Theologie der Lessingzeit* (Halle, 1929), ch. 2.

[19] Reimarus was later to be the subject of a biography by Strauss, the most controversial theologian of his era: David Friedrich Strauss, *Hermann Samuel Reimarus und seine Schutzschrift für die vernünftigen Verehrer Gottes* (Leipzig, 1862).

waters with some innocuous excerpts in 1772, Lessing published a selection of the contentious fragments in 1777, although in his editorial introduction he distanced himself from some of the extreme positions taken up by Reimarus.[20] But even with its publication, Reimarus' version of a revelation-free Christian natural theology would probably not have attracted much attention, being dismissed as the rantings of a madman, had its appearance not been associated with Lessing, and subsequently with the 1785 *Pantheismusstreit* (pantheism controversy),[21] prompted by the revelation by Jacobi that Lessing had considered himself a Spinozist.[22]

Reimarus' advocacy of a form of rational religion fit well with Spinozean politico-theology. Jacobi's opening up, in 1785, of the question of Spinozism raised the stakes in the rationalism/pietism debate. Yet it also muddied the waters. As Beiser has noted, most of this first generation of Spinozists actually began as Lutheran pietists who had lost confidence in the course of the Reformation.[23] They held to the original ideals of the Reformation, notably the universal priesthood of believers, freedom of conscience, and a direct and unmediated relationship to God, and they saw in Spinoza a critical attitude to the Bible, support for democracy, a call for a universal religion, and a separation of church and state. That Lutherans should be taking up Spinoza is not quite as counter-intuitive as it seems, for Spinoza's pantheism was seen, by some at least, as highly religious in orientation. Novalis, for instance, talks of Spinoza as someone 'intoxicated by God' (*Gottrunkener*).[24]

If rationalism was a radical position, the kind of alternative offered by Hamann was no less so. The core argument of Hamann's *Sokratische Denkwürdigkeiten* (1759)[25] is that reason is propositional knowledge, but that there is another kind

[20] Gotthold Ephraim Lessing, *Sämtliche Schriften* (23 vols, Stuttgart, 1886–1924), xii. 62–82. On Lessing's role in the affair see H. B. Nisbet, *Gotthold Ephraim Lessing: His Life, Works, and Thought* (Oxford, 2013), ch. 18. Despite his caution, Lessing was attacked as party to Reimarus' views, beginning with Johann Melchior Goeze, *Etwas vorläufiges gegen des Herrn Hofraths Lessings mittelbare und unmittelbare feindselige Angriffe auf unsre allerheiligste Religion und auf den einigen Lehrgrund derselben, die heilige Schrift* (Hamburg, 1778).

[21] See Beiser, *The Fate of Reason*, chs 2–4; Jonathan Israel, *Democratic Enlightenment*, ch. 25.

[22] See Friedrich Heinrich Jacobi, *Briefe über die Lehre Spinozas* (Leipzig, 1785). On the reception of Spinoza in Germany, see David Bell, *Spinoza in Germany from 1670 to the Age of Goethe* (London, 1984). Jonathan Israel has promoted a view of Spinoza as a key figure in the 'radical Enlightenment' in great detail in a series of books: *Radical Enlightenment and the Making of Modernity, 1650–1750* (Oxford, 2002); *Enlightenment Contested: Philosophy, Modernity, and the Emancipation of Man 1670–1752* (Oxford, 2008); and *Democratic Enlightenment*. On Lessing and Spinozism, see Aner, *Die Theologie der Lessingzeit*, and Henry Allison, *Lessing and the Enlightenment: His Philosophy of Religion and its Relation to Eighteenth-Century Thought* (Ann Arbor, 1966). Allison plausibly questions just how seriously Lessing's comment was meant (72–9), noting the ironical and half-serious nature of his responses to Jacobi (73). Cf. Nisbet, *Gotthold Ephraim Lessing*, ch. 21, *contra* Israel's more literal reading in *Democratic Enlightenment*, 694.

[23] See Beiser, *The Fate of Reason*, 48–52.

[24] *Novalis Schriften*, ed. R. Samuel, H.-J. Mähl, and G. Schultz (6 vols, Stuttgart, 1960–2006), iii. 651.

[25] Johann Georg Hamann, *Sokratische Denkwürdigkeiten für die lange Weile des Publikums zusammengetragen von einem Liebhaber der langen Weile* ('Amsterdam' [Köningsberg], 1759). The

of knowledge, that of faith, which is non-propositional, and this is superior to propositional knowledge. Hamann is concerned here not with non-propositional understanding generally but with non-propositional *knowledge*, and this is rather different, because what motivates his account is not the variety of ways in which we can stand in relation to the world, but rather an epistemological assumption about the priority of unmediated over mediated grasp. The assumption had been crucial in Descartes' *Meditationes*, where the immediacy of the cognitive grasp afforded by the *cogito* plays a crucial role in establishing its standing as a form of understanding secure in itself, and from Descartes onwards there was a general assumption that any mediating factors militate against the security of knowledge.[26] But there had been critics. Whether the *cogito* could even count as a form of knowledge at all had been questioned from an Aristotelian perspective: knowledge, it was argued, was associated with increasing one's understanding, having an explanation or demonstration for something, and it was clear that grasping my own existence could not be knowledge on this conception.[27] Descartes treats the *cogito* as propositional knowledge, as knowledge *that* something is the case, yet it is crucial for his argument that the grasp of the *cogito* must be non-inferential: it is what he elsewhere terms an *intuitus*. For Hamann such things as our own existence and the existence of a physical world cannot be demonstrated, yet we can be certain of these. Like Descartes, he thinks of the immediacy of our grasp in these cases as being the crucial factor in our certainty but, unlike Descartes, he associates such certainty with faith.

Because of the way in which he writes, it is not easy to put one's finger on exactly what Hamann's argument is. In effect, however, he is putting a form of non-propositional understanding, faith, in the place of the *cogito*, treating faith as a *bona fide* form of knowledge which has been unjustly neglected, not so much because it has been denied the status of knowledge but because non-propositional forms of knowledge have illegitimately been rejected in favour of propositional ones. A good deal depends here on how we think of knowledge. Assuming knowledge must involve belief, but that it must go beyond mere belief, we can consider it in philosophical or in psychological terms. If we think of knowledge in philosophical terms, as justified true belief, then it would be at best highly contentious, even in eighteenth-century Germany, to treat faith as a form of knowledge. However, we can also think of knowledge in psychological terms,

false 'Amsterdam' imprint was not designed to avoid censorship as the work already had government permission: it seems to have been a whim of the publisher. There is an extensively annotated English translation: *Hamann's Socratic Memorabilia*, ed. J. C. O'Flaherty (Baltimore, 1967). A very useful selection in English of Hamann's works can be found in Hamann, *Writings on Philosophy and Language*, ed. K. Haynes (Cambridge, 2007).

[26] The exception was Antoine Arnauld, *Des vrais et fausses idées contre ce qu'enseigne l'auteur de la recherche de la verité* (Cologne, 1683). See the discussion in John W. Yolton, *Perceptual Acquaintance from Descartes to Reid* (Oxford, 1984), ch. 3.

[27] See Gaukroger, *Descartes, An Intellectual Biography*, 309–21.

whereby knowing something is believing it with certainty. On this latter understanding, there can be little doubt that faith in eighteenth-century Germany was a form of knowledge, and there can be no doubt that, for many, religious belief had an unassailable degree of certainty. In the *Sokratische Denkwürdigkeiten*, Hamann develops the claims for non-propositional knowledge by examining the contrast between Socrates and Jesus, traditionally taken to be diametrically opposed advocates of reason and faith respectively. Hamann reinterprets Socrates as a critic of reason, a precursor of Jesus. In particular, he takes Socrates' attacks on the sophists as a repudiation of those who believe that everything can be grasped through reason: it is the sophists who are the rationalists, whereas Socrates is presented as the antithesis of the sophists, someone for whom the recognition of one's own ignorance is the highest form of knowledge.[28] On Hamann's reading, the thing about Socrates, by contrast with the sophists, is that he is inspired, and it is his inspiration that enables him to grasp the truth. By assimilating this inspiration to faith, Hamann is able to treat faith as a special kind of knowledge. It stands in contrast to propositional knowledge: it is intuitive as opposed to 'rational' knowledge, and, unlike rational knowledge, it is unmediated.

In a characteristically unorthodox reading, Hamann also draws on Hume to support his defence of the autonomy of faith. Hume, he writes to Kant in a letter of 27 July 1759, needs faith if he is to eat an egg and drink a glass of water, because their existence is something of which we can be certain, although this certainty cannot derive from reason. But if Hume 'needs faith for food and drink, why does he deny faith when he judges matters that are higher than sensuous eating and drinking?'[29] Hamann then quotes a passage from Hume, telling us that it

> will prove that one can preach the truth *in jest*, and without being conscious or desirous of it. . . . Here it is: 'The Christian religion not only was at first attended with miracles, but even at this day cannot be believed by any reasonable person without one. Mere reason is insufficient to convince us of its veracity. And whoever is moved by *Faith* to assent to it, is conscious of a continued miracle in his own person, which subverts all the principles of his understanding, and gives him a determination to believe what is most contrary to custom and experience'.[30]

Hamann treats reason and faith here as legitimate forms of knowledge in their own right, something that makes sense if we construe knowledge in psychological terms, as certain belief. But it is clear that this is not how Hume saw things, for, although he was adamant that forms of non-propositional understanding were important to balance reason, he did not treat religion in such terms. The kinds of

[28] Towards the end of the century, Christian Garve, one of the last representatives of *Popularphilosophie*, would promote a similar view, arguing that Socrates' achievement was to oppose learned pedantry with true love of wisdom: see Zande, 'In the Image of Cicero', 431.

[29] Kant, *Gesammelte Schriften*, x. 14–15; translated in Immanuel Kant, *Correspondence* (Cambridge, 1999), 52–3.

[30] *Gesammelte Schriften*, x. 53; *Correspondence*, 53.

claims that Christianity made were assimilated by him to metaphysics. As he puts it in the *Enquiries Concerning Human Understanding*:

All the philosophy, therefore, in the world, and all religion, which is nothing but a species of philosophy, will never be able to carry us beyond the usual course of experience, or give us measures of conduct and behaviour different from those which are furnished by reflections on common life.[31]

To put it in other terms, for Hume religion must be judged against the standards of justified true belief, not those of certainty of belief.

Hamann's use of Hume nicely illustrates the intricacy of the perceived threat to reason, the complexity of its motivations, and the diversity of the proposed solutions. This is important because reason, along with free will, had traditionally been treated as the distinguishing feature of human beings, so any revision to, or threat to, its standing manifestly had profound consequences for how the human realm was conceived. Above all, it had consequences for what resources were drawn on in the attempts to understand human behaviour.

This, then, is the context within which we should situate Kant's solution to the problem of religious understanding. But it was a solution that only gained a general currency with Reinhold. When Reinhold translated the 'critical philosophy' from an academic treatise into a series of popular essays, his focus was on whether Kant's critical philosophy could show us how belief in the existence of God, personal immortality, and divine providence can be vindicated. In particular, Reinhold set out to show that, through Kant's doctrine of practical faith, it has a unique claim to success in this. His argument was that the critical philosophy was able to save reason from the scepticism/fideism dichotomy,[32] and to help integrate theology and science into a unified comprehensive vision grounded in metaphysics (or that transcendental form of enquiry that has replaced traditional metaphysics for Kant). On the face of it, the account of the conditions of possibility of experience seems to reinforce the sceptical conclusion as far as religion is concerned. Even worse, by attempting to solve the problem of the standing of religion by offering a general account that removes space, time, and causation from the world, and makes them forms of structuring the world, religion seems to follow science along the sceptical path. The general difficulty here arises from the fact that, for Kant, all our knowledge is constrained by the way in which the world must appear to us. There is no way in which we can grasp how things are prior to this structuring, so we cannot ask whether there really is a God, independently of an experience that is the product of our own minds, any more than we can ask whether space and time really exist independently of our experience. Metaphysicians had striven to provide answers to the questions of how

[31] Hume, *Enquiries Concerning the Human Understanding*, 38 (sec. xi).

[32] Reinhold, like Kant's followers more generally, was much more concerned with scepticism than was Kant: see Kuehn, *Kant*, 351–2.

things really are, but Kant believes he has shown this to be impossible.[33] In Kantian terminology, metaphysicians have attempted to use the resources of pure reason to establish how things are in themselves: they have attempted to step outside the boundaries of possible experience. However, we only have access to appearances, so the task should rather be that of understanding how critical reflection on appearances can reveal something about those things that go beyond them.

The way in which experience is generated is crucial here, for Kant's elaborate argument is that this experience, and the conditions under which it is generated, are universal: the world *must* appear to us in a particular way, and it must appear to all of us in this way, because the conditions of possibility of its appearing in this way are universal. This is why Kant can have no compromise with anthropological projects, such as those of Herder and Hamann, which contextualize and relativize experience in various ways. If experience were not constrained by something universal, the critical philosophy would be no better placed than metaphysics in establishing basic truths. But the universal constraints are not all there is to our experience of the world. Equally important for Kant is the freedom and autonomy of the subject. Our experience is organized not only by sensory and intuitive structuring over which we have no control, but also by self-imposed conceptual rules. In other words, this structuring of our experience is both a manifestation of something that we simply have to take as given, and at the same time a manifestation of our freedom from some order in the world that simply discloses itself to passive agents. Here, in this second element, we begin to see the origins of the path to the idealism that will be followed by Fichte, Schelling, and Hegel, who considered themselves to be completing Kant's project. Our concern here however is to understand how the Kantian project was seen to integrate religion and science—or more specifically Christianity and *Aufklärung*—into an account of our understanding of the world.

The best way to understand the attraction of the *Kritik* to Kant's contemporaries is through its most influential exponent, Reinhold. Indeed, Reinhold was so influential that Kant's philosophy was known as Kant–Reinhold philosophy in the second half of the 1780s.[34] Reinhold was concerned above all with the question of the limits of reason in religion, and in his *Briefe über die kantische Philosophie*, he argued that Kant had not only shown that reason and religion dealt with different aspects of reality, but that he had vindicated the latter. More specifically, Kant had shown that knowledge of God cannot be achieved by theoretical reason, but this did not mean that we should be sceptical about the existence of God. Quite the contrary, he had set out to establish the legitimacy of practical faith, that is, to show that there were, on practical grounds, not merely reasons, but necessary reasons, for postulating human freedom and the existence

[33] The large section on 'Transcendental Dialectic' in the *Kritik*, which runs from A293/B249 to A704/B732, is devoted specifically to this question.

[34] See Kuehn, *Kant*, 351.

of a personal God, providence, and personal immortality. It is important to understand that by 1785, when Reinhold began his *Briefe*, the question of the standing of religion was paramount in the minds of German philosophers and theologians, with 'rationalists' pitted against 'pietists'.[35] Reinhold starts by explaining that this is an irresoluble deadlock: the kinds of reasons the rationalists adduce for their case are incapable of convincing the pietists, and vice versa. Since the question on which everything hinged was the limits of reason, it could only be resolved by investigating the nature of reason, and Reinhold points out that such an investigation had already been carried out five years earlier by Kant, in the first *Kritik*, generally ignored up to that point. What Kant had established, Reinhold explains, is a middle way between rationalism and pietism. The rationalist argues that God's existence is knowable through rational demonstration, which makes faith superfluous, whereas the pietist argues that faith in God cannot be justified by reason. Kant, as we have seen, argues that theoretical reason cannot be used to prove the existence of God, but nor does he accept that faith can establish the existence of God. Belief in the existence of God, he argues, is indeed rational: but to understand how this can be, we must recognize that as well as theoretical reason, there is also another form of reason, practical reason.

This novel type of demonstration deploys a form of transcendental argument. It is designed to show that the existence of a personal God is a condition of possibility of our having the kind of experience we have, although what is at issue here is not our experience of the natural world, but our experience of morality.[36] The issues turn in the first instance on the question of freedom, specifically on the undeniability of our freedom. In our free actions and deliberations, we can always distinguish between these and determined behaviour, and this difference can be captured in the idea that free actions are regulated by a normative principle: unlike determined actions, we can say of free ones that they are correct or incorrect, right or wrong. This is so intrinsic to our behaviour that our freedom is a necessary condition of our acting in the way we do. Note that this freedom is not established by theoretical reason—as we have already seen, in the antinomies of pure reason Kant had shown that theoretical reason (in the form of metaphysics) cannot establish whether our behaviour is free or determined—but by practical reason. Our desires and inclinations may be motivated by considerations and principles that have been determined outside our control, but we choose which of these

[35] See Beiser, *The Fate of Reason*, 226–36. More generally, see: Martin Bondeli, 'The Conception of Enlightenment in Reinhold's "Letters on Kantian Philosophy"'; Pierluigi Valenza, 'Reinhold, Lavater et les Lumières'; and Michael Gerten, 'Bügerliche Aufklärung und streng systematische Philosophie. Zum Verhältnis von Leben und Philosophie bei K. L. Reinhold', in George di Giovanni, ed., *Karl Leonhard Reinhold and the Enlightenment* (Dordrecht, 2010), 43–52, 53–64, 65–85 respectively.

[36] The argument is set out in three texts: *Grundlegung zur Metaphysische der Sitten* (1785): *Gesammelte Schriften*, iv. 385–464; *Kritik der praktischen Vernunft* (1788): *Gesammelte Schriften*, v. 1–164; *Die Metaphysik der Sitten* (1797): *Werke*, vi. 203–494. English translations in Kant, *Practical Philosophy*.

principles to act upon, and we unhesitatingly experience this activity as guided by self-imposed principles. I can always distinguish between the normative question of what I should have done in some particular case, and the factual question of what I did do in that case. That I cannot demonstrate that I am in fact free, as opposed to merely believing that I am free, is a limitation of theoretical reason. What practical reason shows is that my standing back from the world, as it were, and taking a position on what to do and what not to do, or, to put it in different terms, being a genuine agent as opposed to merely part of a deterministic world, is only possible on the condition of freedom.

The principles on which we act when we act freely are not merely subjective however. When we distinguish between what we did and what we ought to have done, the latter judgement may reflect two different kinds of practical principle: those that we actually act upon, and those that reflect a general criterion for behaving morally or appropriately. The important difference between the two lies in the fact that in the former case we would not extend the principles to the behaviour of others, whereas in the latter case we would, in the form of a command or imperative. Kant asks whether any of these imperatives could be unconditionally binding, whether there might be something that binds every rational agent independently of any particular tastes, inclinations, or purposes that guide their behaviour, and he concluded that while this cannot be a substantive principle, it can be a formal one, namely the principle of universalizability: to act only on principles that you would want to become universal.

One of the most controversial features of Kant's account of morality was that it appeared to sever the connection between morality and Christianity. But in fact he conceives there to be a form of transcendental connection between morality and God. As it turns out, he would in any case seem to need this connection on independent grounds because of a very problematic aspect of his account: the relation between morality and happiness, or morality and the 'good life', things which are tightly connected in the Aristotelian tradition of virtue ethics for example, seem to have come apart in the Kantian picture. The demands of his principle of universalizability—the 'moral law'—override any concern with happiness, yet if the moral good led us in a completely different direction from happiness, then the plausibility of Kant's account of moral behaviour and its motivations would be severely diminished, as he recognized. Instead of acknowledging this as a failing of his account, however, he turns it into a strength, using it to show the necessity of a religious basis for ethics. His solution is to argue that, given we are in fact motivated by the moral law (among other things), and given that the moral law does not guarantee happiness in this life, then to explain our moral behaviour, at a motivational level, we need to invoke a kind of idealized 'highest good', something on which morality and happiness converge.[37] We need

[37] In the *Kritik der Urtheilskraft* Kant pursues these questions even further, arguing that it is rational to expect there to be a harmony between the phenomenal realm and human interest in

to act as if this convergence were possible, and two things must be the case for it to be possible: we need to postulate continued existence after this life, i.e. personal immortality; and we need to postulate the existence of a personal God who correlates virtue and happiness in due proportion. In short, given the insights into the nature of morality that Kant believes he has provided, we need to invoke the existence of God and personal immortality if we are to make sense of our moral behaviour. The argument does not establish the existence of either personal immortality or God, for it was one of the basic errors of 'pure' or theoretical reason to believe that these were things that could be demonstrated. Rather, it is designed to show that we have no option but to believe in these things: we must act (and in effect always do act, in Kant's view) as if they were true.

If we compare Fichte's reading of Kantian theology in the *Versuch* with Kant's account of 'religion within the limits of reason alone', which appeared one year later, what is striking is that Fichte's account is actually closer to orthodoxy than is Kant's.[38] Fichte follows Lessing's view that revealed religion constitutes what Lessing called 'an education of the human race' towards reason, civilization, and morality, and he thinks of morality as legislated by our own will and the will of a God external to us. Kant by contrast has a much more pragmatic view of revealed religion. Although he thinks that we can argue, through practical reasoning, that our moral life would make little sense unless there were an afterlife in which there were punishment and rewards, it does not follow from this that morality requires religion, an idea he explicitly rejects in the 'Preface' to the first edition of *Die Religion*:

> So far as morality is based on the conception of the human being as one who is free but also, just because of that, binds himself through his reason to unconditional laws, it is in need neither of the idea of another being above him in order that he recognizes his duty, nor, that he observe it, of an incentive other than the law itself. . . . Hence on its own behalf morality in no way needs religion . . . but is rather self-sufficient by virtue of pure practical reason.[39]

Kant goes on to argue in the body of the work that we have a propensity to evil within ourselves, and that our recognition of moral duties as divine commands requires us to join together in religious communities if we are to combat these. The historical form these communities take, the various churches, are based on the acceptance of particular sacred writings. In the response to the letter from the king, he sets out to explain what the standing of his work on religion is:

> Since, in [*Die Religion*], I make no *appraisal* of Christianity, I cannot be guilty of *disparaging* it. In fact, it is only natural religion that I appraise: The only possible basis

natural science, virtue, and happiness, because it is rational to expect the human condition to be in harmony with divine purpose.

[38] I draw here on Allen Wood's introduction to Fichte, *Attempt at a Critique of All Revelation*, xii–xvi.

[39] *Gesammelte Schriften*, vi. 3; *Religion and Rational Theology*, 57.

> for this misunderstanding is the fact that I cite some biblical texts to corroborate certain purely rational teachings in religion. . . . But when reason speaks, in these matters, as if it were sufficient to itself and as if revealed teachings were therefore superfluous (as assertion which, were it to be taken objectively, would have to be considered a real disparagement of Christianity), it is merely expressing its appraisal of itself—not in terms of its [theoretical] ability [but] in terms of what it prescribes us to do insofar as it alone is the source of the *universality*, *unity*, and *necessity* in the tenets of faith that are the essence of any religion as such, which consists in the morally practical (in what we *ought* to do). On the other hand, what we have cause to believe on historical grounds (where '*ought*' does not hold at all)—that is, revelation as contingent tenets of faith—it regards as nonessential. But this does not mean that reason considers it idle and superfluous; for revelation is useful in making up the *theoretical* deficiency which our pure rational belief admits it has (in the questions, for example, of the origin of evil, the conversion from evil to good, the human being's assurance that he has become good, etc.) and helps—more or less, depending on the times and the person concerned—to satisfy a rational need.[40]

As Allen Wood points out, it is 'difficult to resist the temptation that Kant hopes the reliance on such scriptures, and the supposed revelations they contain, will gradually dissipate or even disappear as religion becomes more enlightened, rational, and universal'.[41] If, by contrast with an ahistorical and entirely necessary reason, revelation is historical and contingent, then clearly there are two ways in which religion itself needs to be examined: in terms of reason and in terms of historical contingency.

Kant's concern was with the former but his notion of a morality and religion guided by rationality was not universally accepted. Its most fervent critic was Schleiermacher, for whom religious faith and theology were grounded in a direct and immediate form of personal religious consciousness. 'It can be rightly be said', he writes, 'that in religion everything is immediately true, since nothing at all is expressed in its individual moments except the religious person's own state of mind'.[42] For Schleiermacher, this does not make religion something subjective or purely personal, however, for he argues that human beings experience a feeling of absolute dependence which is relational, and its *relatum* can only be something greater than ourselves.[43] The argument is best construed as transcendental, in that it tells us what things must be like if we are to have the kind of experience we have, and is reminiscent of Kant, whom Schleiermacher had studied carefully, although he rejected Kant's notion of happiness as completely divorced from any earthly state and hence unable to provide any religious motivation. Neither

[40] *Gesammelte Schriften*, vii. 8–9; *Religion and Rational Theology*, 241–2.

[41] Wood in Fichte, *Attempt at a Critique of All Revelation*, xv.

[42] Quoted in Robert Merrihew Adams, 'Faith and Religious Knowledge', in J. Mariña, ed., *The Cambridge Companion to Friedrich Schleiermacher* (Cambridge, 2005), 35–52: 35. The religious state of mind would consist, in the final instance, in an immersion of the self in the rest of the universe: see Schleiermacher's most popular work, *Über die Religion: reden an die Gebildeten unter ihren Verächtern* (Berlin, 1799).

[43] See the discussion in *Über die Religion*, passim.

Schleiermacher nor Kant offer what in Kantian terms is a theoretical argument for the existence of God, but whereas Kantian morality is something that is entirely rational, Schleiermacher rejects such an understanding, and in particular rejects Kantian ethics on the grounds that its formalism does not enable one to derive the content of specific maxims.[44] On his conception of religion it is not 'rational' in Kant's sense, but has a closer connection with the sensibility tradition. Yet like Kant, he offers little that goes beyond the idea of a higher being.[45]

Lessing had tried to separate genuine Christian doctrine completely from what he considered to be its discredited historical foundations, and Schelling saw the account of Jesus' life and death in the gospels as being based on Jewish fables invented after his resurrection on the basis of Old Testament prophecies. But Schleiermacher, like Kant, acknowledges a significant role for contingent historical factors. In fact these now came to assume a central position in religious understanding, and, despite the fact that some theologians used Kant's account to reject historical enquiry,[46] it is clear that Kant's minimalist conception of God was a key factor in establishing the need for historical enquiry in the German philosophical and theological traditions. It is to such contingent historical factors that we now turn. These call for a different type of enquiry from that in which Kant engages, a resolutely empirical type. I shall be concerned with two kinds of historical investigation. These fall within distinct traditions, philosophical history of the kind initiated in France and Britain in writers such as Voltaire and Hume, and philosophical theology of the kind pursued by philosophically minded theologians in Germany in the wake of Schelling's and Hegel's absolute idealism. They focused on very different questions: second-century Christianity in the first case, and Jesus' mission in the second. Yet, in their different ways, the two, represented above all in the work of Gibbon and Strauss respectively, converged on the realization that historical enquiry showed that there was something deeply wrong in the traditional understanding of Christianity.

THE HISTORICIZATION OF CHRISTIANITY

Gibbon's approach to the investigation of the credentials of Christianity could not be more different from that of Kant. The first volume of his *The History of the*

[44] On these questions see Frederick Beiser, 'Schleiermacher's Ethics', in J. Mariña, ed., *The Cambridge Companion to Friedrich Schleiermacher* (Cambridge, 2005), 53–72.

[45] There is a case to be made that he believes that certain core Christian doctrines, such as the Trinity, can receive some kind of legitimation, but it is not a very strong case. See Francis Schüssler Fiorenza, 'Schleiermacher's Understanding of God as Triune', in J. Mariña, ed., *The Cambridge Companion to Friedrich Schleiermacher* (Cambridge, 2005), 171–88.

[46] The most influential such reading was that of the Tübingen theologian Gottlob Christian Storr, *Annotationes quædam theologicæ ad philosophicam Kantii de religione doctrinam* (Tübingen, 1793).

Decline and Fall of the Roman Empire appeared in 1776. Its last two chapters cover the history of the Church before Constantine, and begin with a discussion of 'the Progress of the Christian Religion, and the Sentiments, Manners, Numbers, and Condition of the Primitive Christians'. In these two key chapters, 15 and 16, a number of doctrines are examined, not least the immortality of the soul and miracles, and the history of their development questioned. In defence of the procedure that he adopts, Gibbon writes that a 'candid but rational inquiry into the progress and establishment of Christianity may be considered as a very essential part of the history of the Roman empire'. But there was a problem, namely that:

this inquiry, however useful or entertaining, is attended with two peculiar difficulties. The scanty and suspicious materials of ecclesiastical history seldom enable us to dispel the dark cloud that hangs over the first age of the church. The great law of impartiality too often obliges us to reveal the imperfections of the uninspired teachers and believers of the gospel; and, to a careless observer, *their* faults may seem to cast a shade on the faith which they professed. But the scandal of the pious Christian, and the fallacious triumph of the Infidel, should cease as soon as they recollect not only *by whom*, but likewise *to whom*, the Divine Revelation was given. The theologian may indulge the pleasing task of describing Religion as she descended from heaven, arrayed in her native purity. A more melancholy duty is imposed on the historian. He must discover the inevitable mixture of error and corruption which she contracted in a long residence upon earth, among a weak and degenerate race of beings.[47]

It is worth noting from the outset that Gibbon was not working in the context of fixed doctrine. There was a significant degree of variation in the teachings of various churches, and great variation in the limits of tolerance. In the case of the religious culture of eighteenth-century England in which Gibbon was writing, for example, Pocock notes:

We may simplify this narrative for heuristic purposes, supposing that from time to time there arose a challenge to the Church's position in the state, leading to a willingness to enlarge or minimize doctrine in order to widen its support. There would be a movement along a scale running from Nicene Trinitarianism through Arianism towards unitarianism, and beyond that towards a scepticism of revealed religion altogether. The term 'deism' might connote either a philosophic theism termed 'natural religion', or a pantheism that identified God with nature; and critical scepticism might fasten on any of the main points at which revelation was held to have made itself known.[48]

Nevertheless, there is no doubt from the reaction to Gibbon's two chapters on pre-Constantine Christianity that both Christians and critics of Christianity believed that what he was proposing was radical and threatening. Arnaldo

[47] Edward Gibbon, *The History of the Decline and Fall of the Roman Empire*, with notes by Dean Milman and M. Guizot: a new edition with additional notes by William Smith (8 vols, London, 1881), i. 151–2.
[48] Pocock, *Barbarism and Religion*, v. 216.

Momigliano has remarked that 'we owe it to Gibbon that the problem of the relations between Christianity and the political and social developments of Europe has come to stay in European historiography'.[49] Above all, Gibbon's treatment of Christianity put Christian apologists in a defensive position, forcing them to deal with it as a historical religion whose development was subject to the same contingencies as any other historical movement.[50] But why was such an approach so challenging? Investigation of the historical circumstances under which Christian doctrines were formulated certainly revealed a significant element of contingency—indeed, on Gibbon's analysis what, in the case of Constantine's reforms, might best be termed political contingency—into the heart of core doctrines. But the furious reaction to Gibbon wasn't merely a response to the results of his investigation: much of it attacked the very idea of dealing with Christian doctrine in such a historical fashion. Christianity, it was considered, had a divine origin and as such was atemporal, and therefore simply not the kind of thing that could be elucidated in historical terms. This was not the view of all Christians, and many were seeking a more critical perspective that could be reconciled with Christianity.[51] But there were those who sharply distinguished the secular from the sacred: these were now in a difficult position, and open to the charge of double standards. The author of the First Epistle to Timothy—traditionally attributed to the Apostle Paul (but now usually attributed to an unknown second-century author)—had dismissed the 'godless and silly myths' of the Greeks and Romans, and Christians had routinely treated other religions, including those which had likewise claimed divine origins, in terms that relied on a contextualization, whether historical or otherwise, to disparage their claims. So it is not as if the historical investigation of religions per se was condemned, nor could claims of divine origin in themselves exempt it from such treatment. There were other routes that could be followed, which avoided these problems, and we shall be looking in the next section at attempts to establish the uniqueness of Christianity other than by fiat. The idea, which lies at the core of Christian attempts to formulate a comparative theory of religion in the nineteenth

[49] Arnaldo Momigliano, *Studies in Historiography* (London, 1966), 52. One of the most significant consequences of this is that it raises the question of whether Christianity, very much a minority religion at the time, would have died out, like the numerous other religious cults of late antiquity, had Constantine not given it imperial endorsement. Although he did not pose it in these terms, Gibbon is preoccupied with the question why Christianity persisted.

[50] Cf. Leslie Stephen's comment that Gibbon not merely 'struck a heavy blow at Christianity, but struck by far the heaviest blow which it had yet received from any single hand. What he did was to bring the genuine spirit of historical inquiry for the first time face to face with the facts.' *History of English Thought in the Eighteenth Century* (2 vols, London, 1881), i. 449–50.

[51] In the case of England, for example, see Brian W. Young, *Religion and Enlightenment in Eighteenth-Century England: Theological Debate from Locke to Burke* (Oxford, 1998). Note also that the question of secular versus sacred was not one restricted to theologians: Newton, for example, who if not orthodox was certainly pious, had promoted secular history at the expense of sacred history on a number of issues. See Garry W. Trompf, 'On Newtonian History', in S. Gaukroger, ed., *The Uses of Antiquity: The Scientific Revolution and the Classical Tradition* (Dordrecht, 1991), 231–50.

century,[52] was that religions manifest a form of evolutionary development in which Christianity emerges as the final form of religion.[53] Such an approach harbours its own problems however, and our immediate concern here is with the dividing line between secular history and sacred history: in particular with whether Gibbon was doing something contrary to the spirit of Christianity by treating it in historical terms.

The first point to note is that historical self-reconstruction had been central to Christianity from an early stage of its development, gradually becoming fixed as a form of sacred history. Before the last decades of the second century, which was when Christianity began to forge a history for itself, the attraction of the Pauline religion had lain in its novelty and its radical difference from everything else. But at the same time there had developed a view among its critics that something barely a century and a half old could not make claims to being the true religion. Theophilus, the bishop of Antioch, decided that it needed a longer genealogy, and, believing that the sacred Jewish books had foretold the coming of Christ, he maintained that the history of Christianity was continuous with that of Judaism. Adding up the lives of the patriarchs, he concluded that the origins of Christianity predated those of any other religion. The antiquity of Christianity, its standing as the oldest religion, became a crucial element in defending its legitimacy. The second important strand in establishing a historical standing for Christianity emerged in the work of second- and third-century theologians: notably Clement of Alexandra, Julius Africanus, and Hippolytus of Rome. As Momigliano notes, they 'created the frame for the divine administration of the world; they transformed Hellenistic chronography into a Christian science and added the lists of the bishops of the most important sees to the list of kings and magistrates of the pagan world. They presented history in such a way that the scheme of redemption was easy to perceive.'[54] History, imbued with religious content, became the typical form of Christian instruction, concerned as it was with the pattern of history rather than details.

In short, there is both a historicization of Christianity and a Christianization of history by theologians in the second and third centuries. History becomes in effect an instrument of propaganda: its purpose is to sustain conviction, not to satisfy curiosity. But not all history can be Christianized. There are no Christianized versions of Herodotus or Thucydides for example, and with the serious revival of historical writing in the Renaissance, these became the models for historical

[52] On nineteenth-century comparative religion, see Masuzawa, *The Invention of World Religions*.

[53] The comparative approach to establishing the uniqueness of Christianity does not have to take a straightforward evolutionary form. An alternative approach can be found in Rémi Brague: *Europe, la voie romaine* (Paris, 1992); cf. idem, *La Sagesse du Monde: Histoire de l'expérience humaine de l'univers* (Paris, 1999).

[54] Arnaldo Momigliano, 'Pagan and Christian Historiography in the Fourth Century A.D.', in idem, ed., *The Conflict between Paganism and Christianity in the Fourth Century* (Oxford, 1963), 79–99: 83–4.

enquiry. This meant that there were now two forms of historical writing, a sacred and a secular, but the relation between these two had regularly been fraught. In particular, mixing the sacred and the secular was always a delicate area. In 1573, for example, the Venetian Inquisitors condemned a painting of the Last Supper by Paolo Veronese, complaining that one of the Apostles was shown eating with a fork: it was evidently felt that the Last Supper had been transformed into an exercise in civility, thereby challenging the divide between the social and the spiritual. Nearly two centuries later, the Abbé Edme-François Mallet, in his entry on Noah's ark in the *Encyclopédie*, devotes columns of detail to size specifications, the kind of wood used, the discrepancy between the dimensions reported in the Bible and what was needed to accommodate the animals mentioned, what units of measurement were used, what species were on board, how much fodder would have been needed, the amount of fresh water required as ballast, and so on.[55] Mallet effectively removes the account of Noah's ark from a religious context and treats it in a wholly naturalized way as a problem in the logistics of ship building. Note however that the response to such issues varied on a national level. Mallet's approach was treated as wholly sacrilegious in France. But in Britain, in the context of physico-theological disputes over the universality of the Flood, it was assumed that there was a convergence between natural history and Christian theology, and such calculations were deemed appropriate: Wilkins and others in late seventeenth-century England undertook such calculations in a spirit of orthodoxy.

The stakes were raised, at the end of the seventeenth century, by Bayle's refusal to make any concessions to sacred history. His writings on religion do not look at the development of religious doctrine: rather he adopts a posture of standing back from Christianity, considering it as an object of study without making any assumptions as to the validity of its claims to truth. This was to become a procedure that Hume would employ in the study of religion,[56] and Bayle and Hume were two decisive influences on Gibbon.[57] But it was also the method of philology. From the late fifteenth century, new philological methods based on techniques for studying literary texts were applied to legal ones, but also to the Bible and the Gospels by Valla, Erasmus, Reuchlin, and others.[58] Crucial here were philological methods of establishing texts, developed with great sophistication by biblical scholars.[59] Even among orthodox writers, there was an acute

[55] 'Arche de Noé', *Encyclopédie*, iii. 228–35.

[56] See Gaukroger, *The Collapse of Mechanism and the Rise of Sensibility*, 444–52.

[57] On the influences see the comprehensive account in the first two volumes of Pocock, *Barbarism and Religion*.

[58] See Gaukroger, *The Emergence of a Scientific Culture*, 139–42, for a summary discussion and further references.

[59] Philosophers too used these methods. In the Renaissance, it was not just Platonists but also the Aristotelians of the northern Italian universities who historicized Aristotle, detaching his philosophy in crucial respects from Christian thought, for which his philosophy had provided the foundations

awareness of the need for secular history. As the Catholic theologian Sebastien Le Nain de Tillemont, one of Gibbon's contemporary sources, remarked, 'there is such a close connection between sacred history and secular history that one must study the latter carefully in order to master the former, and in order to resolve its difficulties properly'.[60]

In short, the kinds of probing of sources that Gibbon employs in establishing the origins of doctrines are continuous with a number of contemporary approaches, not least philologically oriented scholarship, including biblical scholarship, that had been practised since the sixteenth century. Nevertheless, the challenge he faced was considerable. As Pocock notes, once Gibbon moved, in the last two chapters of the first volume of the *Decline and Fall*, from Roman history to Christian history, a major conceptual shift was required. Christian writings, as Pocock notes, had a history of their own, which Gibbon was required to study, and which we need to understand if we are to appreciate what Gibbon was doing. The two interwoven strands of this Christian history are sacred history, 'the history of God's actions in and upon the world', and ecclesiastical history, the history of 'the human association charged with continuing and interpreting his intervention and his presence'.[61] The two were connected because the idea of the Church was inseparable from that of God's presence, which put the Incarnation at the centre of ecclesiastical history. Gibbon was writing ecclesiastical history, but this was so intimately tied up with sacred history that he could not avoid encroaching on the latter. As the Catholic theologian Claude Fleury, one of Gibbon's contemporary sources, wrote in the Preface to his thirty-two-volume *Histoire ecclésiastique*:

> The subject of ecclesiastical history is the representation of the continuity of Christianity from its founding. For the true religion has the advantage that its origin is certain, and the tradition continues to our own day without interruption. Its origin is certain because it is undisputed that, even on the evidence of infidels, Jesus Christ came in to the world around seventeen centuries ago. We possess an account of him written by his disciples, who saw him and bore witness; we have prophecies which foretold him long before his coming; and we know the dates and authors of these as far back as Moses, whose books are the most ancient in the world.[62]

There are some contingent features of the development of Christianity that were considered to be integral to its history, such as its persecution, which provided martyrs who offered a kind of living proof of the Word in their rejection of the values of this world (by contrast with Islam, which as Christian writers pointed

from the thirteenth century onwards. See Craig Martin, *Subverting Aristotle: Religion, History, and Philosophy on Early Modern Science* (Baltimore, 2014).

[60] Sebastien Le Nain de Tillemont, *Histoire des empereurs et des autres princes, qui ont regné durant les six premiers siècles de l'eglise* (3 vols, Brussels, 1743), i. p. iii.

[61] Pocock, *Barbarism and Religion*, v. 2.

[62] Claude Fleury, *Histoire ecclesiastique* (32 vols, Brussels, 1716), i. 2a.

out was never persecuted), but the value of investigation of such features is very limited, and anything more is curiosity, which is condemned.[63] By contrast, there was one contingent feature of the development of Christianity that was recognized as such and to which a great deal of attention was paid: heresy. We saw earlier that the doctrinal nature that Christianity came to take on from an early stage made heresy integral to its self-conception. Nevertheless, heresy, unlike martyrdom, could never have been a feature of the *internal* development of Christianity on the traditional account. More than anything else, the history of heresy opens up a space for a kind of 'secular' ecclesiastical history.

What provoked the intense reaction to the first volume of the *Decline and Fall* from Protestant critics was the period of Christianity that it covered. Constantine's joining of Christianity to the Roman state and the attendant establishment of the powers of the priesthood was the subject of much Protestant criticism, and although Gibbon does deal with this question, it is in chapters 20 and 21 of the second volume. The last two chapters of the first volume, chapters 15 and 16, deal with the Church before Constantine's reforms, a period in which sacred history had traditionally driven ecclesiastical history. Gibbon was not the first to subject the period to scrutiny, and eighteenth-century Protestant theologians, in asking to what extent the Nicene theology of post-Constantine Christianity was genuinely a continuation of apostolic teaching, and to what extent it was a matter of historical contingencies, raised questions about the status of Christian teaching in earlier centuries. There emerged serious scholarly reasons for suggesting that the question might be reduced to a historical narrative. But this was a dangerous route. Was it possible, as Pocock asks, to deny that the second-century Fathers 'had continued the mission of the Apostles without nullifying that mission and with it the Gospel narratives, the miracles of Christ himself and even his presence as the incarnate Word?'[64]

An important source for Gibbon was Le Clerc, who believed that in the second century the Church was providentially guided by four principal means: the miracles that continued to the last disciples of the Apostles; the refutation of pagans by Christian writers; martyrdom, by which Christians signalled that they preferred death to abandonment of their faith; and by the superior lives and morality of Christians.[65] Gibbon, while accepting these are causes, parts company with Le Clerc significantly here, because, in opposition to Le Clerc, he denies that each of them is supported and promoted by a divine active agency.[66] This raises the issue of what was contingent in the early history of Christianity, and here another major source that Gibbon drew on, namely the two ecclesiastical histories

[63] Pocock, *Barbarism and Religion*, v. 52. [64] Ibid., 18.

[65] Jean Le Clerc, *Historia Ecclesiastica duorum primorum a Christo nato sæculorum a veteribus monumentis depromta* (Amsterdam, 1716), 533–4.

[66] Pocock, *Barbarism and Religion*, v. 134.

of Mosheim, was crucial.[67] Mosheim was a Lutheran, so considered that salvific truth was in the Scriptures, not in the Church, which was simply a human association subject to geographical, historical, and cultural contingencies like any other association. In particular, the papal ascendancy was for him the outcome of unintended consequences and circumstantial changes, a shift from a democracy of believers to an absolutist monarchy, an issue on which there was unique agreement between *philosophes* and devout Protestant authors.

The ground on which the debate over just what was subject to a secular historical reading and what was considered an essential feature of Christianity that was either sacred or part of a providential design had, then, been prepared along confessional lines. This is the ground on which Gibbon will work, though he is not writing a confessional history. Chapter 15 of *The History of the Decline and Fall* sets out his proposed treatment of pre-Constantine Christianity in these terms:

> Our curiosity is naturally prompted to inquire by what means the Christian faith obtained so remarkable a victory over the established religions of the earth. To this inquiry an obvious but satisfactory answer may be returned; that it was owing to the convincing evidence of the doctrine itself, and to the ruling providence of its great Author. But as truth and reason seldom find so favourable reception in the world, and as the wisdom of Providence frequently condescends to use the passions of the human heart, and the general circumstances of mankind, as instruments to execute its purpose, we may be permitted, though with becoming submission, to ask, not indeed what were the first, but what were the secondary causes of the rapid growth of the Christian church? It will, perhaps, appear that it was most effectually favoured and assisted by the five following causes:—I. The inflexible, and, if we may use the expression, the intolerant zeal of the Christians . . . II. The doctrine of a future life, improved by every additional circumstance which could give weight and efficacy to that important truth. III. The miraculous powers ascribed to the primitive church. IV. The pure and austere morals of the Christians. V. The union and discipline of the Christian republic.[68]

I shall confine my attention to the first three, as it is these that bear most centrally on the question of the secular history of Christianity.

In discussing the first set of causes, 'intolerant zeal of the Christians', Gibbon contrasts both ancient Judaism and early Christianity with ancient forms of polytheism in which the questions of toleration simply do not arise. The theme is a Humean one, though one that is also pursued in other sources for Gibbon's account, such as Warburton and Mosheim.[69] The view that ancient polytheism was a social form of religion that bound together the members of ancient society,

[67] As well as Mosheim's *De Rebus Christianorum*, see also his *Institutionum historiæ ecclesiasticæ antiqui et recentioris quattuor* (Helmstedt, 1755). See the discussion in Pocock, *Barbarism and Religion*, v. 163–212.

[68] Gibbon, *The History of the Decline and Fall*, i. 152.

[69] See Pocock, *Barbarism and Religion*, v. 248–58.

only to be overcome by a form of zealotry, was one that Christians, to the extent to which they might have accepted the historical outline, would have considered a case where a statement of secondary causes obscured and relativized the events.

On the second set of causes, the discussion of the immortality of the soul takes its bearings from Book II of William Warburton's *Divine Legation of Moses*,[70] where the case is made that personal immortality is a distinctively Christian doctrine, unknown in Judaism. Immortality cannot be demonstrated philosophically, Warburton argues, and can only be grasped through revelation, but such revelation was not granted to the Jews, showing the superiority of Christianity. Personal immortality was one of the core doctrines of Christianity, and Warburton's critics worried that, given the continuity between ancient Judaism and Christianity, something so fundamental should have been withheld from the Jews, that life after death, with God in another world, was completely unknown until it was revealed by Jesus. Gibbon comes very close to endorsing Warburton's position in his discussion, in contrast to the view that Jesus had come to perfect earlier religion, not to replace it with something completely new, a doctrine that reflects Augustine's rejection of the Manichaean teaching that Christianity was a completely new religion of a loving God, replacing the vengeful God of the old Jewish religion. A great deal hung on the relation between Christianity and Judaism, and severing the continuity on an important doctrine like immortality was a potentially dangerous move. Moreover, Hume had intimated that a belief supported only by revelation was without foundation and susceptible to sceptical doubt. Gibbon is careful in his treatment of immortality, but from the point of view of a purely historical investigation, one examining 'second causes', immortality becomes reduced to an opinion conducive to the propagation of Christianity.[71]

Gibbon's treatment of the third cause, miracles, was for his critics one of the most problematic sections of the whole book. Even more was at stake than in the case of immortality. Miracles were evidence of divine authority. They convinced through the senses, an analogue to revelation, which convinced through the intellect. In the case of Jesus' miracles, they were evidence that his mission, increasingly construed in moral terms in the eighteenth century, was divine, but miracles had extended beyond Jesus to his Apostles. On the Roman Catholic view, the power to perform miracles may have diminished since that time but it had never ceased, whereas Protestants considered that miracles had stopped at an early stage. A crucial source here for Gibbon was Conyers Middleton's *Free Enquiry into the Miraculous Powers* of 1749, which was devoted largely to a

[70] William Warburton, *The Divine Legation of Moses Demonstrated, on the Principles of a Religions Deist, From the Omission of the Doctrine of a Future State of Reward and Punishment in the Jewish Dispensation* (4 vols, London, 1738–41), i. 309–472.

[71] See Pocock, *Barbarism and Religion*, v. 261.

refutation of claims of the existence of miracles in the second century.[72] This in itself was in accord with much Protestant thinking, but the procedures by which Middleton assessed the evidence, and the scathing conclusions he reached about the reliability of witnesses infatuated with signs and wonders rather than the evidence of the senses, could very easily be extended backwards to the Apostles themselves and even to Jesus. Middleton himself did not do this—by contrast with Reimarus, who as we have seen did ridicule the Apostles' reports of miracles—but it was a possibility that he opened up, and a road that his critics believed he was going down. He located the cessation of the miracles at the end of the first century, as Le Clerc had done, but this was too close to the bone for many theologians. Gibbon believed that, given the nature of miracles and the difficulty of establishing the authenticity of reports, we would never be able to identify a precise time at which they ceased (in contrast to Le Clerc's claim that they ceased with the death of John at the end of the first century). He seemingly had no doubt that Jesus and the Apostles performed miracles, but his beliefs were not at issue. As a historian, his project was neither to affirm nor deny the existence of miracles, but to look at the consequences of a belief in revelation for ecclesiastical history. But such a bracketing of questions of truth and falsity on such foundational issues came close to a denial of the existence of miracles, to active disbelief, especially at a time where there existed such an explicit denial, in the work of Hume, and when Middleton was generally being taken to be denying their existence.

THE HISTORICIZATION OF DOGMA

Gibbon offered historical understanding in place of faith. Turning from Gibbon to the Germany of the 1830s, a different kind of critical approach to Christianity was being developed, but it was one where again history played the key role. The use of secular history in rethinking the nature of religion in general, and Christianity in particular, was one of the most powerful weapons in the reform of Christian theology. As Strauss was to put it, 'the true criticism of dogma is its history'.[73] Note however that what was at issue here, at least initially, was reform of Christianity, not its overthrow. As Bödeker notes, 'the greatest challenge to the churches, and the greatest threat to them, lay in the fact that the *Aufklärung* did not turn away from religion, but instead took over its questions and tried to

[72] Conyers Middleton, *A Free Inquiry into the Miraculous Powers which are supposed to have subsisted in the Christian Church From the Earliest Ages through several successive centuries. By which it is shewn, That we have no sufficient Reason to believe, upon the Authority of the Primitive Fathers, That any such Powers were continued to the Church, after the Days of the Apostles* (London, 1749).

[73] David Friedrich Strauss, *Die christliche Glaubenslehre in ihrer geschichtlichen Entwicklung und im Kampfe mit der modernen Wissenschaft* (2 vols, Tubingen, 1840–1), i. 71.

answer them itself, with the aim of providing a new foundation for religion'.[74] German biblical scholarship from the second decade of the eighteenth century had uncovered various layers in the Bible, which seriously undermined the idea that it was the single document on which Lutheran and Calvinist theology depended. Johann David Michaelis had argued in detail that the Jewish law of the Bible was a result of the imposition of a legal code derived from Egypt on primitive Hebrew customary law, and that elements of Egyptian mythology remained, such as the creation story, originally an Egyptian polytheistic cosmology. At the same time, the Halle theologian Johann Salomo Semler, who had published a response to Reimarus in 1779, had argued that the Bible comprised different books written at different times under different social and historical conditions, and that large sections of both the Old and the New Testaments, despite being God's word, were designed to accommodate local superstitions and prejudices, and as a consequence should be subjected to a secular historical reading. This kind of approach culminated in 1835, in two works that were uncompromising in their criticism of canonical texts: Wilhelm Vatke's *Die Religion des Alten Testaments*,[75] which dealt with the Old Testament, and the one-and-a-half thousand page treatise of the twenty-seven-year-old Strauss, *Das Leben Jesu*.[76] Vatke's treatise was written in dense prose, and was largely ignored, whereas that of Strauss, more elegantly written, immediately became a *cause célèbre*,[77] although, as one reviewer of *Das Leben Jesu* pointed out, the book was not original, but rather a synthesis of many different sources from Kant onwards,[78] including Schleiermacher and Strauss' teacher Ferdinand Christian Baur, who in his *Symbolik und Mythologie* had argued for the crucial role played by myth in religious thought.[79]

[74] Hans Erich Bödeker, 'Die Religiosität der Gebideten', in K. Gründer and K. H. Rengstorf, eds, *Religionskritik und Religiostät in der deutschen Aufklärung* (Heidelberg, 1989), 145–95: 148.

[75] Wilhelm Vatke, *Die Religion des Alten Testaments nach den kanonischen Büchern entwickelt* (2 vols, Berlin, 1835).

[76] Strauss, *Das Leben Jesu, kritische bearbeitet* (2 vols, Tübingen, 1835–6). On Strauss' Christology see Friedrich W. Graf, *Kritik und Pseudo-Spekulation: David Friedrich Strauß al Dogmatiker im Kontext der positionellen Theologie seiner Zeit* (Munich, 1982).

[77] Among its fiercest critics was Strauss' colleague at Tübingen theology faculty, Carl August von Eschenmayer, *Der Ischiotismus unserer Zeit—eine Zugabe zu dem jüngst erscheinen Werke: Das Leben Jesu, von Strauss* (Tübingen, 1835). But it also had supporters. It was translated into English, after two years labour, by George Eliot: *The Life of Jesus: Critically Examined*, trans. Marian Evans (3 vols, London, 1846), and subsequently became influential in England. In France it was taken up by Ernest Renan in his *Vie de Jésus* (Paris, 1863), the first of an eight-volume series on the history of the origins of Christianity (1863–81).

[78] Edgar Quinet, 'Examen de La Vie de Jésus' (1838), in *Œuvres complètes* (10 vols, Paris, 1857) iii. 283–352.

[79] Ferdinand Christian Baur, *Symbolik und Mythologie, oder die Naturreligion des Altherthums* (2 vols, Stuttgart, 1824–5). Cf. Georg Friedrich Creuzer, *Symbolik und Mythologie der alten Völker, besonders der Griechen* (4 vols, Leipzig and Darmstadt, 1810–12), which attempted to reconcile ancient religions with Christianity through the notions of symbol and myth.

Nevertheless, with the appearance of *Das Leben Jesu* it was as if all the fears of the critics of Middleton and Gibbon had been realized. But Vatke and Strauss, while they used philological methods that would have been familiar, in a less sophisticated form, to Middleton and Gibbon, combined these with a philosophical theory, that of Hegel, of a kind utterly different from anything with which Middleton and Gibbon would have been familiar. Hegelianism set the questions of sacred and secular history in a context that was quite distinct from that in which Gibbon had been working. In particular, Hegelianism deploys a broad notion of myth, very close to the modern one that I quoted at the beginning of the chapter, that in itself does not differentiate between Christian sacred history and other forms of engagement with the world that operate through the intermediary of religious and aesthetic means.

The key notion in the Hegelian treatment is that of 'interpretation'. We have seen that Kant's conception of God was indirect, approached via practical rather than theoretical reason, as something that must be the case if we are make sense of our understanding of morality. Because of this, the content of the notion was very open, and while there was no obstacle to identifying Kant's God with the God of Christianity, nor was there any obstacle to identifying it with some other understanding. The move from Fichte's Kantian subjective idealism to the absolute idealism of Schelling and Hegel involved the postulation of an even more abstract notion—the 'Absolute'—which had strong pantheistic overtones.[80] The Absolute, conceived in undeniably Spinozean terms as the totality of things, could be represented, in Schelling and in Hegel, through symbols, myths, dogmas, and historical narratives. Classical and biblical myths, Hegel argued, possess an inner meaning that was distinct from their representations. 'We are directly conscious,' he writes, 'that they are only images, but that they [also] have a significance distinct from that which the image as such primitively expresses—that the image is something symbolic or allegorical and that we have before us something twofold, first the immediate and then what is meant by it, its inner meaning.'[81] In other words, as one commentator sums it up, myth is 'the aesthetic representation of the Absolute in the form of narrative. The narrative shell of myth was ultimately separable from its true meaning and possessed no significance of itself.'[82] In the case of Christianity, the Trinity, Creation, and the Last Judgement fell under the rubric of myth, but not the story of Jesus, which Hegel considered was a historical fact and a direct revelation.

[80] The details of this complex process fall outside our current concerns. See the accounts in Frederick C. Beiser, *German Idealism: The Struggle Against Subjectivism 1781–1801* (Cambridge, Mass., 2002); and Eckart Förster, *The Twenty-Five Years of Philosophy: A Systematic Reconstruction* (Cambridge, Mass., 2012).

[81] George W. F. Hegel, *Lectures on the Philosophy of Religion, One-Volume Edition. The Lecture of 1827*, ed. P. C. Hodgson (Berkeley, 1988), 145–6.

[82] Williamson, *The Longing for Myth in Germany*, 157.

On this conception, both history and myth were forms of representation, and Hegel's ultimate aim was the reconciliation of religion and philosophy into a new synthesis. The task of reconciliation was not straightforward, however, and Hegel himself had avoided questions of the Bible's historical truth. One possible route was to try to reconcile Hegelian philosophy with Lutheran theology, a path taken by conservative opponents of Schleiermacher in the University of Berlin theology faculty who had converted to Hegelianism, such as Daub, Marheineke, and Göschel, and towards the end of his life Hegel sanctioned this way of proceeding.[83] A particularly important development of this approach can be found in Karl Rosencranz. Hegel had considered philosophy and religion as different in form (the two forms were *Vorstellung*, representation, and *Begriff*, concept) but identical in content. Rosencranz, considering Hegelian philosophy as the way of grasping the Christian revelation of the actualization of spirit in the world, attempted to shift theology away from historical studies entirely in favour of an ahistorical philosophically guided systematic theology.[84]

By contrast, there were those who came to consider that the reconciliation of Hegelian philosophy and Christianity merely covered over a fundamental divergence, and it was above all Strauss who opened the path in the opposite direction to the reconciliatory strategy, separating religious experience and philosophical comprehension and bringing them into open conflict.[85] Strauss' aim was to use historical criticism to set out the next spiritual stage after traditional Christianity, to move from a religious to a philosophical understanding. His original intention in *Das Leben Jesu* had been to divide the study into three parts: a study of the traditional view of Jesus as presented in the evangelists; a critical analysis of the historical record of Jesus; and a philosophical reconstruction of what had been subjected to critical analysis in the second part. But he abandoned the first part and reduced the third to a supplement, so that the book dealt almost exclusively with what had originally been intended as the second part.[86]

There are two main theses in *Das Leben Jesu*. The first is the importance of myth. Strauss rejects not only the supernaturalist explanations whereby divine guidance drives the accounts of the Apostles, but also those rationalizing accounts in which everything is the result of natural phenomena. He does not want simply to accept the accounts of the Apostles but nor does he want to explain them away. These were the kinds of limited options facing Gibbon and other writers of ecclesiastical history in England and France. Using the resources of Hegelianism, Strauss is able to introduce a third option: myth. Examining the Gospels' accounts, he concluded that the events reported could not have happened.

[83] See John E. Toews, *Hegelianism: The Path Towards Dialectical Humanism, 1805–1841* (Cambridge, 1980), 141–55.

[84] Karl Rosencranz, *Encyklopädie der theologischen Wissenschaften* (Halle, 1831).

[85] See Toews, *Hegelianism*, 165–75, 255–87.

[86] See William J. Brazill, *The Young Hegelians* (New Haven, 1970), ch. 3.

They were not deliberate falsehoods, but representations in a concrete literary form of the truths inherent in the states of mind and emotions of those reporting the events. They were symbolic representations of truth, not literal ones. Myths were very much products of their time in a way that literal historical narratives were not however, which meant that they were not valid in perpetuity. The time had now come, in Strauss' view, to move beyond the mythical accounts of the Apostles. But historical accounts, crucial as they were to establishing the standing of the Apostles' stories, were just not the right kinds of things to replace the now outdated mythological accounts. This task was rather one for a philosophical understanding of the Christian message.

The second main thesis of *Das Leben Jesu* sets out this philosophical understanding. It is that it was not just Jesus, but humanity as a whole that shared human and divine natures. This was the deep underlying truth of the Gospels, which was not evident in the mythical form that they took, but could finally be revealed once the message was translated into a philosophical form. The basic argument is that the doctrine that Jesus taught was not his own divinity but rather the dual nature of all humanity. However, because of the myths associated with him, what happened was that this message was lost and instead he was deified. By worshipping him, human beings alienated themselves from what was in fact their true spirit.[87] The solution was to cease worshipping Jesus, and instead to follow his example and consciously participate in the life of humanity. This could only be achieved if the traditional moral tasks of Christianity were transferred from the church to the state: 'No peace will be found until eternity is fully absorbed into time, piety has been completely taken up in morality, and the church as been absorbed into the state'.[88]

THE EVOLUTION OF RELIGIONS

Strauss treated Christianity as an 'absolute religion', an end point of religious consciousness. The idea that religious consciousness had evolved through a number of stages was a common theme in many theologians and historians influenced by Hegel, but while the idea of an evolution of 'religious consciousness' is distinctively Hegelian, the general theme of the evolution of religions is not. In fact in neither case does the term 'evolution' necessarily imply a chronological sequence here. It was central to Christian theology that Christianity, through its link with early Judaism, was the oldest religion, so there were theologically speaking no religions that predated it. But there were religions, for example forms of ancient paganism, that were taken to be more primitive than Christianity, and by the eighteenth century it had become a task of Christian

[87] Strauss, *Das Leben Jesu*, ii. 729–30.
[88] Strauss, *Die christliche Glaubenslehre*, ii. 618.

theology to account for other forms of religion and to relate them to Christian teaching, usually on the medieval model of identifying and explaining heresies. There was no uniform approach here however. Consider the question of pagan deities. In Samuel Shuckford's euhemerist *Sacred and Profane History of the World* (1728), Hesiod's theogony becomes political history, with Cronus' devouring of his children meaning simply that he sent them to school in foreign parts, and Zeus was made into a contemporary of Moses who ruled Crete wisely, administered justice through subordinates, and was later divinized.[89] By contrast, Le Clerc notes that early Christians, far from explaining away the Greek gods, fully accepted their reality, believing that they were fallen angels who had become demons.[90] Different yet again, Greek gods were sometimes simply Christianized, as with the sculpture of Jupiter in the guise of a monk on the bell tower of Santa Maria del Fiore, Florence, dating from the late 1330s.[91] While none of these extremes of interpretation would have been generally accepted by theologians in the period with which we are concerned, they give us a sense not only of the range of possibilities, but also of the dangers of dealing with religions in comparative terms. To accept the reality of non-Christian gods, even by fitting them into a biblical story, would have potentially meant a fundamental reassessment of all non-Christian deities, whereas simply explaining them away raises the issue of whether the Christian God could be explained away on similar grounds.

One important form of comparative studies of religions and mythologies was a form of stadial history. As I indicated earlier, for Christian theologians and historians, the assumption behind their stadial histories was that religions manifest a form of evolutionary development in which Christianity emerges as the final form of religion. Pre-Christian Judaism had been treated as a precursor to Christianity from the second century, and writers like Augustine had treated classical philosophy as an incomplete form of Christianity: in effect anything of significance in the pre-Christian era was assimilated to Christianity by the time of Augustine, so the idea of non-Christian forms of thought and worship being provided with a place in the development of Christianity was not something novel. The kind of development we are concerned with here, however, is distinctive in that it is one in which this approach is combined with an independent study of ancient mythologies with a view to providing a more comprehensive understanding of the development of religions. Despite the Jesuit missionary

[89] Samuel Shuckford, *The Sacred and Profane History of the World Connected* (2 vols, London, 1728). Shuckford relies on the main renaissance source for euhemerist readings of Greek mythology: Giacomo Filippo Foresti da Bergamo, *Supplementum chronicarum* (Venice, 1483). Euhemerus was writing in the third century BCE.

[90] Le Clerc, *Historia Ecclesiastica*, 48–95.

[91] See Barbara Graziosi, *The Gods of Olympus: A History* (London, 2013), 178–81. Cf. Dante's reference to Jesus Christ as 'the highest Jupiter': *Purgatorio*, 6.118. In an influential account, Jean Seznec argued that the Greek gods had a continuous history between antiquity and the fourteenth century: *La survivance des dieux antiques. Essai sur le rôle de la tradition mythologique dans l'humanisme et dans l'art de la Renaissance* (London, 1940).

reports and accounts in travel books of the existence of various forms of atheism outside Europe from the sixteenth century onwards, there was a seemingly unassailable assumption that there were forms of religious consciousness in all societies, and the aim was to discover how these had ultimately become (or should ultimately become) focused in the form of Christianity—how they effectively converge on Christianity as the ultimate goal.

In mid-eighteenth-century France, the study of ancient mythologies was institutionalized in the Académie des Inscriptions et Belles-Lettres, where Nicolas Fréret played an important role. In the 1720s, a number of his *mémoires* were published by them, including an important piece on chronology and historical evidence,[92] and it was Fréret who was at the forefront of the scholarly criticisms of Newton's chronologies of the ancient world in the 1740s.[93] In 1760, the academician Charles de Brosses[94] published his *Du culte des dieux fétiches*,[95] whose two most significant features were its historicization of fetishism, and its hierarchical ordering of religions, with fetishism occupying the lowest and most primitive level. In seeking a single origin for religions, de Brosses' work was not unique: the *L'antiquité dévoilée* of his contemporary Boulanger, published posthumously in 1766, offered an elaborate account of how all religious cults were responses to the Flood.[96] But *Du culte des dieux fétiches* opened up enduring questions about the origins of religion in primitive societies.

De Brosses' account of fetishism (a term coined by him[97]) was a solution to the problem of zoolatry, the worship of brute animals, as depicted in the images in

[92] Nicolas Fréret, 'Réflexions générales sur l'étude des anciennes histoires et sur le degré de certitude des différentes preuves historiques (1724)', in idem, *Mémoires académiques* (Paris, 1996), 73–126.

[93] Fréret, *Défense de la chronologie fondée sur les monuments de l'histoire ancienne, contre le système chronologique de M. Newton* (Paris, 1758).

[94] De Brosses' contemporary standing also rested on an anthropologically inspired history of navigation: *Histoire des navigations aux terres australes, contenant ce que l'on sait des moeurs et des productions des contrées découvertes jusqu'à ce jour* (2 vols, Paris, 1756); and a history of languages: *Traité de la formation méchanique des langues et des principes physiques de l'étymologie* (2 vols, Paris, 1765).

[95] De Brosses, *Du culte des dieux fétiches, ou Parallèle de l'ancienne Religion de l'Égypte avec la Religion actuelle de Nigritie* (Geneva, 1760). See Manuel, *The Eighteenth Century Confronts the Gods*, 184–209; Aaron Freeman, 'Charles de Brosses and the French Enlightenment Origins of Religious Fetishism', *Intellectual History Review* 24 (2014), 203–14; Joan-Pau Rubiés, 'Theology, Ethnography, and the Historicization of Idolatry,' *Journal of the History of Ideas* 67 (2006), 571–96; Clelia Martînez Maza, 'De la théologie à l'anthropologie: utilisation du paganisme gréco-romain dans l'Europe du XIXème siècle', in C. Avlami, J. Alvar, and M. R. Recio, eds, *Historiographie de l'antiquité et transferts culturels: Les histoires anciennes de l'Europe des XVIIIe et XIXe siècles* (Amsterdam, 2010), 85–98.

[96] Nicolas-Antoine Boulanger, *L'antiquité dévoilée par ses usages, ou Examen critique des principales opinions, cérémonies et institutions religieuses et politiques des différens peuples de la terre* (3 vols, Amsterdam, 1766).

[97] It is a Gallicization of the Portuguese *fetisso*, a word used by merchants to describe the small ritual statues they found on the coast of Senegal.

Egyptian obelisks, mummies, and papyri.[98] Zoolatry was regarded as the most degrading form of irreligion, far worse than atheism since it reversed the order of nature, in which human beings occupied the pinnacle of the natural realm. Consequently, a number of writers, particularly in the seventeenth century, had attempted to explain away the Egyptian images allegorically.[99] De Brosses by contrast mounted a detailed refutation of allegorical readings, arguing that all Egyptian religion was literally a form of animal worship, something also to be found in contemporary travel reports of the practices of savages. Indeed, his larger claim was that primitive religions were not monotheistic, as Christian theologians believed, nor were they polytheistic as Hume had argued. Rather there was a universal primitive stage of fetishism: this was the original form of religion. The original Greek gods, he suggests, took shape from sacred rocks.[100] Combing through the classical literature, he finds overwhelming evidence of an original enduring fetishism: Diodorus Siculus reports on monkey worship in Egypt; Plutarch records the clandestine 'murder' of fetishes during calamities; Herodotus reports Mycerinus encasing his daughter in a cow in order to have her worshipped; Suetonius describes how Augustus, after a disaster, avenged himself on the statue of Neptune as if it were a fetish; further afield, there are reports of Germanic tribes worshipping trees. But as Manuel points out, de Brosses' most radical thesis was that the scriptural record is awash with fetish objects, providing 'confirmation of his thesis that the practice was universal. Abraham had a sacred grove; Rachel stole her father's teraphim [idols]; Jacob anointing a sacred stone at Bethel was performing a particularly dramatic fetish act.'[101] Although such traces of fetishism can be found in monotheism, de Brosses is somewhat circumspect about drawing the consequences for Christianity. Not so Diderot. In his article 'Oindre' (anointing) in the *Encyclopédie*, he writes that:

> in fetishism—considered the oldest, most widespread, and the first of all religions in conjectural and natural history—those who took a stone as a fetish *anointed* it in order to mark it out; from this arose the custom of *anointing* everything that was endowed with a divine or a sacred character. . . . It is said of Jesus Christ that he is *anointed* of the Lord. The

[98] See Manuel, *The Eighteenth Century Confronts the Gods*, 189–94.

[99] For example, Nicolas Caussin, *Symbolica Aegyptiorum Sapientia* (Paris, 1631), and Athanasius Kircher, *Oedipus Aegyptiacus* (3 vols, Rome, 1652–4). Allegorical readings had a long history. In particular, the Greek gods were interpreted allegorically as early as the sixth century BCE, by Theagenes of Rhegium—see Jean-Pierre Vernant, *Mythe et Société en Grèce ancienne* (Paris, 1971), 212—and in the fifth century by Metrodorus of Lampsacus. They were taken to represent different elements (Poseidon water and Apollo fire) or to represent pyschological qualities (Athena wisdom versus Ares folly, for example). See Giulia Sissa and Marcel Detienne, *La Vie quotidienne des dieux grecs* (Paris, 1989), and Paul Veyne, *Did the Greek Believe in their Myths?: An Essay on the Constitutive Imagination* (Chicago, 1988).

[100] This view goes back at least to the second-century CE historian Pausanius: 'in early times, all the Greeks worshipped unshaped stones instead of statues', *Hellados Periegesis* 7.22.4.

[101] Manuel, *The Eighteenth Century Confronts the Gods*, 202.

Lord said, beware of touching my *anointed* ones: these *anointed* ones are kings, priests, and prophets.[102]

The implication is that, since traces of ancient fetishism remain in Christianity, it is the product of an evolution from the most primitive form of religion, something which could not be further from monotheism. The theme was developed in detail by German theologians in the 1780s and 1790s, starting with Johann Gottfried Eichhorn's *Urgeschichte* of 1779,[103] which introduced a mythical (and indeed historicist) reading of the Old Testament.

The possibility of traces of ancient forms of savage worship remaining in Christianity was not the only problem raised by stadial theories of the development of religious consciousness. Far more challenging was the combination of two developments: a stadial account, in which Christianity stood at the end point of the evolution of religious consciousness, with a Straussian conception of Christianity. This combination turned out to be explosive, yielding a Godless but divine humanity as the final point of the evolution of religious consciousness. It is not surprising that Comte, in 1853, in his *Système de politique positive*, a work whose subtitle tells us that it institutes 'the religion of humanity', should write that 'the characteristic feature of the present century will be the overwhelming importance it assigns to history, by the light of which philosophy, politics, and even poetry will be pursued'.[104] And it was indeed history—even if a largely conjectural history—that provided the tools that enabled Feuerbach, twelve years earlier, to take the development of religious consciousness beyond Christianity to a form of humanism in his *Das Wesen des Christentums* (1841).[105]

Working within a Hegelian framework, Feuerbach's aim in examining the development of religious consciousness was, even more explicitly than that of Strauss, to consider it within the context of an account of the meaning and direction of history, in which religion played a crucial role but, it turns out, not the permanent one that its advocates had assumed. Hegel himself had prevaricated over whether history was ultimately sacred or secular, but that group of his followers called the 'Young Hegelians'—a group that included Strauss, Feuerbach, Bauer, and Stirner—had insisted that it was secular, and they contended that, as one commentator has put it, 'the essential feature of the historical age in which they were living was that it marked the spirit's progress from religion to philosophy. Men must accept that progress in order to be free—that was the meaning of history.'[106] The difference between Strauss and Feuerbach was that,

[102] Diderot and d'Alembert, *Encyclopédie*, xxiii. 570.

[103] Johann Gottfried Eichhorn, *Urgeschichte* (Altdorf and Nüremberg, 1779), appearing in a greatly expanded edition with commentary by Johann Philipp Gabler, 1790–3.

[104] August Comte, *Système de politique positive, ou Traité de sociologie, Instituant la Religion de l'Humanité* (4 vols, Paris, 1851–4), iii. 1.

[105] Ludwig Feuerbach, *Das Wesen des Christentums* (Leipzig, 1841); 2nd edition translated into English by George Eliot as *The Essence of Christianity* (London, 1854).

[106] Brazill, *The Young Hegelians*, 136.

whereas Strauss was concerned to reveal the true content of Christianity through historical criticism of dogma, Feuerbach was concerned rather with how dogma was simply a form of representation of human fears, aspirations, and needs. In Feuerbach, history becomes a process whereby one increasingly comes to know oneself through creating representations of the divine until ultimately one comes to see that these are mere representations.

Feuerbach's first published work after his doctoral dissertation, a work that led to his being dismissed from his academic position, was his 'thoughts on death and immortality',[107] which focused on personal immortality, in effect offering a form of Averroism. Averroes had argued that individuals have a separate identity through their bodies while alive, but on the death and corruption of the body there is no longer any way of individuating minds, so that they become one and the same.[108] Feuerbach's argument works in a similar way, except that he construes the individuality that one has while alive very much in terms of one's contribution to humanity as a whole, so that death becomes a natural extension of this: one becomes part of humanity in general. This then allows him to move to a programme of criticism of Hegel's speculative idealism, transforming the theological notion that the idea of God is universal into the claim that there is a universal urge to create God.

One of the core arguments of Feuerbach's *Das Wesen des Christentums* is that human beings had created ideals which they had projected on to the world, reified, and made properties of a divine being. The ostensible objects of worship, whether these be the rocks and animals of primitive religions or the anthropomorphic God of Christianity, are only ever images or representations, devised by human beings and invested with special qualities. For Feuerbach, 'the true meaning of theology is anthropology', and religion, in the qualities it assigns to God 'only defines or represents the true human nature of the word'.[109] The arguments turn on a rejection of the idea that we have access to some extra-human perspective from which human nature could somehow be regarded as defective or limited. The supposed entities that religion postulates are 'nothing but the subject's own nature taken objectively'.[110] This being the case, the goal of history cannot be the reconciliation of God and man, but must rather be the actualization and self-comprehension of human essential capacities and powers.[111] This is an essentially collective activity, not one that it makes sense to ascribe to an individual. Moreover the historical dimension of this collective activity is crucial: religious consciousness goes through various stages of development and self-awareness until finally, in Christianity, God is identified with a man, Jesus, and

[107] Feuerbach, *Gedanken über Tod und Unsterblichkeit* (Nürnberg, 1830).
[108] See Gaukroger, *The Emergence of a Scientific Culture*, 101–6 on the Averroist doctrine of one mind.
[109] Feuerbach, *The Essence of Christianity*, xxxvii.
[110] Ibid., 12. [111] See Toews, *Hegelianism*, 340–55, which I am indebted to here.

it remains only to take the final step, namely that of accepting that God has been man, in the form of humanity, all along.[112] The key here lies in the interpretation of the Incarnation. On the Christian reading, God is led by love of humanity to alienate his divinity. For Feuerbach, the 'natural reduction' of the Incarnation reveals the deep meaning of all religion, that God is not indifferent to those who worship him. But such a God is already human:

It is the consciousness of love by which man reconciles himself with God, or rather with his own nature as represented in the moral law. The consciousness of the divine love, or what is the same thing, the contemplation of God as human, is the mystery of the Incarnation. The Incarnation is nothing else than the practical, material manifestation of the human nature of God. God did not become man for his own sake; the need, the want of man—a want which still exists in the religious sentiment—was the cause of the Incarnation. God became man out of mercy: thus he was in himself already a human God before he became an actual man; for human want, human misery, went to his heart. . . . If in the Incarnation we stop short at the fact of God becoming man, it certainly appears a surprising, inexplicable, marvellous event. But the incarnate God is only the apparent manifestation of deified man; for the descent of God to man is necessarily preceded by the exaltation of man to God. Man was already in God, was already God himself, before God became man, i.e. showed himself as man.[113]

What Feuerbach, building on the efforts of Strauss and others, has done here is to reconfigure Christianity in such a way as to reveal what he argues is its real content—the work is, after all, entitled the *essence* of Christianity—and to discard those parts of it which obscure that content. In the process of stripping it down and rebuilding it, he has transformed it into a humanist anthropology that has no connection with Christianity as traditionally conceived. Nevertheless, if one wished to question the legitimacy of such an exercise, it must be borne in mind that in many respects it mirrors the process by which Augustine set out to establish the universal scope of Christianity. Christianity, on Augustine's view, had always existed, not just from before the Incarnation but from the beginning of time. He believed that this could be glimpsed in ancient Platonist philosophy, but that this worked in purely intellectual terms, whereas the superiority of Christianity over ancient philosophies and over the contemporary rivals of Christianity lay in the institution of the sacraments. The claim was not that Christianity was ancient philosophy plus the sacraments; it was that ancient philosophy was Christianity minus the sacraments. Ancient philosophy was essentially incomplete. Augustine reworked ancient thought in such a way as to establish Christianity as the culmination of all previous philosophical reflection and religious belief.[114] Although the details of how they accomplish their tasks are very different, both Augustine on the one hand, and Strauss and Feuerbach on the

[112] See the detailed discussion in Marx W. Wartofsky, *Feuerbach* (Cambridge, 1977), ch. 8.
[113] Feuerbach, *The Essence of Christianity*, 50.
[114] See Gaukroger, *The Emergence of a Scientific Culture*, 51–3.

other, are engaged in the same kind of exercise: radically reworking one tradition so as to incorporate it into another, by showing how, despite itself, it leads to something that can be seen as its culmination.

AESTHETIC HUMANISM AND THE CULTIVATION OF THE SELF

Nevertheless, we can still ask: if, as Feuerbach argues, religious consciousness is the alienated form of human self-consciousness, what happens when Christianity has been transcended? How does one come to terms with the world without the representational forms that religion provides? For *philosophes* such as Diderot and Condorcet, the transcendence of religion meant that society could finally achieve a harmonious balance through a combination of individual freedom and an education based on rational, humanitarian principles, which would replace the fear- and ignorance-driven religious conformity of the Christian era. The shaping of character was also a core concern of German Romanticism, but the mechanism was different: here everything hinged on *Bildung*, the cultural formation of character. As Toews notes in the case of Feuerbach:

> [his] continued use of the word *Bildung* to describe the defining characteristic of the new order was an expression of a vision of cultural liberation that would include the whole man, and not just the human spirit. . . . The Christian religion provided an illusory compensation for individual subjectivity through an imaginary union with an omniscient God. In [*Bildung*] the subjective deficiencies of the individual were overcome through progressive elevation to 'species consciousness'.[115]

German Romanticism engaged with three themes in Christianity: its concern with moral guidance, its aesthetic and literary qualities, and its antiquity. On the question of morality, it was deeply indebted to a Kantian notion of absolute freedom, which replaced a Christian ethics with an overtly humanist one. Morality was a matter of something generated by a process of coming to full self-consciousness, rather than something inculcated by an outside authority. But morality did not exhaust Christianity. Much of its standing derived from the aesthetic appeal not only of its architectural and pictorial heritage but above all of its sacred texts. The aesthetic appeal of its sacred texts was naturally construed within Christianity as a product of their divine nature, so that any account of their content had to be in religious rather than aesthetic terms, which had a strictly secondary importance. The Romantics, however, not only separated aesthetic appeal and made it part of an autonomous realm, but reversed the relation

[115] Toews, *Hegelianism*, 352.

between the divine and the aesthetic. As Schiller was to put it: 'In those days, nothing but the Beautiful was divine.'[116]

At the same time, much of the authority of Christianity lay in its claims to antiquity, that is, in the idea of its being the only religion that dated back to the beginnings of recorded history, but it also lay, as we have seen, in the postulated existence of an unbroken continuity between its origins and its present state. What the continuity consisted in was disputed among Catholics and Protestants, but each confession claimed continuity. For Protestants, for example, a 'figural interpretation' of the Bible, in which it offered a literal and self-sufficient universal narrative of creation, the Fall, and ultimate redemption, lay at the core of Christian belief.[117] In the wake of Winckelmann, both the Romantics and the Idealists took continuity with ancient roots as crucial to the standing of one's own culture, but these ancient roots were now, at least in large part, those of classical antiquity rather than those of Christianity. Classical antiquity was treated as a unique culture in which language, art, and politics merged into a coherent and harmonious whole. Although there was also a movement, associated with Herder and Jacob Grimm, to trace contemporary German culture back to an authentic Nordic culture of the original German *Volk*,[118] this did not displace the centrality of classical culture in the early decades of the nineteenth century, and in any case was not strictly in competition with it. Hegel, in a speech delivered on 29 September 1809 to the Nüremberg Gymnasium, stresses the continuity with classical antiquity:

> The spirit and purpose of our foundation is preparation for learned study, a preparation grounded on Greece and Rome. For more than a thousand years this has been the soil on which all civilization has stood, from which it has sprung, and with which it has been in continuous connection. Just as the natural organisms, plants and animals, struggle to free themselves from gravitation without being able to renounce this element of their own nature, so the fine arts and the sciences have grown up on that soil, and, while they have attained a self-subsistence of their own, they have not yet emancipated themselves from the recollection of that older culture. As Antaeus renewed his energies by touching his mother-earth, so every new impetus and invigoration of science and learning has emerged into the daylight from a return to antiquity.[119]

[116] Quoted in Marchand, *Down from Olympus*, 3. The most ardent defender of aesthetics over religion was Karl Philipp Moritz, a member of Goethe's Weimar circle: see his *Götterlehre oder Mythologische Dictungen der Alten* (Berlin, 1790).

[117] See Frei, *The Eclipse of Biblical Narrative.*

[118] See Williamson, *The Longing for Myth in Germany*, ch. 2. For Herder, the German language was a corrective to the decadent language of French courtiers and salons: see Anthony La Volpa, 'Herder's *Publikum*: Language, Print, and Sociability in Eighteenth-Century Germany', *Eighteenth-Century Studies* 29 (1995), 5–24.

[119] Hegel, 'On Classical Studies', in *Early Theological Writings*, trans. T. M. Knox and R. Korner (Philadelphia, 1971), 321.

One feature that binds questions of aesthetics and questions of antiquity is a larger concern with myth. In particular, once religion was thought of in mythological terms, then aesthetics became crucial, for myth was above all an aesthetic means of expression of religious consciousness. This held for those Romantics who remained within Christianity—such as Schleiermacher, for whom the object of religious intuition was conceived in explicitly aesthetic terms—as much as for those who rejected Christianity and called for a new mythology as a maieutic to the birth of a more thoroughly Romantic poetry, such as Friedrich Schlegel.[120]

Herder is a key figure here. Following on from the work of Heyne, he treated myth as something that bound together the language, literature, customs, and religion of the classical world into a coherent whole. The contrast was between the beauty and the public nature of Greek religion, and the hieratic and divided Christian churches.[121] Unlike stadial theories that saw development in largely teleological terms, where the interest in the 'primitive' was confined to where it led, Herder (like Vico) viewed the early stages of human history as its most aesthetically vital and religiously creative, becoming progressively debased as civilization and luxury came to predominate. Primitive language, in particular, had a richness lacking in modern forms, and Herder's pressing task was to recapture the vitality of the language of ancient fables, which dated from the time of creation and as a consequence reflected humanity's relation to the natural and human realms in a pure and fertile fashion.

We saw in Chapter 4 how aesthetics played a crucial role in shaping Herder's anthropology. His concern with language was above all with a means of expression, rather than simply a means of representation, and it was in large part aesthetically motivated. This aesthetic concern is immediately evident in his anthropology, for, in the practical realm, philosophical anthropology was directed towards *Bildung*.[122] Herder's conception of the cultivation of character was one based on an aesthetic model, as opposed to the traditional religious, metaphysical, or humanist models, or the newer socio-political and medical ones.

Some of the more significant features of Herder's understanding of the shaping of character can best be seen by contrast with a more traditional view, which goes back to Aristotle. Like Aristotle, Herder thinks of life in terms of the realization of purpose, but whereas the Aristotelian conception fostered an idea of this realization as establishing an order and harmony which is constantly threatened by disorder and disharmony, Herder thinks of it in terms of a force or striving against a world that opposes and would seek to shape one to its own ends. So far, this sounds very much like Spinoza, and Herder was indeed influenced by the revival

[120] See Richards, *The Romantic Conception of Life*, 94–113; Williamson, *The Longing for Myth in Germany*, 56–71.

[121] Williamson, *The Longing for Myth in Germany*, 9.

[122] In the tradition of German classical philology, clear classical precedents for this could be identified. See, for example, Werner Jaeger, *Paideia: The Ideals of Greek Culture* (3 vols, Oxford, 1939–45).

of Spinozism. There are however two significant features of Herder's account which marked it out from Spinoza's. The first is the view that one's ends are not perspicacious in their own right. They are not accessible simply through introspection. The attempt to realize these ends is a crucial part of understanding them, for struggling to realize one's ends is a way of clarifying just what those ends are: it is through action that we come to know ourselves. As Charles Taylor puts it: 'man comes to know himself by expressing and hence clarifying what he is and recognizing himself in his expression. The specific property of human life is to culminate in self-awareness through expression.'[123] Second, one does not act in the abstract, but rather in a particular historical context, which shapes what one can do and how. This may resist one in various ways, a resistance that one can strive to overcome, or it may offer paths by which one can come to know oneself. Whichever it is, we are always acting within a particular cultural context—typified in some respects by one's language—and within a particular type of physical context. In short, to the extent to which we come to discover and realize our purposes, that is, come to know ourselves, this is never *ab initio*, but always using the resources of a particular culture.

Bildung is, however, not just a matter of self-expression, but above all a question of moral striving. Moral striving as a feature of self-expression is a central theme from Lessing onwards. Lessing himself sums up the idea in terms of truth: 'The true value of a man is not determined by whether, in reality or not, he possesses the truth, but rather by his sincere striving to reach the truth. It is not possession of the truth, but rather the pursuit of truth that allows him to extend his powers, and in which his constantly increasing capacity for perfection is to be found.'[124] The idea of striving became reinforced in the 1790s when many philosophers, dramatists, and others took up Kant's idea of freedom as radical self-determination by the rational will. This complemented the idea, broadly shared, of one's responsibility for one's own inner development. No external factor, not even God, could constrain our freedom. This equation of one's moral life with absolute freedom would play an especially significant role in the development of German idealism, but it was also crucial as a complement to the notion of cultural self-expression, for which freedom of self-determination was a core objective.

Note however that the idea of 'moral striving' was a general one at the time, and not restricted to those who were to follow the Kantian path. Goethe for example had no place for the Kantian conception of absolute freedom and its associated

[123] Charles Taylor, *Hegel* (Cambridge, 1975), 17. Taylor sees a central connection between Herder's expressivism and Hegelian idealism, but this is questionable: see Michael Rosen, *Hegel's Dialectic and its Criticism* (Cambridge, 1982), 122–43.

[124] Anti-Goetze: Eine Duplik (1778) in: Gotthold Ephraim Lessing, *Werke* ed. H. G. Gölpert et al. (8 vols, Berlin, 1978), viii. 32. As Nietzsche commented: 'Therefore Lessing, that most honest thinker, dared to announce that he cared more for the search after truth than for truth itself': Friedrich Nietzsche, *Die Geburt der Tragödie aus dem Geiste der Musik* (Leipzig, 1872), §15.

notion of radical subjectivity. He had worked closely with Herder in the 1780s, and was particularly concerned with the idea of *Bildung*, philosophically but also in a literary context: his novel *Wilhelm Meister* was the first *Bildungsroman*—the plot is driven by the idea that the hero comes to discover himself as he progresses through the world—and set a model for the subsequent development of German literature.[125] Moral striving is as crucial to Goethe's idea of *Bildung* as it was to those who construed this striving in Kantian terms. Of the more explicitly philosophical accounts of how one should attempt to achieve self-understanding, however, the most influential was one that attempted to integrate the Kantian conception of morality and freedom into a picture of self-expression drawn from Herder and Goethe. This was Schiller's *Über die ästhetische Erziehung des Menschen* (1794): 'on the aesthetic education of man'. Schiller departs from Kant on a number of questions, and the most important for our purposes hinge on the problem of moral agency.[126] First, he argues that moral worth springs not just from the agent's intention, as Kant had argued in the 1780s, but that it is also necessary that the intention derive from the moral character of the agent. It is this that makes the formation of character the central concern of his anthropology. Second, freedom for Schiller is an autonomous development not just of reason but of reason and sensibility. In contrast to Kant, freedom could not be a matter of reason alone. Third, moral agency is something which is shaped by history and sensibility, rather than just being a product of reason, insulated from the natural world. In these respects, Schiller is far closer to Herder than to Kant.

The centrality of an *aesthetic* education to Schiller's anthropology is due to his conception of beauty as reconciling the two natures of human beings: sensibility and rationality, what in Kant are respectively our sensory receptive powers and our spontaneity, and what in Letter X of *Über die ästhetische Erziehung* take on a moral tone: 'So you are at one with me about this, and are convinced . . . that man can be drawn aside from his destiny in two opposite ways, that our age is actually travelling along both these false roads, and has fallen prey to coarseness on the one hand, and to ennervation and perversity on the other. From this twofold straying it must be brought back by means of Beauty.'[127] It is in aesthetic perception that we ultimately discover our unity and integrity.[128] The significance of this cannot be overestimated. For Schiller, life has an aesthetic dimension that is crucial to our

[125] See William H. Bruford, *Culture and Society in Classical Weimar, 1755–1806* (Cambridge, 1962).

[126] I take the following summary points from Frederick C. Beiser, *Schiller as a Philosopher: A Re-Examination* (Oxford, 2005), 3–4. Beiser develops them in detail in the course of the book.

[127] Friedrich Schiller, *On the Aesthetic Education of Man, in a Series of Letters*, ed. and trans. Elizabeth Wilkinson and L. A. Willoughby (Oxford, 1967), 63. This edition contains facing German text and translation.

[128] Cf. the other two landmark defences of art as a means of overcoming the fragmentation of humanity: 'Jean Paul' [Friedrich Richter], *Vorschule der Aesthetik* (Hamburg, 1804); and Samuel Taylor Coleridge, *Biographia Literaria, or Biographical Sketchs of My Literary Life and Opinions* (2 vols, London, 1817).

understanding of who we are, just as his predecessors and contemporaries had considered that life has a religious dimension that is crucial to our understanding of who we are. What Schiller is proposing is in effect a replacement of a religious by an aesthetic model of human development.

In Letter X, after proposing aesthetics as the solution to reconciling sensibility and reason, Schiller asks how it can achieve this, and the remainder of the essay is devoted to this question. Actually, there are two questions: how human nature can realize itself only as beauty, and—what he calls a brief excursion into the realm of speculation—what effect the different kinds of beauty have in developing the ideal human being. In the first, art is treated as a means to an end, whereas in the second it is treated as an end in itself.[129] Schiller's aim is to establish systematic connections between the two, building up a general account of the necessity for aesthetic education.

It is classical antiquity that provides his starting point:

> The Greeks put us to shame not only by the simplicity to which our age is a stranger; they are at the same time our rivals, indeed often our models, in those very excellences with which we are wont to console ourselves for the unnaturalness of our manners. In fullness of form no less than of content, at once philosophic and creative, sensitive and energetic, the Greeks combined the first youth of imagination with the manhood of reason in a glorious manifestation of beauty. At that first fair awakening of the powers of the mind, sense and intellect did not as yet rule over strictly separate domains.[130]

In classical times, there was a harmony of reason and sensibility, because the artist was still in unity with nature. The correspondence between thought and feeling has been lost in the modern age, however, so has to be re-established, albeit in a way that surpasses its original unreflective naive form, for it is now something that exists not inside oneself but outside, in the form of an idea that one strives to realize.[131] In classical antiquity, there was no specialization of knowledge, whereas in the modern era there are 'whole classes of men developing but one part of their potentialities'.[132] With the division of labour, not only a fragmentation of skills, but a fragmentation of the human being arose. This is associated by Schiller with a

[129] See Beiser, *Schiller as Philosopher*, 135–68.

[130] Schiller, *On the Aesthetic Education of Man*, 31. Goethe shared this view: see Humphry Trevelyan, *Goethe and the Greeks* (Cambridge, 1941). This was to be an assessment of the Greeks that shaped the German educational system. As Lionel Gossman notes: 'To the generation of [Wilhelm] Humboldt the study of antiquity was no narrow technical or scholarly exercise: it was conceived as an essential stage in the education of German youth and in the transformation and reconstruction of the German nation; and for this reason it was built into the new educational system that Humboldt and his collaborators devised for the new Prussia.' *Orpheus Philologus: Bachofen versus Mommsen on the Study of Antiquity*, Transactions of the American Philosophical Society, 73/5 (Philadelphia, 1983), 1–89: 76–7.

[131] See Philip J. Kain, *Schiller, Hegel, and Marx: State, Society, and the Aesthetic Ideal of Ancient Greece* (Kingston, Ontario, 1982), 15–19.

[132] Schiller, *On the Aesthetic Education of Man*, 33.

number of dysfunctions of the modern age.[133] The first is the rigorous separation of ranks and occupations.[134] Second, there is a set of connected fragmentations:

> That zoophyte character of the Greek States, in which every individual enjoyed an independent existence but could, where need arose, grow into the whole organism, now made way for an ingenious clock-work, in which, out of the piecing together of innumerable but lifeless parts, a mechanical kind of collective life ensued. State and Church, laws and customs, were now turned asunder; enjoyment was divorced from labour, the means from the end, the effort from the reward.[135]

The separation of enjoyment from labour, and of means from ends, are the crucial forms of fragmentation here as far as *Bildung* is concerned. Third, one's activity or labour or occupation no longer expresses one's humanity: instead, one merely becomes defined by one's occupation. It is true that the 'outstanding individual will never let the limits of his occupation dictate the limits of his activity', but this is unavoidable in the usual case.[136] Finally, the state becomes alien to its citizens.[137]

Reconciliation of reason and sensibility cannot occur under these conditions. One's activity (*Tätigkeit*) is what constitutes one and defines what one is, so that when, as a result of the division of labour for example, this activity becomes mere toil (*Arbeit*), when it results in a fragmentation of oneself rather than in a unification, then it must be transformed. The model for this transformation is artistic production. This is because, in the case of artistic labour, we can see most clearly how labour could, in changed circumstances, be pleasurable and, above all, genuinely fulfilling. On the conception of self-knowledge that Schiller is offering, one comes to know oneself not introspectively, for 'man without perception or sensations is nothing but form and empty potential';[138] prior to one's doing anything there can be no self-knowledge because there is simply nothing to know about oneself. Rather, one comes to understand oneself through what one has produced. The paradigm form of expression here is artistic production, which for Schiller is unique among forms of production because the product is not alienable in the way that other products are.

The aesthetic model for a philosophical anthropology was to prove exceptionally fruitful in nineteenth- and twentieth-century culture. But it was always fraught with problems. Among the many variants, two different developments bring to light an inherent tension in the model, one jettisoning the idea of the

[133] See the discussion in Kain, *Schiller, Hegel, and Marx*, ch. 1.
[134] Schiller, *On the Aesthetic Education of Man*, 33.
[135] Ibid., 35. [136] Ibid., 37. [137] Ibid., 37.
[138] Ibid., 77. Compare the way Fichte, in his *Grundlage der gesamten Wissenschaftslehre* of 1794, combines, in his inimitably uncompromising way, a distinctively Herderian view of what is distinctive about humans with a Kantian conception of radical freedom, to produce a view that goes beyond either of them when he writes that 'All animals are perfect and complete: man however, is merely suggested . . . Every animal is what it is; man alone is originally nothing at all': Johann Gottlieb Fichte, *The Science of Rights*, trans. A. E. Kroeger (New York, 1889), 118–19.

uniqueness of artistic production, the other effectively aestheticizing life as a whole.

The first route was that followed most influentially by Marx in the 1840s.[139] Marx rejected the idea that artistic production is wholly distinct from other forms of production. It is not a non-alienated area outside normal production, but as subject to alienation as any other form of production. The aesthetic aspects of his conception are manifest in his statements about labour under socialism in his comments on Mill's political economy:

> In my *production* I would have objectified my *individuality*, its specific *character*, and therefore enjoyed not only an individual *manifestation of my life* during the activity, but also when looking at the object I would have the individual pleasure of knowing my personality to be *objective*, visible to the senses and hence a power *beyond all doubt* . . . My work would be a *free manifestation of life*, hence an *enjoyment of life*.[140]

For Marx the alienation of labour was at the basis of all other forms of alienation, and with the overcoming of alienated labour all other forms of alienation would be overcome. But the aesthetic/anthropological model (one that Marx gradually moved away from, it should be said), rather than providing a goal for politics in fact leads to a disconnection from politics. We are not being given an account of how social and political problems are to be recognized, or to be posed, or the mechanisms by which we might hope to resolve them. Rather, what we are presented with is a picture of a state of affairs in which there are effectively no social problems: if there is no alienation from nature, oneself, the human species, and others, then we effectively have an end of politics.

The second route, by contrast, celebrated what it took to be the liberation from politics offered by an aesthetically modelled anthropology. It led to the valuing of cultural achievement over everything else, as it became a substitute for religion and politics. This was primarily a German phenomenon, and it ran in different directions. On the one hand, the cosmopolitan nature of culture was attractive to assimilated German Jews, who, cut off from their own religious tradition but denied full participation in civic life, embraced and immersed themselves in Germany's cultural life, and were able to forge an identity for themselves independently of national politics.[141] On the other hand, it also led to a scorn for politics among intellectuals and artists, and, in the Nazi era, a substitution of

[139] See the discussion in Gaukroger, 'Romanticism and Decommodification: Marx's Conception of Socialism'.

[140] Karl Marx and Friedrich Engels, *Werke* (43 vols, Berlin, 1956–68), Ergänzungsband i (vol. 42), 462–3; Karl Marx and Friedrich Engels, *Collected Works* (50 vols, London and New York, 1975–2005), iii. 227–8.

[141] Cassirer is a good case in point. See Edward Skidelsky, *Ernst Cassirer: The Last Philosopher of Culture* (Princeton, 2008). See also Ned Curthoys, 'Redescribing the Enlightenment: The German-Jewish Adoption of *Bildung* as a Counter-Normative Ideal', *Intellectual History Review* 23 (2013), 365–86.

culture for politics, or even an apolitical aestheticization of Nazism.[142] But although these developments are most clearly seen in Germany, they are also evident elsewhere. In Britain from the late eighteenth century, for example, German philosophical anthropology was seen by Coleridge and his circle as something purely intellectual, unconnected with any political programme, and consequently something that freed one from deciding between French revolutionary thought, with its materialism, and British reaction. It offered a politically neutral mode of engaging with the world. As Mike Jay remarks of the British reception of German idealism, 'the Kantian philosophies were revolutionary, but it was a revolution that pointed the way to inner freedoms and the primacy of the life of the mind in forms that could evolve under the political lockdown of the absolutist Prussian state'.[143]

The problematic standing of philosophical anthropology with regard to political questions was one of its most challenging aspects, either because, in the one case, it failed to yield the political understanding to which it aspired through its programme of uncovering how one discovered one's humanity; or because, in the other, its celebration of liberation from politics was itself in many cases a taking of a political stance. Nevertheless, despite this it offered powerful new understandings that shaped nineteenth- and twentieth-century views of how one might come to terms with the human condition. Philosophical anthropology, in the form that it came to take in Herder and Schiller and their successors, was, of all the forms of naturalization of the human, the only one to offer a plausible secular world view on a par with the Christian one, and the only one in a position to displace the Christian one. Its power lay in its ability to replace a Christian world view with something that was not merely non-reductionist, by contrast with the programmes of the *philosophes* for example, but more importantly was able to bear the weight of the non-propositional forms of engaging with the world: in terms of aspirations, goals, and desires, as well as fears and anxieties. It was able to do this because it provided the basis for a programme of naturalization of a range of human aspirations, including the rich dimension of human experience that had been articulated in a mythical or religious form. It encountered deep and in some cases insurmountable problems. But these were above all an indication not only of its ambition, but also of its refusal simply to reduce or ignore features of human experience that appeared to offer resistance to the direction of its enquiry.

142 See Edith May Butler, *The Tyranny of Greece over Germany* (Cambridge, 1935); and Wolf Lepenies, *The Seduction of Culture in German History* (Princeton, 2006). See also, specifically on how this culture bore on the pursuit of science, the area with the most dangerous consequences: Alan D. Beyerchen, *Scientists under Hitler: Politics and the Physics Community in the Third Reich* (New Haven, 1977); and Philip Ball, *Serving the Reich: The Struggle for the Soul of Physics under Hitler* (London, 2013).

143 Jay, *The Atmosphere of Heaven*, 193–4.

Conclusion

There was no single dimension of human experience to which the forms of naturalization that we have identified were exclusively directed. The attempt to understand human psychology through its medicalization; the attempt to investigate language as an empirical manifestation of states of mind; the attempt to consider human nature in comparative terms, whether anatomical, geographical, or historical; the attempt to understand individual behaviour on the model of collective or aggregate behaviour: all of these took aspects of core questions about human nature that had previously been treated as purely conceptual, stripped them down to basics, and reformulated them so as to open them up to empirical investigation. But the motivations behind the exercise, and consequently the means used to carry it out, varied radically. It is not just that in many cases there was no overlap, but that in those cases where there was overlap the accounts often conflicted. One thing that is clear is that we cannot think of naturalization as being primarily a relationship between the natural sciences and other disciplines. Materialism and reduction to the natural sciences play a very small and inessential role in the forms of naturalization that shaped the transformation of the study of human psychology and conduct into empirical disciplines in the second half of the eighteenth century. Indeed, the most powerful naturalizing resource, in the sense of the one that led most effectively to the replacement of basic traditional beliefs about the world and our place in it, was history. The power of history lay in its ability to do something that was wholly outside the resources of reductive forms of naturalization, namely to engage non-propositional forms of understanding.

At the same time, the historicization of the world was closely linked with the dynamization of a world that had been rendered inert by seventeenth-century mechanism. And the means by which this dynamization was achieved reveals an important feature of naturalization that is obscured in its reductive forms: naturalization has a counterpart, the humanization of nature, whereby there is an explicit awareness that human interests guide inquiry. This took a range of forms, from the claim that the world is not already 'carved up' into genera and species, to the claim that religious discourse is at bottom about something human rather than something divine. This was a characteristic feature of the exercises in naturalization with which we have been concerned. Underlying all of these is the single most important outcome of the combination of naturalization and humanization: the insertion of the human being into the world in such a way that empirical evidence—whether historical, medical, linguistic, anatomical, or economic—can be deployed in exploring and accounting for the relationship between the human and the natural world. As Friedrich Schlegel put it, 'Man is nature looking back on itself'.

Bibliography of Works Cited

Aarsleff, Hans. 'The Tradition of Condillac: The Problem of the Origin of Language in the Eighteenth Century and the Debate in the Berlin Academy before Herder', in idem, *From Locke to Saussure: Essays on the Study of Language and Intellectual History* (London, 1982), 146–209.

Abou-Nemeh, Catherine. 'Réaumur's Crayfish Experiments in Hartsoeker's *Système*: Regeneration and the Limits of Mechanism', in O. Nachtomy and J. E. H. Smith, eds, *The Life Sciences in Early Modern Philosophy* (Oxford, 2014), 157–80.

Ackermann, Jacob Fidelis. *Über die körperliche Verschiendenhiet des Mannes vom Weibe ausser Geschlechtstheilen* (Koblenz, 1788).

Ackermann, Jacob Fidelis. *Versuch einer physikalischen Darstellung der Lebenskräfte organischer Körper* (2 vols, Frankfurt-on-Main, 1797–1800).

Ackernecht, Erwin. *Medicine at the Paris Hospital, 1794–1848* (Baltimore, 1967).

Adams, George. *An Essay on Electricity* (2nd edn, London, 1785).

Adams, Robert Merrihew. 'Faith and Religious Knowledge', in J. Mariña, ed., *The Cambridge Companion to Friedrich Schleiermacher* (Cambridge, 2005), 35–52.

Adanson, Michel. *Familles des plantes* (2 vols, Paris, 1763).

Adler, Hans. *Die Prägnanz des Dunklen: Gnoseologie, Ästhetik, Geschichtsphilosophie bei J. G. Herder* (Hamburg, 1990).

Aït-Touati, Frédérique and Stephen Gaukroger. *Le monde en images: Voir, représenter, savoir, de Descartes à Leibniz* (Paris, 2015).

Albury, William Randall. 'The Logic of Condillac and the Structure of French Chemical and Biological Theory, 1780–1801', unpublished PhD dissertation, Johns Hopkins University, 1972.

Albury, William Randall. 'Experiment and Explanation in the Physiology of Bichat and Magendie', *Studies in History of Biology* 1 (1977), 47–131.

Aldini, Giovanni. *Essai théorique et expérimental sur le galvanisme* (Paris, 1804).

Aldini, Giovanni. *General Views on the Application of Galvanism to Medical Purposes; Principally in Cases of Suspended Animation* (London, 1819).

Alembert, Jean Baptiste d'. *Traité de Dynamique, dans lequel les Loix de l'Equilibre & du Mouvement des Corps sont réduites au plus petit nombre possible, & démontrées d'une manière nouvelle, & où l'on donne un Principe général pour trouver le Mouvement de plusieurs Corps qui agissent les uns sur les autres, d'une manière quelconque* (Paris, 1743).

Alembert, Jean Baptiste d'. *Opuscules mathématiques* (8 vols, Paris, 1761–80).

Alembert, Jean Baptiste d'. *Œuvres* (5 vols, Paris, 1821–2).

Alibert, Jean-Louis. 'Discours sur les rapports de la médecine avec les sciences physiques et morales', *Mémoires de la Société Médicale d'Émulation* 2 (1798), i–cxii.

Allison, Henry. *Lessing and the Enlightenment: His Philosophy of Religion and its Relation to Eighteenth-Century Thought* (Ann Arbor, 1966).

Allison, Henry. 'Kant's Critique of Spinoza', in Richard Kennington, ed., *The Philosophy of Baruch Spinoza* (Washington, DC, 1980), 199–227.

Anderson, Lorin. *Charles Bonnet and the Order of the Known* (Dordrecht, 1982).

Aner, Karl. *Die Theologie der Lessingzeit* (Halle, 1929).

Anon. 'Review of Quetelet, *Sur l'Homme*', *Athenaeum* 8 August 1835, 593–5; 15 August 1835, 611–13; 29 August 1835, 658–61.

Arbuthnot, John. *Proposals for Printing A Very Curious Discourse... intitled ΨΕΥΔΟΛΟΓΓΑ ΠΟΛΙΤΙΚΗ; or, a Treatise on the Art of Political Lying* (London, 1712).

Arbuthnot, John. *Essay Concerning the Effects of Air on Human Bodies* (London, 1733).

Archibald, Raymond C. 'A Rare Pamphlet of De Moivre and Some of His Discoveries', *Isis* 8 (1926), 671–84.

Arnauld, Antoine. *Des vrais et fausses idées contre ce qu'enseigne l'auteur de la recherche de la verité* (Cologne, 1683).

Arnsperger, Walther. *Christian Wolffs Verhältnis zu Leibniz* (Weimar, 1897).

Asad, Talal. *Genealogies of Religion: Discipline and Reasons of Power in Christianity and Islam* (Baltimore, 1993).

Bachmann-Medick, Doris. *Die ästhetische Ordnung des Handelns: Moralphilosophie und Ästhetik in der Popularphilosophie des 18. Jahrhunderts* (Stuttgart, 1989).

Baker, Keith Michael. *Condorcet: From Natural Philosophy to Social Mathematics* (Chicago, 1975).

Balan, Bernard. *L'ordre et le temps: l'anatomie comparée et l'histoire des vivants au XIXe siècle* (Paris, 1979).

Ball, Philip. *Serving the Reich: The Struggle for the Soul of Physics under Hitler* (London, 2013).

Barclay, John. *The Anatomy of the Bones of the Human Body* (Edinburgh, 1829).

Barthez, Paul-Joseph. *Oratio academica de principio vitali hominis* (Montpellier, 1773).

Barthez, Paul-Joseph. *Nova doctrina de functionibus naturae humanae* (Montpellier, 1774).

Barthez, Paul-Joseph. *Nouveaux éléments de la science de l'homme* (Montpellier, 1778).

Barthez, Paul-Joseph. *Nouveaux éléments de la science de l'homme, seconde édition* (2 vols, Paris, 1806).

Batteux, Charles. *Histoire des causes premières: ou, Exposition sommaire des pensées des philosophes sur les principes des êtres* (Paris, 1769).

Baumé, Antoine. *Manuel de Chymie, ou exposé des opérations et des produits d'un cours de chymie* (Paris, 1763).

Baumgarten, Alexander Gottlieb. *Meditationes philosophicae de nonnullis ad poema pertinentibus* (Halle, 1735).

Baumgarten, Alexander Gottlieb. *Metaphysica* (Halle, 1739).

Baumgarten, Alexander Gottlieb. *Ethica philosophica* (Halle, 1740).

Baumgarten, Alexander Gottlieb. *Aesthetica* (2 vols, Frankfurt an der Oder, 1750–8).

Baumgarten, Alexander Gottlieb. *Initia philosophiae practicae primae* (Halle, 1760).

Baumgarten, Alexander Gottlieb. *Ius Naturae* (Halle, 1763).

Baur, Ferdinand Christian. *Symbolik und Mythologie, oder die Naturreligion des Altherthums* (2 vols, Stuttgart, 1824–5).

Beauchêne, Edmé-Pierre Chauvot de. *De l'Influence des affections de l'âme dans les maladies nerveuses des femmes, avec le traitement qui convient à ces maladies* (Montpellier and Paris, 1781).

Beccaria, Cesare. *Dei delitti e delle pene* ([Milan], 1784).

Beck, Lewis White. *Early German Philosophy: Kant and his Predecessors* (Cambridge, Mass., 1969).

Beddoes, Thomas. *On the Nature of Demonstrative Evidence* (London, 1793).
Beddoes, Thomas. *Notice of Some Observations Made at the Medical Pneumatic Institution* (Bristol, 1799).
Beiser, Frederick C. *The Fate of Reason: German Philosophy from Kant to Fichte* (Cambridge, Mass., 1987).
Beiser, Frederick C. *German Idealism: The Struggle Against Subjectivism 1781–1801* (Cambridge, Mass., 2002).
Beiser, Frederick C. *The Romantic Imperative: The Concept of Early German Romanticism* (Cambridge, Mass., 2003).
Beiser, Frederick C. *Schiller as a Philosopher: A Re-Examination* (Oxford, 2005).
Beiser, Frederick C. 'Schleiermacher's Ethics', in J. Mariña, ed., *The Cambridge Companion to Friedrich Schleiermacher* (Cambridge, 2005), 53–72.
Beiser, Frederick C. *Diotoma's Children: German Aesthetic Rationalism from Leibniz to Lessing* (Oxford, 2009).
Beiser, Frederick C. *Late German Idealism: Trendelenburg and Lotze* (Oxford, 2013).
Bell, David. *Spinoza in Germany from 1670 to the Age of Goethe* (London, 1984).
Belon, Pierre. *L'Histoire de la nature des oyseaux* (Paris, 1555).
Bentham, Jeremy. *An Introduction to the Principles of Morals and Legislation* (London, 1789).
Bentham, Jeremy. *The Works of Jeremy Bentham*, ed. J. Brownring (11 vols, Edinburgh, 1838–43).
Bentham, Jeremy. *An Introduction to the Principles of Morals and Legislation*, ed. J. H. Burns and H. L. A. Hart (Oxford, 1970).
Berdoe, Marmaduke. *An Enquiry into the Influence of the Electric Fluid in the Structure and Formation of Animated Beings* (Bath, 1771).
Beretta, Marco. 'Between the Workshop and the Laboratory: Lavoisier's Network of Instrument Makers', *Osiris* 29 (2014), 197–214.
Bergman, Torbern. *Traité des affinités chimiques ou attractions électives* (Paris, 1788).
Berman, Harold J. *Law and Revolution: The Formation of the Western Legal Tradition* (Cambridge, Mass., 1983).
Berman, Harold J. *Law and Revolution* ii.: *The Impact of the Protestant Reformations on the Western Legal Tradition* (Cambridge, Mass., 2003).
Bernoulli, Daniel. 'Essai d'une nouvelle analyse de la mortalité causée par la petite vérole et des avantages de l'inoculation pour le prévenir', *Mémoires de l'Académie Royale des Sciences, 1760* (Paris, 1765), Part 2, 1–79.
Bernoulli, Jacob. *Ars Conjectandi, opus posthumum, accedit Tractatus de seriebus infinitis, et Epistola Gallicè scripta de ludo pilae reticularis* (Basel, 1713).
Berthollet, Claude Louis. *Recherches sur les Lois de l'Affinité* (Paris, 1801).
Berthollet, Claude Louis. *Essai de statique chimique* (2 vols, Paris, 1803).
Bertholon, Pierre. *De l'électricité du corps humain dans l'état de santé et de maladie* (Paris, 1780).
Bertholon, Pierre. *De l'électricité des végétaux* (Paris, 1782).
Bertin, Exupère-Joseph. *Traité d'ostéologie* (4 vols, Paris, 1754).
Bertoletti, Stefano Fabbri. *Impulso, formazione e organismo. Per una storia del concetto di Bildungstrieb nella cultura tedesca* (Florence, 1990).
Bertoletti, Stefano Fabbri. 'The Anthropological Theory of Johann Friedrich Blumenbach', in S. Poggi and M. Bossi, eds, *Romanticism in Science* (Dordrecht, 1994), 103–25.

Bertrand, Philippe. *Nouveaux principes de géologie, comparés et opposés à ceux des philosophes anciens et modernes, notamment de J.-C. Delamétherie, qui les a tous analysés dans sa théorie de la terre* (Paris, 1797).

Beyerchen, Alan D. *Scientists under Hitler: Politics and the Physics Community in the Third Reich* (New Haven, 1977).

Bichat, Xavier. *Anatomie générale appliquée à la physiologie et à la médecine* (Paris, 1801).

Bilfinger, Bernhard. *Dilucidationes philosophiae de Deo, anima humana, mundo et generalibus rerum affectionibus* (Tübingen, 1725).

Biot, Jean-Baptiste. *Essai sur l'histoire générale des sciences pendant la révolution française* (Paris, 1803).

Black, Joseph. 'Experiments Upon Magnesia Alba, Quicklime and some other Alcaline Substances', *Essays and Observations, Physical and Literary, read before a Society in Edinburgh* 2 (1756), 157–225.

Black, Joseph. *Lectures on the Elements of Chemistry*, ed. John Robinson (Edinburgh, 1803).

Blanckaert, Claude. 'L'*Anthropologie* en France: Le mot et l'histoire (XVI–XIXe siècle)', *Bulletin et mémoires de la Société d'Anthropologie de Paris* 1 (1989), 13–44.

Blewitt, George. *An Enquiry whether a General Practice of Virtue tends to the Wealth or Poverty, Benefit or Disadvantage of a People?* (London, 1725).

Blumenbach, Johann Friedrich. *De generis humani varietate nativa* (Göttingen, 1775).

Blumenbach, Johann Friedrich. *Über den Bildungstrieb und das Zeugungsgeschäfte* (Göttingen, 1781).

Blumenbach, Johann Friedrich. *De generis humani varietate nativa* (3rd edn, Göttingen, 1795).

Blumenbach, Johann Friedrich. *Handbuch der Naturgeschichte* (Göttingen, 1797).

Blumenbach, Johann Friedrich. *Handbuch der vergleichenden Anatomie* (Göttingen, 1805).

Blumenbach, Johann Friedrich. *Beyträge zur Naturgeschichte* (2 vols, 2nd edn, Göttingen, 1806).

Blumenberg, Hans. *Arbeit am Mythos* (Frankfurt am Main, 1979).

Boantza, Victor and Ofer Gal. 'The "Absolute Existence" of Phlogiston: The Losing Party's Point of View', *British Journal for the History of Science* 44 (2011), 317–42.

Bödeker, Hans Erich. 'Die Religiosität der Gebideten', in K. Gründer and K. H. Rengstorf, eds, *Religionskritik und Religiostät in der deutschen Aufklärung* (Heidelberg, 1989), 145–95.

Bödeker, Hans Erich. 'Von der "Magd der Theologie" zur "Leitwissenschaft": Vorüberlegungen zu einer Geschichte der Philosophie des 18. Jahrhunderts', *Das achtzehnte Jahrhundert* 14 (1990), 19–57.

Bodemer, Charles W. 'Regeneration and the Decline of Preformationism in Eighteenth Century Embryology', *Bulletin of the History of Medicine* 38 (1962), 21–31.

Bodin, Jean. *Methodus ad facilem historiarum cognitionem* (Paris, 1566).

Boerhaave, Hermann. *Elementa chemiae, quae anniversario labore docuit in publicis privatisque scholis* (2 vols, Leyden, 1732).

Bondeli, Martin. 'The Conception of Enlightenment in Reinhold's "Letters on Kantian Philosophy"', in George di Giovanni, ed., *Karl Leonhard Reinhold and the Enlightenment* (Dordrecht, 2010), 43–52.

Bonnet, Charles. *Essai de psychologie; ou considerations sur les operations de l'ame, sur l'habitude et sur l'education* (London, 1755).
Bonnet, Charles. *Essai analytique sur les facultés de l'ame* (Copenhagen, 1760).
Bonnet, Charles. *Mémoires autobiographiques de Charles Bonnet de Genève*, ed. Raymond Savioz (Paris, 1948).
Bordeu, Théophile de. *Recherches anatomiques sur la position des glands et leur action* (Paris, 1752).
Bordeu, Théophile de. *Recherches sur le tissu muqueux* (Paris, 1767).
Bordeu, Théophile de. *Recherches sur les maladies chroniques* (Paris, 1775).
Bordeu, Théophile de. *Œuvres complètes*, ed. A. Richerand (2 vols, Paris, 1818).
Boscovich, Roger Joseph. *Theoria philosophiae naturalis* (Vienna, 1758).
Boscovich, Roger Joseph and Christopher Maire. *De Litteraria Expeditione per Pontificiam ditionem ad dimetiendas duas Meridiani gradus* (Rome, 1755).
Bossuet, Jacques-Bénigne. *Discours sur l'histoire universelle a monseigneur le dauphin pour expliquer la fuite de la religion et les changemens des empires* (Paris, 1681).
Botero, Giovanni. *Della Ragion di Stato* (Venice, 1589).
Boulanger, Nicolas-Antoine. *L'Antiquité dévoilée par ses usages, ou Examen critique des principales opinions, cérémonies & institutions religieuses & politiques des différens peuples de la terre* (3 vols, Amsterdam, 1766).
Boyarin, Daniel. *Border Lines: The Partition of Judeo-Christianity* (Philadelphia, 2004).
Bradley, Richard. *A Philosophical Account of the Works of Nature. Endeavouring to set forth the several Gradations Remarkable in the Mineral, Vegetable, and Animal Parts of the Creation. Tending to the Composition of a Scale of Life* (London, 1721).
Brague, Rémi. *Europe, la voie romaine* (Paris, 1992).
Brague, Rémi. *La Sagesse du Monde: Histoire de l'expérience humaine de l'univers* (Paris, 1999).
Brandis, Joachim Dietrich. *Versuch über die Lebenscraft* (Hannover, 1795).
Brandt, Reinhard. *Kritischer Kommentar zu Kants Anthropologie in pragmatischer Hinsicht (1789)* (Hamburg, 1999).
Brazill, William J. *The Young Hegelians* (New Haven, 1970).
Breidbach, Olaf. 'The Culture of Science and Experiments in Jena Around 1800', in R. M. Brain, R. S. Cohen, and O. Knudsen, eds, *Hans Christian Ørsted and the Romantic Legacy in Science* (Dordrecht, 2007), 177–216.
Brendel, Otto J. *Prologomena to the Study of Roman Art* (New Haven, 1979).
Bressy, Joseph. *Recherches sur les vapeurs* (London and Paris, 1789).
The Bridgewater Treatises, on the Power, Wisdom, and Goodness of God as Manifested in the Creation (8 vols, London, 1834–7).
Brissot de Warville, Jacques-Pierre. *De la verité, ou Méditations sur les moyens de parvenir à la vérité dans toutes les connaissances humaines* (Neuchâtel, 1782).
Broberg, Gunnar. '*Homo sapiens*: Linnaeus's Classification of Man', in Tore Frängsyr, ed., *Linnaeus: The Man and his Work* (Berkeley, 1983), 156–94.
Broce, Gerald. 'Herder and Ethnography', *Journal of the History of the Behavioural Sciences* 22 (1986), 150–70.
Brockliss, Laurence W. B. 'Before the Clinic: French Medical Teaching in the Eighteenth Century', in Caroline Hannaway and Ann La Berge, eds, *Constructing Paris Medicine* (Amsterdam, 1998), 71–116.

Broman, Thomas. 'Metaphysics for an Enlightened Public: The Controversy over Monads in Germany, 1746–1748', *Isis* 103 (2012), 1–23.

Brosses, Charles de. *Histoire des navigations aux terres australes, contenant ce que l'on sait des moeurs et des productions des contrées découvertes jusqu'à ce jour* (2 vols, Paris, 1756).

Brosses, Charles de. *Du culte des dieux fétiches, ou Parallèle de l'ancienne Religion de l'Égypte avec la Religion actuelle de Nigritie* (Geneva, 1760).

Brosses, Charles de. *Traité de la formation méchanique des langues et des principes physiques de l'étymologie* (2 vols, Paris, 1765).

Brown, John. *Essays on the Characteristics of Lord Shaftesbury* (London, 1751).

Brown, John. *Estimate of the Manners and Principles of the Times* (2 vols, London, 1757–8).

Browne, Janet. 'Botany for Gentlemen: Erasmus Darwin and *The Loves of Plants*', *Isis* 80 (1989), 593–621.

Browne, Janet. 'Biogeography and Empire', in N. Jardine, J. A. Secord, and E. C. Spary, eds, *Cultures of Natural History* (Cambridge, 1996), 305–21.

Bruford, William H. *Culture and Society in Classical Weimar, 1755–1806* (Cambridge, 1962).

Buchenau, Stefanie. *The Founding of Aesthetics in the German Enlightenment: The Art of Invention and the Invention of Art* (Cambridge, 2013).

[Bucher, Urban Gottfried and Röschel, Johann Baptist]. *Zweyer Guten Freunde vertrauter Brief-Wechsel vom Wesen der Seelen* (The Hague, 1713).

Buchez, Philippe. *Essai d'un traité complet de philosophie du point de vue du catholicisme et du progrès* (3 vols, Paris, 1838–40).

Buck, Peter. 'Seventeenth-Century Political Arithmetic: Civil Strife and Vital Statistics', *Isis* 68 (1977), 67–84.

Buck, Peter. 'People Who Counted: Political Arithmetic in the Eighteenth Century', *Isis* 73 (1982), 28–45.

Buffon, George Louis Leclerc, Comte de. *Histoire naturelle, générale et particulière* (15 vols, Paris, 1749–67).

Buffon, George Louis Leclerc, Comte de. *Histoire naturelle des oiseaux* (9 vols, Paris, 1770–83).

Buffon, George Louis Leclerc, Comte de. *Histoire naturelle, générale et particulière. Supplément* (7 vols, Paris, 1774–89).

Buffon, George Louis Leclerc, Comte de. *Époches de la Nature* (Paris, 1778).

Buffon, George Louis Leclerc, Comte de. *Histoire naturelle des minéraux* (5 vols, Paris, 1783–8).

Buffon, George Louis Leclerc, Comte de. *Histoire naturelle, générale et particulière, par Leclerc de Buffon. Nouvelle édition, accompagné de notes, dans lesquelles les supplémens sont insérés dans la première texte, à la place qui leur convient. L'on y a ajouté l'histoire naturelle des quadrupèdes et des oiseaux découverts depuis la mort de Buffon, celle des reptiles, des poissons, des insects et des vers; enfin, l'histoire des plantes dont ce grand naturaliste n'a pas eu de tempts de s'occuper. Ouvrage formant un cours complet d'histoire naturelle. Rédigé par C. S. Sonnini* (127 vols, Paris, 1789–1808).

Buffon, George Louis Leclerc, Comte de. *Œuvres complètes de Buffon, mises en ordres et précedées d'une notice historique par M. A. Richard . . . suivies de deux volumes sur le progrès des sciences physiques et naturelles depuis la Mort de Buffon par M. Le Baron Cuvier* (34 vols, Paris, 1827).

Burke, Edmund. *The Correspondence of Edmund Burke*, ed. T. H. Copeland (10 vols, Cambridge, 1958–78).
Burns, Chester R. 'Nonnaturals: A Paradox in the Western Concept of Health', *Journal of Medicine and Philosophy* 1 (1976), 202–11.
Buschman, Cornelia. 'Die philosophische Preisfragen und Preisschriften der Berliner Akademie der Wissenschaften in 18. Jahrhundert', in W. Förster, ed., *Aufklärung in Berlin* (Berlin, 1989), 165–228.
Butler, Charles. *The Feminine Monarchy, or, A Treatise concerning bees, and the due ordering of them* (Oxford, 1609).
Butler, Edith May. *The Tyranny of Greece over Germany* (Cambridge, 1935).
Buxton, Richard. *Imaginary Greece: The Contexts of Mythology* (Cambridge, 1994).
Bynum, William F. 'The Anatomical Method, Natural Theology, and the Functions of the Brain', *Isis* 64 (1973), 444–68.
Bynum, William F. 'Time's Noblest Offspring: The Problem of Man in the British Natural Historical Sciences, 1800–1863', unpublished PhD dissertation, University of Cambridge, 1974.
Cabanis, Pierre Jean Georges. *Observations sur les hôpitaux* (Paris, 1790).
Cabanis, Pierre Jean Georges. *Du degré de certitude de la médecine* (Paris, 1798).
Cabanis, Pierre Jean Georges. *Rapports du physique et du moral de l'homme* (2 vols, Paris, 1802).
Cabanis, Pierre Jean Georges. *Coup d'oeil sur les Révolutions et sur la réforme de la médecine* (Paris, 1804).
Cabanis, Pierre Jean Georges. *Œuvres complètes de Cabanis* (5 vols, Paris, 1823–5).
Cabanis, Pierre Jean Georges. *Œuvres philosophiques de Cabanis*, ed. C. Lehec and J. Cazeneuve (2 vols, Paris, 1956).
Calan, Ronan de. *Généalogie de la sensation: Physique, physiologie et psychologie en Europe, de Fernel à Locke* (Paris, 2012).
Canguilhem, Georges. *The Normal and the Pathological* (New York, 1991).
Canz, Israel Theophil. *Philosophiae Leibnitianae et Wolffianae usus in theologia* (Frankfurt and Leipzig, 1733).
Carey, Daniel. *Locke, Shaftesbury, and Hutcheson* (Cambridge, 2006).
Carra, Jean Louis. *Nouveaux principes de physique ornés de planches* (2 vols, Paris, 1781).
Carra, Jean Louis. *Systême de la raison, ou le Prophete philosophe* (Paris, 1782).
Carra, Jean Louis. *Examen physique du magnétisme animal* (Paris, 1785).
Carrithers, David. 'The Enlightenment Science of Society', in C. Fox, R. Porter, and R. Wokler, eds, *Inventing Human Science: Eighteenth-Century Domains* (Berkeley, 1995), 232–70.
Casmann, Otto. *Psychologia anthropologica sive animae humanae doctrina* (Hanau, 1594).
Casmann, Otto. *Nucleus mysteriorum Naturae enucleatus* (Hamburg, 1605).
Cassirer, Ernst. *The Philosophy of the Enlightenment* (Princeton, 1951).
Caussin, Nicolas. *Symbolica Aegyptiorum Sapientia* (Paris, 1631).
Cavallo, Tiberius. *An Essay on the Theory and Practice of Medical Electricity* (London, 1780).
Cavendish, Henry. 'Three Papers, containing Experiments on Factitious Air', *Philosophical Transactions* 56 (1766), 141–84.
Cavendish, Henry. 'An Account of Some Attempts to Imitate the Effects of the Torpedo by Electricity,' *Philosophical Transactions* 66 (1776), 196–225.

Chalmers, Alan. *The Scientist's Atom and the Philosopher's Stone: How Science Succeeded and Philosophy Failed to Gain Knowledge of Atoms* (New York, 2011).

Chang, Hasok. *Inventing Temperature: Measurement and Scientific Progress* (New York, 2004).

Chang, Hasok. *Is Water H_2O? Evidence, Realism and Pluralism* (Dordrecht, 2012).

Chardin, Jean-Baptiste. *Journal du Voiage du Chevalier Chardin en Perse & aux Indes Orientales, par la Mer Noire & par la Colchide* (Amsterdam/Paris/London, 1686).

Chardin, Jean-Baptiste. *The Travels of Sir John Chardin in Persia* (2 vols, London, 1720).

Châtelet, Gabrielle Émilie Le Tonnier de Breteuil, Marquise de. *Institutions de physique* (Paris, 1740).

Cheseldon, William. *Osteographia or the Anatomy of the Bones* (London, 1733).

Cheyne, George. *The English Malady; or, A Treatise of Nervous Diseases of All Kinds, as Spleen, Vapours, Lowness of Spirits, Hypochondriacal and Hysterical Distempers* (Dublin, 1733).

Christensen. Dan C. *Hans Christian Ørsted: Reading Nature's Mind* (Oxford, 2013).

Clark, Robert T. Jr. 'Herder, Cesarotti and Vico', *Studies in Philology* 44 (1947), 645–71.

Cohen, Alix. *Kant and the Human Sciences: Biology, Anthropology and History* (London, 2009).

Coleman, William. *Georges Cuvier Zoologist: A Study in the History of Evolution Theory* (Cambridge, Mass., 1962).

Coleman, William. 'Health and Hygiene in the *Encyclopédie*: A Medical Doctrine for the Bourgeoisie', *Journal of the History of Medicine and Allied Sciences* 29 (1974), 399–421.

Coleridge, Samuel Taylor. *Biographia Literaria, or Biographical Sketchs of My Literary Life and Opinions* (2 vols, London, 1817).

Collini, Stefan, Donald Winch, and John Burrow. *That Noble Science of Politics: A Study in Nineteenth-Century Intellectual History* (Cambridge, 1983).

Comte, August. *Cours de Philosophie Positive* (6 vols, Paris, 1830–42).

Comte, August. *Système de politique positive, ou Traité de sociologie, Instituant la Religion de l'Humanité* (4 vols, Paris, 1851–4).

Condillac, Etienne Bonnot de. *Essai sur l'origine des connoissances humaines: ouvrage où l'on réduit à un seul principe tout ce qui concerne l'entendement* (2 vols, Amsterdam, 1746).

Condillac, Etienne Bonnot de. *Traité des sistêmes, où l'on démêle les inconvéniens et des avantages* (The Hague, 1749).

Condillac, Etienne Bonnot de. *Traité des sensations* (Paris, 1754).

Condillac, Etienne Bonnot de. *An Essay on the Origin of Human Knowledge. Being a Supplement to Mr. Locke's Essay on the Human Understanding. Translated... by Mr. Nugent* (London, 1756).

Condillac, Etienne Bonnot de. *Œuvres de Condillac, revues, corrigées par l'auteur* (23 vols, Paris, 1798).

Condorcet, Marie Jean Antoine Nicolas de Caritat, Marquis de. *Le Marquis de Condorcet à Mr. D'Alembert sur le Système du Monde et sur le Calcul Intégral* (Paris, 1768).

Condorcet, Marie Jean Antoine Nicolas de Caritat, Marquis de. *Essai sur l'application de l'analyse à la probabilité des décisions rendues à la pluralité des voix* (Paris, 1785).

Condorcet, Marie Jean Antoine Nicolas de Caritat, Marquis de. *Esquisse d'un tableau historique des progrès de l'esprit humain* (Paris, 1795).

Condorcet, Marie Jean Antoine Nicolas de Caritat, Marquis de. *Outlines of an Historical View of the Progress of the Mind* (London, 1795).

Condorcet, Marie Jean Antoine Nicolas de Caritat, Marquis de. *Œuvres de Condorcet*, ed. A. Condorcet-O'Connor and F. Arago (12 vols, Paris, 1847–9).
Condren, Conal. *Argument and Authority in Early Modern England: The Presupposition of Oaths and Offices* (Cambridge, 2006).
Contant, Jean-Paul. *Enseignment de la chimie au Jardin Royal des Plantes de Paris* (Cahors, 1952).
Corneanu, Sorana. *Regimens of the Mind: Boyle, Locke, and the Early Modern* Cultura Animi *Tradition* (Chicago, 2012).
Corr, Charles. 'Christian Wolff and Leibniz', *Journal of the History of Ideas* 36 (1975), 241–62.
Corry, Leo. *David Hilbert and the Axiomatization of Physics (1898–1918): From* Grunglagen der Geometrie *to* Grundlagen der Physik (Dordrecht, 2010).
Corsi, Pietro. *The Age of Lamarck: Evolutionary Theories in France 1790–1830* (Berkeley, 1988).
Coulomb, Charles-Augustin. 'Sur l'électricité et le magnetisme. [I] Construction et usage d'une balance électrique', *Académie Royale des Sciences: Mémoires de mathematique et de physique, présentés . . . par divers savans* (1785), 569–77.
Coulomb, Charles-Augustin. 'Sur l'électricité et le magnetisme. [II] Ou l'on détermine suivant quelles lois le fluide magnétique ainsi que le fluide électrique agissent', *Académie Royale des Sciences: Mémoires de mathematique et de physique, présentés . . . par divers savans* (1785), 578–611.
Creuzer, Georg Friedrich. *Symbolik und Mythologie der alten Völker, besonders der Griechen* (4 vols, Leipzig and Darmstadt, 1810–12).
Curthoys, Ned. 'Redescribing the Enlightenment: The German-Jewish Adoption of *Bildung* as a Counter-Normative Ideal', *Intellectual History Review* 23 (2013), 365–86.
Cuvier, Frédéric. 'Description d'un orangoutang, et observations sur ses facultés intellectuelles', *Annales du Muséum d'histoire naturelle* 16 (1810), 46–65.
Cuvier, Georges, Baron. 'Mémoire sur les espèces d'élephants vivantes et fossiles. Lu le premier pluvoîse an IV', *Mémoires de l'Institut national des sciences et des arts. Sciences mathématiques et physiques* 2 (1799), 1–22.
Cuvier, Georges, Baron. *Leçons d'anatomie comparée* (5 vols, Paris, 1800–5).
Cuvier, Georges, Baron. 'Histoire de la classe des sciences mathématiques et physiques', *Mémoires de l'Institut national des sciences et arts. Sciences mathématiques et physiques* 7 (1806), 1–79.
Cuvier, Georges, Baron. *Le règne animal distribué d'après son organisation pour servir de base à l'histoire naturelle des animaux et l'introduction à l'anatomie comparée* (5 vols, Paris, 1829–30).
Cuvier, Georges, Baron. *The Animal Kingdom arranged after its Organization; forming a Natural History of Animals, and an Introduction to Comparative Anatomy*, new edn (London, 1863).
Dalton, John. *A New System of Chemical Philosophy* (3 vols, Manchester, 1808–27).
Darnton, Robert. *Mesmerism and the End of the Enlightenment in France* (Cambridge, Mass., 1968).
Darnton, Robert. 'History of Reading', in P. Burke, ed., *New Perspectives on Historical Writing* (University Park, Penn., 1992), 140–67.
Darwin, Charles. *The Formation of Vegetable Mould, Through the Action of Worms* (London, 1881).

Darwin, Erasmus. *Zoonomia; or, The Laws of Organic Life* (2 vols, Dublin, 1794–1806).
Darwin, Francis, ed. *The Life and Letters of Charles Darwin* (3 vols, London, 1887).
Daston, Lorraine. *Classical Probability in the Enlightenment* (Princeton, 1988).
Daston, Lorraine. 'The Ideal and Reality of the Republic of Letters in the Enlightenment', *Science in Context* 4 (1991), 367–86.
Daston, Lorraine. 'Unruly Weather: Natural Law Confronts Natural Variability', in L. Daston and M. Stolleis, eds, *Natural Laws and Laws of Nature in Early Modern Europe* (Aldershot, 2008), 233–48.
Daston, Lorraine. 'The Humboldtian Gaze', in M. Epple and C Zittel, eds, *Science as Cultural Practice. Volume 1: Cultures and Politics of Research from the Early Modern Period to the Age of Extremes* (Berlin, 2010), 45–60.
Daston, Lorraine and Peter Galison, *Objectivity* (New York, 2007).
Daston, Lorraine and Michael Stolleis. 'Introduction: Nature, Law and Natural Law in Early Modern Europe', in L. Daston and M. Stolleis, eds, *Natural Laws and Laws of Nature in Early Modern Europe* (Aldershot, 2008), 1–12.
Daubenton, Louis-Georges-Marie. 'Introduction à l'histoire naturelle', *Encyclopédie méthodique. Histoire naturelle des animaux* (4 vols, Paris, 1782–9).
Davies, Martin L. 'Karl Philipp Moritz's *Erfahrungsseelenkund*: Its Social and Intellectual Origins', *Oxford German Studies* 16 (1985), 13–35.
Davies, Martin L. *Identity or History! Marcus Herz and the End of Enlightenment* (Detroit, 1995).
Davy, Humphry. *Researches Chemical and Philosophical, Chiefly concerning Nitrous Oxide, or Dephlogisticated Nitrous Air, and its Respiration* (London, 1800).
Davy, Humphry. *The Collected Works* (9 vols, London, 1839–40).
De Angelis, Simone. *Anthropologien: Genese und Konfiguration einer 'Wissenschaft vom Menschen' in der Frühen Neuzeit* (Berlin, 2010).
De Moivre, Abraham. *The Doctrine of Chances* (2nd edn, London, 1738).
Defoe, Daniel. *Essays upon Several Projects, or Effectual Ways, for Advancing the Interests of the Nation* (London, 1702).
Delamétherie, Jean-Claude. *Théorie de la terre* (3 vols, Paris, 1795).
Delamétherie, Jean-Claude. 'Discours préliminaire', *Journal de physique* 48 (1798), 3–134.
Demandt, Alexander. 'Winckelmann und die alte Geschichte', in Thomas W. Gaehtgens, ed., *Johann Joachim Winckelmann, 1717–1768* (Hamburg, 1986), 301–13.
Denby, David. 'Herder: Culture, Anthropology and the Enlightenment', *History of the Human Sciences* 18 (2005), 55–76.
Dent, Nicholas. *Rousseau* (London, 2005).
Des Chene, Dennis. 'Life and Health in Cartesian Natural Philosophy', in S. Gaukroger, J. Schuster, and J. Sutton, eds, *Descartes' Natural Philosophy* (London, 2000), 723–35.
Des Chene, Dennis. '"Animal" as a Category in Bayle's *Rorarius*', in Justin E. H. Smith, ed., *The Problem of Animal Generation in Early Modern Philosophy* (Cambridge, 2006), 215–32.
Descartes, René. *Œuvres de Descartes*, ed. Charles Adam and Paul Tannery (2nd edn, 11 vols, Paris, 1974–86).
Deschamps, Jean. *Cours abrégé de la philosophie wolfienne* (2 vols, Leipzig and Amsterdam, 1743–7).

Deslon, Charles. *Observations sur les deux Rapports de MM. les Commissaires nommés par Sa Majesté pour l'Examen du Magnétisme animal* (Paris, 1784).

Destutt de Tracy, Antoine-Louis-Claude. 'Mémoire sur la faculté de penser', *Mémoires de l'Institut national des sciences et arts. Classe de sciences morales et politiques*, 1 (1798), 283–451.

Detienne, Marcel. *L'invention de mythologie* (Paris, 1981).

Detienne, Marcel. *Maîtres de vérité dans la Grèce archaïque* (Paris, 1990).

Dettlebach, Michael. 'Humboldtian Science', in N. Jardine, J. A. Secord, and E. C. Spary, eds, *Cultures of Natural History* (Cambridge, 1996), 287–304.

Diderot, Denis. *Lettre sur les aveugles, á l'usage de ceux qui voient* ('London' [Paris], 1749).

Diderot, Denis. *Œuvres philosophiques et dramatiques* (6 vols, 'Amsterdam' [Paris], 1772).

Diderot, Denis. *Œuvres complètes*, ed. J. Assézat and M. Tourneux (20 vols, Paris, 1875–9).

Diderot, Denis. *Eléments de physiologie*, ed. Jean Mayer (Paris, 1964).

Diderot, Denis and d'Alembert, Jean le Rond. *Encyclopédie ou Dictionnaire raisonné des sciences, des arts et des métiers par une société des gens de Lettres, mis en ordre et publié par Diderot et quant à la Partie mathématique par d'Alembert* (2nd edn, 40 vols, Geneva, 1777–9).

Digby, Kenelm. *Two Treatises: In the One of Which, The Nature of Bodies; In the Other, The Nature of Mans Soule, is Looked into: In Way of Discovery of the Immortality of Reasonable Soules* (Paris, 1644).

Dijksterhuis, Eduard Jan. *The Mechanization of the World Picture* (Oxford, 1961).

Dobbin, Leonard. 'A Cullen Chemical Manuscript of 1753', *Annals of Science* 1 (1936), 138–56.

Dobbs, Betty Jo Teeter. *The Janus Faces of Genius: The Role of Alchemy in Newton's Thought* (Cambridge, 1991).

Domenech, Jacques. *L'éthique des Lumières. Les fondements de la morale dans la philosophie française du XVIIIe siècle* (Paris, 1989).

Donovan, Arthur L. *Philosophical Chemistry in the Scottish Enlightenment: The Doctrines and Discoveries of William Cullen and Joseph Black* (Edinburgh, 1975).

Donovan, Arthur L. 'Pneumatic Chemistry and Newtonian Natural Philosophy in the Eighteenth Century: William Cullen and Joseph Black', *Isis* 67 (1976), 217–28.

Donovan, Arthur L. *Antoine Lavoisier: Science, Administration and Revolution* (Cambridge, 1993).

Doody, Aude. *Pliny's Encyclopedia: The Reception of the* Natural History (Cambridge, 2010).

Dove, Heinrich Wilhelm. 'Einige meteorologische Untersuchungen über der Wind', *Annalen der Physik und Chemie* 11 (1827), 545–90.

Dove, Heinrich Wilhelm. *Meteorologische Untersuchungen* (Berlin, 1837).

Drake, James. *Antropologia nova; or, a New System of Anatomy, Describing the Animal Oeconomy* (2 vols, London, 1707).

Dreike, Beate. *Herders Naturauffassung in ihrer Beeinflussung durch Leibniz' Philosophie* (Wiesbaden, 1973).

Du Pont de Nemours, Pierre-Samuel. *Idées sur les secours à donner aux pauvres dans une grande ville* (Paris, 1786).

Dubois d'Amiens, Frédéric. 'Discussion sur la statistique médicale', *Bulletin de l'Académie royale de médecine* 13 (1837), 684–714.

Dubos, Jean-Baptiste. *Réflexions critiques sur la poésie et sur la peinture: Ut Pictura Poesis* (2 vols, Paris, 1719).

Duchesneau, François. *La physiologie des lumières: Empirisme, modèles et théories* (The Hague, 1982).

Duchesneau, François. 'Diderot et la physiologie de la sensibilité', *Dix-Huitième Siècle* 31 (1999), 195–216.

Duffy, Eamon. *The Stripping of Altars: Traditional Religion in England, 1400–1580* (New Haven, 1992).

Dulieu, Luis. 'Les articles d'Henri Fouquet dans l'*Encyclopédie*', *Revue d'histoire des sciences et de leurs applications* 5 (1952), 18–25.

Dumas, Charles Louis. *Principes de physiologie, ou introduction à la science expérimentale, philosophique et médicale de l'homme vivant* (3 vols, Paris, 1800–3).

Dunbar, James. *Essays on the History of Mankind in Rude and Cultivated Ages* (London, 1780).

Dutrochet, René Joachim Henri. *Mémoires pour servir à l'histoire anatomique et physiologique des végétaux et des animaux* (Paris, 1824).

Eberhard, Johann August. *Neue Apologie des Sokrates oder Untersuchung der Lehre von der Seligkeit der Heiden* (Berlin, 1772).

École, Jean. 'De la connaissance qu'avait Kant de la métaphysique wolffienne, ou Kant avait-il lu les ouvrages métaphysiques de Wolff?', *Archiv für Geschichte der Philosophie* 73 (1991), 161–76.

Eichhorn, Johann Gottfried. *Urgeschichte* (Altdorf and Nüremberg, 1779).

Emch-Dériaz, Antoinette. *Tissot: Physician of the Enlightenment* (New York, 1992).

Engelsing, Rolf. *Der Bürger als Leser: Lesergeschichte in Deutschland 1500–1800* (Stuttgart, 1974).

Engfer, Hans-Jürgen. 'Von der Leibnizischen Monadologie zur empirischen Psychologie Wolffs', in S. Carbocin and L. Madonna, eds, *Nuovi studi sul pensiero di Christian Wolff* (Hildesheim, 1992), 193–215.

Erikson, Paul, et al., *How Reason Almost Lost its Mind: The Strange Career of Cold War Rationality* (Chicago, 2013).

Eschenmayer, Carl August von. *Der Ischiotismus unserer Zeit—eine Zugabe zu dem jüngst erscheinen Werke: Das Leben Jesu, von Strauss* (Tübingen, 1835).

Euler, Leonard. 'Réflexions sur l'espace et le temps', *Histoire de l'Académie Royale des sciences et belles lettres de Prusse* (Berlin, 1748) vol. 4: 324–33.

Euler, Leonard. *Lettres à une princesse d'Allemagne sur divers sujets de physique et de philosophie* (3 vols, St Petersburg, 1768–72).

Euler, Leonard. *Letters of Euler on Different Subjects in Physics, Addressed to a German Princess* (2 vols, London, 1802).

Euler, Leonard. *Mechanica sive motus scientia analytice exposita* (1736), in *Leonhardi Euleri opera omnia*, series 2, vols i, ii (Leipzig & Berlin, 1912).

Euler, Leonard. *Theoria motus corporum solidorum seu rigidorum* (1765) in *Leonhardi Euleri opera omnia*, series 2, vols iii, iv (Bern, 1948).

Falconer, William. *Remarks on the Influence of Climate, Situation, Nature of Country, Population, Nature of Food, and Way of Life, on the Disposition and Temper, Manners and Behaviour, Intellects, Laws and Customs, Form of Government, and Religion, of Mankind* (London, 1781).

Feder, Johann Georg Heinrich. *Untersuchungen über den menschlichen Willen* (4 vols, Göttingen, 1779–93).

Feijóo y Montenegro, Benito Jerónimo. *Teatro Critico Universal* (8 vols, Madrid, 1726–36).

Feil, Ernst. *Religio: Die Geschichte eines neuzeitlichen Grundbegriffs vom Frühchristentum bis zum Reformation* (Göttingen, 1986).

Feke, John. *An Essay to Shew the Cause of Electricity* (London, 1746).

Ferguson, Adam. *An Essay on the History of Civil Society* (4th edn, London, 1773).

Ferguson, Adam. *Institutes of Moral Philosophy. For the Use of Students in the College of Edinburgh* (2nd edn, Edinburgh, 1773).

Feuerbach, Ludwig. *Gedanken über Tod und Unsterblichkeit* (Nürnberg, 1830).

Feuerbach, Ludwig. *Das Wesen des Christentums* (Leipzig, 1841).

Feuerbach, Ludwig. *The Essence of Christianity*, trans. Mary Evans (London, 1854).

Feuerbach, Ludwig. *Das Wesen des Christentums*, ed. W. Schuffenhauer (2 vols, Berlin, 1956).

Fichte, Johann Gottlieb. *Versuch einer Kritik aller Offenbarung* (Königsberg, 1792).

Fichte, Johann Gottlieb. *Grundlage der gesamten Wissenschaftslehre* (2 vols, Jena, 1794–5).

Fichte, Johann Gottlieb. *Ueber den Grund unseres Glaubens an eine göttliche Weltregierung* (Jena, 1798).

Fichte, Johann Gottlieb. *The Science of Rights*, trans. A. E. Kroeger (New York, 1889).

Fichte, Johann Gottlieb. *Attempt at a Critique of All Revelation*, ed. Allen Wood (Cambridge, 2010).

Fielding, Henry. *Some papers proper to be read before the R____L SOCIETY, concerning the terrestrial chrysipus, golden-foot or guinea; an insect, or vegetable, resembling the polypus, which hath this surprising property, that being cut into several pieces, each piece becomes a perfect animal, or vegetable, as complete as that of which it was originally only a part* (London, 1743).

Figlio, Karl. 'Theories of Perception and the Physiology of Mind in the Late Eighteenth Century', *History of Science* 12 (1975), 177–212.

Fiorenza, Francis Schüssler. 'Schleiermacher's Understanding of God as Triune', in J. Mariña, ed., *The Cambridge Companion to Friedrich Schleiermacher* (Cambridge, 2005), 171–88.

Fischer, Rotraut and Gabriele Stumpp, 'Die Allegorisierung des Individuums in der Physiognomik Johann Caspar Lavaters und Carl Gustav Carus', in R. Fischer, G. Schrader, and G. Stumpp, eds, *Natur nach Mass: Physiognomik zwischen Wissenschaft une Ästhetik* (Marburg, 1989), 11–58.

Fleury, Claude. *Histoire ecclesiastique* (32 vols, Brussels, 1716).

Flögel, Carl Friedrich. *Geschichte des menschlichen Verstandes* (2nd edn, Breslau, 1773).

Fontenelle, Bernard le Bovier de. *Entretiens sur la pluralité des mondes* (Paris, 1686).

Fontenelle, Bernard le Bovier de. *Digressions sur les anciens et les modernes* (Paris, 1688).

Förster, Eckart. *The Twenty-Five Years of Philosophy: A Systematic Reconstruction* (Cambridge, Mass., 2012).

Forster, Georg. *A Voyage Round the World in His Majesty's Sloop, Resolution, commanded by Capt. James Cook, during the Years 1772, 3, 4, 5* (London, 1777).

Forster, Georg. 'Noch etwas über die Menschenraßen,' *Teutsche Merkur* (Fourth Quarter, 1786), 57–86, 150–66.

Forster, Georg. *Werke: Sämtliche Schriften, Tagesbücher, Briefe* (Berlin, 1958–).
Forster, Johann Reinhold. *Observations Made During a Voyage Round the World* (London, 1778).
Forster, Michael N. *After Herder: Philosophy of Language in the German Tradition* (Oxford, 2010).
Foucault, Michel. *La Naissance de Clinique* (Paris, 1963).
Fourcroy, Antoine François de. *Elements of Natural History and Chemistry* (3rd edn, 3 vols, London, 1790).
Fox-Genovese, Elizabeth. *The Origins of Physiocracy* (Ithaca, NY, 1976).
Franck, Manfred, *Vorlesungen über die neue Mythologie* (2 vols, Frankfurt am Main, 1982–8).
Franke, Ursula. *Kunst als Erkenntnis: die Rolle d. Sinnlichkeit in d. Ästhetik d. Alexander Gottlieb Baumgarten* (Wiebaden, 1972).
Franklin, Benjamin. *Works*, ed. John Bigelow (12 vols, New York, 1904).
Franks, Paul W. *All or Nothing: Systematicity, Transcendental Arguments, and Skepticism in German Idealism* (Cambridge Mass., 2005).
Fraser, Craig G. 'Lagrange's Analytical Mathematics, its Cartesian Origins and Reception in Comte's Positive Philosophy', *Studies in History and Philosophy of Science* 21 (1990), 243–56.
Frazer, Michael L. *The Enlightenment of Sympathy: Justice and the Moral Sentiments in the Eighteenth Century and Today* (Oxford, 2010).
Freeman, Aaron. 'Charles de Brosses and the French Enlightenment Origins of Religious Fetishism', *Intellectual History Review* 24 (2014), 203–14.
Frei, Hans W. *The Eclipse of Biblical Narrative: A Study in Eighteenth and Nineteenth Century Hermeneutics* (New Haven, 1974).
Fréret, Nicolas. *Défense de la chronologie fondée sur les monuments de l'histoire ancienne, contre le système chronologique de M. Newton* (Paris, 1758).
Fréret, Nicolas. 'Réflexions générales sur l'étude des anciennes histoires et sur le degré de certitude des différentes preuves historiques (1724)', in idem., *Mémoires académiques* (Paris, 1996), 73–126.
Fricke, Christel. 'Adam Smith: The Sympathetic Process and the Origin and Function of Conscience', in C. J. Berry, M. P. Paganelli, and C. Smith eds, *The Oxford Handbook of Adam Smith* (Oxford, 2013), 177–200.
Friedman, Michael. *Kant and the Exact Sciences* (Cambridge, Mass., 1992).
Fuchs, Eckhardt. 'Nature and *Bildung*: Pedagogical Naturalism in Nineteenth-Century Germany', in Lorraine Daston and Fernando Vidal, eds, *The Moral Authority of Nature* (Chicago, 2004), 155–81.
Gagliano, Monica et al. 'Experience Teaches Plants to Learn Faster and Forget Slower in Environments Where It Matters', *Oecologia* 175/1 (2014), 63–72.
Galvani, Luigi. *De viribus electricitatis in motu musculari* (Bologna, 1791).
Galvani, Luigi. *Memorie ed esperimenti di Luigi Galvani con la iconografia di lui e un saggio di bibliografia degli scritti* (Bologna, 1937).
Garrett, Aaron. 'Human Nature', in Knud Haakonssen, ed., *The Cambridge History of Eighteenth-Century Philosophy* (Cambridge, 2006), 160–233.
Garve, Christian. *Versuche über verschiedene aus der Moral, der Litteratur und dem gesellschaftlichen Leben* (5 vols, Breslau, 1792–1802).
Gaub, Jerome. *Sermo academicus de regimine mentis quod medicorum est* (Leiden, 1747).

Gaukroger, Stephen. 'Romanticism and Decommodification: Marx's Conception of Socialism', *Economy and Society* 15 (1986), 287–333.

Gaukroger, Stephen. 'Vico and the Maker's Knowledge Principle', *History of Philosophy Quarterly* 3 (1986), 29–44.

Gaukroger, Stephen. *Descartes, An Intellectual Biography* (Oxford, 1995).

Gaukroger, Stephen. *Francis Bacon and the Transformation of Early-Modern Philosophy* (Cambridge, 2001).

Gaukroger, Stephen. *Descartes' System of Natural Philosophy* (Cambridge, 2002).

Gaukroger, Stephen. *The Emergence of a Scientific Culture: Science and the Shaping of Modernity, 1210–1685* (Oxford, 2006).

Gaukroger, Stephen. *The Collapse of Mechanism and the Rise of Sensibility: Science and the Shaping of Modernity, 1680–1760* (Oxford, 2010).

Gaukroger, Stephen. 'Descartes' Theory of Perceptual Cognition and the Question of Moral Sensibility', in John Cottingham and Peter Hacker, eds, *Mind, Method and Morality: Essays in Honour of Anthony Kenny* (Oxford, 2010), 230–51.

Gaukroger, Stephen. *Objectivity: A Very Short Introduction* (Oxford, 2012).

Gawlick, Günter. 'Über einige Charakteristika der britischen Philosophie des 18. Jahrhundert', *Studia Leibnitiana* 15 (1983), 30–41.

Gehler, Johann Samuel. *Physikalisches Wörterbuch* (5 vols, Leipzig, 1787–92).

Geoffroy, Étienne François. 'Observations sur les analyses du corail & de quelques autres plantes pierreuses, faites par M. le Comte Marsigli', *Mémoires de l'Académie royale des sciences* (1709), 102–5.

Geoffroy, Étienne François. 'Tables des différens rapports observés en chymie entre différentes substances', *Mémoires de l'Académie royale des sciences* (1718), 202–12.

Gerten, Michael. 'Bügerliche Aufklärung und streng systematische Philosophie. Zum Verhältnis von Leben und Philosophie bei K. L. Reinhold', in George di Giovanni, ed., *Karl Leonhard Reinhold and the Enlightenment* (Dordrecht, 2010), 65–85.

Gesner, Conrad. *Historia animalium* (5 vols, Zurich, 1551–87).

Giacomo Filippo Foresti da Bergamo, *Supplementum chronicarum* (Venice, 1483).

Gibbon, Edward. *The History of the Decline and Fall of the Roman Empire*, with notes by Dean Milman and M. Guizot: a new edition with additional notes by William Smith (8 vols, London, 1881).

Gibson, Susannah C. 'The Pursuit of Nature: Defining Natural Histories in Eighteenth-Century Britain', unpublished PhD thesis, University of Cambridge, 2011.

Giglioni, Guido. 'Jean-Baptiste Lamarck and the Place of Irritability in the History of Life and Death', in S. Normandin and C. Wolfe, eds, *Vitalism and the Scientific Image in Post-Enlightenment Life Science, 1800–2010* (Dordrecht, 2012), 19–49.

Gill, Michael B. *Humean Moral Pluralism* (Oxford, 2014).

Girtanner, Christoph. 'Abhandlungen über die Irritabilität', *Journal der Physik* (1791) 3.2: 317–51, 3.3: 507–37.

Girtanner, Christoph. *Über das Kantische Prinzip für die Naturgeschichte. Ein Versuch diese Wissenschaft philosophisch zu behandeln* (Göttingen, 1796).

Glacken, Clarence J. *Traces on the Rhodian Shore: Nature and Culture in Western Thought from Ancient Times to the End of the Eighteenth Century* (Berkeley, 1967).

Gmelin, Johann Friedrich. *Grundriß der allgemeinen Chemie zum Gebrauch bei Vorlesungen* (2 vols, Göttingen, 1789).

Gode-von Aesch, Alexander Gottfried Friedrich. *Natural Science in German Romanticism* (New York, 1941).

Godwin, William. *Enquiry Concerning Political Justice and its Influence on Morals and Happiness* (3rd edn, London, 1797).

Goeze, Johann Melchior. *Etwas vorläufiges gegen des Herrn Hofraths Lessings mittelbare und unmittelbare feindselige Angriffe auf unsre allerheiligste Religion und auf den einigen Lehrgrund derselben, die heilige Schrift* (Hamburg, 1778).

Golinski, Jan. *Science and Public Culture: Chemistry and Enlightenment in Britain, 1760–1820* (Cambridge, 1992).

Golinski, Jan. *British Weather and the Climate of Enlightenment* (Chicago, 2007).

Gossman, Lionel. *Orpheus Philologus: Bachofen versus Mommsen on the Study of Antiquity*, Transactions of the American Philosophical Society, 73/5 (Philadelphia, 1983), 1–89.

Gottsched, Johann Christoph. *Erste Gründe der Gesamten Weltweisheit, darinn alle Philosophische Wissenschaften in ihrer natürlichen Verknüpfung abgehandelt werden* (2 vols, Leipzig, 1733–4).

Gould, Stephen Jay. 'Chimp on a Chain', *Natural History* 92/12 (1983), 18–26.

Gould, Stephen Jay. *The Mismeasure of Man* (rev. edn, New York, 1996).

Graf, Friedrich W. *Kritik und Pseudo-Spekulation: David Friedrich Strauß al Dogmatiker im Kontext der positionellen Theologie seiner Zeit* (Munich, 1982).

Graham, John. *Lavater's Essays on Physiognomy: A Study in the History of Ideas* (Bern, 1979).

Granger, Gilles-Gaston. *La mathématique sociale du Marquis de Condorcet* (Paris, 1956).

Grau, Conrad. *Die Preußische Akademie der Wissenschaften zu Berlin* (Heidelberg, 1993).

Graunt, John. *Natural and Political Observations mentioned in a following Index, and made upon the Bills of Mortality* (London, 1662).

Gray, Richard T. *About Face: German Physiognomic Thought from Lavater to Auschwitz* (Detroit, 2004).

Graziosi, Barbara. *The Gods of Olympus: A History* (London, 2013).

Gregory, John. *A Comparative View of the State and Faculties of Man with those of the Animal World* (London, 1765).

Griswold, Charles L. Jr. *Adam Smith and the Virtues of Enlightenment* (Cambridge, 1999).

Gusdorf, Georges. *La Conscience révolutionnaire, les idéologues* (Paris, 1978).

Habermas, Jürgen. *Erkenntnis und Interesse* (Berlin, 1968).

Habermas, Jürgen. *Knowledge and Human Interests* (London, 1972).

Hacking, Ian. 'Jacques Bernoulli's *Art of Conjecturing*', *British Journal for the Philosophy of Science* 22 (1971), 209–29.

Hacking, Ian. *The Taming of Chance* (Cambridge, 1990).

Haigh, Elizabeth. *Xavier Bichat and the Medical Theory of the Eighteenth Century* (London, 1984).

Halbwachs, Maurice. *La théorie de l'homme moyen: Essai sur Quetelet et la statistique morale* (Paris, 1912).

Haldane, Jack B. S., et al. 'Discussion: Future of the Mind', in Gordon E. W. Wolstenholme, ed., *Man and his Future*, Ciba Foundation Seminar (London, 1963), 322–7.

Hales, Stephen. *Vegetable Staticks: Or, An Account of Some Statical Experiments on the Sap of Vegetables: being an Essay towards a Natural History of Vegetation. Also, a Specimen of an Attempt to Analyse the Air, By a great Variety of Chymio-Statical Experiments* (London, 1727).

Hales, Stephen. *Statical Essays: Containing Haemastatics; or An Account of Some Hydraulic and Hydrostatical Experiments Made on the Blood and Blood-vessels of Animals* (2 vols, London, 1740).

Hales, Stephen. *Vegetable Staticks* (London, 1969).

Haller, Albrecht von. *Hermanni Boerhaave Praelectiones academicae in proprias institutiones rei medicae edidit*... (7 vols, Göttingen, 1739–44).

Haller, Albrecht von. *De partibus corporis humani sensibilibus et irritabilibus* (Göttingen, 1752).

Haller, Albrecht von. *A Dissertation on the Sensible and Irritable Parts of Animals* (London, 1755).

Haller, Albrecht von. *Mémoires sur la nature sensible et irritable des parties du corps animal* (4 vols, Lausanne, 1756–60).

Haller, Albrecht von. *Ad Roberti Whyttii nuperum scriptum Apologia* (Göttingen, 1764).

Hamann, Johann Georg. *Sokratische Denkwürdigkeiten für die lange Weile des Publikums zusammengetragen von einem Liebhaber der langen Weile* ('Amsterdam' [Köningsberg], 1759).

Hamann, Johann Georg. *Hamann's Socratic Memorabilia*, ed. and trans. J. C. O'Flaherty (Baltimore, 1967).

Hamann, Johann Georg. *Writings on Philosophy and Language*, ed. K. Haynes (Cambridge, 2007).

Hamann, Julian. '*Bildung* in German Human Sciences: The Discursive Transformation of a Concept', *History of the Human Sciences* 24 (2011), 48–72.

Hammerstein, Notker. 'Zur Geschichte der deutschen Universitäten im Zeitalter der Aufklärung', in Hellmuth Rössler and Günther Franz, eds, *Universität und Gelehrtenstand 1400–1800* (Limburg, 1970), 145–82.

Hammerstein, Notker. 'Die Universitätsgründungen im Zeichen der Aufklärung', in Peter Baumgarten and Notker Hammerstein, eds, *Universität und Gelehrenstand 1400–1800* (Nendeln, 1978), 263–98.

Hammerstein, Notker. 'Christian Wolff und die Universitäten', in Werner Schneiders, ed., *Christian Wolff 1679–1754* (Hamburg, 1983), 266–77.

Hammerstein, Notker. 'Die deutschen Universitäten im Zeitalter der Aufklärung', *Zeitschrift für historische Forschung* 19 (1983), 73–89.

Hannaway, Caroline. 'The Société Royale de Médecine and Epidemics in the Ancien Régime', *Bulletin of the History of Medicine* 46 (1972), 257–73.

Hannaway, Caroline. 'Vicq d'Azyr, Anatomy and a Vision of Medicine', in Ann La Berge and Mordechai Feingold, eds, *French Medical Culture in the Nineteenth Century* (Amsterdam, 1994), 280–95.

Harnack, Adolf. *Geschichte der Königlich Preussischen Akademie der Wissenschaften zu Berlin* (4 vols, Berlin, 1900).

Harris, Henry. *The Birth of the Cell* (New Haven, Conn., 1999).

Harrison, Peter. *'Religion' and Religions in the English Enlightenment* (Cambridge, 1990).

Harrison, Peter. 'Experimental Religion and Experimental Science in Early Modern England', *Intellectual History Review* 21 (2011), 413–33.

Harrison, Ross. *Bentham* (London, 1983).

Hatfield, Gary. 'Remaking the Science of the Mind', in C. Fox, R. Porter, and R. Wokler, eds, *Inventing Human Science: Eighteenth-Century Domains* (Berkeley, 1995), 184–231.

Heath, Eugene. 'Adam Smith and Self-Interest', in C. J. Berry, M. P. Paganelli, and C. Smith eds, *The Oxford Handbook of Adam Smith* (Oxford, 2013), 241–64.

Hecht, Harmut. 'In the Spirit of Leibniz—Two Approaches from 1742', in Ruth Hagengruber, ed., *Emilie du Châtelet Between Leibniz and Newton* (Dordrecht, 2012), 61–76.

Hegel, George W. F. *Early Theological Writings*, trans. T. M. Knox and R. Korner (Philadelphia, 1971).

Hegel, George W. F. *Lectures on the Philosophy of Religion, One-Volume Edition. The Lecture of 1827*, ed. P. C. Hodgson (Berkeley, 1988).

Heilbron, John L. 'The Measure of Enlightenment', in T. Frängsmyr, J. Heilbron, and R. Rider, eds, *The Quantifying Spirit in the Eighteenth Century* (Berkeley, 1990), 207–42.

Heilbron, John L. *Electricity in the Seventeenth and Eighteenth Centuries* (Mineola, NY, 1999).

Heinz, Marion. *Sensualisticher Idealismus: Untersuchungen zur Erkenntnistheorie des jungen Herder* (Hamburg, 1994).

Heinz, Marion. 'Grundzüge von Herders Psychologie. "Uebers Erkennen und Empfinden der Menschlichen Seele" (1774)', in W. Koepke, ed., *J. G. Herder: Academic Disciplines of Knowledge and the Pursuit of Knowledge* (New York, 1996), 137–51.

Helvétius, Claude-Adrien. *De l'esprit: or, Essays on the Mind, and its several faculties* (London, 1759).

Herder, Johann Gottfried. *Fragmente über die neuere deutsche Literatur* (Riga, 1766–7).

Herder, Johann Gottfried. *Ueber deu Ursprung der Sprache* (Berlin, 1772).

Herder, Johann Gottfried. *Herders Briefe*, ed. Wilhelm Dobbek (Weimar, 1959).

Herder, Johann Gottfried. *Briefe: Gesamtausgabe 1763–1803*, ed. Wilhelm Dobbek and Günter Arnold (10 vols, Weimar, 1977).

Herder, Johann Gottfried. *Werke in zehn Bänden*, ed. Günther Arnold et al. (9 vols in 10 parts, Frankfurt am Maine, 1985–2000).

Herder, Johann Gottfried. *Philosophical Writings*, ed. and trans. Michael N. Forster (Cambridge, 2002).

Hill, Christopher. *The World Turned Upside Down: Radical Politics During the English Revolution* (London, 1972).

Hilton, Phillip. *Bitter Honey: Recuperating the Medical and Scientific Context of Bernard Mandeville* (Bern, 2011).

Hinske, Norbert. 'Wolffs Stellung in der deutschen Aufklärung', in Werner Schneiders, ed., *Christian Wolff 1679–1754* (Hamburg, 1983), 306–20.

Hobbes, Thomas. *Leviathan or The Matter, Forme and Power of a Commonwealth Ecclesiasticall and Civil* (London, 1651).

Hoffmann, Paul. *La Femme dans la Pensée des Lumières* (Paris, 1977).

[Holbach, Paul Henri Thiery, Baron d']. *Système de la Nature ou Des Loix du Monde Physique* (2 vols, [Paris], 1770).

Holmes, Frederic Lawrence. *Eighteenth-Century Chemistry as an Investigative Enterprise* (Berkeley, 1989).

Holzhey, Helmut. 'Philosophie als Eklektik', *Studia Leibniziana* 15 (1983), 19–29.

Home, Roderick W. *Aepinus's Essay on the Theory of Electricity and Magnetism*, introd. monograph and notes by R. W. Home, trans. P. J. O'Connor (Princeton, 1979).

Hont, Istvan. *Jealousy of Trade: International Competition and the Nation-State in Historical Perspective* (Cambridge, Mass., 2005).

Hooke, Robert. *Micrographia: or some Physiological Descriptions of Minute Bodies made by Magnifying Glasses* (London, 1665).
Hoppit, Julian. 'Political Arithmetic In Eighteenth-Century England', *Economic History Review* 49 (1996), 516–40.
Howard, Thomas Albert. *Protestant Theology and the Making of the Modern German University* (Oxford, 2006).
Hübner, Johann. *Kurtze Fragen aus der politischen Historie biß auf gegenwärtige Zeit continuiret* (10 vols, Leipzig, 1715).
Hübner, Kurt. *Die Wahrheit des Mythos* (Munich, 1985).
Hufbauer, Karl. *The Formation of the German Chemical Community (1720–1795)* (Berkeley, 1982).
Hufeland, Christoph Wilhelm. *Ideen über Pathologie und Einfluß der Lebenskraft auf Entstehung und Form der Krankheit* (Jena, 1795).
Humboldt, Alexander von. *Florae Fribergensis specimen plantas cryptogamicas praesertim subterraneas exhibens* (Berlin, 1793).
Humboldt, Alexander von. *Versuche über der Gereizte Muskel- und Nervenfaser* (2 vols, Posen and Berlin, 1797).
Humboldt, Alexander von. 'Des lignes isothermes et de la distribution de la chaleur sur le globe', *Mémoires de physique et de chimie de la Société d'Arcueil* 3 (1817), 462–602.
Humboldt, Alexander von. *Essai géognostique sur le gisement des roches dans les deux hémisphères* (Paris, 1823).
Humboldt, Alexander von. 'Ueber der Mittel, die Ergründung einiger Phänomene des tellurischen Magnetismus zu erleichtern', *Annalen der Physik und Chemie* 91 (1829) 319–36.
Humboldt, Alexander von. *Kosmos: Entwurf einer physischen Weltbeschreibung* (5 vols, Stuttgart, 1845–62).
Humboldt, Alexander von. *Ansichten der Natur: mit wissenschaftlichen Erläuterungen* (Stuttgart, 1849).
Humboldt, Alexander von. *Cosmos: A Sketch of a Physical Description of the Universe*, trans. E. C. Otte (5 vols, London, 1849–58).
Hume, David. *A Treatise of Human Nature: Being an Attempt to Introduce the Experimental Method of Reasoning in Moral Subjects* (3 vols, London, 1739).
Hume, David. *Essays, Moral and Political* (2 vols, Edinburgh, 1742).
Hume, David. *The History of England, from the Invasion of Julius Ceasar until the Revolution in 1688 . . . a new edition, corrected* (8 vols, Dublin, 1788).
Hume, David. *Essays and Treatises on Several Subjects* (2 vols, Edinburgh, 1793).
Hume, David. *Enquiries concerning the Human Understanding and Concerning the Principles of Morals*, ed. L. A. Selby-Bigge, 2nd edn (Oxford, 1962).
Humphrey, Nicholas, John. R. Skoyles, and Roger Keynes. 'Human Hand-Walkers: Five Siblings Who Never Stood Up', *LSE Research Online*, 2005. Available at: <http://eprints.lse.ac.uk/463/>.
Hunard, Pierre. *Dissertation sur les vapeurs et les pertes de sang* (Paris, 1765).
Hundert, Edward G. *The Enlightenment's* Fable: *Bernard Mandeville and the Discovery of Society* (Cambridge, 1994).
Hundt, Magnus. *Anthropologia, de hominis dignitate, nature et proprietatibus, de elementis, partibus et membris humani corporis etc., de spiritu humano eiusque natura et partibus et operibus, de anima humana et ipsius appendiciis* (Leipzig, 1501).

Hunter, Ian. *Rival Enlightenments: Civil and Metaphysical Philosophy in Early Modern Germany* (Cambridge, 2011).
Hunter, Ian. 'Secularization: The Birth of a Modern Combat Concept', *Modern Intellectual History*, 12 (2015), 1–32.
Hunter, John. *The Works of John Hunter*, ed. J. F. Palmer (4 vols, London, 1835–7).
Hutcheson, Francis. *An Inquiry into the Original of our Ideas of Beauty and Virtue* (London, 1726).
Hutcheson, Francis. *An Essay on the Nature and Conduct of the Passions and Affections, with Illustrations on the Moral Sense* (Glasgow, 1729).
Hutton, James. *Theory of the Earth, with Proofs and Illustrations* (2 vols, Edinburgh, 1795).
Iacono, Alfonzo Maurizio. 'The American Indians and the Ancients of Europe: The Idea of Comparison and the Construction of Historical Time in the Eighteenth Century', in W. Haase and M. Reinhold, eds, *The Classical Tradition and the Americas, Volume One: European Images of the Americas and the Classical Tradition* (Berlin, 1994), 658–81.
Irmscher, Hans Dieter. 'Beobachtungen zur Funktion der Analogie im Denken Herders'. *Deutsche Vierteljahrshrift fu r Literaturwissenschaft und Geschichte* 55 (1981), 64–98.
Irwing, Karl Franz von. *Erfahrungen und Untersuchungen über den Menschen* (2 vols, Berlin, 1777–9).
Isambert, François-André. *Politique, religion et science de l'homme chez Philippe Buchez (1796–1865)* (Paris, 1967).
Iselin, Isaak. *Über die Geschichte der Menscheit* (Zürich, 1764).
Israel, Jonathan. *Radical Enlightenment and the Making of Modernity, 1650–1750* (Oxford, 2002).
Israel, Jonathan. *Enlightenment Contested: Philosophy, Modernity, and the Emancipation of Man 1670–1752* (Oxford, 2008).
Israel, Jonathan. *Democratic Enlightenment: Philosophy, Revolution, and Human Rights, 1750–1790* (Oxford, 2011).
Itard, Jean. *De l'éducation d'un homme sauvage, ou, Des premiers développemens physiques et moraux du jeune sauvage de l'Aveyron* (Paris, 1801).
Itard, Jean. *Rapport fait à son excellence le ministre de l'interieur sur les nouveaux développements de l'état actuel du sauvage de l'Aveyron* (Paris, 1807).
Jaeger, Werner. *Paideia: The Ideals of Greek Culture* (3 vols, Oxford, 1939–45).
Jakob, Ludwig Heinrich. *Prüfung der Mendelssohnischen* Morgenstunden *oder aller spekulativen beweise für des Daseyn Gottes* (Leipzig, 1786).
Jacobi, Friedrich Heinrich. *Briefe über die Lehre Spinozas* (Leipzig, 1785).
Jacobi, Friedrich Heinrich. *David Hume über den Glauben, oder Idealismus und Realismus* (Breslau, 1787).
Jacobs, Brian. 'Kantian Character and the Science of Humanity', in Brian Jacobs and Patrick Kain, eds, *Essays on Kant's Anthropology* (Cambridge, 2003), 105–34.
Jardine, Nicholas. '*Naturphilosophie* and the Kingdoms of Nature', in N. Jardine, J. A. Secord, and E. C. Spary, eds, *Cultures of Natural History* (Cambridge, 1996), 230–45.
Jaumann, Herbert. 'Rousseau in Deutschland, Forschungsgeschichte und Perspektiven', in Herbert Jaumann, ed., *Rousseau in Deutschland: Neue Beiträge zur Erforschung seiner Rezeption* (Berlin, 1994), 1–22.
Jay, Mike. *The Atmosphere of Heaven: The Unnatural Experiments of Dr Beddoes and His Sons of Genius* (New Haven, 2009).

'Jean Paul' [Friedrich Richter]. *Vorschule der Aesthetik* (Hamburg, 1804).
Jouard, Gabriel. *Nouvel Essai sur la femme considerée comparativement à l'homme* (Paris, 1804).
Jungnickel, Christa and McCormach, Russell. *Cavendish: The Experimental Life* (Lewisburg, 1999).
Jurin, James. 'A Letter to the Learned Caleb Cotesworth, MD, FRS, of the College of Physicians, and Physician to St. Thomas' Hospital, Containing a Comparison between the Mortality of the Natural Small Pox, and That Given by Inoculation', *Philosophical Transactions* 32 (1722–3), 213–24.
Jussieu, Antoine-Laurent de. *Genera plantorum, secundum ordines naturales disposita, juxta methodum in Horto regio parisiensi exaratum* (Paris, 1789).
Kain, Philip J. *Schiller, Hegel, and Marx: State, Society, and the Aesthetic Ideal of Ancient Greece* (Kingston, Ontario, 1982).
Kaitaro, Timo. *Diderot's Holism* (Frankfurt am Main, 1997).
Kames, Henry Home, Lord. *Sketches of a History of Man* (4 vols, Edinburgh, 1788).
Kant, Immanuel. *Gesammelte Schriften* ('Akademie' edition, 29 vols, Berlin, 1900 onwards).
Kant, Immanuel. *Bermerkungen in den 'Beobachten über das Gefühl des Schönen und Erhabenen'*, ed. Marie Rischmüller (Hamburg, 1991).
Kant, Immanuel. *Lectures on Logic* (Cambridge, 1992).
Kant, Immanuel. *Theoretical Philosophy, 1755–1770* (Cambridge, 1992).
Kant, Immanuel. *Practical Philosophy* (Cambridge, 1996).
Kant, Immanuel. *Religion and Rational Theology* (Cambridge, 1996).
Kant, Immanuel. *Critique of Pure Reason* (Cambridge, 1998).
Kant, Immanuel. *Correspondence* (Cambridge, 1999).
Kant, Immanuel. *Critique of the Power of Judgement* (Cambridge, 2000).
Kant, Immanuel. *Theoretical Philosophy after 1781* (Cambridge, 2002).
Kant, Immanuel. *Anthropology, History, and Education* (Cambridge, 2007).
Kant, Immanuel. *Lectures on Anthropology* (Cambridge, 2012).
Kant, Immanuel. *Natural Science* (Cambridge, 2012).
Karsten, Wenceslaus Johann Gustav Renger. *Entleitung zur gemeinnützlichen Kenntniß der Natur* (Halle, 1783).
Karsten, Wenceslaus Johann Gustav Renger. *Physische-chemische Abhandlung, durch neuere Schriften von hermetischen Arbeiten und andere neue Untersuchungen veranlaßet* (2 vols, Halle, 1786–7).
Keel, Othmar. 'Was Anatomical and Tissue Pathology a Product of the Paris Clinical School or Not?', in Caroline Hannaway and Ann La Berge, eds, *Constructing Paris Medicine* (Amsterdam, 1998), 117–84.
Kepler, Johannes. *Ad vitellionem paralipomena, quibus Astronomiae pars optica traditvr* (Frankfurt, 1605).
Kielmeyer, Carl Friedrich. *Über die Verhältnisse der organischen Kräfte* (Stuttgart, 1793).
Kim, Mi Gyung. *Affinity, That Elusive Dream: A Genealogy of the Chemical Revolution* (Cambridge, Mass., 2003).
Kim, Soo Bae. *Die Entstehung der Kantischen Anthropologie und ihre Beziehung zur empirischen Psychologie der Wolffschen Schule* (Frankfurt am Main, 1995).
Kircher, Athanasius. *Oedipus Aegyptiacus* (3 vols, Rome, 1652–4).
Kirwan, Richard. *An Essay on Phlogiston, and the Constitution of Acids* (London, 1789).

Klein, Jacob Theodor. *Summa Dubiorum Circa Classes Quadrupedum et Amphibium in Linnaei Systemate Naturae* (Danzig, 1743).

Klein, Ursula. 'E. F. Geoffroy's Table of Different "Rapports" Observed Between Different Chemical Substances—A Reinterpretation', *Ambix* 42 (1995), 251–87.

Klein, Ursula. *Experiments, Models, Paper Tools: Cultures of Organic Chemistry in the Nineteenth Century* (Stanford, 2003).

Klein, Ursula and Lefèvre, Wolfgang. *Materials in Eighteenth-Century Science: A Historical Ontology* (Cambridge, Mass., 2007).

Klose, Carl Ludwig. *Über den Einfluss des Geschlechts-Unterschiedes auf Ausbildung und Heilung von Krankenheiten* (Stendal, 1829).

Knight, Gowin. *An Attempt To Demonstrate, That all the Phenomena in Nature, May be Explained by Two Simple Active Principles, Attraction and Repulsion* (London, 1754).

[Knoblauch, Carl von.] *Taschenbuch für Aufklärer und Nichtaufklärer auf das Jahr 1791* (Berlin, 1790).

[Knoblauch, Carl von.] *Euclides Anti-Thaumaturgicus, oder demonstrantiver Beweis von der Unmöglichkeit hyperphysischer Begebenheiten* ('Germany' [i.e. Berlin], 1791).

Koerner, Lisbet. *Linnaeus: Nature and Nation* (Harvard, Mass., 1999).

Koestlin, Karl Heinrich. *Dissertatio Physica Experimentalis de Effectibus Electricitatis in Quaedam Corpora Organica* (Tübingen, 1775).

Köhler, Johann David. *Sculptura Historiarum et Temporum Memoratrix: Das ist, Gedächtnüß-Hülfiche Bilder-Lust der Merwürdigsten Welt-Geschichten aller Zeiten von Erschaffung der Welt biß auf gegenwärtige* (Nüremberg, 1726).

Köhler, Johann David. *Erneurter Entwurf eines Collegii über den gegenwärtigen Zustand von Europa und die jetzigen Welt-Händel* (Göttingen, 1736).

Kraus, Andreas. 'Die Bedeutung der deutschen Akademien des 18. Jahrhunderts für historische und naturwissenschaftliche Forschungen', in F. Hartmann and R. Vierhaus, eds, *Der Akademiegedanke im. 17 und 18. Jahrhundert* (Bremen, 1977), 139–70.

Krüger, Johann Gottlob. *Naturlehre: Nebst Kupfern und vollständigem Register* (3 vols, Halle, 1740–9).

Krüger, Johann Gottlob. *Versuch einer Experimental-Seelenlehre* (Halle/Helmstädt, 1756).

Kuehn, Manfred. *Kant: A Biography* (Cambridge, 2001).

Kuehn, Manfred. 'The Reception of Hume in Germany', in Peter Jones, ed., *The Reception of David Hume in Europe* (London, 2005), 98–138.

Kusukawa, Sachiko. *Picturing the Book of Nature: Image, Text, and Argument in Sixteenth-Century Human Anatomy and Medical Botany* (Chicago, 2012).

La Berge, Ann and Caroline Hannaway. 'Paris Medicine: Perspectives Past and Present', in Caroline Hannaway and Ann La Berge, eds, *Constructing Paris Medicine* (Amsterdam, 1998), 1–70.

[La Caze, Louis de]. *Idée de l'homme physique et moral, pour servir d'introduction à un traité de médecine* (Paris, 1755).

La Condamine, Charles-Marie de. *Histoire d'une jeune fille sauvage trouvée dans les bois à l'âge de dix ans* (Paris, 1755).

La Créquinière, M. de. *Conformité des coutumes des Indiens orientaux, avec celles des Juifs & des autres peuples de l'antiquité* (Brussels, 1704).

La Méthrie, Jean-Claude de. 'Discours préliminaire, contenant un précis des nouvelles découvertes', *Observations sur la physique, l'histoire naturelle et les arts* 28 (1787), 1–53.

[La Mettrie, Julien Offray de]. *L'Homme-machine* (Leiden, 1747).

[La Mettrie, Julien Offray de]. *L'Homme plante* (Potsdam/Berlin, 1748).

La Mettrie, Julien Offray de. *Œuvres philosophiques* (3 vols, Berlin, 1796).

La Vopa, Anthony. 'Herder's *Publikum*: Language, Print, and Sociability in Eighteenth-Century Germany', *Eighteenth-Century Studies* 29 (1995), 5–24.

La Vopa, Anthony. *Fichte: The Self and the Calling of Philosophy, 1762–1799* (Cambridge, 2001).

Laënnec, René-Théophile-Hyacinthe. *Traité de l'auscultation médiate ou Traité diagnostic des maladies des poumons et du cœur, fondé principalement sur ce nouveau moyen d'exploration* (2 vols, Paris, 1819).

Lafitau, Joseph-François. *Mœurs des sauvages ameriquains, comparées aux mœurs des premiers temps* (2 vols, Paris, 1724).

Lagrange, Joseph Louis de. *Mécanique analytique* (Paris, 1788).

Lamarck, Jean-Baptiste-Pierre-Antoine de Monet de. *Flore Françoise, ou description succincte de toutes les plantes qui croissent naturellement en France, disposée selon une nouvelle méthode d'analyse à laquelle on a joint la citation de leurs vertus les moins équivoques en médecine, et de leur utilité dans les arts* (3 vols, Paris, 1779).

Lamarck, Jean-Baptiste-Pierre-Antoine de Monet de. *Recherches sur les causes des principaux faits physiques* (2 vols, Paris, 1794).

Lamarck, Jean-Baptiste-Pierre-Antoine de Monet de. *Mémoires de physique et d'histoire naturelle, établis sur des bases de raisonnement indépendantes de toute théorie* (Paris, 1797).

Lamarck, Jean-Baptiste-Pierre-Antoine de Monet de. *Recherches sur l'organisation des corps vivans et particulièrement sur leur origine, sur la cause de ses développemens et des progrès de sa composition, et sur celle qui, tendant continuellement à la détruire dans chaque individu, amène nécessairement sa mort; précédé du discours d'ouverture du cours de zoologie, donné dans le Muséum national d'Histoire Naturelle* (Paris, 1802).

Lamarck, Jean-Baptiste-Pierre-Antoine de Monet de. *Philosophie zoologique, ou exposition des considérations relatives à l'histoire naturelle des animaux; à la diversité de leur organisation et des facultés qu'ils en obtiennent; aux causes physiques qui maintiennent en eux la vie et donnent lieu aux mouvemens qu'ils exécutent; enfin, à celles qui produisent les unes le sentiment et les autres l'intelligence de ceux qui en sont doués* (2 vols, Paris, 1809).

Landmann, Michael. *Einführung in die philosophische Anthropologie. Menschliche Selbstdeutung in Geschichte und Gegenwart* (Berlin, 1964).

Lane, Harlan. *The Wild Boy of Aveyron* (London, 1977).

Laplace, Pierre-Simon. *Œuvres complètes* (14 vols, Paris, 1878–1912).

Laqueur, Thomas. 'Organism, Generation, and the Politics of Reproductive Biology', in C. Gallagher and T. Laqueur, eds, *The Making of the Modern Body: Sexuality and Society in the Nineteenth Century* (Berkeley, 1987), 1–41.

Larmore, Charles. *Patterns of Moral Complexity* (Cambridge, 1987).

Larson, James L. *Interpreting Nature: The Science of Living Form from Linnaeus to Kant* (Baltimore, 1994).

Lavater, Johann Caspar. *Physiognomische Fragmente, zur Beförderung der Menschenkenntniß und Menschenliebe* (4 vols, Leipzig and Winterthur, 1775–8).

Lavoisier, Antoine-Laurent de. *Traité Élémentaire de Chimie, présenté dans un ordre nouveau et d'Après les Découvertes Modernes* (Paris, 1789).

Le Bègue de Presle, Achille. *Conservateur de la santé, ou avis sur les dangers qu'il importe à chacun d'éviter, pour se conserver en bonne santé et prolonger sa vie* (Paris, 1763).

Le Camus, Antoine. *La Médecine de l'esprit* (2 vols, Paris, 1753).

Le Camus, Antoine. *La Médecine de l'esprit* (2nd edn, 3 vols, Paris, 1769).

Le Cat, Claude Nicolas. *Traité de la couleur de la peau humaine en général et de celle des nègres en particulier, et de la métamorphose d'une de ces couleurs en l'autre, soit de naissance, soit actuellement* (Amsterdam, 1765).

Le Clerc, Jean. *Historia Ecclesiastica duorum primorum a Christo nato sæculorum a veteribus monumentis depromta* (Amsterdam, 1716).

Le Roy, Loys. *De la vicissitude ou variété des choses en l'univers, et concurrence des armes et des lettres par les premieres et plus illustres nations du monde, depuis le temps où a commencé la ciuilité, & mémoire humaine iusques à présent* (Paris, 1575).

Lécuyer, Bernard-Pierre. 'Probability in Vital and Social Statistics: Quetelet, Farr, and the Bertillons', in L. Krüger, L. Daston, and M. Heiderlberger, eds, *The Probabilistic Revolution, Volume 1: Ideas in History* (Cambridge, Mass., 1987), 317–35.

Ledru, Nicolas-Philippe. *Rapport de MM. Cosnier, Maloet, Darcet, [et al] sur les avantages reconnus de la nouvelle méthode d'administrer l'électricité dans les maladies nerveuses, particulièrement dans l'épilepsie, & dans la catalepsie, par M. Ledru, connu sous le nom de Comus* (Paris, 1783).

Leeuwenhoek, Anton von. *Arcana Naturae* (Leiden, 1695).

Legendre, Adrien Marie. *Nouvelles méthodes pour la détermination des orbites des comètes* (Paris, 1805).

Leifer, Walter. *Hellas in deutschen Geistesleben* (Bad Herrenalb, 1963).

Lemery, Nicolas. *Cours de Chimie* (Paris, 1675).

Lepenies, Wolf. *The Seduction of Culture in German History* (Princeton, 2006).

Lespinasse, Julie de. *Lettres inédites de Mlle de Lespinasse*, ed. Charles Henry (Paris, 1887).

Lessing, Gotthold Ephraim. *Laokoon: Oder, Über die Grenzen der Malerei und Poesie* (Berlin, 1766).

Lessing, Gotthold Ephraim. *Die Erziehung des Menschengeschlechts* (Berlin, 1780).

Lessing, Gotthold Ephraim. *Sämtliche Schriften* (23 vols, Stuttgart, 1886–1924).

Lessing, Gotthold Ephraim. *Werke*, ed. H. G. Gölpert et al. (8 vols, Berlin, 1978).

Lestition, Steven. 'Kant's Philosophical Anthropology: Texts and Historical Contexts, Continuity and Change', unpublished PhD dissertation, University of Chicago, 1985.

Lévesque de Pouilly, Louis-Jean. *Théorie des sentimens agréables* (Geneva, 1747).

Lidner, Herbert. *Das Problem des Spinozismus im Schaffen Goethes und Herders* (Weimar, 1960).

Lifschitz, Avi. *Language and Enlightenment: The Berlin Debates of the Eighteenth Century* (Oxford, 2012).

Linden, Mareta. *Untersuchungen zum Anthropologiebegriff des 18. Jahrhunderts* (Bern, 1976).

Linnaeus, Carl. *Systema Naturæ per Regna Tria Naturæ, secundum Classes, Ordines, Genera, Species* (Leiden, 1735).

Linnaeus, Carl. *Fauna Svecica* (Stockholm, 1746).

Linnaeus, Carl. *Genera morborum, in auditorum usum* (Uppsala, 1763).

Linnaeus, Carl. *The Animal Kingdom or Zoological System . . . being a translation of that part of the Systema Naturae, as lately published . . . by Professor Gmelin* (Edinburgh 1792).

Livi, Jocelyne. *Vapeurs de femmes: Essai historique sur quelques fantasmes médicaux et les pertes de sang* (Dijon, 1984).

Livingston, Donald W. *Philosophical Melancholy and Delirium: Hume's Pathology of Philosophy* (Chicago, 1998).

Lloyd, Geoffrey E. R. 'Fuzzy natures?', in idem, *Aristotelian Explorations* (Cambridge, 1996), 67–82.

Locke, John. *An Essay concerning Humane Understanding* (London, 1690).

Lohr, Charles H. 'Metaphysics', in Charles B. Schmitt, Quentin Skinner, and Eckhard Kessler, eds, *The Cambridge History of Renaissance Philosophy* (Cambridge, 1988), 537–638.

Lord, Beth. *Kant and Spinozism: Transcendental Idealism and Immanence from Jacobi to Deleuze* (London, 2011).

Lott, Frans van der. 'Kort bericht van den Conger-all, ofte drilvisch; getrokken uit eenen brief van Frans van der Lott, gedateerd Rio Essequebo den 7 Juny 1761', *Verhandelingen uitgegeeven door de Hollandsche Maatschappye der Weetenschappen, te Haarlem, 6* (1762), part II, supp., 87–95.

Louden, Robert B. 'The Second Part of Morals', in Brian Jacobs and Patrick Kain, eds, *Essays on Kant's Anthropology* (Cambridge, 2003), 60–84.

Louis, Pierre Charles Alexandre. *Recherches Sur Les Effets De La Saignée Dans Quelques Maladies Inflammatoires Et Sur L'action De L'émétique Et Des Vésicatoires Dans La Pneumonie* (Paris, 1835).

Loveland, Jeff. 'Panckouck and the Circle Squarers', *Eighteenth-Century Studies* 37 (2004), 215–36.

Lovett, Richard. *The Subtil Medium prov'd: or, That Wonderful Power of Nature, so Long ago Conjectur'd by the most Ancient and Remarkable Philosophers, which they call'd sometimes Aether, but oftener Elementary Fire, verify'd* (London, 1756).

Lovett, Richard. *The Electrical Philosopher. Containing a new System of Physics founded upon the Principle of an Universal Plenum of Elementary* Fire (Worcester, 1774).

Lyell, Charles. *Scientific Journals on the Species Question*, ed. L. G. Wilson (New Haven, 1970).

Lyon, John. *Experiments and Observations Made with a View to Point Out the Errors of the Present Received Theory of Electricity* (London, 1780).

Lyons, David. *In the Interest of the Governed: A Study of Bentham's Philosophy of Utility and Law* (Oxford, 1973).

MacCulloch, Diarmaid. *A History of Christianity* (London, 2009).

Mackintosh, James. *Dissertation on the Progress of Ethical Philosophy* (Edinburgh, 1836).

Macquer, Pierre-Joseph. *Elémens de chimie-théorique* (Paris 1749).

Macquer, Pierre-Joseph. *Elémens de chimie-pratique, contenant la description des opérations fondamentales de la chymie, avec des explications & des remarques sur chaque opération* (2 vols, Paris, 1751).

Macquer, Pierre-Joseph. *Dictionnaire de Chymie* (2 vols, Paris, 1766).

Magendie, François. *Précis élémentaire de physiologie* (2 vols, Paris, 1816–17).

Maier, Hans. 'Die Lehre der Politik an den deutschen Univeritäten vornehmlich vom 16. bis 18. Jahrhundert', in Dieter Oberndorfer, ed., *Wissenschaftliche Politik: Eine Einführung in Grundfragen ihrer Tradition und Theorie* (Freiburg, 1967), 59–116.

Mairan, Jean-Jacques Dortous de. *Dissertation sur la glace* (4th edn, Paris, 1749).

Malherbe, Michel. 'Hume's Reception in France', in Peter Jones, ed., *The Reception of David Hume in Europe* (London, 2005), 43–96.

Malthus, Thomas. *On the Principle of Population, as it affects the Future Improvement of Society. With Remarks on the Speculations of Mr. Godwin, M. Condorcet, and other writers* (London, 1798).

[Mandeville, Bernard]. *The Grumbling Hive: Or, Knaves turn'd honest* ([London], 1705).

[Mandeville, Bernard]. *The Virgin Unmask'd: or, Female Dialogues betwixt an Elderly Maiden Lady, and her Niece* (London, 1709).

[Mandeville, Bernard]. *Treatise of the Hypochondriack and Hysterick Passions, Vulgarly Call'd the Hypo in men and Vapours in Women* (London, 1711).

[Mandeville, Bernard]. *Free Thoughts on Religion, the Church and National Happiness* (London, 1720).

[Mandeville, Bernard]. *A Modest Defence of Publick Stews: Or, an Essay upon Whoring as it is now Practis'd in these Kingdoms, Written by a Layman* (London, 1724).

[Mandeville, Bernard]. *An Enquiry into the Origin of Honour, and the Usefulness of Christianity in War* (London, 1732).

[Mandeville, Bernard]. *A Letter to Dion, Occasion'd by his Book call'd Alciphron, or the Minute Philosopher, by 'The Author of the Fable of the Bees'* (London, 1732).

[Mandeville, Bernard]. *The Fable of the Bees: Or, Private Vices, Publick Benefits. By Bernard Mandeville. With a Commentary Critical, Historical, and Explanatory by F. B. Kaye* (2 vols, Oxford, 1924).

Mangin, Abbé de. *Histoire générale et particulière sur l'électricité* (Paris, 1782).

Manuel, Frank E. *The Eighteenth Century Confronts the Gods* (Cambridge, Mass., 1959).

Marat, Jean-Paul. *Recherches physiques sur la feu* (Paris, 1780).

Marat, Jean-Paul. *Les Charlatans modernes; ou, Lettres sur le charlatanisme académique publiées par M. Marat, l'Ami du peuple* (Paris, 1793).

Marchand, Suzanne L. *Down from Olympus: Archaeology and Philhellenism in Germany, 1750–1970* (Princeton, 1996).

Marks, Jonathan. *Perfection and Disharmony in the Thought of Jean-Jacques Rousseau* (Cambridge, 2005).

Marsais, César Chesneau, sieur du. *Œuvres* (7 vols, Paris, 1797).

Marshall, Dorothy. *The English Poor in the Eighteenth Century: A Study in Social and Administrative History* (London, 1926).

Martens, Wolfgang. *Botschaft der Tugend: Die Aufklärung im Spiegel der deutschen Moralischen Wochenschriften* (Stuttgart, 1968).

Martin, Craig. *Subverting Aristotle: Religion, History, and Philosophy on Early Modern Science* (Baltimore, 2014).

Martins, Roberto de Andrade. 'Ørsted, Ritter and Magnetochemistry', in R. M. Brain, R. S. Cohen, and O. Knudsen, *Hans Christian Ørsted and the Romantic Legacy in Science: Ideas, Disciplines, Practices*, (New York, 2007), 339–85.

Martyn, John. *Elements of Natural History* (Cambridge, 1775).

Marx, Karl and Engels, Friedrich. *Werke* (43 vols, Berlin, 1956–68).

Marx, Karl and Engels, Friedrich. *Collected Works* (50 vols, London and New York, 1975–2005).

Masuzawa, Tomoko. *The Invention of World Religions* (Chicago, 2005).

Matthews, J. Rosser. *Quantification and the Quest for Medical Certainty* (Princeton, 1995).

Mauduyt de la Varenne, Pierre Jean Claude. *Mémoire sur les différentes manières d'administrer l'électricité: et observations sur les effets qu'elles ont produits* (Paris, 1784).

Maupertuis, Pierre-Louis Moreau de. *Réflexions philosophiques sur l'origine des langues et la signification des mots* (Paris, 1748).

Maupertuis, Pierre-Louis Moreau de. *Œuvres de Mr de Maupertuis, nouvelle edition* (4 vols, Lyon, 1756).

Maupertuis, Pierre-Louis Moreau de. 'Dissertation sur les différens moyens dont les hommes se sont servis pour exprimer leurs idées', *Histoire de l'Academie, 1754* (Berlin, 1756), 349–64.

Mayhew, Robert J. *Malthus: The Life and Legacies of an Untimely Prophet* (Cambridge, Mass., 2014).

Maza, Clelia Martînez. 'De la théologie à l'anthropologie: utilisation du paganisme grécoromain dans l'Europe du XIXème siècle', in C. Avlami, J. Alvar, and M. R. Recio, eds, *Historiographie de l'antiquité et transferts culturels: Les histoires anciennes de l'Europe des XVIIIe et XIXe siècles* (Amsterdam, 2010), 85–98.

McCormick, Ted. *William Petty: And the Ambitions of Political Arithmetic* (Oxford, 2009).

McLaughlin, Peter. 'Blumenbach und der Bildunstrieb: Zum verhältnis von epigenetischer Embryologie und typologischem Artbegriff', *Medizinhistorisches Journal* 17 (1982), 357–72.

Meek, Ronald L. *Social Science and the Ignoble Savage* (Cambridge, 1976).

Mehta, Pratap Bhanu. 'Self-Interest and Other Interests', in Knud Haakonssen, ed., *The Cambridge Companion to Adam Smith* (Cambridge, 2006), 246–69.

Meiners, Christoph. *Grundriss der Geschichte der Menschheit* (Lemgo, 1785).

Mendelssohn, Moses. *Abhandlung über die Evidenz in metaphysischen Wissenschaften* (Berlin, 1764).

Mendelssohn, Moses. *Morgenstunden: oder, Vorlesungen über das daseyn Gottes* (Berlin, 1785).

Mensch, Jennifer. *Kant's Organicism: Epigensis and the Development of Critical Philosophy* (Chicago, 2013).

Menzer, Paul. *Kants Lehre von der Entwicklung in Natur und Geschichte* (Berlin, 1911).

Mesmer, Franz. *Mémoire sur la découverte du magnétisme animal* (Geneva, 1779).

Metzger, Hélène. *Newton, Stahl, Boerhaave et la doctrine chimique* (Paris, 1930).

Metzger, Johann Daniel. *Medizinisch-philosophische Anthropologie fiir Aerzte und Nichtarzte. Zum Gebrauch akademischer Vorlesungen* (Weissenfels and Leipzig, 1790).

Michaelis, Johann David. *Mosaisches Recht* (6 vols, Frankfurt, 1770–5).

Middleton, Conyers. *A Free Inquiry into the Miraculous Powers which are supposed to have subsisted in the Christian Church, From the Earliest Ages through several successive centuries. By which it is shewn, That we have no sufficient Reason to believe, upon the Authority of the Primitive Fathers, That any such Powers were continued to the Church, after the Days of the Apostles* (London, 1749).

Millar, John. *Observations concerning the Distinction of Ranks in Society* (2nd enlarged edn. London, 1773).

Miller, Geneviève. *The Adoption of Inoculation for Smallpox in England and France* (Philadelphia, 1957).

Momigliano, Arnaldo. 'Pagan and Christian Historiography in the Fourth Century A.D.', in idem., ed., *The Conflict between Paganism and Christianity in the Fourth Century* (Oxford, 1963), 79–99.

Momigliano, Arnaldo. *Studies in Historiography* (London, 1966).
Monboddo, Lord (James Burnett). *Of the Origin and Progress of Language* (2nd edn, 6 vols, Edinburgh, 1774–1809).
Monboddo, Lord (James Burnett). *Antient Metaphysics* (6 vols, Edinburgh, 1779–99).
Monro, Alexander. *The Anatomy of Human Bones* (Edinburgh, 1726).
Monro, Alexander. *Traité de ostéologie* (Paris, 1759).
Montesquieu, Charles de Secondat, Baron de. *Lettres persanes* ('Cologne' [i.e. Amsterdam], 1721).
Montesquieu, Charles de Secondat, Baron de. *De l'esprit des lois: ou Du rapport que les loix doivent avoir avec la constitution de chaque gouvernement, les moeurs, le climat, la religion, le commerce, &c.* (Geneva, 1748).
Montesquieu, Charles de Secondat, Baron de. *Œuvres de Montesquieu* (7 vols, Lyon, 1805).
Montesquieu, Charles de Secondat, Baron de. *Œuvres complètes*, ed. R. Caillois (2 vols, Paris, 1949–51).
Moravia, Sergio. 'Philosophie et médecine en France à la fin du XVIIIe siècle', *Studies on Voltaire and the Eighteenth Century* 89 (1972), 1089–151.
Moravia, Sergio. *Il Pensiero degli idéologues: scienza e filosofia in Francia (1780–1815)* (Florence, 1974).
Morgagni, Giovanni B. *De sedibus et causis morborum per anatomen indagatis libri V, dissectiones et animadversiones nunc primum editas, complectuntur* (2 vols, Venice, 1761).
Morgan, George Cadogan. *Lectures on Electricity* (2 vols, Norwich, 1794).
Moritz, Karl Philipp. *Götterlehre oder Mythologische Dictungen der Alten* (Berlin, 1790).
Morris, Robert J. 'Lavoisier and the Caloric Theory', *British Journal for the History of Science* 6 (1972), 1–38.
Mortier, Roland. *Diderot en Allemagne, 1760–1850* (Paris, 1954).
Morus, Iwan Rhys. *When Physics Became King* (Chicago, 2005).
Mosheim, Johann Lorenz von. *De Rebus Christianorum ante Constantinum Magnum Commentarii* (Helmstedt, 1753).
Mosheim, Johann Lorenz von. *Institutionum historiæ ecclesiasticæ antiqui et recentioris quattuor* (Helmstedt, 1755).
Mücke, Dorothea E. von. 'Language as the Mark of the Soul: Herder's Narcissistic Subject', in Kurt Mueller-Vollmer, ed., *Herder Today* (Berlin, 1990), 331–44.
Muldrew, Craig. *Food, Energy, and the Creation of Industriousness: Work and Material Culture in Agrarian England, 1550–1780* (Cambridge, 2011).
Muller, Richard A. 'The Debate over Vowel Points and the Crisis in Orthodox Hermeneutics', *Journal of Medieval and Renaissance Studies* 10 (1980), 53–72.
Mulligan, Lotte. '"Reason", "Right Reason", and "Revelation" in Mid-Seventeenth-Century England', in Brian Vickers, ed., *Occult and Scientific Mentalities in the Renaissance* (Cambridge, 1984), 375–402.
Mulligan, Lotte. 'Robert Boyle, Right Reason, and the Meaning of Metaphor', *Journal of the History of Ideas* 55 (1994), 235–57.
Muthu, Sankar. *Enlightenment against Empire* (Princeton, 2003).
Nassar, Dalia. 'Sensibility and Organic Unity: Kant, Goethe, and the Plasticity of Cognition', *Intellectual History Review* 25 (2015), 311–26.

Nicholson, William. 'Account of the New Electrical or Galvanic Apparatus of Sig. Alessandro Volta, and Experiments Performed with the Same', *Journal of Chemistry, Natural Philosophy, and the Arts* 4 (1800), 179–87.
Niebyl, Peter H. 'The Non-Naturals', *Bulletin of the History of Medicine* 45 (1971), 486–92.
Nietzsche, Friedrich. *Die Geburt der Tragödie aus dem Geiste der Musik* (Leipzig, 1872).
Nisbet, H. Barry. *Gotthold Ephraim Lessing: His Life, Works, and Thought* (Oxford, 2013).
Nitsch, Friedrich A. *A General and Introductory View of Kant's Principles concerning Man, the World, and the Deity* (London, 1796).
Norton, Robert E. *Herder's Aesthetics and the European Enlightenment* (Ithaca, NY, 1991).
Novalis (Friedrich von Hardenberg). *Novalis Schriften*, ed. R. Samuel, H.-J. Mähl, and G. Schultz (6 vols, Stuttgart, 1960–2006).
O'Neal, John C. *The Authority of Experience: Sensationalist Theory in the French Enlightenment* (University Park, Penn., 1996).
Olson, Michael J. 'The Camera Obscura and the Nature of the Soul: On the Tension between the Mechanics of Sensation and the Metaphysics of the Soul', *Intellectual History Review* 25 (2015), 279–92.
Ormoy, Abée d'. 'De l'influence de l'électricité sur la végétation, prouvée par de nouvelles expériences', *Observations sur la Physique, sur l'histoire naturelle et sur les arts* 35 (1789), 161–76.
Outram, Dorinda. *Georges Cuvier: Vocation, Science and Authority in Post-Revolutionary France* (Manchester, 1984).
Palaia, Roberto. 'Berlino 1747: Il dibattito occasione del concorso dell'Academia della scienze', *Nouvelles de la République des Lettres* (1993), 91–119.
Palmer, Robert R. *The Improvement of Humanity: Education and the French Revolution* (Princeton, 1985).
Pancaldi, Giuliano. *Volta: Science and Culture in the Age of Enlightenment* (Princeton, 2003).
Parekh, Bhikho, ed. *Bentham's Political Thought* (New York, 1973).
Parent, André. 'Giovannio Aldini: From Animal Electricity to Human Brain Stimulation', *Canadian Journal of Neurological Sciences* 31 (2004), 576–84.
Parker, Geoffrey. *Global Crisis: War, Climate Change and Catastrophe in the Seventeenth Century* (New Haven, Conn., 2013).
Parslow, Christopher Charles. *Rediscovering Antiquity: Karl Weber and the Excavation of Herculaneum, Pompeii and Stabiae* (Cambridge, 1995).
Patrin, Eugène-Louis-Melchior. *Histoire naturelle des minéraux, contenant leur description, celle de leur gîte, la théorie de leur formation, leurs rapports avec la géologie ou l'histoire de la terre, le détail de leurs propriétés et de leurs usages, leur analyse chimique* (5 vols, Paris, 1801).
Pauw, Cornelius de. *Recherches philosophiques sur les Américains ou Mémoires intéressants pour servir à l'histoire de l'espèce humaine* (2 vols, Berlin, 1768–9).
Pera, Marcello. *The Ambiguous Frog: The Galvani–Volta Controversy on Animal Electricity* (Princeton, 1992).
Pestalozzi, Karl. 'Physiognomische Methodik', in A. Finck and G. Gréciano, eds, *Germanistik aus interkultureller Perspektive* (Strasbourg, 1988), 137–53.
Petersen, Peter. *Geschichte der aristotelianischen Philosophie im protestantischen Deutschland* (Leipzig, 1921).

Petty, William. *A Treatise of Taxes and Contributions* (London, 1662).
Petty, William. *A Discourse Made Before the Royal Society... Concerning the Use of Duplicate Proportion* (London, 1674).
Petty, William. *An Essay Concerning the Multiplication of Mankind: together with another Essay in Political Arithmetick, concerning the growth of the city of London* (London, 1683).
Petty, William. *Political Arithmetick* (London, 1690).
Pfaff, Christoph Heinrich. *Über thierische Elektricität und Reizbarkeit. Ein Beytrag zu den neuesten Entdeckungen über diese Gegenstände* (Leipzig, 1795).
Phillipson, Nicholas. 'Adam Smith as Civic Moralist', in I. Hont and M. Ignatieff, eds, *Wealth and Virtue: The Shaping of Political Economy in the Scottish Enlightenment* (Cambridge, 1983), 179–202.
Pick, Daniel. *War Machine: The Rationalization of Slaughter in the Modern Age* (New Haven, Conn., 1993).
Pierdziwol, Annette. 'Extending Nature: Rousseau on the Cultivation of Moral Sensibility' in M. Lenz and A. Waldow, eds, *Contemporary Perspectives on Early Modern Philosophy: Nature and Norms in Thought* (Dordrecht, 2013), 135–58.
Pilloud, Séverine. 'Mettre les maux en mots, médiations dans la consultation épistolaire au XVIIIe siècle: Les malades du Dr Tissot (1728–1797)', *Bulletin Canadien d'Histoire de la Médecine* 16 (1999), 215–45.
Pinel, Philippe. 'Mémoire sur la manie périodique ou intermittente', *Mémoires de la Société médicale d'émulation* 1 (1796–7), 94–119.
Pinel, Philippe. 'Recherches et observations sur le traitement moral des aliénés', *Mémoires de la Société médicale d'émulation* 2 (1797–8), 215–55.
Pinel, Philippe. 'Observations sur les aliénés et leur division en espèces distinctes', *Mémoires de la Société médicale d'émulation* 3 (1798–9), 1–26.
Pinkard, Terry. *German Philosophy 1760–1860: The Legacy of Idealism* (Cambridge, 2002).
Platner, Ernst. *Anthropologie für Ärzte und Weltweise, erster Theil* (Leipzig, 1772).
Pocock, John G. A. *Virtue, Commerce, and History* (Cambridge, 1985).
Pocock, John G. A. *Barbarism and Religion* (6 vols, Cambridge, 1999–2015).
Pomeranz, Kenneth. *The Great Divergence: China, Europe, and the Making of the Modern World Economy* (Princeton, 2000).
Pomme, Pierre. *Traité des affections vaporeuses des deux sexes* (Lyon, 1760).
Popelinière, Henri Lancelot du Voisin de la. *L'Histoire des Histoires, avec l'idée de l'Histoire accomplie* (Paris, 1599).
Porta, Giambattista Della. *De humana physiognomonia Libri III* (Naples, 1586).
Porter, Theodore M. *The Rise of Statistical Thinking 1820–1900* (Princeton, 1986).
Pressavin, Jean-Baptiste. *Nouveau traité des vapeurs* (Lyon, 1770).
Price, Richard. *Observations on reversionary payments on schemes for providing annuities for widows, and for persons in old age; on the method of calculating the values of assurances on lives; and on the national debt* (London, 1771).
Priestley, Joseph. *Experiments and Observations on Different Kinds of Air* (3 vols, London 1774–7).
Priestley, Joseph. *Experiments and Observations Relating to the Analysis of Atmospherical Air* (London, 1796).
Prothero, Stephen. *God is Not One* (New York, 2010).

Pufendorf, Samuel. *De Jure Naturae et Gentium Libri Octo* (London, 1672).
Puységur, Armand Marie-Jacques de Chastenet, Marquis de. *Mémoires pour servir à l'histoire et à l'establissement du magnétisme animal* (London, 1786).
Quesnay, François. *Le Tableau économique* (Paris, 1758).
Quetelet, Lambert Adolphe Jacques. *Sur l'homme et le développment de ses facultés, ou essai de physique sociale* (2 vols, Paris, 1835).
Quetelet, Lambert Adolphe Jacques. *A Treatise on Man and the Development of His Faculties* (Edinburgh, 1842).
Quetelet, Lambert Adolphe Jacques. *Recherches statistiques* (Brussels, 1844).
Quetelet, Lambert Adolphe Jacques. *Anthropométrie ou mesure des différentes facultés de l'homme* (Brussels, 1871).
Quinet, Edgar. *Œuvres complètes* (10 vols, Paris, 1857).
Rackstrow, Benjamin. *Miscellaneous Observations, together with a Collection of Experiments on Electricity* (London, 1748).
Rashed, Roshdi. *Condorcet: Mathématique et société* (Paris, 1974).
Raspail, François-Vincent. *Nouveau système de physiologie végétale et de botanique fondé sur les méthodes d'observation, qui ont été dévelopées dans le nouveau système de chimie organique* (Paris, 1833).
Ratcliffe, Marc J. *The Quest for the Invisible: Microscopy in the Enlightenment* (Farnham, 2009).
Rather, Lelland J. *Mind and Body in Eighteenth Century Medicine: A Study Based on Jerome Gaub's De regimine mentis* (London, 1965).
Raunlin, Joseph. *Traité des affections vaporeuses du sexe* (Paris, 1758).
Réaumur, René Antoine Ferchault de. 'Sur la reproduction de quelques parties des écrevisses', *Histoire de l'Académie des sciences* (1712), 35–7.
Réaumur, René Antoine Ferchault de. 'Observations sur la formation du corail, & des autres productions appellées plantes pierreuses', *Mémoires de l'Académie royale des sciences* (1728), 269–81.
Redlhamer, Josef. *Philosophiae Naturalis. Pars prima, seu Physica generalis ad praefixam in scholis nostris normam concinnata* (Vienna, 1755).
Reill, Johann Christian. 'Von der Lebenskraft', *Archiv für die Physiologie* 1 (1795), 8–162.
Reill, Peter Hanns. *The German Enlightenment and the Rise of Historicism* (Berkeley, 1975).
Reill, Peter Hanns. *Vitalizing Nature in the Enlightenment* (Berkeley, 2005).
Reinhold, Karl Leonhard. *Briefe über die kantische philosophie* (2 vols, Leipzig 1790–2).
Reinhold, Karl Leonhard. *Beyträge zur Berichtung bisheriger Missverständnisse der Philosophen* (2 vols, Jena, 1790–4).
Reinhold, Karl Leonhard. *Ueber das Fundament des philosophischen Wissens* (Jena, 1791).
Reinhold, Karl Leonhard. *Letters on Kantian Philosophy*, ed. K. Ameriks (Cambridge, 2005).
Renan, Ernest. *Vie de Jésus* (Paris, 1863).
Rey, Roselyne. *Naissance et développement du vitalisme en France de la deuxième moitié du 18ᵉ siècle à la fin du Premier Empire* (Oxford, 2000).
Richards, Robert J. 'Christian Wolff's Prolegomena to Empirical and Rational Psychology: Translation and Commentary', *Proceedings of the American Philosophical Society*, 124/3 (1980), 227–39.

Richards, Robert J. *The Romantic Conception of Life: Science and Philosophy in the Age of Goethe* (Chicago, 2002).
Riskin, Jessica. *Science in the Age of Sensibility: The Sentimental Empiricists of the French Enlightenment* (Chicago, 2002).
Ritchie, Jack. *Understanding Naturalism* (Stocksfield, 2008).
Ritter, Johann Wilhelm. *Beweis, dass ein beständiger Galvanismus den Lebensprozess in Therreich begleite* (Weimar, 1798).
Ritterbush, Philip C. *Overtures to Biology: The Speculations of Eighteenth-Century Naturalists* (New Haven, 1964).
Robertson, John. *The Case for Enlightenment: Scotland and Naples 1680–1760* (Cambridge, 2005).
Robertson, William. *The History of Scotland, During the reigns of Queen Mary and of King James VI, till his accession to the crown of England* (2nd edn, 2 vols, London, 1759).
Robertson, William. *The History of America, volume the first* (Dublin, 1777).
Robinson, Bryan. *A Dissertation on the Aether of Sir Isaac Newton* (Dublin, 1746).
Roe, Shirley A. *Matter, Life, and Generation: Eighteenth-Century Embryology and the Haller–Wolff Debate* (Cambridge, 1981).
Roger, Jacques. *Buffon: A Life in Natural History* (Ithaca, NY, 1997).
Ronalds, Francis. *Catalogue of Books and Papers Relating to Electricity, Magnetism, the Electric Telegraph, Etc.* (Cambridge, 1880).
Rosen, Michael. *Hegel's Dialectic and its Criticism* (Cambridge, 1982).
Rosencranz, Karl. *Encyklopädie der theologischen Wissenschaften* (Halle, 1831).
Rouelle, Guillaume François. *Cours d'experience chymique* (Paris, 1759).
Rousseau, Jean-Jacques. *Discours qui a remporté le prix a l'Academie de Dijon en l'année 1750. Sur cette Question proposée par la même Académie: Si le rétablissement des sciences et des arts a contributé à épurer les mœurs* (Geneva, 1750).
Rousseau, Jean-Jacques. *Discours sur l'origine et les fondements de l'inégalité parmi les hommes* (Amsterdam, 1755).
Rousseau, Jean-Jacques. *Du Contrat Social, ou Principes du Droit Politique* ('Amsterdam' [i.e. Germany], 1762).
Rousseau, Jean-Jacques. *Émile, ou de l'éducation* (The Hague, 1762).
Rousseau, Jean-Jacques. *Essai sur l'origine des langues* (Geneva, 1781).
Rousseau, Jean-Jacques. *Œuvres Complètes* (4 vols, Paris, 1883).
Roussel, Pierre. *Système physique et moral de la Femme, ou tableau philosophique de la Constitution, de l'Etat organique, du Tempérament, des Mœurs, et des Fonctions propres au Sexe* (Paris, 1775).
Rubiés, Joan-Pau. 'Theology, Ethnography, and the Historicization of Idolatry,' *Journal of the History of Ideas* 67 (2006), 571–96.
Rudwick, Martin J. S. *Bursting the Limits of Time: The Reconstruction of Geohistory in the Age of Revolution* (Chicago, 2005).
Rudwick, Martin J. S. *Worlds before Adam: The Reconstruction of Geohistory in the Age of Reform* (Chicago, 2008).
Rusnock, Andrea A. *Vital Accounts: Quantifying Health and Population in Eighteenth-Century England and France* (Cambridge, 2002).
Saumarez, Richard. *A New System of Physiology, comprehending the laws by which animated beings in general, and the human species in particular, are governed, in their several states of health and of disease* (2nd edn, 2 vols, London, 1799).

Saumarez, Richard. 'On Generation and the Principle of Life', *Medical and Physical Journal* 2 (1799), 242–7, 321–6.

Saussure, Horace-Benédict de. 'Agenda, ou tableau général des observations et des recherches dont les résultats doivent servir de base à la théorie de la terre', *Journal des mines* 4/20 (1795), 1–70.

Sauvages de la Croix, François Boissier. *Nouvelles classes de maladies, qui dans un ordre semblable à celui des botanistes, comprennent les genres et les especes de toutes les maladies, avec leurs signes et leurs indications* (Avignon, 1737).

Sauvages de la Croix, François Boissier. *Nosologia methodica sistens morborum classes, genera et Species juxta Sydenhami mentem et botanicorum ordinem* (2 vols, Amsterdam, 1763).

Scheele, Carl Wilhelm. *Chemische Abhandlung von der Luft und dem Feuer* (Uppsala and Leipzig, 1777).

Scheele, Carl Wilhelm. *Chemical Observations and Experiments on Air and Fire... with a Prefatory Introduction by Torbern Bergman, translated from the German by J. R. Forster* (London, 1780).

Schelling, Friedrich Wilhelm Joseph. *Ideen zu einer Philosophie der Natur als Einleitung in das Studium dieser Wissenschaft* (Jena & Leipzig, 1797).

Schelling, Friedrich Wilhelm Joseph. *Einleitung zu dem Entwurf eines Systems der Naturphilosophie. Oder über den Begriff der Speculativen Physik und die innere Organisation eines Systems dieser Wissenschaft* (Jena & Leipzig, 1799).

Schelling, Friedrich Wilhelm Joseph. *Vorlesungen über die Methode des Akademischen Studiums* (Leipzig, 1803).

Schelling, Friedrich Wilhelm Joseph. 'Introduction to the Outlines of a System of Natural Philosophy; or, on the Idea of Speculative Physics and the Internal Organization of a System of this Science', trans. Tom Davidson, *The Journal of Speculative Philosophy* 1 (1867), 193–220.

Schiebinger, Londa. *The Mind Has No Sex? Women in the Origins of Modern Science* (Cambridge, Mass., 1989).

Schiebinger, Londa. *Nature's Body: Gender in the Making of Modern Science* (Boston, Mass., 1993).

Schiller, Friedrich. *On the Aesthetic Education of Man, in a Series of Letters*, ed. and trans. Elizabeth Wilkinson and L. A. Willoughby (Oxford, 1967).

Schleiermacher, Friedrich. *Über die Religion: reden an die Gebildeten unter ihren Verächtern* (Berlin, 1799).

Schmauß, Johann Jacob. *Vorstellung des wahren Begriffs von einem Recht der Nature* (Göttingen, 1748).

Schneiders, Werner. 'Zwischen Welt und Weisheit: Zur Verweltlichung der Philosophie in der frühen Moderne', *Studia Leibnitiana* 15 (1983), 2–18.

Schneiders, Werner. 'Der Philosophiebegriff des philosophischen Zeitalters: Wandlungen im Selbstverständnis der Philosophie von Leibniz bis Kant', in Rudolf Vierhaus, ed., *Wissenschaft im Zeitalter der Aufklärung* (Göttingen, 1985), 58–92.

Schneiders, Werner. 'Vernünftiger Zweifel und wahre Eklektik: Zur Entstehung des modern Kritikbegriffes', *Studia Leibnitiana* 17 (1985), 143–61.

Schönfeld, Martin. *The Philosophy of the Young Kant: The Precritical Project* (Oxford, 2000).

Schofield, Robert E. *Mechanism and Materialism: British Natural Philosophy in an Age of Reason* (Princeton, 1970).

Schofield, Robert E. *The Enlightened Joseph Priestley: A Study of His Life and Work from 1773 to 1804* (University Park, Penn., 2004).

Schulte-Sasse, Jochen. *Der Kritik an der Trivialliteratur seit der Aufklärung: Studien zur Geschichte des modernen Kitschbegriffs* (Münich, 1971).

Schwann, Theodor. *Mikroskopische Untersuchungen über die Uebereinstimmung in der Struktur und den Wachstum der Thiere und Pflanzen* (Berlin, 1839).

Schwann, Theodor. *Microscopical Researches into the Accordance in the Structure and Growth of Animals and Plants* (London, 1847).

Selwyn, Pamela Eve. *Everyday Life in the German Book Trade: Friedrich Nicolai as Bookseller and Publisher in the Age of Enlightenment, 1750–1810* (University Park, Penn., 2000).

Sen, Amartya. 'Adam Smith's Prudence', in S. Lall and F. Stewart, eds, *Theory and Reality in Development* (London, 1986), 28–37.

Sepper, Dennis. *Understanding Imagination: The Reason of Images* (Dordrecht, 2013).

Seznec, Jean, *La survivance des dieux antiques. Essai sur le rôle de la tradition mythologique dans l'humanisme et dans l'art de la Renaissance* (London, 1940).

Shaffer, Elinor S. 'Romantic Philosophy and the Organization of the Disciplines: The Founding of the Humboldt University of Berlin', in A. Cunningham and N. Jardine, eds, *Romanticism and the Sciences* (Cambridge, 1990), 38–54.

Shaftesbury, Anthony Ashley Cooper, Earl of. *Characteristicks of Men, Manners, Opinions, Times* (3 vols, London, 1737).

Sheynin, Oscar B. 'On the Mathematical Treatment of Astronomical Observations', *Archive for History of Exact Sciences* 11 (1973), 17–126.

Sheynin, Oscar B. 'Laplace's Theory of Error', *Archive for History of Exact Sciences* 17 (1977), 1–61.

Shuckford, Samuel. *The Sacred and Profane History of the World Connected* (2 vols, London, 1728).

Sidgwick, Henry. *The Methods of Ethics* (London, 1874).

Siegfried, Robert. 'Lavoisier's View of the Gaseous State and Its Early Application to Pneumatic Chemistry', *Isis* 63 (1972), 59–78.

Siegrist, Christoph. '"Letters of the Divine Alphabet": Lavater's Concept of Physiognomy', in Ellis Schookman, ed., *The Faces of Physiognomy: Interdisciplinary Approaches to Johann Caspar Lavater* (Columbia, South Carolina, 1993), 25–39.

Sikka, Sonia. 'Herder's Critique of Pure Reason', *Review of Metaphysics* 61 (2007), 31–50.

Singy, Patrick. 'The Popularization of Medicine in the Eighteenth Century: Writing, Reading, and Rewriting Samuel Auguste Tissot's *Avis au peuple sur sa santé*, *Journal of Modern History* 82 (2010), 769–80.

Sissa, Giulia and Marcel Detienne. *La Vie quotidienne des dieux grecs* (Paris, 1989).

Skelton, Philip. *Ophiomaches; or, Deism Revealed* (2 vols, London, 1749).

Skidelsky, Edward. *Ernst Cassirer: The Last Philosopher of Culture* (Princeton, 2008).

Smellie, William. *The Philosophy of Natural History* (Edinburgh, 1790).

Smith, Adam. *The Theory of Moral Sentiments* (London, 1759).

Smith, Adam. *The Glasgow Edition of the Works and Correspondence of Adam Smith* (6 vols, Oxford, 1976–83).

Smith, Justin E. H. 'Language, Bipedalism, and the Mind–Body Problem in Edward Tyson's Orang-Outang (1699),' *Intellectual History Review* 17 (2007), 291–304.

Smith, Justin E. H. *Divine Machines: Leibniz and the Sciences of Life* (Princeton, 2011).
Smith, Wilfred Cantwell. *The Meaning and End of Religion* (London, 1978).
Snelders, Henricus Adrianus Marie. 'The Influence of the Dualistic System of Jakob Joseph Winterl (1732–1809) on the German Romantic Era', *Isis* 61 (1970), 231–40.
Snelders, Henricus Adrianus Marie. 'Romanticism and Dutch Scientists', in S. Poggi and M. Bossi, eds, *Romanticism in Science* (Dordrecht, 1994), 175–88.
Soemmerring, Samuel Thomas. *Ueber die körperliche Verschiedenheit des Negers vom Europäer* (Frankfurt am Main, 1785).
Soemmerring, Samuel Thomas. *Vom Baue des menschlichen Körpers* (5 vols, Frankfurt am Main, 1791–6).
Soemmerring, Samuel Thomas. *Tabula sceleti feminini juncta descriptione* (Frankfurt am Main and Mainz, 1796).
Sombart, Werner. 'Beiträge zur Geschichte der wissenschaftliche Anthropologie,' *Sitzunesberichte der Preussische Akademie der Wissenschaften philosophisch-historische Klasse* 13 (1938), 96–130.
Spallanzani, Lazzaro. *Prodromo di un opera da imprimersi sopra le riproduzioni animali* (Modena, 1768).
Stafford, Barbara Maria. *Artful Science: Enlightenment Entertainment and the Eclipse of Visual Education* (Cambridge, Mass., 1994).
Stahl, Georg Ernst. *Disquisitio de mechanismi et organismi diversitate* (Halle, 1706).
Staum, Martin S. *Cabanis: Enlightenment and Medical Philosophy in the French Revolution* (Princeton, 1980).
Stavenhagen, Kurt. *Kant und Königsberg* (Göttingen, 1949).
Steffens, Henrich. *Beyträge zur inneren Naturgeschichte der Erde* (Freiberg, 1801).
Steffens, Henrich. *Grundzüge der philosophischen Naturwissenschaft* (Berlin, 1806).
Steigerwald, Joan. 'Rethinking Organic Vitality in Germany at the Turn of the Nineteenth Century', in S. Normandin and C. Wolfe, eds, *Vitalism and the Scientific Image in Post-Enlightenment Life Science, 1800–2010* (Dordrecht, 2013), 51–75.
Stengers, Jean. 'Buffon et la Sorbonne', *Études sur le XVIIIe Siècle* 1 (1974), 97–127.
Stephen, Leslie. *History of English Thought in the Eighteenth Century* (2 vols, London, 1881).
Stewart, Dugald. *The Collected Works of Dugald Stewart*, ed. William Hamilton (11 vols, Edinburgh, 1854–60).
Stigler, George J. 'Smith's Travels on the Ship of State', *History of Political Economy* 3 (1971), 265–77.
Stigler, Stephen M. *The History of Statistics: The Measurement of Uncertainty before 1900* (Cambridge, Mass., 1986).
Stocking, George W. *Race, Culture and Evolution: Essays in the History of Anthropology* (New York, 1968).
Storost, Jürgen. *Langue française—Langue universelle? Die Diskussion über die Universalität des Französichen an der Berliner Akademie der Wissenschaften* (Bonn, 1994).
Storr, Gottlob Christian. *Annotationes quædam theologicæ ad philosophicam Kantii de religione doctrinam* (Tübingen, 1793).
Strangeland, Charles E. *Pre-Malthusian Doctrines of Population: A Study in the History of Economic Though*t (New York, 1904).
Strauss, David Friedrich. *Das Leben Jesu, kritische bearbeitet* (2 vols, Tübingen, 1835–6).

Strauss, David Friedrich. *Die christliche Glaubenslehre in ihrer geschichtlichen Entwicklung und im Kampfe mit der modernen Wissenschaft* (2 vols, Tubingen, 1840–1).
Strauss, David Friedrich. *The Life of Jesus: Critically Examined*, trans. Marian Evans (3 vols, London, 1846).
Strauss, David Friedrich. *Hermann Samuel Reimarus und seine Schutzschrift für die vernünftigen Verehrer Gottes* (Leipzig, 1862).
Stroumsa, Guy. *Barbarian Philosophy: The Religious Revolution of Early Christianity* (Tübingen, 1999).
Struve, Wilhelm Friedrich. *Anthropologia naturalis sublimior* (Jena, 1754).
Sturm, Thomas. *Kant und die Wissenschaften vom Menschen* (Paderborn, 2009).
Süßmilch, Johann Peter. *Die Göttliche Ordnung in den Veränderungen des menschlichen Geschlechts, aus der Geburt, dem Tode und der Fortpflanzung desselben erwiesen* (Berlin, 1741).
Süßmilch, Johann Peter. *Versuch eines Beweises, daß die erste Sprache ihren Ursprung nicht vom Menschen, sondern allein vom Schöpfer erhalten habe* (Berlin, 1766).
Suisky, Dieter. 'Leonhard Euler and Emilie du Châtelet: On the Post-Newtonian Development of Physics', in Ruth Hagengruber, ed., *Emilie du Châtelet Between Leibniz and Newton* (Dordrecht, 2012), 113–56.
Sutton, Geoffrey. 'Electric Medicine and Mesmerism', *Isis* 72 (1981), 375–92.
Swoboda, Gudrun. 'Die Sammlung Johann Caspar Lavater in Wien', in G. Mraz and U. Schölg, eds, *Das Kunstkabinett des Johannes Caspar Lavater* (Vienna, 1999), 75–95.
Taminaux, Jacques. *La nostaglie de la Grèce à l'aube de l'idéalisme allemand* (The Hague, 1967).
Tarin, Pierre. *Ostéographie, ou Description des os de l'adult* (Paris, 1753).
Taylor, Charles. *Hegel* (Cambridge, 1975).
Tenon, Jacques. *Mémoires sur les hôpitaux de Paris* (Paris, 1788).
Thiel, Udo. *The Early Modern Subject: Self-Consciousness and Personal Identity from Descartes to Hume* (Oxford, 2011).
Thomas, Keith. *The Ends of Life: Roads to Fulfilment in Early Modern England* (Oxford, 2009).
Tillemont, Sebastien Le Nain de. *Histoire des empereurs et des autres princes, qui ont regné durant les six premiers siècles de l'eglise* (3 vols, Brussels, 1743).
Timm, Herman. *Gott und die Freiheit: Studien zur Religionsphilosophie*, i. *Die Spinozarenaissance* (Frankfurt, 1974).
Timoni, Emanuele. 'An Account, or History, of the Procuring of the Small Pox by Incision, or Inoculation; As It Has For Some Time Been Practised at Constantinople. Being the Extract of a Letter from Emanuel Timonius, Oxon. & Patav. M.D. F.R.S. Dated at Constantinople, December 1713. Communicated to the Royal Society by John Woodward, M.D. Profes. Med. Gresh. and F.R.S.', *Philosophical Transactions* 29 (1714), 72–82.
Tinland, Franck. *L'Homme sauvage: Homo ferus et Homo sylvestris, de l'animal à l'homme sauvage* (Paris, 2003).
Tissot, Samuel August. *Avis au peuple sur sa santé* (Lausanne, 1767).
Tissot, Samuel August. *De la santé des gens de lettres* (Lausanne, 1768).
Tissot, Samuel August. *Essai sur les maladies des gens du monde* (Lausanne, 1770).
Toews, John E. *Hegelianism: The Path Towards Dialectical Humanism, 1805–1841* (Cambridge, 1980).

Tomaselli, Sylvana. 'Reflections on the History of Science of Woman', *History of Science* 29 (1991), 185–205.

Tonelli, Giorgio. 'Der Streit über die mathematische Methode in der Philosophie in der ersten Hälfte des 18. Jahrhunderts und die Entstehung von Kants Schrift Über die "Deutlichkeit"', *Archiv für Philosophie* 9 (1959), 37–66.

Topham, Jonathan. 'Science and Popular Education in the 1830s: The Role of the *Bridgewater Treatises*', *British Journal for the History of Science* 25 (1992), 397–430.

Trembley, Abraham. *Mémoires pour servir à l'histoire d'un genre de polypes d'eau douce* (Leiden, 1744).

Trevelyan, Humphry. *Goethe and the Greeks* (Cambridge, 1941).

Trompf, Garry W. *The Idea of Historical Recurrence in Western Thought: From Antiquity to the Reformation* (Berkeley, 1979).

Trompf, Garry W. 'On Newtonian History', in S. Gaukroger, ed., *The Uses of Antiquity: The Scientific Revolution and the Classical Tradition* (Dordrecht, 1991), 231–50.

Trotter, Thomas. *A View of the Nervous Temperament: being a Practical Treatise on Nervous, Bilious, Stomach, and Liver Complaints* (London, 1807).

Truesdell, Clifford. 'Rational Fluid Mechanics, 1687–1765', *Leonhardi Euleri opera omnia* series 2, vol. xii, section 1 (Zurich, 1954).

Truesdell, Clifford. 'Rational Fluid Mechanics, 1765–1788', *Leonhardi Euleri opera omnia* series 2, vol. xiii (Lausanne, 1955).

Truesdell, Clifford. 'A Program Towards Rediscovering the Rational Mechanics of the Age of Reason', *Archive for History of Exact Sciences* 1 (1960), 3–36.

Truesdell, Clifford. 'The Rational Mechanics of Flexible or Elastic Bodies, 1638–1788', *Leonhardi Euleri opera omnia* series 2, vol. xi, section 2 (Zurich, 1960).

Tuck, Richard. *Philosophy and Government, 1572–1651* (Cambridge, 1993).

Tully, James. *A Discourse on Property* (Cambridge, 1980).

Tulp, Nicolaes. *Observationum medicarum* (Amsterdam, 1641).

Tulp, Nicolaes. *Observationum medicarum. Editio Nova* (Amsterdam, 1672).

Turgot, Anne Robert Jacques, Baron de l'Aulne. *Œuvres de Turgot*, ed. Eugène Daire (2 vols, Paris, 1844).

Turnbull, George. *The Principles of Moral Philosophy: An Enquiry into the Wise and Good Government of the Moral World* (London, 1740).

Turner, James. *Philology: The Forgotten Origins of the Modern Humanities* (Princeton, 2014).

Tyson, Edward. *Orang-Outang, sive Homo Sylvestris: or, The Anatomy of a Pymgie Compared with that of a Monkey, an Ape, and a Man* (London, 1699).

Ueding, Gert. 'Populärphilosophie', in R. Grimminger, ed., *Deutsche Aufklärung bis zur Französischen Revolution 1680–1789* (Munich, 1984), 605–34.

Valenza, Pierluigi. 'Reinhold, Lavater et les Lumières', in George di Giovanni, ed., *Karl Leonhard Reinhold and the Enlightenment* (Dordrecht, 2010), 53–64.

Vandermonde, Charles Augustin. *Essai sur la manière de perfectionner l'espèce humaine* (Paris, 1756).

Vatke, Wilhelm. *Die Religion des Alten Testaments nach den kanonischen Büchern entwickelt* (2 vols, Berlin, 1835).

Vernant, Jean-Pierre. *Mythe et Société en Grèce ancienne* (Paris, 1971).

Veyne, Paul. *Did the Greek Believe in their Myths?: An Essay on the Constitutive Imagination* (Chicago, 1988).

Vico, Giambattista. *De Nostri temporis studiorum ratione* (Naples, 1709).

Vico, Giambattista. *Principj di una Scienza Nuova intorno alla natura delle nazioni, per la quale si ritruovano i principj di altro sistema del Diritto Naturale delle genti* (2 vols, Naples, 1725).

Vico, Giambattista. *The New Science of Giambattista Vico*, trans. T. G. Bergin and M. H. Fisch (Ithaca, NY, 1968).

Vico, Giambattista. *Le Orazioni Inaugurali I–IV*, ed. G. G. Visconti (Bologna, 1982).

Vicq d'Azyr, Félix. *Traité d'anatomie et de physiologie, avec des planches coloriées* (Paris, 1786).

Vicq d'Azyr, Félix. 'Nouveau Plan de constitution pour la médecine en France', *Histoire et mémoires de la Société royale de médecine, 1787–1788*, ix (Paris, 1790), 1–201.

Vidal, Fernando. 'Psychology in the 18th Century: A View from Encyclopedias', *History of the Human Sciences* 6 (1993), 89–119.

Vierhaus, Rudolph. 'Bildung', in O. Brunner, W. Conze, and R. Koselleck, eds, *Geschichtliche Grundbegriffe* (5 vols, Stuttgart, 1972–89), i. 508–51.

Vila, Anne C. *Enlightenment and Pathology: Sensibility in the Literature and Medicine of Eighteenth-Century France* (Baltimore, 1998).

Virey, Julien-Joseph. *Histoire naturelle du genre humain, ou Recherches sur les principaux fondemens physiques et moraux: précédées d'un discours sur la nature des êtres organiques, et sur l'ensemble de leur physiologie. On y a joint une dissertation sur le sauvage d'Aveyron* (Paris, 1800).

Viroli, Maurizio. *From Politics to Reason of State: The Acquisition and Transformation of the Language of Politics, 1250–1600* (Cambridge, 1992).

Volney, Constantin-François Chassebœuf de Boisgiray de. *La Loi naturelle, ou Catéchisme du citoyen français, ou Principes physiques et de la morale déduits de l'organisation de l'homme et de l'univers* (Paris, 1793).

Voltaire, François-Marie Arouet de. *Œuvres complètes de Voltaire* (72 vols, Gotha, 1784–90).

Waldow, Anik. 'Back to the Facts: Herder on the Normative Role of Sensibility and Imagination', in Martin Lenz and Anik Waldow, eds, *Contemporary Perspectives on Early Modern Philosophy* (Dordrecht, 2013), 115–34.

Wallace, Robert. *Dissertation on the Numbers of Mankind in Ancient and Modern Times, in which the superior populousness of Antiquity is maintained* (Edinburgh, 1753).

Wallerius, Johan Gottschalk. *Decades Binae Thesium Medicarum* (Uppsala, 1741).

Warburton, William. *The Divine Legation of Moses Demonstrated, on the Principles of a Religious Deist, From the Omission of the Doctrine of a Future State of Reward and Punishment in the Jewish Dispensation* (4 vols, London, 1738–41).

Wartofsky, Marx W. *Feuerbach* (Cambridge, 1977).

Weber, Ernst Heinrich. *Allgemeine Anatomie des menschlichen Körpers* (Brunswick, 1830).

Weber, Moritz. *Die Lehre von den Ur- und Racen-Formen der Schädel und Becken des Menschen* (Düsseldorf, 1830).

Weil, Hans. *Die Entstehung des deutsches Bildungsprinzips* (Bonn, 1930).

Werner, Abraham Gottlob. *Von den äusserlichen Kennzeichen der Fossilien* (Leipzig, 1774).

Werner, Abraham Gottlob. *Kurze Classification und Beschreibung der Gebirgsarten* (Dresden, 1787).

Werner, Abraham Gottlob. *Neue Theorie über Entstehung der Gänge* (Freiberg, 1791).

White, Charles. *An Account of the Regular Gradation in Man, and in Different Animals and Vegetables, and from the Former to the Latter* (London, 1799).
Whytt, Robert. *An Essay on the Vital and other Involuntary Motions of Animals* (Edinburgh, 1751).
Wiegleb, Johann Christian. 'Doctrine de Stahl sur le Phlogistique, rectifiée et appuyée par des preuves, en opposition au nouveau Systême chimique des François, dont on cherche en même tems à démontrer le peu de solidité, Extrait des Annales de Chimie de Crell', *Observations sur la physique* 41 (1792), 81–5.
Willey, Thomas E. *Back to Kant: The Revival of Kantianism in German Social and Historical Thought, 1860–1914* (Detroit, 1978).
Williams, Elizabeth A. 'The French Revolution, Anthropological Medicine, and the Creation of Medical Authority', in Bryant T. Ragan Jnr and Elizabeth A. Williams, eds, *Re-Creating Authority in Revolutionary France* (Westport, Conn., 1992), 79–92.
Williams, Elizabeth A. *The Physical and the Moral: Anthropology, Physiology, and Philosophical Medicine in France, 1750–1850* (Cambridge, 1994).
Williamson, George S. *The Longing for Myth in Germany: Religion and Aesthetic Culture from Romanticism to Nietzsche* (Chicago, 2004).
Willis, Thomas. *Cerebri anatome: cui accessit nervorum descriptio et usus* (London, 1664).
Wilson, Alexander. *Some Observations relative to the Influence of Climate on Vegetable and Animal Bodies* (London, 1780).
Wilson, Catherine. *The Invisible World: Early Modern Philosophy and the Invention of the Microscope* (Princeton, 1995).
Wilson, Lindsay. *Women and Medicine in the French Enlightenment: The Debate over 'Maladies des Femmes'* (Baltimore, 1993).
Winch, Donald. *The Emergence of Economics as a Science, 1750–1870* (London, 1971).
Winckelmann, Johann Joachim. *Sendschreiben von den herculanischen Entdeckungen* (Dresden, 1762).
Winckelmann, Johann Joachim. *Die Geschichte der Kunst des Altertums* (Dresden, 1764).
Winckelmann, Johann Joachim. *Monumenti antichi inediti spiegati ed illustrati* (2 vols, Rome, 1767).
Windsor, Mary. 'The Development of Linnaean Insect Classification', *Taxon* 25 (1976), 57–67.
Winterl, Jakob Joseph. *Prolusiones ad chemiam saeculi decimi noni* (Buda, 1800).
Winterl, Jakob Joseph. *Accessiones novae ad Prolusiones suam primam et secundam* (Buda, 1803).
Wise, M. Norton. 'What's in a Line?', in M. Epple and C Zittel, eds, *Science as Cultural Practice. Volume 1: Cultures and Politics of Research from the Early Modern Period to the Age of Extremes* (Berlin, 2010), 61–102.
Withers, Charles W. J. *Placing the Enlightenment: Thinking Geographically about the Ages of Reason* (Chicago, 2007).
Wizenmann, Thomas. *Die Resultate der Jacobischen und Mendelsohnschen Philosophie: kritische untersucht* (Leipzig, 1786).
Wokler, Robert. 'Anthropology and Conjectural History in the Enlightenment', in C. Fox, R. Porter, and R. Wokler, eds, *Inventing Human Science: Eighteenth-Century Domains* (Berkeley, 1995), 31–52.

Wolfe, Charles T. 'Sensibility as Vital Force or Property of Matter in Mid-Eighteenth Century Debates', in Henry Martin Lloyd, ed., *The Discourse of Sensibility: The Knowing Body in the Enlightenment* (Dordrecht, 2013), 147–70.

Wolfe, Charles T. and Terada, Motoichi. 'The Animal Economy as Object and Program in Montpellier Vitalism', *Science in Context* 21 (2008), 537–80.

Wolff, Christian. *Disquisitio philosophica de loquela* (Leipzig, 1703).

Wolff, Christian. *Vernünftige Gedanken von den Kräften des menschlichen Verstandes* (Halle, 1713).

Wolff, Christian. *Anmerckungen über die Vernünfftige Gedancken von Gott, der Welt und der Seele des Menschen, auch alle Dinge überhaupt, zu besserem Verstande und bequemerem Gebrauch derselben* (Frankfurt am Main, 1724).

Wolff, Christian. *Psychologia empirica, methodo scientifica pertractata, qua ea, quae de anima humana indubia experientiae fide constant, continentur* (2nd edn, Frankfurt and Leipzig, 1732).

Wolff, Christian. *Ausführliche Nachricht von seinen eigenen Schriften, die er in deutscher Sprache von den verschiedenen Teilen der Weltweisheit ans Licht gestellet* (Frankfurt am Maine, 1733).

Wolff, Christian. *Theologia naturalis pertractata* (2 vols, Frankfurt and Leipzig, 1736–7).

Wolff, Christian. *Elementa Matheseos Universae Tomvs IV, Qvi, Geographiam cum Hydrographia, Chronologiam, Gnomonicam, Pyrotechnicam, Architectvram Militarem atqve Civilem Complectitur* (Magdeburg, 1738).

Wolff, Christian. *Psychologia rationalis, methodo scientifica pertractata, qua ea, quae de anima humana indubia experientiae fide innotescunt, per essentiam et naturam animae explicantur* (2nd edn, Frankfurt and Leipzig, 1740).

Wolff, Christian. *Vernünfftige Gedancken von Gott, der Welt und der Seele des Menschen, auch allen Dinge überhaupt* (2 vols, Halle, 1751).

Wolff, Christian. *Discursus praeliminaris de philosophia in genere. Einleitende Adhandlung über Philosophie im allgemeinen; Historische-kritische Ausgabe*, ed. Gunther Gawlick and Lothar Kreimendahl (Stuttgart, 1996).

Wood, Allen W. 'Kant and the Problem of Human Nature', in Brian Jacobs and Patrick Kain, eds, *Essays on Kant's Anthropology* (Cambridge, 2003), 38–59.

Wood, Paul B. 'The Natural History of Man in the Scottish Enlightenment', *History of Science* 27 (1989), 89–123.

Wrigley, Edward Anthony. *Continuity, Chance and Change: The Character of the Industrial Revolution* (Cambridge, 1988).

Wundt, Max. *Die deutsche Schulphilosophie im Zeitalter der Aufklärung* (Tübingen, 1945).

Yolton, John W. *Perceptual Acquaintance from Descartes to Reid* (Oxford, 1984).

Yolton, John W. *Locke and French Materialism* (Oxford, 1991).

Young, Brian W. *Religion and Enlightenment in Eighteenth-Century England: Theological Debate from Locke to Burke* (Oxford, 1998).

Zammito, John H. *The Genesis of Kant's Critique of Judgment* (Chicago, 1992).

Zammito, John H. *Kant, Herder, and the Birth of Anthropology* (Chicago, 2002).

Zammito, John H. '"The Most Hidden Conditions of Men of First Rank"—The Pantheist Current in Eighteenth-Century Germany "Uncovered" by the Spinoza Controversy', *Eighteenth-Century Thought* 1 (2003), 335–68.

Zammito, John H. '*Médecin-philosophe*: *Persona* for Radical Enlightenment', *Intellectual History Review* 18 (2008), 427–40.
Zande, Johan van der. 'In the Image of Cicero: German Philosophy between Wolff and Kant', *Journal of the History of Ideas* 56 (1995), 419–42.
Zelbstein, M. U. 'Mauduyt de la Varenne (1733–1792), un médecin au siècle des "Lumières"', *Histoire des sciences médicales* 18 (1984), 295–306.
Zimmermann, Johann Georg. *Von der Erfahrung in der Arzneykunst* (2 vols, Zürich, 1763–4).
Zöller, Günther. 'Kant's Political Anthropology', *Kant Jahrbuch* 3 (2011), 131–61.
Zuckert, Rachel. *Kant on Beauty and Biology: An Interpretation of the* Critique of Judgment (Cambridge, 2007).

Index

Printed and bound by CPI Group (UK) Ltd, Croydon, CR0 4YY